Lecture Notes in Physics

Edited by H. Araki, Kyoto, J. Ehlers, München, K. Hepp, Zürich
R. Kippenhahn, München, H. A. Weidenmüller, Heidelberg
J. Wess, Karlsruhe and J. Zittartz, Köln
Managing Editor: W. Beiglböck

266

The Physics of Accretion onto Compact Objects

Proceedings of a Workshop Held in Tenerife, Spain
April 21–25, 1986

Edited by K.O. Mason, M.G. Watson and N.E. White

Springer-Verlag

Berlin Heidelberg New York London Paris Tokyo

Editors

K.O. Mason
Mullard Space Science Laboratory, University College London
Holmbury St. Mary, Dorking, RH5 6NT, England

M.G. Watson
Physics Department, University of Leicester
Leicester LE1 7RH, England

N.E. White
Space Science Department, ESTEC
Postbus 299 2200 AG Noordwijk ZH, The Netherlands

ISBN 3-540-17195-9 Springer-Verlag Berlin Heidelberg New York
ISBN 0-387-17195-9 Springer-Verlag New York Berlin Heidelberg

Printing: Druckhaus Beltz, Hemsbach/Bergstr.; Bookbinding: J. Schäffer OHG, Grünstadt
2153/3140-543210

Foreword

The Tenerife workshop had its genesis in The Hague in November 1984 during the XVIII ESLAB symposium. That meeting was devoted to new results from the European Space Agency's EXOSAT X-ray observatory, but we realised that a full assimilation of the EXOSAT results would eventually require a different sort of meeting in which they were set in the context of the subject as a whole. A major part of the work done with EXOSAT concerned sources in which accretion is the dominant energy generation process, so it was natural to choose this topic as the focus of the Tenerife Workshop. We deliberately organised sessions around physical, rather than observational, topics to avoid the schism that often occurs between people with primarily Galactic and Extragalactic interests, we hope to the benefit of both groups. We have followed the organisation of the meeting as far as possible in these proceedings which contain the main review papers that were presented.

Considerable credit for the success of the meeting must go to the scientific organising committee. Not only did we follow their suggestions when constructing the main program, but individual members were responsible for the conduct and content of the discussion intervals that made up approximately half of each session. Their enthusiasm, and that of the audience, established the characteristic flavour of the meeting.

In choosing to hold the conference on Tenerife, we were mindful that the Canary Islands are a rapidly expanding center for ground-based astronomical research. The observatories at Izana, and at El Roque de los Muchachos on the nearby island of La Palma are among the best observing sites in the world and are home for the telescopes of several European nations. We received considerable encouragement in setting up the meeting from Brian Taylor of ESA, and from the staff of the Royal Greenwich Observatory on La Palma, and the Instituto de Astrofisica de Canarias on Tenerife under their director, Professor F. Sanchez. Particular mention should be made of Paul Murdin of the RGO, and Campbell Warden of the IAC who found us a pleasant, congenial site at the Maritim Hotel in Puerto dela Cruz, and who smoothed the way for the conference participants. The local organisation was the responsibility of David Pike (RGO), Evencio Mediavilla (IAC), and Carmen Rosa Santos (IAC), ably assisted by Sandra Andrews (ESOC) and Lindsay Simpson (RGO). The local organisers put in a great deal of hard work behind the scenes to make the whole thing work. The best tribute that we can pay to their efforts is to say that none of the disasters that we imagined could have happened, did happen! We also thank the Patronato Insular de Turismo of the Cabildo of Santa Cruz de Tenerife for administrative support.

K. O. Mason
M. G. Watson
N. E. White
September 1986

Scientific Organising Committee

H.V. Bradt - *Massachussetts Institute of Technology*
J. Buitrago - *Instituto de Astrofisica de Canarias*
F.A. Córdova - *Los Alamos National Laboratory*
M.S. Elvis - *Harvard-Smithsonian Center for Astrophysics*
A.C. Fabian - *Institute of Astronomy, Cambridge*
S.M. Kahn - *Physics Department, U.C. Berkeley*
K.O. Mason - *Mullard Space Science Laboratory, U.C. London*
P.G. Murdin - *Royal Greenwich Observatory*
R.F. Mushotzky - *Goddard Space Flight Center*
J.A. van Paradijs - *University of Amsterdam*
F. Verbunt - *Max Plank Institut, Garching*
M.G. Watson - *University of Leicester*
N.E. White - *EXOSAT Observatory, SSD–ESA*

Local Organising Committee

E. Mediavilla - *Instituto de Astrofisica de Canarias*
C.D. Pike - *Royal Greenwich Observatory*
C. Rosa Santos - *Instituto de Astrofisica de Canarias*

Workshop Participants

Atrio F.	Hjellming R. M.	Penston M. V.
Barr P.	Horne K.	Peran C.
Battaner E.	Ilovaisky S. A.	Perez Fournon I.
Belli B. M.	Jones D. H. P.	Pike C. D.
Belvedere G.	Kahabka P.	Pietsch W.
Bender R.	Kahn S. M.	Pollock A.
Berman N. M.	Kaisig M.	Pounds K. A.
Binette L.	Kallman T. R.	Priedhorsky W.
Boyle C. B.	Kawai N.	Reichert G.
Bradt H.	Kendziorra E.	Robba N.
Branduardi-Raymont G.	Kidger M.	Rosen S. R.
Breedon L. M.	King A. R.	Shalinski C.
Brooker J. R. E.	Krautter J.	Schnopper H. W.
Brunner H.	Lamb D. Q.	Shafer R.
Buitrago J.	Lamb P. A.	Smale A. P.
Burm H. M. G.	Lambert A.	Smith M. D.
Cabezas-Flores J. A.	Larsson S.	Solheim J. E.
Canal R.	Lasota J.-P.	Spencer R. E.
Coe M.	Lawrence A.	Staubert R.
Cook M. C.	Leahy D. A.	Stella L.
Corbet R.	Lewin W.	Stewart G.
Córdova F. A.	Madau P.	Stollman G. M.
Courvoisier T. J.-L.	Makishima K.	Svensson R.
Damen E. M. F.	Malkan M.	Sztajno M.
Day C.	Maraschi L.	Szuszkiewicz E.
De Kool M.	Marsh T. R.	Taylor B. G.
Elvis M.	Mason K. O.	Tennant A.
Eulderink F.	Matsuoka M.	Truemper J.
Fabian A. C.	McClintock J.	Turner T. J.
Foster A. J.	Mediavilla E.	Urry C. M.
George I. M.	Meyer H.	Van Amerongen S. F.
Ghisellini G.	Molnar L.	Van der Klis M.
Giommi P.	Morini M.	Van Paradijs J. A.
Gonzalez Serrano J. I.	Mouchet M.	Verbunt F.
Gottwald M.	Mukai K.	Vrtilek S. D.
Gregory P. C.	Mushotzky R.	Warwick R. S.
Haberl F.	Novick R.	Watson M. G.
Hanson C.	Ogelman H.	White N. E.
Hasinger G.	Osborne J.	
Hessman F. V.	Parmar A. N.	

Table of Contents

Spectral Formation

Indirect Imaging of Accretion Disks in Binaries

Keith Horne
Space Telescope Science Institute
3700 San Martin Drive
Baltimore, MD 21218, USA

T. R. Marsh
Royal Greenwich Observatory
Herstmonceaux Castle
Hailsham, East Sussex BN27 1RP, UK

Abstract

We review two indirect imaging techniques that are now being used to form spatially-resolved images of accretion disks in mass-transfer binaries, mainly the cataclysmic variable stars. The *eclipse mapping method* reconstructs the continuum brightness distribution on the surface of a disk from the shape of the light curve during eclipses of the disk by the secondary star. *Doppler tomography* uses emission line velocity profiles observed over the full range of binary phases to recover the line emissivity distribution in a 2-dimensional *velocity space*. The doppler tomogram may be interpreted directly in velocity space or transformed to a spatial image of the disk with the assumption of a Keplerian flow. We use the *maximum entropy method* with *default level steering* to adjust the disk images.

1 Introduction

Accretion disks offer solutions to several astrophysical puzzles: the flattened structure of our solar system, the shedding of angular momentum during star formation, the fueling of the central engines in active galactic nuclei and quasars, the collimation of jets from these and such systems as SS 433, and the moderation of mass-transfer in binary systems.

The theoretical treatment of accretion disks currently uses a viscous fluid approximation, although the physical processes that produce viscosity in disks are unknown. Friction causes an outward transfer of angular momentum, which permits the inward migration of the material, and heats the disk to produce a strong radial temperature gradient, $T_{\text{eff}} \propto \dot{M} R^{-3/4}$ for a steady-state disk around a compact object. Disks around the white dwarf primaries of cataclysmic variable stars may have temperatures ranging from 5,000 K at the outer rim to over 50,000 K near the center, where a hot boundary layer may form. Emission lines may arise from an outer region that becomes optically-thin, or in chromospheric layers above opaque zones of the disk.

What we can observe directly is the superposition of spectra from these diverse regions, and the interpretation of disk observations is hindered by the ambiguity always associated with composite spectra. The physics of disks is so uncertain at present that disk modellers have the freedom to make a disk spectrum look like almost anything.

These difficulties could be largely overcome with spatially-resolved data. The angular size of a disk with a diameter of 1 solar radius at a distance of 100 parsecs is only 0.05 milli-arcseconds, thus spatial resolution is beyond the reach of any direct imaging technique short of optical VLBI. This paper describes two indirect imaging techniques, *eclipse mapping* and *doppler tomography*, that can be used to map the accretion disks in binary systems in the continuum and in the emission lines.

2 Eclipse Mapping

Eclipse mapping is described in detail by Horne (1985). The method aims to recover a 2-dimensional image of the surface brightness on the face of the accretion disk. The observations that constrain the desired image are eclipse light curves. Deep eclipses occur in high inclination systems ($i > 70°$) as the secondary star, which is cool and has a low surface brightness compared with that of the disk, passes between the observer and the disk. The secondary stars in mass-transfer binaries precisely fill their Roche lobes, thus the eclipses can be accurately simulated with numerical techniques developed by Mochnacki (1971) for contact binary stars (see also Mochnacki & Doughty 1972, Young & Schneider 1980).

The eclipse geometry is completely specified (apart from a scale factor) by the mass ratio q and inclination i. For deep eclipses the measured phase width provides one tight constraint between these two parameters (Bailey 1979). A second constraint is available if the light curves show eclipses of the bright-spot, which forms where the gas stream from the secondary star hits the outer rim of the disk (Cook & Warner 1984, Wood *et al.* 1986). In most cases the bright-spot eclipses are indistinct features and a mass ratio must be assumed.

Synthetic light curves are provided by the numerical eclipse simulation for any image of the disk that we might wish to consider. The disk image must then be adjusted until the synthetic light curve fits the observed light curve. With a full image of free parameters, a close fit is almost always achieved.

We must contend with the problem of indeterminacy, however, since the 2-dimensional disk image is constrained by a 1-dimensional light curve. Many different disk images will be equally successful in fitting a given light curve. The

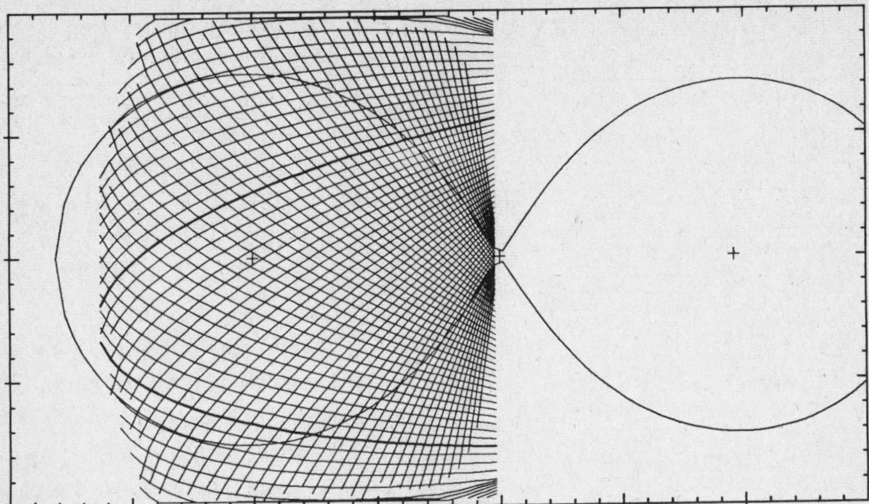

Figure 1: The geometry of an accretion disk eclipse is illustrated by the network of ingress/egress arches on the face of the disk. Each arch is the boundary of the disk region that is occulted by the secondary star at a particular binary phase.

maximum entropy method (MEM) copes neatly with the indeterminacy by selecting the unique image that has the largest *entropy* among images that fit the light curve at a specified level of χ^2. MEM also ensures that the disk surface brightness is positive, a distinct advantage over linear reconstruction methods such as those used in geophysics (Parker 1977). The image adjustment is accomplished by iterative corrections made with the general purpose maximum entropy fitting program MEMSYS, developed by Skilling and Bryan (1984). MEMSYS makes it easy to apply MEM to any reconstruction problem. One need only code a subroutine to calculate predicted data from an arbitrary image.

In its simplest form, MEM seeks the most *uniform* image that is consistent with the data. The entropy of the image I is given by

$$S = -\sum_i p_i \ln p_i \ ,$$

where $p_i = I_i / \sum_j I_j$. The entropy is an information-like measure of the uniformity of the image values; maximizing S makes the image values as uniform as possible within the constraints of the light curve data.

This simple version of MEM is unsuitable in the eclipse mapping problem. The MEM images generated by Horne (1985) achieve good fits to the light curves but are criss-crossed by artifacts along the grid lines formed by the arch-shaped occultation boundaries (see Figure 1). These artifacts arise through a competition between the data constraints and the entropy. MEM tends to suppress peaks in the image, since this makes the image more uniform and increases the entropy (Bryan & Skilling 1980, Horne 1985, Narayan & Nityananda 1986). However, flux can be redistributed along the occultation arches without greatly changing the predicted light curve. Thus flux is removed from bright regions (to increase the entropy) by migrating along the arches (to maintain the light curve shape and the required χ^2).

The artifacts generated by simple MEM are suppressed by a technique known as *default level steering*. The entropy of the image I can be defined with respect to a prior or *default image*, D. The modified definition is

$$S = - \sum_i p_i \ln (p_i/q_i) \ ,$$

where $q_i = D_i/\sum_j D_j$, and D_j is the default level at pixel j. The entropy then measures how closely the image I resembles the default image D. In simple MEM the default is a uniform image. If instead we make D an axi-symmetric image with the same radial profile as I, the entropy becomes a measure of axi-symmetry, and loses its sensitivity to the radial profile. MEM then seeks the *most axi-symmetric* image that fits the data. The light curve freely determines the radial profile, and the entropy suppresses as much azimuthal structure as the data will permit. Artifacts stemming from the bright region at the center of the disk are thereby greatly reduced.

Simulation tests of the eclipse mapping method, both with and without radial profile defaults, are presented by Horne (1985). The need for default level steering makes the method most accurate for images with only moderate deviations from azimuthal symmetry. The compact bright-spot near the outer rim of the disk gets smeared in the azimuthal direction. However, accurate brightness temperature profiles are recovered in the simulation tests even when a strong bright-spot is present.

Eclipse maps have been formed from UBR eclipse light curves of the nova-like variable RW Tri (Horne and Stiening 1985) and from UBV light curves of the dwarf nova Z Cha during an outburst (Horne and Cook 1985). Data from several eclipse cycles were averaged to reduce flickering noise to a suitably low level. The independent images at three different wavelengths demonstrate that these two accretion disks are optically thick. The color-color and color-magnitude relationships of the emission from different parts of the disk fall between those of blackbodies and stars, implying that the vertical temperature gradients in disk atmospheres are not as steep as in stars. The eclipse maps also give a direct observational verification of the $T_{\mathrm{eff}} \propto R^{-3/4}$ law predicted for steady accretion. The eclipse light curve and reconstructed brightness temperature profile of the Z Cha disk are presented in Figure 2. The eclipse mapping experiments yield mass transfer rates accurate to factors of 2-3, and distances to $20 - 40$ %.

In three quiescent dwarf novae, Z Cha (Wood *et al.* 1986), HT Cas (Horne, Wood & Stiening 1986), and OY Car (Wood *et al.* 1987), the brightness temperature distribution is much flatter than the $R^{-3/4}$ law. Optical colors indicate that these quiescent disks are optically-thin. Further analysis should yield temperatures and column densities, giving new constraints on dwarf nova outburst

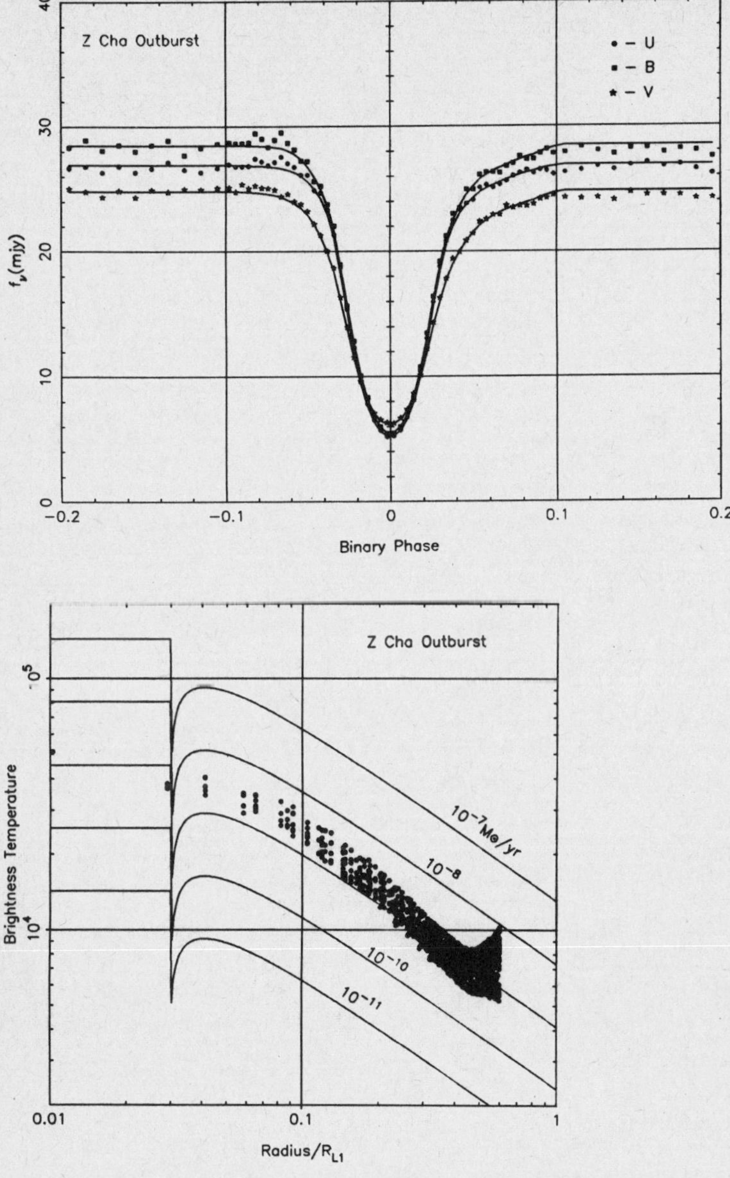

Figure 2: Light curves of the dwarf nova Z Cha in outburst show deep eclipses of the accretion disk. The brightness temperature profile of the disk is reconstructed from the eclipses and compared with steady disk models.

models and perhaps clues to the nature of the accretion disk viscosity.

3 Doppler Tomography

Doppler tomography generates an image of the accretion flow in the light of an emission line from observations of the line's velocity profile at different binary phases. The velocity profiles of emission lines from disks result primarily from doppler shifts due to supersonic velocities in the accretion flow (Smak 1969, Huang 1972, Smak 1981), although radiative transfer effects due to velocity gradients in the disk also affect the line profile (Horne and Marsh 1986). A single line profile carries enough information to determine the radial distribution of the emission if we assume azimuthal symmetry and a Keplerian velocity field (Smak 1981). However, the observed emission lines often have asymmetric profiles that vary with binary phase. The procedure described below generates a two-dimensional image of the disk that accounts for the line profile variations over the full range of binary phases.

3.1 Decomposition of the accretion flow into S-wave components

We assume that the emission line profile of the accretion flow as a whole can be built up by superimposing doppler-shifted contributions from each of its parts. To each part of the accretion flow we can attach a *velocity vector* that is stationary in the rotating frame of the binary. The velocity field for a pure Keplerian flow around a mass M_1 is given by

$$\begin{pmatrix} V_x \\ V_y \\ V_z \end{pmatrix} = \sqrt{\frac{GM_1}{R}} \begin{pmatrix} -y/R \\ x/R \\ 0 \end{pmatrix} - \begin{pmatrix} 0 \\ K_1/\sin i \\ 0 \end{pmatrix} .$$

Here the first term is the circular orbital motion of the disk material around the central object and the second term accounts for the orbital motion of the central object around the center of mass of the binary system. The position coordinates x, y, z and $R = \sqrt{x^2 + y^2}$ are for a cartesian frame co-rotating with the binary, with its origin at the center of the disk, the $+x$ axis in the direction from the primary toward the secondary star, and the $+z$ axis in the direction of the orbital angular momentum. We are interested mainly in accretion disks, but rather than assume a Keplerian velocity field at the outset, we initially retain a more general framework.

The velocity vectors are all stationary in the rotating frame of the binary, but in the observer's inertial frame they revolve around the rotation axis passing through the center of mass of the binary. At each binary phase ϕ, the observed radial velocity $V(\phi)$ is a simple projection of the velocity vector onto the instantaneous line of sight. The rotation of the binary thus gives rise to sinusoidal radial velocity curves of the form

$$V(\phi) = \gamma + K \sin\left(2\pi\phi + \theta\right) \ ,$$

where the three parameters of the sinusoid are given by

$$\begin{aligned}
\gamma &= V_z \cos i \\
K^2 &= [V_x^2 + V_y^2] \sin^2 i \\
\tan\theta &= -V_x/V_y \ .
\end{aligned}$$

The sinusoid's velocity semi-amplitude K and phase θ give the speed and direction of the velocity components in the plane of the disk; the γ velocity gives the component perpendicular to the disk plane. We may think of γ, K and θ as cylindrical coordinates on the three-dimensional velocity space (V_x, V_y, V_z).

When the accretion flow takes the form of a disk, $V_z = 0$, the velocity vectors all lie in the disk plane and the γ velocity is the same for all parts of the flow. For other types of accretion flow, such as the magnetically-controlled flows in AM Her binaries, the velocity component perpendicular to the orbital plane is generally not zero, and the γ velocity can be different in different parts of the flow.

If a small region of the accretion flow has a high line emissivity compared with adjacent regions, that region produces a sharp "S-wave" emission component following a specific sinusoidal velocity curve. Such S-wave emission components are frequently observed in cataclysmic variable stars. In many cases the S-wave emission can be interpreted as emission from the gas stream or bright-spot region (Smak 1985), but in several systems the phase of the S-wave precludes such an interpretation (Young, Schneider & Shectman 1981; Shafter & Szkody 1984).

3.2 Doppler tomograms in velocity space

The observed line profile $f(V, \phi)$ is of course a superposition of many such S-wave components, each arising from a different position in the accretion flow. This superposition is expressed by

$$f(V, \phi) = \int_{-\infty}^{\infty} d\gamma \int_0^{2\pi} d\theta \int_0^{\infty} K \ dK \ I_V(\gamma, K, \theta) \ \delta\left(\gamma + K \sin\left(2\pi\phi + \theta\right) - V\right)$$

where I_V gives the intensity of the emission from each of the S-wave components. The Dirac delta function $\delta(x)$ is used here to select those parts of the accretion flow where the radial velocity $V(\phi)$ matches the desired velocity V. (Strictly speaking, the delta function should be replaced by a local line profile, with a velocity width of order the local sound speed. The delta function is a good approximation, however, since the velocities in a disk are highly supersonic.)

By observing line profiles at numerous binary phases around the cycle, one obtains information that can be used to disentangle the emission contributions from each of the S-waves components. This information is incomplete in general,

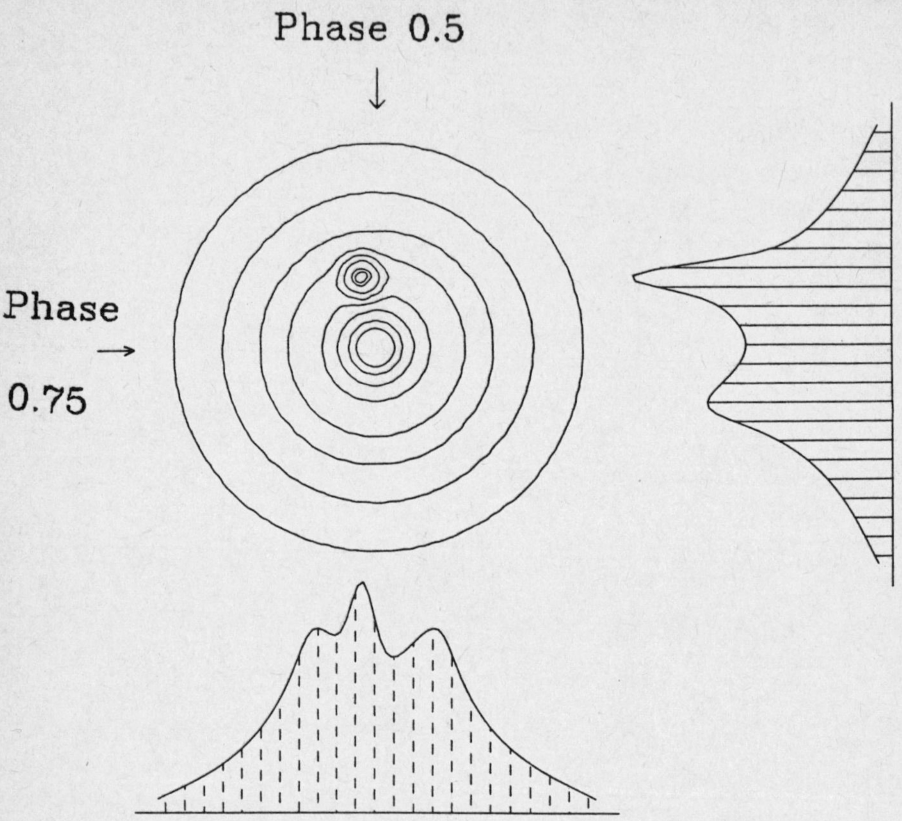

Figure 3: The emission line profiles at different binary phases are simple projections of the two-dimensional emission distribution in velocity space.

since the three-dimensional image $I_V(\gamma, K, \theta)$ is constrained by a two-dimensional dataset $f(V, \phi)$. However, for a disk geometry the γ velocity is the same for all the S-waves. The remaining two-dimensional image $I_V(K, \theta)$ is then completely determined by the observed line profiles.

Figure 3 illustrates how the line profiles at different binary phases are simple projections of the disk image in velocity space. A formula for direct inversion of the line profiles, which can be derived by substituting a fourier expansion for the delta function, is

$$I_V(K, \theta) = \frac{K^6}{2} \int_0^1 2\pi d\phi \int_0^\infty ds\, s\, g(s) \int_{-\infty}^\infty dV\, f(V, \phi) e^{-i2\pi s[V - K\sin(2\pi\phi + \theta)]}.$$

Here $i = \sqrt{-1}$ is not the inclination. The fourier filter $g(s)$ can be used to control the propogation of noise from the data to the image, at the expense of some loss of resolution. Practical numerical algorithms for reconstructing a discretely sampled image from a finite number of projections were developed by Bracewell (1956) for mapping radio sources from measurements with a fan-beam interferometer. The subject of tomography then exploded with the advent of CAT scanners in medicine, which form high-resolution images of a slice through the human body from x-ray profiles taken at a number of different angles. Rowland (1979) gives a thorough mathematical review of the linear techniques that have been developed for tomographic reconstruction.

We employ MEM for doppler tomography of accretion disks. This guarantees us a positive image, and curtails the propogation of noise without unduly degrading resolution of the image. With MEM we can also take into account eclipses by the secondary star, and local radiative transfer effects that produce anisotropic emission. These effects would be difficult to incorporate into a linear reconstruction scheme.

3.3   Transforming a doppler tomogram back to the face of the disk

The velocity image $I_V(K, \theta)$ is of course closely related to the spatial emission distribution $I_X(x, y)$ on the face of the disk, since the velocity field of the disk is highly structured. For a Keplerian disk, the transformation is

$$\begin{pmatrix} x \\ y \end{pmatrix} = \frac{GM_1 \sin^2 i}{[K_x^2 + (K_y + K_1)^2]^{3/2}} \begin{pmatrix} K_y + K_1 \\ -K_x \end{pmatrix} ,$$

where $K_x = -K \sin\theta$ and $K_y = K \cos\theta$. The transformation is illustrated in Figure 4, where a Keplerian disk is partitioned into regions that contribute to different velocity bins of the line profile at a specific phase. The lines of constant radial velocity form a dipole pattern that remains stationary with respect to the observer as the emission distribution of the disk revolves. The radial velocity is of course constant along straight lines in velocity space, where the line profiles at each phase are obtained by a simple projection.

If the true velocity field deviates from the assumed Keplerian flow, the transformation from velocity space to image space will yield a distorted image of the disk. For example, the gas stream has a velocity larger than the local Keplerian velocity and is directed inward. Emission from the gas stream is therefore mapped to a smaller radius and at a larger azimuth, where the local Keplerian velocity vector is the same as the velocity vector of the stream. The inclination must also be known to set the radial scale of the transformed image. For this reason it can be advantageous to interpret doppler tomograms in velocity space rather than in position space.

3.4   Applications

An extensive series of tests with simulated line profiles is presented by Marsh (1985). Doppler tomograms are far better constrained than eclipse maps because both the image $I_V(K, \theta)$ and the data $f_\nu(V, \phi)$ are two-dimensional. Under such circumstances MEM finds virtually the same image regardless of the

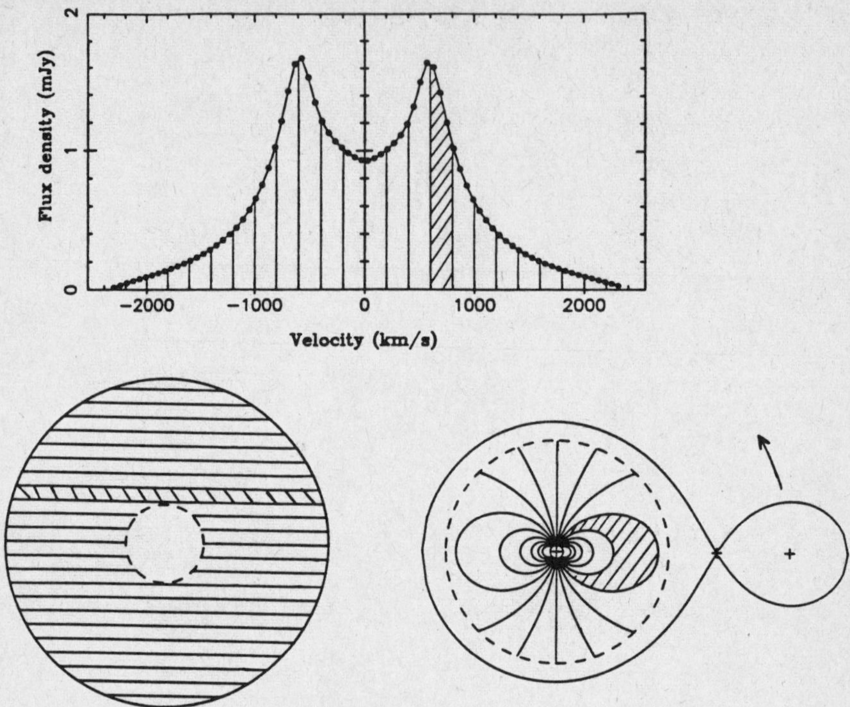

Figure 4: The loci of constant radial velocity form a dipole pattern on the face of a Keplerian accretion disk. The radial velocity contours are straight lines in velocity space.

form of default image. Figure 5 attests to the method's ability to reconstruct a complicated emission pattern from simulated line profile data.

Doppler tomography can be extended to treat three-dimensional accretion flows such as are present in AM Her stars. Each pixel of the image then corresponds to an S-wave emission component with a different γ, K, and θ. (In the case of a disk, the γ velocities are all the same.) However, as the three-dimensional image is constrained by two-dimensional data, such reconstructions must rely heavily on defaults to resolve ambiguity.

Doppler tomograms of the quiescent disk in the dwarf nova Z Cha show that the Balmer emission is highly concentrated toward the center of the disk (Marsh 1985). Smak (1981) obtains similar results for several other systems based on power-law fits to the emission line profiles and one-dimensional reconstructions assuming azimuthal symmetry. The concentrated emission distribution contradicts current models in which emission lines come from an optically-thin outer disk region while the inner disk remains optically-thick (Williams 1980, Tylenda 1981). We evidently do not yet understand the excitation mechanism for these lines.

We also find evidence in the Z Cha maps for enhanced emission along the

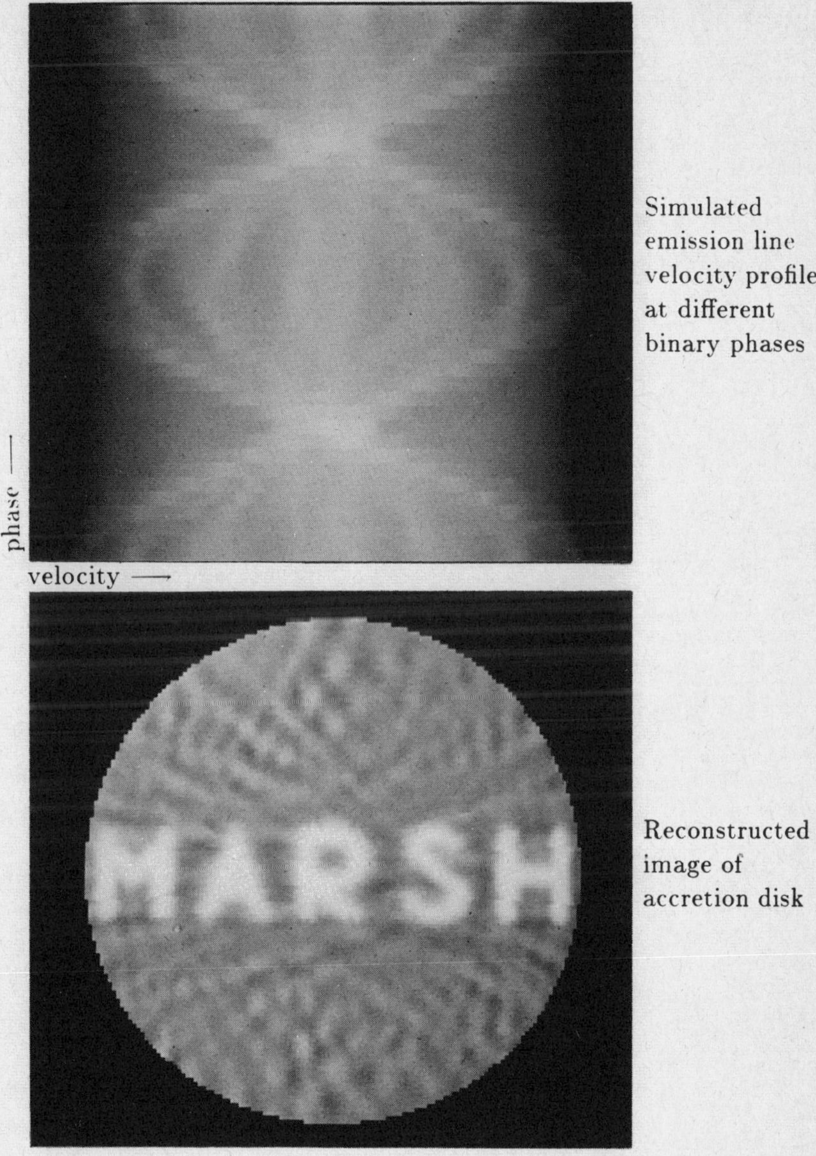

Figure 5: **A complicated emission distribution on the face of an accretion disk is reconstructed from synthetic emission line velocity profiles at different binary phases.**

trajectory of the gas stream. If confirmed, this result would show that the stream interacts directly with material well inside the outer rim of the disk. The emission feature in Z Cha is from material with velocities close to the Keplerian velocities at positions along the stream trajectory. This could be material from the stream that has just been entrained by the disk, or disk material that becomes excited as it passes under the stream. The gradual stripping of gas from the stream is investigated with a phenomenological model by Bath, Edwards & Mantle (1983). The finite vertical width of the stream might cause part of it to clear the outer rim and skim over the surface of the disk, as in the numerical simulations of Livio, Soker and Dgani (1986).

4   Conclusion

An important virtue of our image reconstruction approach is the absence of assumptions about the physics of disks. Model-fitting techniques yield results that are limited to the context of a specific model of the disk. Eclipse maps in different bandpasses and doppler tomograms for different emission lines give in effect the spectrum emitted by each part of the disk. Such model-independent results are particularly valuable given our present ignorance of the viscosity mechanism and the structure of accretion disk atmospheres.

The data requirements of indirect imaging are severe. An ideal experiment would have a time resolution of a few seconds, spectral resolution of ~ 1 Å, a wide wavelength coverage, spectrophotometric accuracy of a few percent, and numerous orbital cycles to average out flickering noise. These requirements can be met by observing simultaneously with a spectrograph on a large telescope and with a multi-color photometer on a smaller telescope. The photometry may alternatively be obtained at the large telescope by placing a comparison star on the slit alongside the variable star.

Eclipse mapping and doppler tomography are not yet widespread in the study of accretion disks. The present situation may be compared to the era in radio astronomy when aperture synthesis techniques began to improve the spatial resolution of radio measurements. Efforts to make more careful observations with higher time- and spectral-resolution, signal-to-noise ratio, and photometric accuracy can now be rewarded by improved resolution of the accretion disk images. Although the work has only just begun, we look forward with optimism to an era of imaging studies of accretion disks in binaries.

References

Bailey J., 1979 *M.N.R.A.S.*, **187**,, 645.

Bath G.T. Edwards,A.C. & Mantle,V.J., 1983, in *IAU Colloquium 72, Cataclysmic Variables and Related Objects*, ed. M.Livio & G.Shaviv (Dordrecht:D.Reidel), p 55.

Bracewell R.N., 1956 *Aust.J.Phys.* **9**, 198.

Bryan R.K. & Skilling J., 1980 *M.N.R.A.S.*, **191**, 69.

Cook M.C. & Warner B., 1984 *M.N.R.A.S.*, **207**, 705.

Horne K., 1985 *M.N.R.A.S.*, **213**, 129.

Horne K. & Cook M.C., 1985 *M.N.R.A.S.*, **214**, 307.

Horne K. & Marsh T.R., 1986 *M.N.R.A.S.*, **218**, 761.

Horne K. & Stiening R.F., 1985 *M.N.R.A.S.*, **216**, 933.

Horne K., Wood J. & Stiening R.F., 1986 *Astrophys.J.*, , in preparation.

Huang S., 1972 *Astrophys.J.*, **171**, 549.

Livio M., Soker N. & Dgani R., 1986 *Astrophys.J.*, **305**, 267.

Marsh T.R., 1985 Ph.D. thesis, University of Cambridge, England.

Marsh T.R. & Horne K., 1986 *M.N.R.A.S.*, , in preparation.

Mochnacki S.W., 1971 M.S. thesis, University of Canterbury, New Zealand.

Mochnacki S.W. & Doughty N.A., 1972 *M.N.R.A.S.*, **156**, 51.

Narayan R. & Nityananda R., 1986 *Annu.Rev.Astr.Ap.*, **24**, 127.

Parker R.L., 1977 *Annu.Rev.Earth Planet.Sci.*, **5**, 35.

Rowland, 1979 in *Topics in Applied Physics 32, Image Reconstruction from Projections*, ed. G.T.Herman (Springer-Verlag).

Shafter A.W. & Szkody P., 1984 *Astrophys.J.*, **276**, 305.

Skilling J. & Bryan R.K., 1984 *M.N.R.A.S.*, **211**, 111.

Smak J., 1969 *Acta Astron.*, **19**, 155.

Smak J., 1981 *Acta Astron.*, **31**, 395.

Smak J., 1985 *Acta Astron.*, **35**, 351.

Tylenda R., 1981 *Acta Astron.*, **31**, 127.

Williams R.E., 1980 *Astrophys.J.*, **235**, 939.

Wood J.H., Horne K.D., Berriman G., Wade R., O'Donoghue D & Warner B., 1986 *M.N.R.A.S.*, **219**, 629.

Wood J.H., Horne K.D., Berriman G. & Wade R., 1987 *M.N.R.A.S.*, , in preparation.

Young P. & Schneider D.P., 1980 *Astrophys.J.*, **238**, 955.

Young P., Schneider D.P. & Shectman,S.A., 1980 *Astrophys.J.*, **245**, 1035.

SPECTRUM AND POLARIZATION OF THE CONTINUUM FROM
ACCRETION DISKS IN ACTIVE GALAXIES AND QUASARS

Wayne Webb

Analytic Decisions Incorporated

A Subsidiary of Ball Corporation;

and Department of Astronomy, UCLA

and

Matthew Malkan

Department of Astronomy

University of California,

Los Angeles

§1. Continuum Observations of Active Galactic Nuclei

Accretion onto a supermassive black hole has long been the most attractive explanation for energy generation in active galaxies and quasars. Observational tests of this hypothesis have only recently become possible, requiring several instruments to measure the spectral energy distribution over a wide range of wavelengths. Spectral shape, polarization, and variability are all vital pieces of information in understanding how the continuum is produced.

The continuous spectra of the nonstellar nuclei of active galaxies and quasars are quite similar over an enormous range of luminosities. From optical to mid-infrared wavelengths, the nuclear continuum resembles a power law with $f_\nu \propto \nu^{-1.2}$ (Malkan and Filippenko 1983). New data from the Infrared Astronomical Satellite confirm that this power law extends at least six octaves to the far infrared with constant slope, until it has a sharp long-wavelength cut-off around 60 μm (Figure 1, from Edelson and Malkan, 1986, hereafter EM). A simple high-frequency extrapolation of this infrared/optical power law with a slope of -1.15 is a remarkably accurate predictor of the observed X-ray flux at 2 keV (Malkan 1984). These observations suggest that the X-ray and infrared power laws may have a common nonthermal origin. Two exceptions to this rule are the low-z quasars, PKS 1004+131 and PG 1351+64. Their spectra have extremely large blue bumps dominating the optical spectrum out to the near infrared. Extrapolation of their mid-infrared continua with a slope of -1.15 overpredicts their observed X-ray fluxes by a factor of 10. Another puzzle is why the variability detected in the X-rays in some Seyfert galaxies appears to have larger amplitudes and shorter timescales than any yet seen in the infrared.

We have tried to avoid studying those quasars whose continuum measurements could be contaminated by the effects of dust near the active galactic nucleus. The dust grains would preferentially absorb ultraviolet

MKN 335

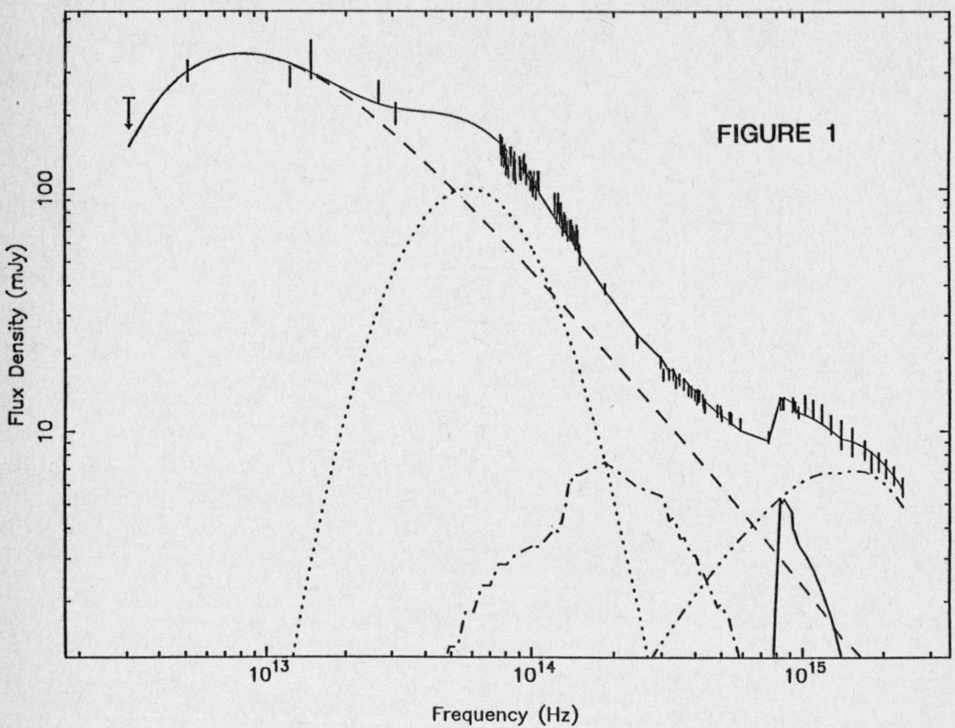

radiation, reddening the observed ultraviolet spectrum. Since the dust will
not survive at temperatures above 1000 K, its thermal reradiation must have an
exponential high-frequency cut off in the mid or near-infrared. Thus the
signature of strong contamination from thermal dust emission is an infrared
spectrum with a steep slope ($\alpha < -1.3$) and strong downward curvature from 20
to 3 μm. EM showed that strong thermal contamination in the IR is only found
in dusty AGNs with independent indicators of substantial emission line and
continuum reddening. Fortunately, the well defined blue limit to the optical/
ultraviolet colors of AGNs and quasars strictly constrains the amount of
internal continuum reddening present in many quasars (Malkan 1984).

The optical and ultraviolet continuum is much stronger than the
extrapolation of the infrared power law would predict. This excess flux is
noticeable as a sharp upward inflection in the spectrum shortward of 1 μm.
The spectral slope from optical to ultraviolet wavelengths is quite flat,
typically $f_\nu \propto \nu^{-0.5}$, and in some cases f_ν is even constant down to 0.25 μm.
Malkan and Sargent (1982, hereafter MS) measured this excess flux component by
subtracting the infrared/red power law from the combined infrared/optical/
ultraviolet spectral energy distributions of eight active galactic nuclei.
The high-frequency extrapolation of the power law accounts for less than half

of the observed optical continuum and virtually none of the ultraviolet. The remaining flux (per unit frequency) has a broad peak around 1500–2000Å. The short-wavelength fall-off of the "blue bump" becomes increasingly difficult to determine at continuum wavelengths shorter than 1216 Å, 1025 Å, and 912 Å, where intervening absorptions of Lyα, Lyβ, and the Lyman continuum begin to depress the apparent continuum level (Malkan 1983; Bechtold et al 1984). IUE observations of high-redshift quasars indicate that the continuum falls with a spectral slope of roughly −1 down to 500Å. The characteristic shape of this ultraviolet excess strongly suggests that it is optically thick thermal emission. It was first quantitatively fitted by the simplest possible function: a single temperature blackbody (MS). In objects spanning an enormous range of luminosities the fitted temperature is almost always within a few thousand degrees of 26,000 K (MS, EM).

The majority of the total energy output from most quasars is in the thermal ultraviolet component. This establishes that the ultraviolet bump is a primary energy component. It is not produced, for example, by optically thick clouds which reradiate the continuum absorbed at some other wavelength. There is not nearly enough energy available for them to intercept. This conclusion is especially strong since the area covering factor of such clouds over the nuclear continuum-emitting region is limited to less than 5 or 10% (Kinney et al. 1985).

We emphasize that the ultraviolet excess cannot be described by a simple flat power law (e.g., $f_\nu \propto \nu^{-0.5}$). As MS pointed out, the addition of such a component to the steep infrared power law cannot produce the sudden inflection in the spectra observed near 1 µm. The data could only be fitted if the infrared component always steepened very sharply at just around 1 µm in all objects. Not only would this explanation be extremely contrived, but there is also no evidence for any break in the $\nu^{-1.2}$ power law down to 0.4 µm in Seyfert galaxies with weak ultraviolet excesses.

The observed continuum level between 3600 and 2200 Å is raised 20–30% by emission from the Balmer continuum and many blended Fe II lines. The low contrast of these features makes them difficult to measure with precision, but the integrated flux of the ultraviolet Fe II lines and Balmer continuum is typically comparable to that of Lyman α or at most twice its flux (MS; Wills, Netzer and Wills 1985). This contribution is too large to ignore in detailed continuum analyses. It is small enough, however, so that uncertainties in the shape of the Balmer continuum plus FeII emission do not lead to significant uncertainties in the determination of the underlying continuous energy distribution.

§2. Accretion Disk Continua

Not surprisinglyobservations extending to higher frequencies revealed that the thermal ultraviolet component is actually broader than a single Planck function. It must arise from gas at a range of temperatures, and Malkan (1983) examined the simple case of a geometrically thin steady state accretion disk. If the disk is assumed to be optically thick, in the steady-state the local surface flux is determined by the radial equations of conservation of mass, angular momentum, and energy in cylindrical coordinates. The disks' spectrum can be calculated without knowledge of its viscosity or vertical structure.

The most common, if somewhat arbitrary, assumption about viscosity is that it is proportional to the total pressure with constant α. This produces the "α-disk" models of Shakura and Sunyaev (1972). The assumption that the viscosity is proportional to the radiation pressure has been questioned by Callahan (1977), Sakimoto and Coroniti (1981), and others who argue that, when applied to very massive black holes, the model predicts disks which are optically thin in their inner regions, and thermally and secularly unstable. The α-disk model for a radiation-pressure dominated accretion disk also predicts that the balance between radiation pressure and the vertical component of gravity will produce a disk scale height proportional to radius, resulting in an accretion wedge.

The thickness of an accretion flow depends on the efficiency with which it can radiate away energy generated by viscous stresses. If the accreting gas is hot, or the disk heats up due to instabilities, the efficiency of energy radiation will fall and the disk will thicken and become a radiation pressure-supported torus (Jaroszynski et al. 1980). At lower accretion rates, if plasma processes are significant enough to keep the electrons and ions in equipartition, then the disk will cool by synchrotron self-Compton emission and the torus will collapse. If, however, equipartition cannot be maintained, then the electrons will cool and the ions will remain at the virial temperature and the disk will remain thick, as an ion-pressure supported torus (Rees, Begelmann, Blandford & Phinney, 1982).

Relativistic effects must be accounted for in the calculation of accretion disk spectra, especially for rapidly rotating Kerr black holes, which have disks which can extend in to a few gravitational radii. Novikov and Thorne (1973) provide an important general relativistic treatment of this disk model resulting in corrections of order 10 to 20 percent at small radii. As illustrated in Figure 2, the inner disk temperatures do not continue to

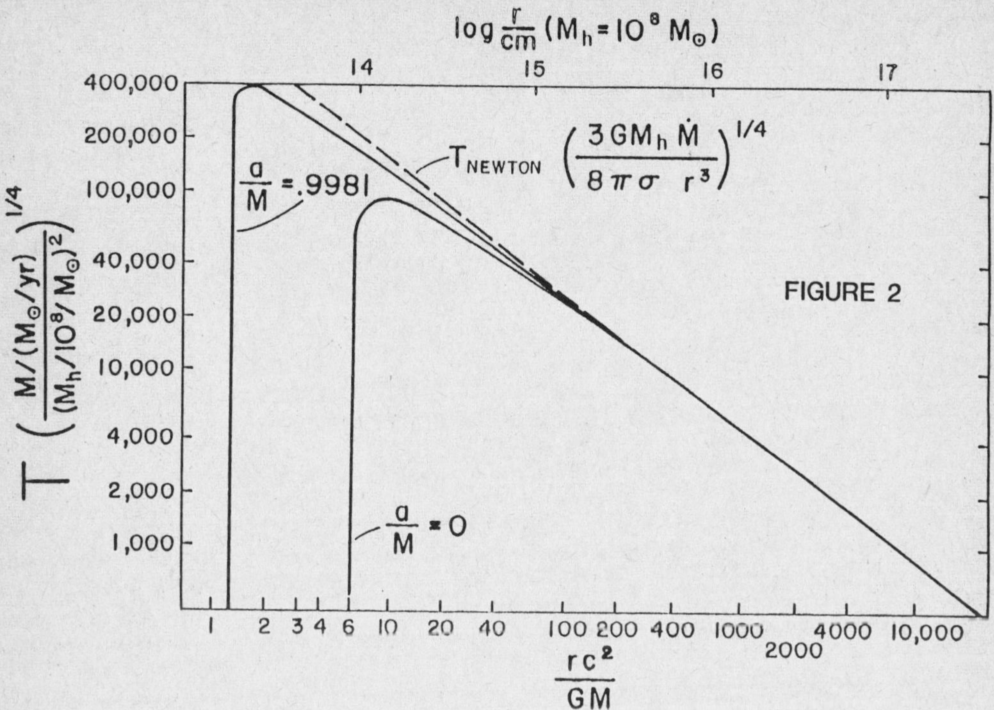

rise like $r^{-3/4}$, as they would in a Newtonian disk. Instead they peak and actually decrease rapidly as the last marginally stable orbit is approached. The effect of doppler shifts, gravitational redshifts, and gravitational lensing on photons leaving the disk must also be included (Cunningham 1975; Malkan 1983). The spectrum of a Kerr disk depends sensitively on the observer's zenith (or polar) angle. An equatorial observer sees large blueshifts, due to "forward peaking" and gravitational focusing of the radiation emitted in the inner regions of the disk. A polar observer sees all light redshifted; the cooler outer regions of the disk are focused and the hot center defocused. While both observers receive nearly the same integrated flux, the equatorial observer sees photons with average energies nearly 10 times those seen along the pole.

The thermal ultraviolet spectrum of the optically thick, geometrically thin accretion disk is determined by only two independent parameters, the mass of the black hole and its accretion rate. These simple models produce excellent fits to the multiwavelength spectral energy distributions of quasars down to 500 Å (Malkan 1983). The fits to the spectra of luminous quasars implied that many had black holes accreting at nearly their Eddington limits. However, two systematic shortcomings of the models probably led to overestimates of the accretion rates. Since most of the disks are probably

not viewed face-on, gravitational focusing would amplify the far ultraviolet
radiation from their inner parts. Also, since their atmospheres are dominated
by electron scattering opacity, the locally emerging disk spectrum will
appear harder in the ultraviolet than would a blackbody of the same effective
temperature. The simple models did not account for either efffect, and thus
required larger mass accretion rates to produce spectra as hard as those
observed.

In addition to studying its spectral shape, we can also learn about the
origin of the continuum by examining its variability and polarization. We and
several other groups are in the process of analyzing AGN variability; here we
consider the polarization.

§3. Polarization of Thermal Continuum Emission from Accretion Flows

In the inner regions of massive accretion disks, as in the atmospheres of
hot stars, the dominant source of opacity is free-electron scattering. The
scattered radiation is linearly polarized perpendicular to the plane formed by
the incident polarization and outgoing direction vectors. When the
distribution of scatterers is not azimuthally symmetric about the observer's
line of sight, as in an inclined disk, there will be a net polarization. Thus
polarization measurements can be used to test models of accretion flows.

The magnitude and direction of the polarization is determined by the
scattering and absorption optical depths, and the geometry of the photon
source, the scatterers, and the observer. By symmetry the orientation of the
polarization can only be parallel to the disk's symmetry axis, or parallel to
the plane of the disk, as projected in the observer's plane of the sky. In
general, disk polarizations range from zero for disks viewed face-on, to
nearly 12% when seen edge-on. The polarizations are parallel to the symmetry
axis (positive) at low optical depths, reach a maximum, decline through zero,
and become parallel to the disk's equatorial plane (negative) at high optical
depths. This 90° rotation in polarization position angle with increasing
depth arises from changes in the predominant photon directions just prior to
the last scattering. At low optical depths photons traveling parallel to the
disk's surface have the best chance of scattering, and they are most likely to
emerge with polarizations parallel to the disk symmetry axis. At high optical
depths, those photons traveling perpendicular to the disk's surface are most
likely to escape, and to have polarizations parallel to the disk symmetry
axis.

The emergent polarization from an α–disk depends principally on how the temperature and optical depth vary throughout the visible luminous region of the disk. The dependence of temperature and optical depth on radius in an α–disk (Figures 2 and 3) are such that most of the visible light arises in regions which are marginally to very optically thick. The dependence of temperature and optical depth on radius, as well as the shape of the torus, is more difficult to obtain agreement upon. Most models postulate that radiation pressure supported tori will be barely optically thick to very optically thin. These tori will also be fat, with comparatively little empty space in their centers. The α–disk and radiation-pressure (or ion-pressure) torus models predict quite different optical depths over the radii that emit most of the visible light. This difference, combined with the different ways in which disks and tori change their cross sections with inclination, suggests the polarization arising from these two accretion flows will be quite different.

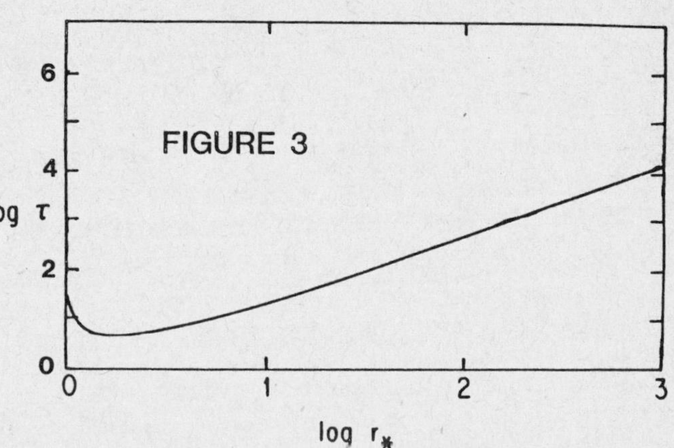

§4. Polarization Model Results for Flat Disks, Wedges and Tori

The polarization produced in the optically thin limit, or for very special optically thick geometries, can be computed semi-analytically. Numerical simulations are best for computing the polarization arising from bodies of arbitrary shape and of intermediate optical depth. In practice, the intermediate optical depth regime, where the optically thin and thick approximations break down, may encompass optical depths from more than a few tenths to over one hundred.

We developed a scattering code to compute the polarization of light originating in disks of arbitrary shape and optical depth. The code was easily modified to include absorptions, and illumination sources outside the disk. Since the polarization varies continuously with the observer's

inclination angle to the disk's symmetry axis, results are binned in .01 and
.10 intervals of the cosine of the zenith angle. Statistics are kept on the
number of scatterings and absorptions, and the number of emerging photons at
each zenith angle (i.e. the limb darkening). The escape paths of one million
photons can be calculated in a few hours of CPU time on the UCLA Astronomy
Department's VAX 11/750.

We calculated polarizations for flat disks, wedges, and tori over a range
of optical depths (Webb and Malkan 1986). We also presented polarizations
for different photon source distributions, calculated the effects of non-zero
absorption optical depths (i. e., albedos less than unity), and predicted the
amounts of limb darkening. Here we plot the polarizations calculated for a
flat disk and a wedge with a 17° opening half-angle, of optical depth six, and
a torus, of optical depth one half (Figure 4).

We found the polarization of wedges deviates only slightly from the flat disk, and only at high zenith angles. This deviation from the flat disk model increases with the opening angle of the wedge. The wedge geometry could account for the apparent deficiency of normal quasars with high polarizations if the typical half-angle were at least 10 degrees. Presumably the wedge would block our direct view of any highly inclined quasar disks, which would otherwise appear strongly polarized.

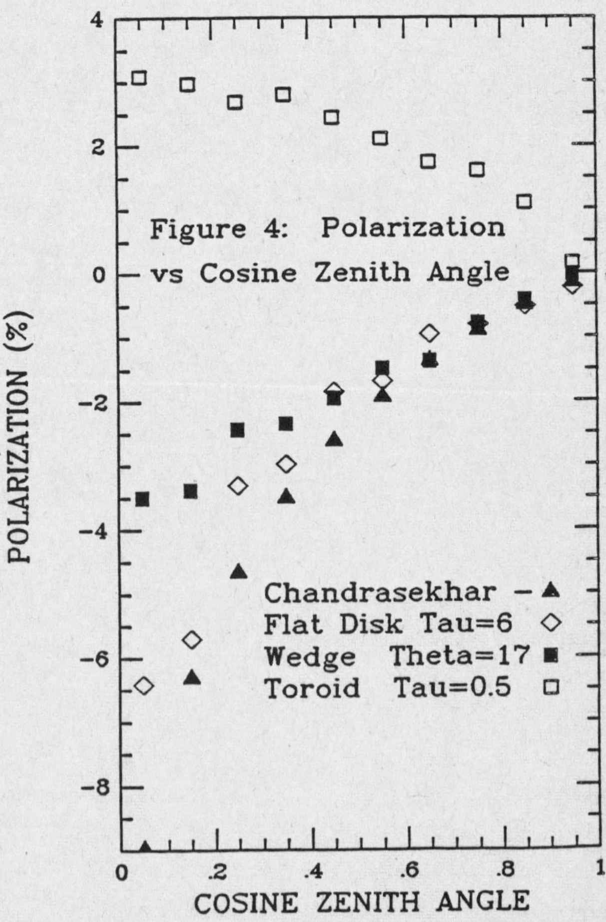

The polarization calculated for the torus is markedly lower than that of the flat or wedge-shaped disks. We averaged the polarizations for the torus over zenith angles of 60 to 90 degrees and investigated their range with optical depth (Webb and Malkan 1986). Peak polarizations of two to three percent were obtained for a narrow range of optical depths around $\tau = 0.5$; the very optically thick torus had a polarization approaching 0.5%.

A general relativistic treatment of the frame dragging around a rapidly rotating black hole predicts rotations in the plane of polarization of light emitted from the very innermost regions of a disk (Connors, Piran, & Stark 1980). The rotations are largest for those high-frequency photons originating at the smallest radii, resulting in a position angle rotation with wavelength. Detection of such a rotation of the polarization plane, perhaps in the far ultraviolet, would indicate the existence of a massive central object.

§5. Polarization Observations of Quasars

Polarization surveys and wavelength-dependence measurements can set constraints on the geometry and physical conditions in the continuum-emitting region that probably could not be obtained in any other way. From multiwavelength polarimetry we can determine if all the observed polarization arises from a single thermal ultraviolet continuum. If the "normal" quasar polarization originates from electron scattering in an accretion disk, then the position angle and magnitude of the polarization are diagnostics of the disk's optical thickness, its inclination, and its shape.

"Normal" low polarization quasars (LPQs) and Seyfert 1 galaxies have white light polarizations of 2% or less, at a fixed position angle indicating a single polarizing process. Our models of tori, flat disks of moderate optical depth, or wedges with opening angles of 10° or more have polarization magnitudes most consistent with the observations. Antonucci and Miller (1985) report that for many Seyfert galaxies, the optical polarization position angle is either parallel or perpendicular to the nuclear symmetry axis defined by the radio morphology. The orientation of this observed polarization is consistent with the interpretations of electron scattering in optically thin and thick disks, respectively.

The net polarized flux per unit frequency in both Mkn 290 (= PG 1543+580) and I ZW 1 (= PG 0051+12) rises steadily from the red into the blue, at a fixed position angle. Of all the continuum components only the thermal ultraviolet excess is increasing in intensity over this wavelength interval. In fact, the polarization of Mkn 290 (Figure 5) at all wavelengths could be attributed to a constant polarization of 3.5% of the ultraviolet excess component alone. The wavelength-dependence of polarization in I ZW 1 has also been measured in the U,B,V,R, and I bands. Again the observed polarization increases into the blue in exactly the same manner as the fitted ultraviolet excess continuum component. A paper further detailing the wavelength-dependence of polarization in quasars is in preparation.

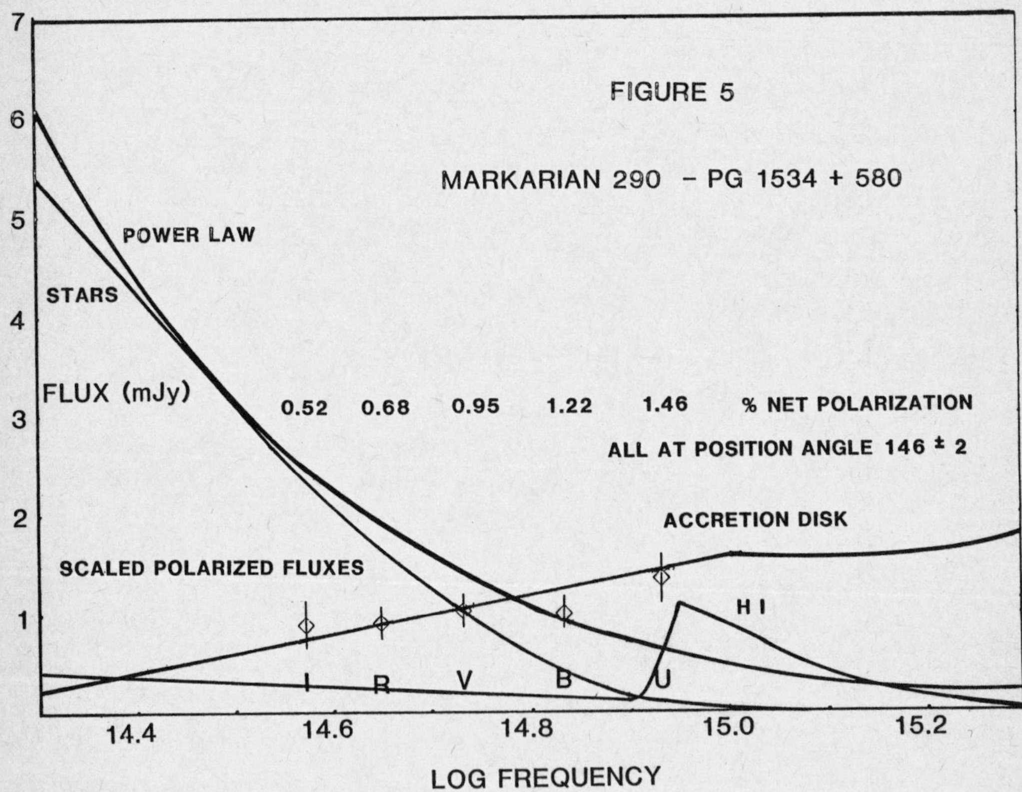

Stockman, Moore, and Angel (SMA 1984) measured the white light polarizations of 137 normal quasars, shown as a histogram in Figure 6. The smooth curve is the derived polarization distribution corrected for the bias of noisy measurements of a positive definite quantity. If there are no observational selection effects, then the distribution of inclination angles of inferred accretion disks should be random. The number of randomly distributed disks one expects to see at each inclination angle is simply proportional to the sine of the zenith angle. Using the predicted polarization at each angle for a given disk model, along with the expected distributions of inclination angles, one can predict the number of sources in a sample having each polarization.

SMA measured the combined polarization of all the continuum components in the optical spectrum. If all the polarization originates in the disk (thermal) component, then the measured polarization is less than that of the intrinsic disk polarization by the ratio of disk light to total light. The fraction of white light attributable to a disk will vary between objects, depending on their relative strengths of starlight, the Balmer continuum, the power law, and the thermal continuum.

We have estimated the fraction of light originating in the disk of a typical source, Mkn 290. Its multiwavelength spectrum was corrected for small amounts of reddening and starlight, and the red power law was extrapolated into the blue. The ultraviolet excess was fitted with a two-parameter model of an optically thick, geometrically thin, relativistic accretion disk. Mkn 290's measured white light polarization of 0.9%, divided by the fraction of its visible light originating in the disk (0.3), implies its disk light is actually about 3% polarized. Adopting Mkn 290's fraction of disk light as typical, and the predicted distribution of inclination angles for a random sample of disks, we calculated the number of quasars that would be seen with each polarization (Figure 7).

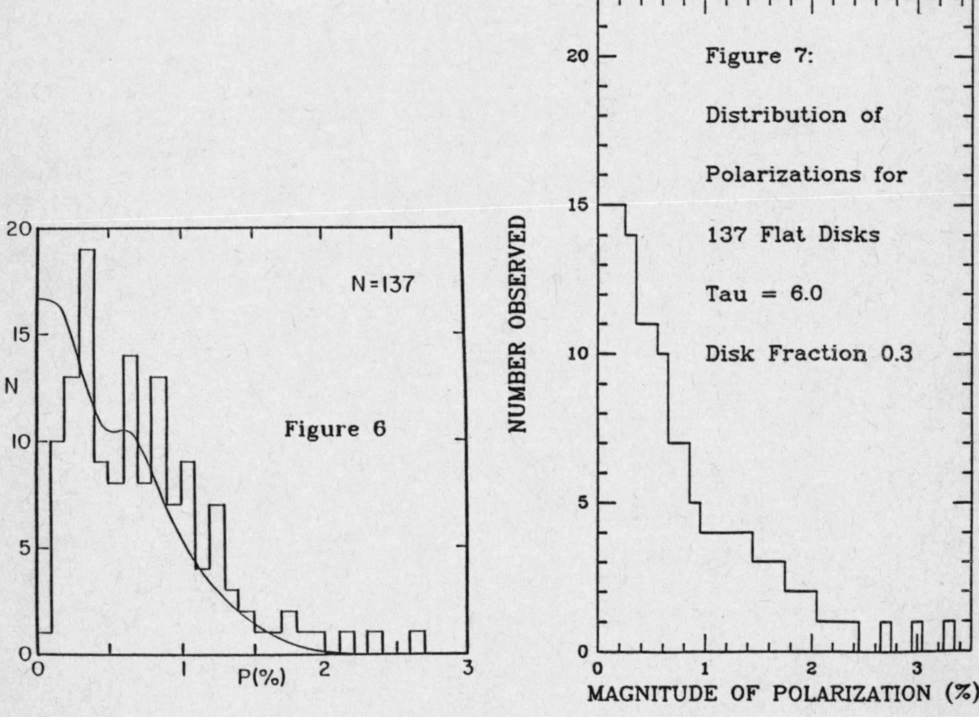

The flat disk polarization distribution is remarkably consistent with the observations. The calculated polarization distribution of tori (Figure 8) is dramatically different. Thin disks only appear highly polarized at high inclinations, hence their peak polarizations are rarely seen. Toroids show nearly their maximum polarization over a much broader range of angles; therefore too many relatively large polarizations are predicted.

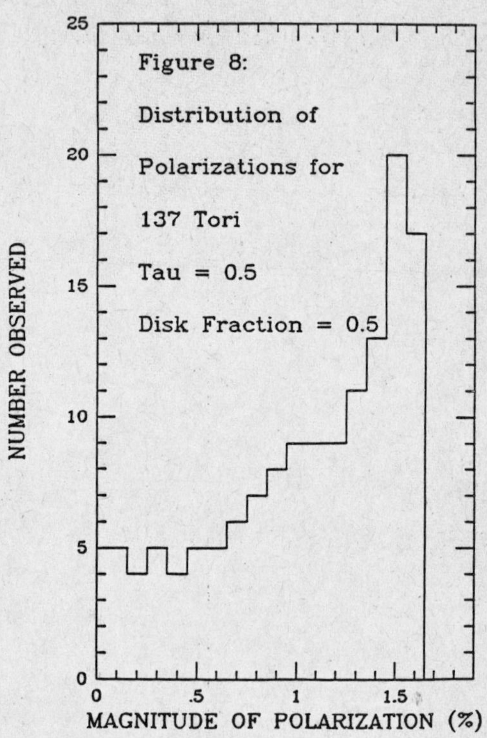

The different redshifts of quasars in the SMA sample cause the disks in otherwise identical objects to produce different fractions of the optical light. As a result, the polarization dilution should decrease with redshift. In many of the high-redshift SMA quasars, the disk would be expected to produce up to 70% of the white light continuum. If these disks were flat and optically thick, some high-redshift quasars ought to be observed with white light polarizations of over 5%. That such high polarizations are not observed suggests the unobscured flat disk may be too simple a model. It is still unclear why, if polarization is proportional to the fraction of continuum produced by disk light, no correlation of polarization with redshift has yet been detected. Perhaps lower optical depths, high absorption coefficients, an appreciable wedge opening angle, or other observational selection effects are suppressing high polarizations.

Janine LaPerch helped in preparing this manuscript.

§7. REFERENCES

Bechtold, J. et al. 1984, ApJ 281, 76

Callahan, P. S. 1977, A&A 59, 127

Connors, P. A., Piran, T. and Stark, R. F. 1980 ApJ 235, 224

Cunningham, C. T. 1975 ApJ 202, 788

Edelson, R. and Malkan, M. 1986 ApJ 306, 1

Jaroszynski, M., Abramowicz, M. A. and Paczynski, B. 1980, Acta Astr., 30, 1

Kinney, A.L.,Huggins, P.J.,Bregman, J.N. and Glassgold, A.E. 1985, ApJ 291, 128

Malkan, M.A. and Filippenko, A.V. 1983, ApJ 275, 477

Malkan, M.A. and Sargent, W.L. 1982, ApJ 254, 22

Malkan, M.A. 1984, Proceedings of Garching Conference on X-Ray and UV Emission
     from Active Galactic Nuclei. MPE Report 184, eds. W. Brinkmann and J.
     Trumper, p. 121-128.

Malkan, M. A. 1983, ApJ 268, 582

Novikov, I.D. and Thorne, K.S. 1973, Black Holes, ed. De Witt, C. and
     De Witt, B. S., Gordon and Breach, New York

Rees, M. J., Begelman, M. C., Blandford, R.D. and Phinney, E.S. 1982,
     Nature 295, 17

Sakimoto, P. J. and Coroniti, F. V. 1981, ApJ 247, 19

Shakura, N. I. and Sunyaev, R. A. 1972, A & A 24, 337

Stockman, H. S., Moore, R. L. and Angel, J. R. 1984, ApJ 279, 485

Webb, W. and Malkan, M. 1986, Proceedings of a Conference on Active Galactic
     Nuclei, N.O.A.O., Tucson, ed. M. Sitko.

Wills, B. J., Netzer, H., and Wills, D. 1985, ApJ 288, 94

Accretion Disks in Low-Mass X-ray Binaries

Keith O. Mason

Mullard Space Science Laboratory
University College London
Holmbury St.Mary, Dorking
Surrey, U.K.

1. INTRODUCTION

The bright, accretion-driven X-ray sources in the Galaxy can be divided into two main types depending on the nature of the mass-donating star in the system. The present paper concerns binary systems in which a compact star (a neutron star or possibly a black hole) accretes matter from a low-mass companion which is overflowing its Roche lobe. These are known as the Low-Mass X-ray Binaries, or LMXRB. The compact star is usually the more massive of the two components in these binaries, and in most cases reprocessed X-radiation completely dominates the intrinsic energy output of the mass-donating companion. The other type of luminous Galactic X-ray source, the high-mass systems, have large, early-type stellar companions. The compact star in the high-mass systems, usually a spinning, magnetised neutron star, is thought to accrete matter primarily from the stellar wind of its companion. The high mass systems are the subject of the contribution by Stella, White and Rosner in these proceedings.

Considerable progress has been made in recent years in the study of the Low-Mass X-ray binaries. Prior to 1980, the orbital periods of only a few (possibly atypical) members of the class were known. An increase in the sensitivity of optical and X-ray observations, occasioned in large part by the advent of CCD detectors and the availability of highly sensitive X-ray observatories such as EXOSAT, has now allowed us to measure the orbital periods of a substantial number of these systems. Knowledge of the orbital parameters of LMXRB is vital if we are to understand their make-up and origin; but of equal interest are the unexpected ways in which the orbital signature is seen, particularly in the X-ray flux. These orbital signatures convey knowledge of previously unsuspected structure in the accretion disks of these systems.

In this paper I review our present knowledge of LMXRB, concentrating on those phenomena which we believe owe their origin to complex structure in the outer parts of their accretion disks. There is not space to cover all aspects of LMXRB research; in particular, X-ray bursts, the structure of the inner accretion disk, and optical studies are only briefly mentioned, if at all. Further information on these topics can be obtained from the recent reviews by Lewin and Joss (1983), van Paradijs (1983), Tanaka (1985) and White and Mason (1985).

In section 2 I introduce the various kinds of orbital modulation seen in LMXRB before reviewing observations of individual sources in section 3. In section 4 I examine the orbital period distribution of LMXRB, while sections 5 and 6 focus in more detail on the disk structure that is required to explain the data. Finally, sections 7 and 8 consider what can be learnt about the mass and chemical composition of the Roche lobe-filling star in LMXRB.

2. X-RAY ORBITAL MODULATION IN LMXRB

Three kinds of orbital modulation can be identified in the X-ray light curves of LMXRB whose orbits are suitably inclined to the line of sight: (i) eclipses of the X-ray source by the companion; (ii) smooth, quasi-sinusoidal modulations of the X-ray flux; and (iii) irregular dips in the X-ray flux which are very variable in width, depth and morphology, but nevertheless occur periodically.

Eclipses of the X-ray source by the companion were an obvious phenomenon to look for in LMXRB. The other two types of periodic modulation, though, were not anticipated. In order to understand them, it has been necessary to move away from the idea that accretion disks are thin and symmetric. Instead a picture has evolved in which accretion disks in LMXRB have substantial thickness, at least at certain radial distances from the accreting star. Variations in the thickness of the disk around its circumference, fixed in the frame of reference of the binary, are believed to give rise to the orbital modulations seen.

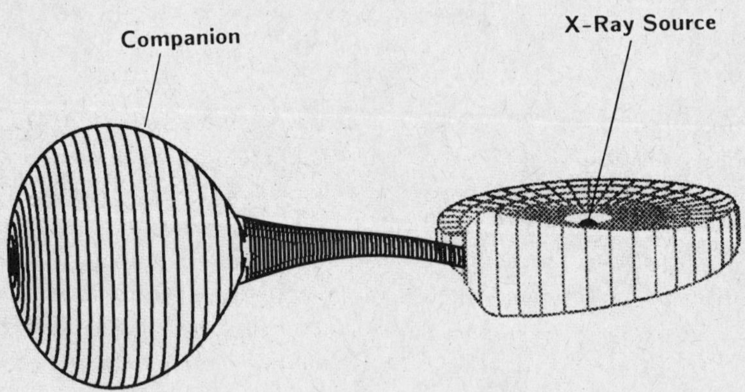

Figure 1 Sketch-model of a Low Mass X-ray Binary seen at about phase 0.8. The orbital inclination in this diagram is such that the X-ray source is obscured by structure on the disk for part of the orbital cycle, producing X-ray dips.

The arguments that lead to the above model are reviewed in Section 5. However, the basic concept can be appreciated readily by reference to Figure 1 which shows the main components of an LMXRB viewed at an orbital inclination of about 80 degrees (i.e.

10 degrees above the orbital plane). There is an obscuring rim in the accretion disk at a distance r from the X-ray source whose thickness varies around the disk. The maximum height of the disk rim above the orbital plane is h. Figure 1 is drawn with the obscuring rim at the outer edge of the disk, but the underlying concept is not sensitive to the exact radial location of the obscuring material within the disk. For orbital inclinations, i, such that $(\pi/2-i) < h/d$, the highest parts of the accretion disk rim will, for part of the orbital cycle, obscure the point-like X-ray source that lies at the center of the disk. The X-ray photons are absorbed by the material in the disk rim to produce dips that recur with the orbital period. The dips are highly structured and irregular because of turbulence and inhomogeneities in the disk material that is responsible for the absorption.

The quasi-sinusoidal X-ray modulations observed in some LMXRB are produced in a similar way, but in systems where the point-like X-ray source is permanently obscured from view by the disk rim. In these binaries we see only X-rays that are scattered into the line of sight by hot, optically thin material above the disk plane. This material is usually referred to as an Accretion Disk Corona, or ADC. The ADC has dimensions comparable to that of the accretion disk, and much larger than the X-ray emitting compact star. The fraction of the ADC that is visible is modulated by the structure on the disk to give a smooth, quasi-sinusoidal, X-ray flux variation. Eclipses of the X-radiation by the mass-donating star in ADC sources are expected to be gradual, and quite probably partial, because the effective size of the X-ray source is comparable to that of the companion star. This expectation is borne out by the available data, as discussed in Section 3.1.

3. REVIEW OF OBSERVATIONAL MATERIAL

Table 1 summarises data on LMXRB with known orbital periods. The periods range from less than an hour to almost ten days. The techniques used to establish these periodicities are indicated in the Table, and include X-ray photometric observations, optical spectroscopy and photometry, and, in the case of the two X-ray pulsars in the group, pulse timing studies. Table 1 also lists the X-ray flux or flux-range of each source, and its optical V-band magnitude where known, together with the depth of the orbital modulation in each band. The latter is expressed as a ratio of minimum to maximum flux in the X-ray range, and as a full amplitude in magnitudes in the optical band. The X-ray modulation depth of the dipping sources changes with time, so the maximum depth recorded to date is quoted in the Table. An estimate of the distance of each source is included where known. The reference numbers refer to the list at the end of this paper. The prefix 'X' is used to identify X-ray sources referred to by equatorial coordinates, coupled with a 'B' if it is a known burst source, or 'T' if it is a transient source.

In the remainder of this section I discuss in more detail the observational data on individual LMXRB sources, concentrating on those which provide evidence for an accretion disk corona (ADC) and/or disk structure. Readers who are not interested in such details are invited to skip to section 4. I also discuss briefly some X-ray observations of cataclysmic variables which may indicate that similar accretion disk structure exists in these systems. Sources within each category are dealt with in order of increasing orbital period.

3.1 LMXRB ADC Sources

3.1.1 X2129+470 (V1727 Cyg) $P=5.2$ hr

This source has a 5.2 hour period which was first noticed as an energy dependent modulation in optical light (Thorstensen et al. 1979). The optical light-curve has a full amplitude of about 2 magnitudes in the U band, decreasing to 1.2 magnitudes in V, and is quasi-sinusoidal, but with a narrower minimum than maximum (McClintock, Remillard, and Margon 1981; Thorstensen et al. 1979). A substantial fraction of the optical radiation of X2129+470 is believed to come from the X-ray heated face of the companion star whose changing aspect with orbital phase gives rise to the observed modulation. Departures of the light curve from a pure sinusoid may be caused by the eclipse of a bright accretion disk (McClintock, Remillard, and Margon 1981; Mason and Córdova 1982b).

The X-ray emission of X2129+470 suffers a partial eclipse, lasting about one hour in total, which is coincident with optical minimum (McClintock et al. 1982; White and Holt 1982; Ulmer et al. 1981). The depth of eclipse is independent of X-ray energy and at minimum the flux drops to about 20% of its peak value. The partial X-ray eclipse together with the low L_x/L_{opt} ratio of the source is taken to mean that the compact X-ray emitting star is always blocked from direct view by the accretion disk, and that X-rays are seen by virtue of scattering off extended, hot material above the disk plane (the ADC). There is some evidence for structure in the X-ray light curve outside of eclipse (White and Holt 1982), but this does not appear to be stable in phase. Mass transfer is not continuous in X2129+470; Since 1983 the source has not been detected in the X-ray band, and has been faint optically (V~18; Pietsch et al. 1986).

3.1.2 Cyg X-3 $P=4.8$ hr

Cygnus X-3 is a spectacular flaring radio and infrared source, a luminous X-ray emitter, and probably a copious ultra-high energy gamma ray source (see Mason, Córdova, and White 1986, and references therein). One of the principle X-ray characteristics of Cyg X-3 is a 4.8 hour, quasi-sinusoidal, periodic modulation. Because of its high stability, this is believed to be the orbital period of the binary. White and Holt (1982) have suggested that the X-ray modulation is the result of obscuration of an extended ADC by structure on the edge of the accretion disk. The height required of the disk structure is extreme, but is perhaps in keeping with the other extreme characteristics of this system. It should be emphasised that no eclipse of the ADC by the companion star is seen in Cyg X-3, so the validity of the model in this case is less secure than for the other ADC sources discussed.

3.1.3 X1822-371 (V691 CrA) $P=5.57$ hr

As with X2129+470, the periodic modulation of radiation from X1822-371 was first detected in the optical band (Mason et al. 1980). The optical light curve of X1822-371 consists of a minimum about 1 magnitude deep which lasts in total about one hour, preceeded by a gradual decline of about 0.5 magnitudes that lasts for about one third of the 5.57 hour cycle. The shape of the light curve is significantly different from that of

TABLE 1: Orbital Periodicities in LMXRB

Source	Period (hours)	Type[1]	F_x (μJy)	F_x^{min}/F_x^{max} (orbital)	V	ΔV (orbital)	d_{kpc}	Reference
XB1820–303	0.19	x	61–410	0.97			8.5	64
X1627–673	0.70	p	10–30		18.5			43
XB1916–053	0.83	d	30	0.0			~10	81,88,65,63
XB1323–618	2.9	d	5	0.4				76
XB1636–536	3.8	o	140–390		17.5	0.25		56
XBT0748–676	3.8	e,d,o	~0.03–50	0.04	17–R>23	0.6	~10	52,54,80
XB1254–690	3.9	d,o	50	0.1	19.1	0.5	10–14	13,47
X1755–338	4.4	d,o	30–120	0.7	18.6	0.5	2–9	87,38
XB1735–444	4.6	o	110–290		17.5	0.16		11
Cyg X-3	4.8	a,i	90–430	0.4	11.5^2	0.3^2	11	53,60,35
X2129+470	5.2	a,e,o	0–30	0.2	16.4–18	1.2		71,41,39,85,57
X1822–371	5.6	a,e,o,s	30	0.5	15.4	1.0	1–5	84,36
LMC X-2	6.4	o	9–44		18.8	0.5	50	46
XBT1658–298	7.2	e,d	<5–80	~0	18.3			10
XT0620–003	7.3	s	<0.02–5E4		11–18		1	40
X1957+115	9.3	o	16–78		18.8	0.23		70
Sco X–1	19.2	o,s	2E4–7E4		12.0-13.4	0.2	~0.7	23,14
X1624–490	21.	d	20–90	0.75				6
Her X–1	40.8	p,e,o,s,d	4–110	0.05	13.2	1.2	~5	69,22,19,2
X0921–630	216.3	a,e,o,s	2	0.3	16	1.5	~8	17,34,31
Cyg X–2	235.2	s,d	220–750	0.7	14.6		~8	15,79

Notes: 1. *Type of Orbital Modulation* — a = ADC Source; e = Eclipses; d = X-ray Dips; o = Optical modulation; x = X-ray modulation (type uncertain); i = Infrared modulation; s = Optical Line Doppler shifts; p = Pulsar Doppler shifts.

2. Infrared (K-band) magnitudes.

X2129+470 and HZ Her/Her X-1. Also unlike the latter stars, the amplitude of the modulation in X1822–371 does not change significantly with wavelength in the optical band, although it does become noticeably deeper in the UV shortward of about 2000A (Mason and Córdova 1982a). The optical continuum spectrum is blue, and has superposed on it emission lines of HeII 4686Å and the CIII/NIII blend at 4640Å, together with broad, dish-shaped H and HeI absorption lines that are characteristic of a disk (Mason et al. 1982). The strength and profile of the HeII 4686Å emission line are modulated with the 5.57 hour period. The line flux varies by a factor of two and peaks at phase 0.8 with respect to the continuum minimum (Mason et al. 1982). The He II line emission is probably produced in material that is recombining in the shadow of the accretion disk, and both the flux and profile changes of the line suggest there is a concentration of emission in the direction of phase 0.8, consistent with the idea that the disk is thicker here. The radial velocity of the base of the He II line is modulated at the 5.57 hour period with a semi-amplitude of $\sim$70 km s^{-1} (Cowley, Crampton and Hutchings, 1982a; Mason et al 1982). The phasing of these velocity changes suggests that this is the orbital velocity of the disk.

The X-ray light curve of X1822–371 is made up of a narrow, partial eclipse that lasts about 15 minutes, and is coincident with the optical minimum, together with a broad, quasi-sinusoidal modulation which has a minimum about 0.2 cycles earlier than the narrow eclipse (White et al. 1981; White and Mason 1985). The X-ray flux drops to $\sim$50% of its maximum value during the minima, independent of photon energy. The narrow minimum in the X-ray band is interpreted as an eclipse of an ADC by the companion star, whereas the corresponding minimum in the optical band is thought to be an eclipse of radiation from the accretion disk by the companion. The broad modulation in both bands is believed to result from variations in the height of the accretion disk rim with azimuth (White and Holt 1982; Mason and Córdova 1982b). In the optical band this is supplemented by a small component due to the X-ray heated face of the companion.

3.1.4 Her X-1 (low state) $P=40.8\ hr$

The well-studied system Her X-1 (HZ Her) is an eclipsing binary containing a magnetised neutron star that is spinning with a 1.24 second period. Her X-1 exhibits high and low X-ray states that usually recur regularly every 35 days, but in the summer of 1983 an extended low state of the source was observed with EXOSAT (Parmar et al. 1985a). During these low states the central X-ray source in Her X-1 is believed to be obscured by the accretion disk. The small residual X-ray flux recorded during the low state has for a long time been explained it terms of scattering by material in the binary system. Parmar et al. (1985a) find that the low-state X-ray flux of Her X-1 suffers a partial eclipse by the companion star, and that the eclipse egress and ingress are gradual, the same characteristics exhibited by the eclipsing ADC sources. The size implied for the scattering region is similar to that of the accretion disk, and these observations provide the most compelling evidence thus far that the scattering material in Her X-1 is associated with the disk. The case of Her X-1 adds weight to the idea that the ADC sources are essentially the same as other LMXRB except that we cannot see the central X-ray source directly.

3.1.5 X0921–630 $P=216.3$ hr

The orbital period of X0921–630 is the second longest known among the LMXRB, and was determined from optical photometric and spectroscopic measurements of its $16^{th.}$ magnitude optical counterpart (Branduardi-Raymont et al. 1981,1983; Chevalier and Ilovaisky 1981, 1982; Cowley, Crampton and Hutchings 1982b; Mason et al. 1985; Thorstensen, Charles and Tuohy 1983). Observation of a persistent, narrow, optical minimum near the predicted time of inferior conjunction of the mass donating companion suggested that this was an eclipsing system (Thorstensen, Charles and Tuohy 1983; Mason et al. 1985). This idea was confirmed by Mason et al. (1986) who recorded a partial X-ray eclipse at the same orbital phase using EXOSAT. The X-ray eclipse lasts for over a day and the flux drops to 30% of its maximum value at eclipse center. The partial nature of the eclipse and the fact that its depth is independent of X-ray energy strongly favours a model in which the X-rays we see are scattered off an extended ADC. The X-ray emitting compact star in this system is presumed to be hidden from direct view by the accretion disk, which is in accord with the star's unusually low X-ray to optical flux ratio ($\sim$1).

3.2 LMXRB Dip Sources

3.2.1 XB1916–053 $P=0.83$ hr

This source has the shortest dip recurrence period known and was the first short-period source in which dips were observed (Walter et al. 1982; White and Swank 1982). The dips vary considerably in length and depth, and are sometimes entirely absent; but like most X-ray dip sources, they are always accompanied by a hardening of the spectrum. "Anomalous" dips often occur approximately mid-way between the "main" dips. The anomalous dips exhibit the same spectral signature as the main dips, and can be just as prominent, so that it is not always possible to distinguish them. Examples of the light curve of XB1916–053 are shown in Figure 2 based on data taken with EXOSAT (Smale, Mason, and White, 1986). Note that on some occasions both the main dips and the anomalous dips can last for about half the 50 minute cycle, so that there is then no part of the light curve that can be claimed to be dip-free. There is uncertainty about the optical counterpart of XB1916–053. There is a $\sim$17th magnitude star within the 3 arcsec radius HRI error circle, but its optical spectrum indicates that it is a normal K star. Walter et al. (1982) suggested that the counterpart was a $\sim$$22nd.$ magnitude star about 3 arcsec away from the K star, but this suggestion lacks confirmation.

3.2.2 XB1323–618 $P=2.9$ hr

A series of twelve consecutive dips, each lasting about one hour and recurring with a 2.9 hour period, have been reported by van der Klis et al. (1985b) in a 1.4 day EXOSAT observation of XB1323–618. A single dip was also observed near the center of an earlier 4 hour observation (van der Klis et al. 1985a). The source flux shows considerable variability on timescales less than one hour during the dips, but does not reach zero intensity. The X-ray spectrum hardens during the dips.

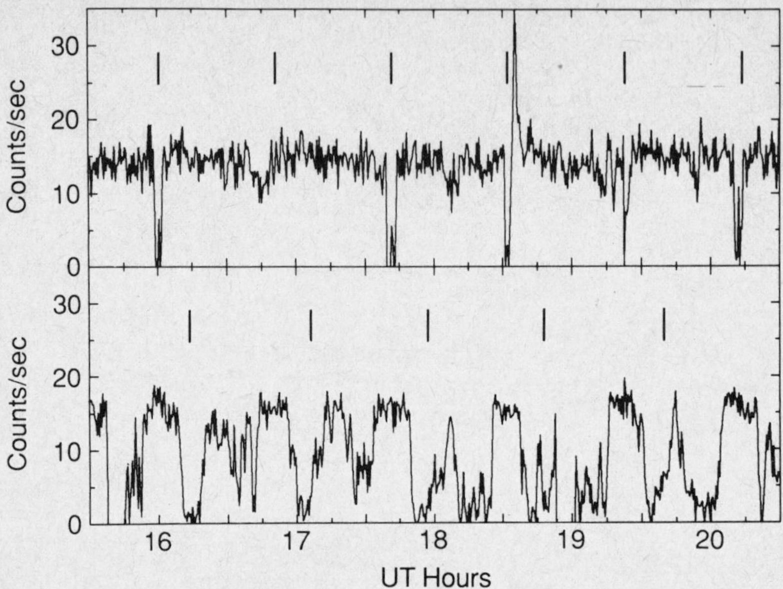

Figure 2 Part of two EXOSAT 1.5-10.0 keV observations of XB1916–053 illustrating changes in dip behaviour. Vertical bars are placed every 50 minutes to mark the underlying periodicity of the source. The *top panel* of data were taken in 1983 September when narrow dips were occuring approximately every 50 minutes. The dip at ~16:50 U.T. was early, however, and shallow. A shallow 'anomalous' dip is seen at ~18:10 U.T., approximately 180° out of phase with the rest. A burst, truncated in the figure, occured just after the dip at 18:30 U.T. The data in the *bottom panel* were taken in 1985 May. Both primary and anomalous dips were much broader during this observation.

3.2.3 XBT0748–676 *P=3.8 hr*

This transient source is one of only two known dip sources in which the X-ray flux is eclipsed by the companion star. The other, also a transient, is XB1658–298 (section 3.2.6). XBT0748–676 was discovered and studied in the X-ray band by Parmar et al. (1986) using EXOSAT, who sampled mean source intensities that ranged over a factor of ten during its outburst. An example of the light curve of XBT0748–676 is shown in Figure 3. The eclipse lasts for 8.3 minutes. The ingress and egress of the eclipse are rapid, but are resolved with the EXOSAT ME detector, and last for about 6 seconds. This time may reflect the scale height of the companion star's atmosphere. The eclipses are not completely total, the average residual flux being about 4% of the un-eclipsed signal in the 2-6 keV range. The residual flux must come from an extended source, and may be scattered photons from an ADC.

The dips in XBT0748–676 show characteristics (variability and energy dependence) that are typical of the class, and the deepest recorded extinguish ~80% of the 2-10 keV

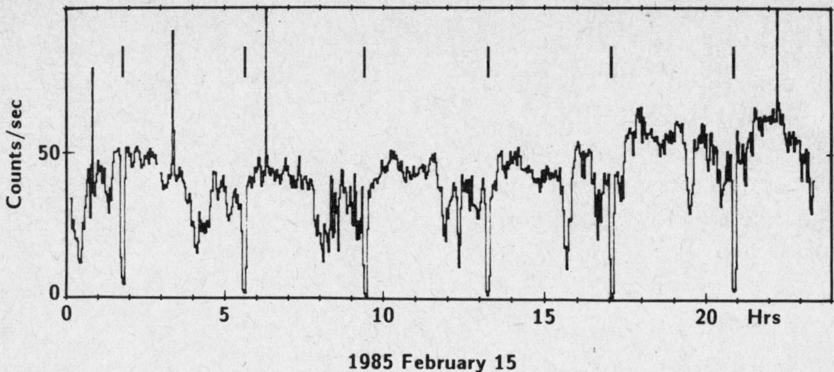

Figure 3 An example of the 1-10 keV light curve of XBT0748–676 (taken from Parmar et al. 1986) which shows periodic eclipses of the X-ray source by the companion star, as well as the irregular dips. The eclipses, which occur every 3.8 hours, are marked with vertical bars. Four X-ray bursts occured during the observation.

X-ray flux. Measured from the time of eclipse, the dips occur primarily between phases 0.8 and 1.2 and near phase 0.65. The structure of the dips changes with time and/or the overall intensity of the source. There is a tendency, for instance, for the dip at phase 0.65 to be more prominent when the overall source intensity is highest.

XBT0748–676 is also a transient object in the optical band, and in February 1985 was observed by Pedersen and Mayer (1985) as a V=$17^{th.}$ magnitude star. Optical spectroscopy reveals a blue emission line spectrum characteristic of the LMXRB (Pedersen et al. 1985; Mouchet, Angebault, and Ilovaisky 1985; Crampton et al. 1986). The optical light is modulated by ~0.6 magnitude on the 3.8 hour period (Pedersen et al. 1985; Wade et al. 1985; Crampton et al. 1986). Minimum light is coincident with X-ray eclipse and the optical light curve suffers erratic variations in shape from cycle to cycle.

3.2.4 XB1254–690 $P=3.9$ hr

X-ray dips in XB1254–690 were reported by Courvoisier et al. (1986). The dips are erratic and energy dependent, varying in depth between an upper limit of 20% (i.e. no significant dip detected) to 95% extinction of the 1-10 keV signal. The dips last for about 0.8 hours, or about 20% of the cycle. As with all dip sources studied, there is considerable variability within the dips, in this case at least down to one second.

XB1254–690 has a V=19 optical counterpart. The 3.9 hour periodicity of the system was independently discovered in the optical band by Motch et al. (1984) and takes the form of a quasi-sinusoidal modulation with a semi-amplitude of 0.3 magnitudes in V. The optical minimum occurs ~0.2 cycles after the center of the X-ray dips.

3.2.5 X1755–338 $P=4.4$ hr

X1755–338 is unusual in that, with the possible exception of XB1837+049 (see below), it is the only dip source known in which the depth of the dips is independent of X-ray energy (White et al. 1984; Parmar et al. 1985b; Mason, Parmar, and White 1985). The dips in X1755–338 are also consistently shallow, with a maximum source extinction of about 30%. In all other respects, though, the characteristics of the dips in X1755–338 are typical of the class; they last for about 20% of the cycle and have an irregular morphology with flux variability within the dips down to timescales of at least 30 seconds. Like XB1254–690, the 4.4 hour period of X1755–338 is also seen as a modulation in the light of its V=18.5 magnitude optical counterpart, with a full amplitude of 0.4 magnitudes (Mason, Parmar, and White 1985). The phasing of the X-ray and optical modulations is also the same in the two stars, the center of the dips preceeding optical minimum by about 0.2 of a cycle (Figure 4).

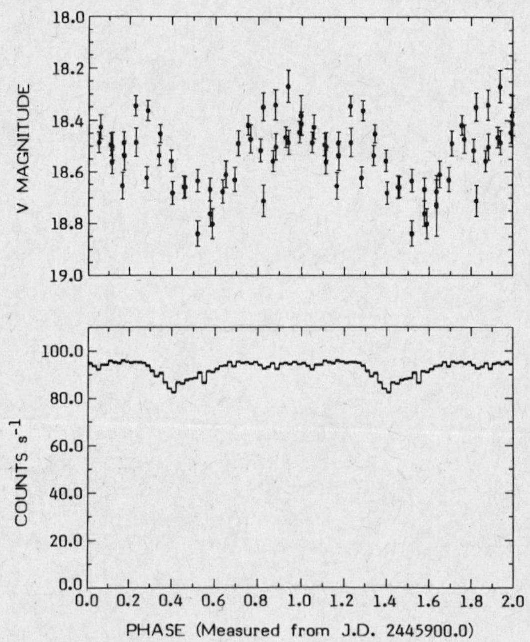

Figure 4 X-ray (1.5-3.5 keV; *below*) and optical (V band; *above*) photometry of X1755–338 folded on the 4.46 hour period of the source. Phase is measured from J.D. 2445900.0. Note the shallowness of the dips, and their phasing with respect to the optical minimum. From Mason, Parmar, and White (1985).

The peculiar energy independence of the dips in X1755–338 has been explained as a deficiency of heavy elements in the material that is obscuring the X-ray source. These are the elements that absorb photoelectrically in the 1-10 keV energy range and their absence leaves scattering by electrons, which is energy independent, as the main source of opacity in the obscuring gas. The required abundance deficiency is $< 1/600^{th}$ of the solar value. In support of this hypothesis we note that (i) the column density required to produce the shallow dips in X1755–338 by electron scattering ($\sim 5 \times 10^{23} cm^{-2}$) is similar to the values necessary to produce the deep dips in other sources by photoelectric absorption; (ii) X1755–338 is one of the few sources where there is no evidence for iron emission lines in the X-ray spectrum; (iii) no X-ray bursts have yet been observed from X1755–338 — low CNO abundance can increase the mean time between bursts (Taam 1980) reducing the probability that one will be detected.

3.2.6 XBT1658–298 $P=7.1$ hr

XBT1658–298 is, like XBT0748–676, a transient X-ray source that exhibits both dips and eclipses. Data on this object taken with the *SAS-3* and *HEAO-1* satellites during an outburst have been examined by Cominsky and Wood (1984). They find that the source exhibits erratic variations over 25% of its 7.1 hour cycle. At the end of this period there is a stable dip in the flux which is interpreted as an eclipse by the companion star. The length of the eclipse is estimated to be ~ 15 minutes.

3.2.7 XB1837+049 (Ser X-1) $P=13$ hr?

Ilovaisky and Chevalier (1986) have reported detecting two dips separated by about 13 hours during a 15 hour EXOSAT observation of XB1837+049. A previous 7.5 hour observation of the source did not record a dip, so the period of this system may be 13 hours. The two dips observed were both shallow (15% depth), lasted 30-40 minutes, and showed considerable time structure. Interestingly, like X1755–338, the depth of the dips in XB1837+049 was independent of energy.

3.2.8 X1624–490 $P=21$ hr

Three consecutive dips recurring with a period of 21 hours were found by Watson et al. (1985) during a 2.5 day EXOSAT observation of X1624–490. Part of another dip was also recorded during a previous, shorter observation. The dips last for between six and eight hours and are characterised by considerable structure on time scales of seconds to tens of minutes. During the dips the 1-10 keV flux never falls below $\sim 25\%$ of the mean level outside the dips. This, together with the behaviour of the source spectrum during the dips, has led Breedon and Watson (1986) to interpret the data in terms of two emission components, one point like, and the other more extended. It is extinction of the point-like source which causes the rapidly variable dip structure, and this is seen against the relatively constant extended emission that makes up about one quarter of the total flux. There is evidence, however, that the extended emission is also partially extinguished before and after the dips.

3.2.9 Her X-1 $P=40.8$ hr

It has been known since the earliest X-ray observations with the *Uhuru* satellite that dips occur in the X-ray light-curve of Herculis X-1 (Giacconi et al. 1973), a facet of the extensive and still poorly understood phenomenology of that source. Dips occur prior to the eclipse by the companion star (phases 0.6-0.9; the so-called 'pre-eclipse dips') and also between phases 0.3 and 0.6 (the 'anomalous' dips), and are accompanied by an increase in the absorption turn-over of the spectrum (Vrtilek and Halpern 1985; Voges et al. 1985). The occurence and orbital phase of the dips is related to phase in the 35-day cycle of the source (Crosa and Boynton 1980). Vrtilek and Halpern (1985) report narrow spikes of emission during the dips that recur with a period of $\sim$108 minutes, but these were not seen by Voges et al. (1985).

3.2.10 Cyg X-2 $P=235.2$ hr

Recent analysis by Vrtilek et al. (1986) of data on Cyg X-2 taken with the *OSO–8* and *Einstein* satellites has revealed dips in the X-ray flux that occur prior to the time of inferior conjunction of the companion star with respect to the X-ray source. The dips, which have an irregular morphology and extinguish up to 30% of the source flux, are associated with an increase in absorbing column density, and sometimes a reduction in spectral normalization as well. Other dips observed in Cyg X-2, which involve a change in normalization of the spectrum without any change in absorbing column, occur at all binary phases and may be an unrelated phenomenon.

3.3 CV Dip Sources

Cataclysmic Variables (CV) are low-mass accreting binary systems that contain a white dwarf as the accreting star, rather than a neutron star or black hole. Because of the parallels between CV and LMXRB, it is instructive to search for disk structure in CV that corresponds to that seen in LMXRB. A difficulty with doing this is that many of the phenomena we regard as diagnostic of disk structure in LMXRB rely on there being a bright X-ray source at the center of the disk to act as a probe, and this is rarely available in the CV. Three cases of dip-like phenomena have been recorded among the brightest X-ray emitting CV, and these are described below. However, it should be added that there is also evidence from optical studies for asymmetries in CV disks in the form of bright spots (e.g. Warner 1976; see also Horne and Marsh, this volume), and evidence for variations in the disk thickness in some magnetic CV systems (e.g. Alpar 1979; Chester 1979; Hassall et al. 1981). It is unclear at present whether these optical phenomena in CV have any direct bearing on the disk structure observed in the LMXRB.

3.3.1 EX Hya $P=1.63$ hr

The orbital period of EX Hya is firmly established by the presence of narrow, partial eclipses of the optical light by the companion star. A study of the X-ray orbital light curve of EX Hya is complicated by the fact that it is a magnetic system (see for example Watson, this volume) and a substantial fraction of the emission is modulated at the spin period of the white dwarf which is 67 minutes, or approximately two thirds of the 98 minute orbital period. Nevertheless, recent EXOSAT data make it clear that the X-ray flux also suffers a partial occultation coincident with the optical eclipse (Beuermann and Osborne, 1985; Córdova, Mason, and Kahn 1985; Mason 1985). Further, and of particular interest here, Córdova, Mason and Kahn (1985) have demonstrated that, prior to the eclipse, there is a broad, partial dip in the low-energy X-ray light curve (0.04-2.0 keV) that lasts for about one third of the orbital cycle. This is not present at higher X-ray energies, suggesting that it is caused by photoelectric absorption. Because of the strong 67-minute modulation, it is not possible to look at the broad orbital variation in

EX Hya on a cycle by cycle basis. However, its phase and energy dependence are very reminiscent of the dip phenomenon in LMXRB.

3.3.2 BG CMi (3A0729+103) *P=3.24 hr*

3A0729+103 is, like EX Hya, a magnetic CV and has a spin period of 15.2 minutes. 3A0729+103 exhibits a broad dip in its orbital light curve whose amplitude decreases with energy (McHardy et al. 1986; see also Watson, this volume). As with EX Hya, the parallel with the dips in LMXRB is suggestive.

3.3.3 U Gem *P=4.25 hr*

An even more striking case of a dip-like phenomenon in cataclysmic variables is afforded by U Gem. This dwarf nova is a very bright *soft* X-ray source during its outbursts. A series of EXOSAT 0.04-2.0 keV band observations of U Gem were made by the author in collaboration with F.A.Córdova, M.G.Watson and A.R.King, towards the end of a long outburst that occured in 1985. These data are shown in Figure 5. A series of dip-like phenomena, very similar in appearance to those found in LMXRB, are found in the U Gem data. The most prominent dip is initially seen at about phase 0.8. Later in the sequence, the dip at phase 0.8 becomes less prominent, and a second dip appears near phase 0.2. This behaviour is very reminiscent of the "anomalous" dips seen in LMXRB.

A word of caution when comparing the LMXRB dips to those in U Gem: The effective energy of the soft X-ray emission recorded from U Gem is of order 10 eV only, compared to a few keV for the LMXRB. Thus the column density of material required to produce the dips in U Gem by photoelectric absorption is much less than that required in the LMXRB.

4. ORBITAL PERIOD DISTRIBUTION OF LMXRB

Figure 6 illustrates the distribution of orbital periods in LMXRB. There is a concentration of systems with periods between about 3 and 8 hours. Analogy has been made between this group of LMXRB and certain types of cataclysmic variable which have a similar period distribution (e.g. White and Mason 1985); these are the classical novae and U Gem group of dwarf novae. There is a notable absence of LMXRB with periods between 1 and 2 hours, a region populated chiefly by AM Her variables and SU UMa-type dwarf novae among the cataclysmic variables.

In Figure 6 I distinguish those systems which exhibit X-ray dips and those where emission from an ADC dominates the X-ray flux. Although the total number in each group is small, there is no obvious distinction between the various sub-types in this plot, supporting the view that the dipping and ADC LMXRB are typical members of the class viewed at favourable inclination.

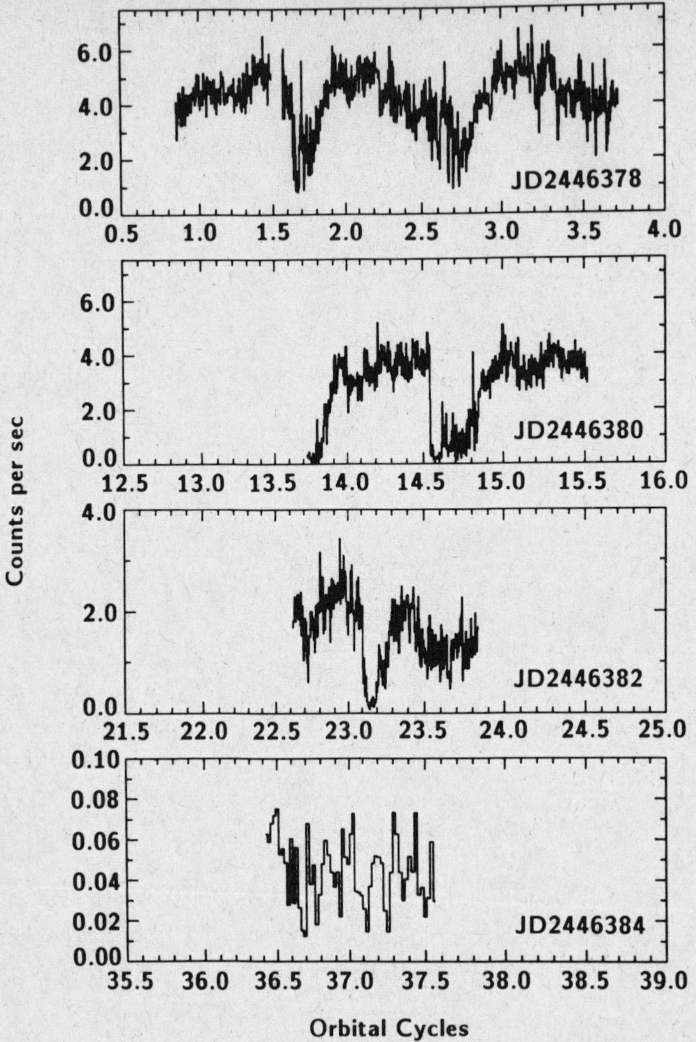

Figure 5 EXOSAT 0.04-2.0 keV observations of U Geminorum made during the decline from a long outburst in November 1985. The time axis is marked in units of orbital cycles from a common origin, zero phase being the time of a well established eclipse of the optical emission by the companion star.

5. INTERPRETATION OF THE DATA

In section 2 I outlined a framework within which the X-ray light curves of LMXRB could be understood. The picture that emerges is radically different from that which our previous understanding of accretion disk physics would have led us to believe. It is therefore especially important to trace critically the logic by which the model has been

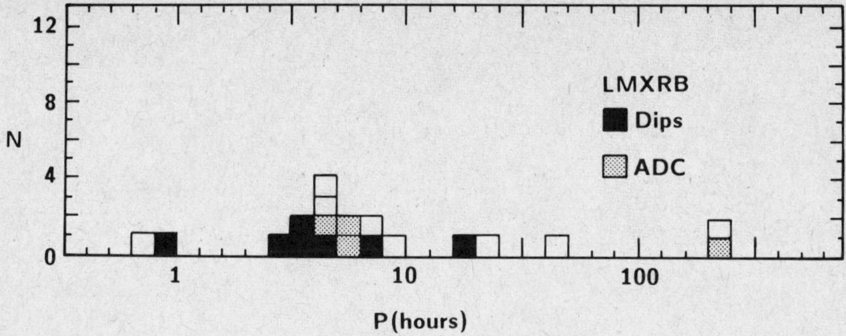

Figure 6 The orbital period distribution of LMXRB plotted on a logarithmic period scale. Accretion Disk Corona (ADC) sources, and those that exhibit dips, are indicated separately.

arrived at. The quantitative evidence is derived largely from the study of one source, X1822–371. In this section I review the interpretation of the X1822–371 light curve, and examine the basis for extending this interpretation to other LMXRB.

5.1 The light curve of X1822–371

The observational data on X1822–371 were summarised in section 3.1.3. Briefly, the X-ray light curve consists of a broad, quasi-sinusoidal modulation, together with a narrow minimum which is offset in phase from the minimum of the broad modulation. The fall and rise of the X-ray flux during the narrow minimum is gradual, and its depth is about 50% of the peak X-ray flux.

The narrow minimum in the X-ray light curve of X1822–371 is interpreted as a partial eclipse of an ADC by the companion star. This assumption is central to the model of X1822–371, since it establishes both the relative and absolute scale of the different components of the accretion disk. The assumption is based on the high stability of the narrow minimum in phase, depth and width over the time-span of almost a decade during which it has been monitored. The stability of the feature is illustrated in Figure 7 which shows 1.5-6.0 keV data from three EXOSAT observations of X1822–371, during which the narrow minimum was monitored a total of seven times. The times of these X-ray minima together with those measured with the *HEAO* satellites (White et al. 1981) are all consistent with a single recurrence period. Based on experience with other LMXRB and cataclysmic variable systems, it is difficult to imagine what could cause a stable feature of this kind other than an eclipse by the companion.

The mass-donating star in X1822–371 is assumed to fill its Roche limiting surface, and as such its size is only weakly dependent on its mass and that of the compact star in the binary. The width and depth of the dip produced when the mass-donating star eclipses the X-ray source thus allows us to measure the effective size of the ADC in this system, with an additional weak dependence on the orbital inclination. The value

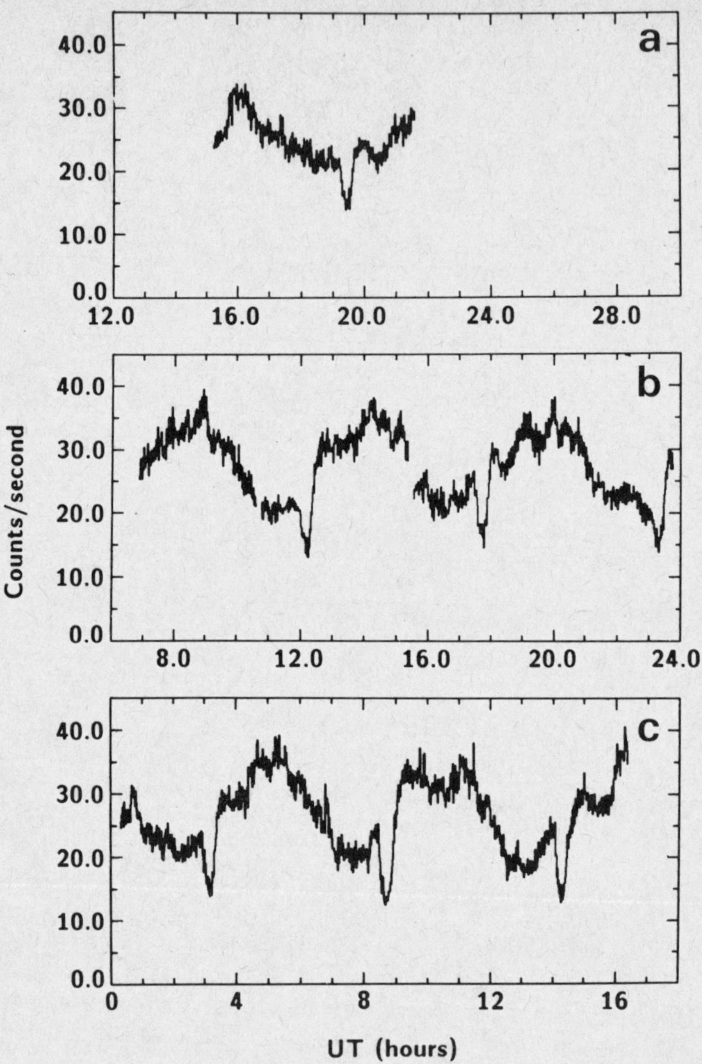

Figure 7 EXOSAT X-ray light curves of X1822–371 taken in the 1.5-6.0 keV energy range. The observations were made on (a) Day 280, 1983, (b) Day 263, 1984, and (c) Day 126, 1985.

obtained for the ADC radius is $\sim$0.3 $R_\odot$ (White and Holt 1982). In Figure 8 I have superposed the mean X-ray light curve of X1822–371 on the optical (B-band) light curve of the star (Mason et al. 1980). The Figure illustrates that the narrow X-ray dip is about half the width of the corresponding feature in the optical light curve. This establishes that the effective size of the ADC in X1822–371 is about half that of the optical emission region. A number of factors suggest that the optical light source being eclipsed by the companion is a bright accretion disk: (i) its size, based on the length of the dip, is ~0.6 $R_\odot$, which is about what is expected for a disk (ii) the optical light curve shows no short timescale flickering suggesting that the whole source is extended (i.e. there is no evidence for significant emission from a hot spot or other localised emission region), and (iii) broad dish-shaped optical absorption lines indicate a high rotational velocity.

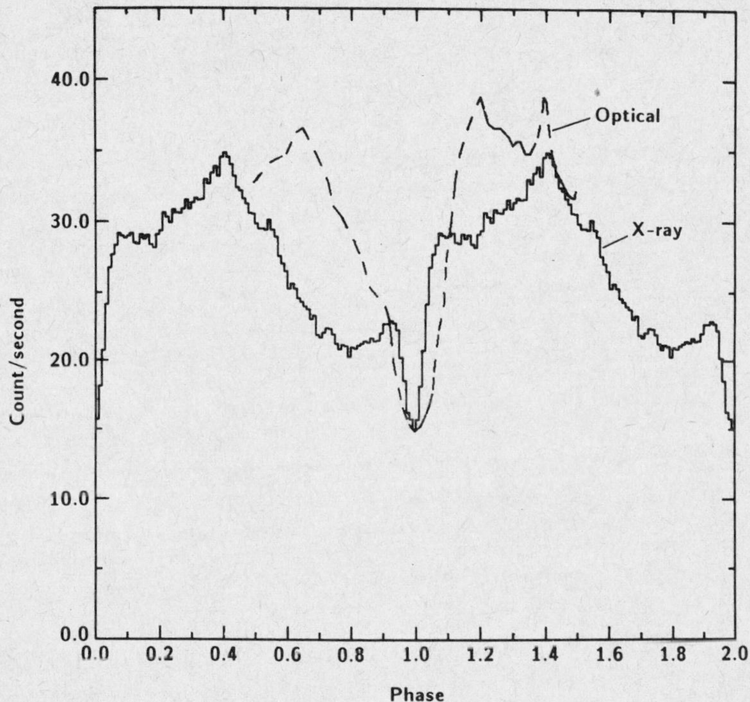

Figure 8 X-ray and optical light curves of X1822–371 aligned in phase and superposed. The scale on the ordinate refers to the X-ray curve which is the mean 1.5–6.0 keV light curve derived by folding all available EXOSAT data together. The optical light curve was measured in the B-band by Mason et al. (1980), and is also plotted on a linear ordinate scale. The ordinate scale of the optical light curve has been adjusted do that the minimum flux points in the two bands coincide.

It remains to establish the cause of the quasi-sinusoidal component of the X1822–371 light curve. The only viable mechanism that has been proposed to account for it is obscuration of the ADC. Of necessity, therefore, the obscuring material must be further from the X-ray source than the boundary of the ADC; i.e. at a distance of greater than ~0.3 $R_\odot$. The obscuring material must have an area that is comparable to the size of the ADC so as to occult ~half the X-rays from it, and this obscuring region will in turn be heated by the X-rays it intercepts. A simple calculation based on the observed X-ray flux of X1822–371 suggests that material exposed to X-rays in the outer part of the disk will be at a temperature of $\sim 10^4$K (e.g. Mason and Córdova 1982b), so will contribute to the optical light of the system. An analysis of the optical light curve indeed suggests that a substantial fraction of the optical emission we see from X1822–371 comes from the material that obscures the ADC (Mason and Córdova, 1982b). The effective size of this optical emission source is required to be similar to the size of the disk; i.e. the data are consistent with the obscuring material being on the outside of the luminous accretion disk.

In order to produce a minimum in the X-ray light curve at phase 0.8, the disk rim must be thickest at this point. The height of the rim above the disk plane at this phase is found to be ~ 0.15 $R_\odot$ at the most likely orbital inclination of 80°, or about a quarter the radius of the disk (Mason and Córdova 1982b; White and Holt 1982). Both the X-ray and optical light curves of X1822–371 are better fit if a second, smaller thickening of the disk rim is included in the model at about phase 0.2 (Mason and Córdova 1982b; White and Holt 1982).

5.2 Relationship to the Dipping Sources

As indicated in section 2, if we take the model of X1822–371 and replace the extended X-ray source by one which is point-like, one would expect a phenomenon very similar to the irregular dips that are seen in many LMXRB. The effect of the extended X-ray source in systems like X1822–371 is to smooth out any irregularities in the obscuring medium, and these become apparent only when the observed X-ray source is point-like. An excellent illustration of this is afforded by a comparison of the X-ray light curve of X1822–371 with that of the eclipsing, dipping X-ray source XBT0748–676 (section 3.2.3). This is shown in Figure 9. The eclipse in XBT0748–676 is sharp and the dips are highly irregular, whereas the eclipse in X1822–371 is smooth and partial, and the irregular dips are replaced by a smooth, broad modulation. It is striking that the dips in XBT0748–676 occur over the same phase interval as the broad dip in X1822–371.

We can list a number of other points of comparison between the dipping and ADC sources:

(i) In those dip sources where there is an indication of the phase of the dips with respect to the line of centers of the two stars in the binary (XBT0748–676, XB1254–690, X1755–338, XBT1658–298, and Cyg X-2), there is major dip activity at about phase 0.8, similar to the time of minimum of the broad dip in X1822–371.

(ii) A number of dip sources show dips occuring twice (or more times) per cycle (e.g. Figure 2). As noted in section 5.1, there is evidence from modelling of both the X-ray and optical light curves of X1822–371 for two maxima in the height of the accretion disk rim in that system.

(iii) Only two of eight confirmed X-ray dippers also exhibit eclipses by the companion star, and no cases are known of systems that exhibit sharp eclipses without dips. This means that the material causing the dips must on average subtend at the X-ray source a solid angle above the disk plane that is larger than the angle subtended by the companion star; i.e. the disks in dipping sources must have substantial thickness.

(iv) Figure 10 shows a number distribution of those ADC and dip sources that

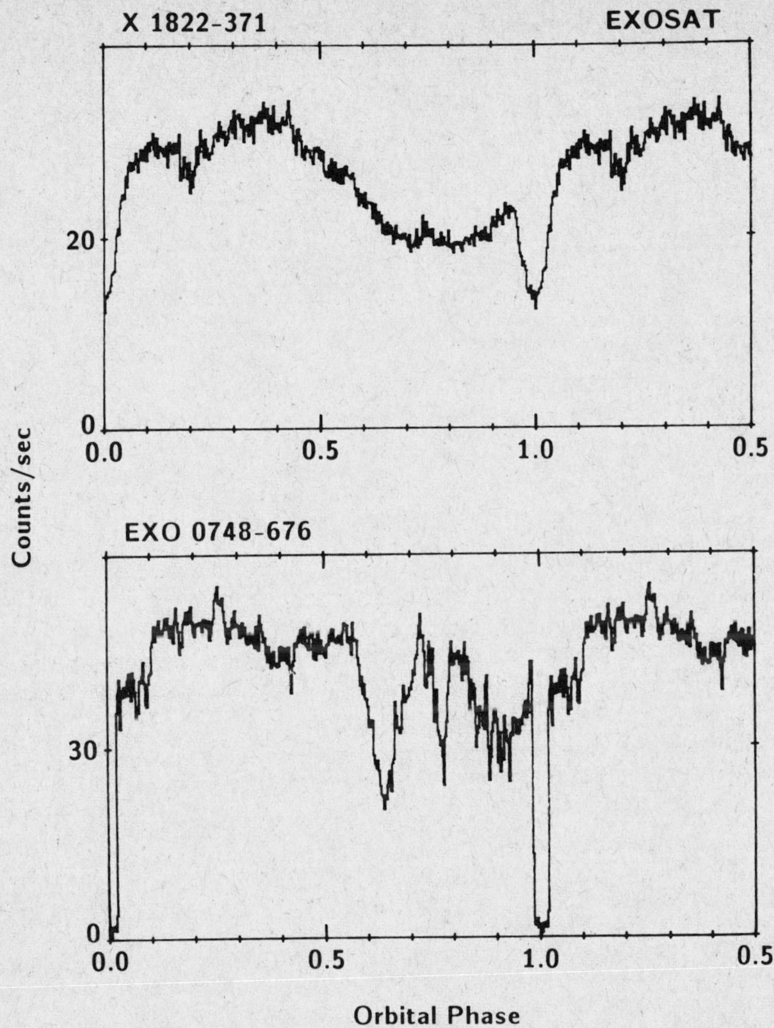

Figure 9 A comparison of the mean, folded X-ray light curves of X1822–371 and XBT0748–676 (taken from Parmar et al. 1986). One and a half orbital cycles are shown in each case, phase being measured from the time of mid-eclipse by the companion star.

have optical counterparts as a function of X-ray to optical flux ratio. There is a clear separation of the two types of LMXRB, the ADC sources having the lower values of L_x/L_{opt}. This distinction is consistent with the idea that ADC's are optically thin and re-direct only a small fraction of the intrinsic X-ray emission of a source into the line of sight. In contrast, the optical emission of an ADC source comes from the X-ray heated

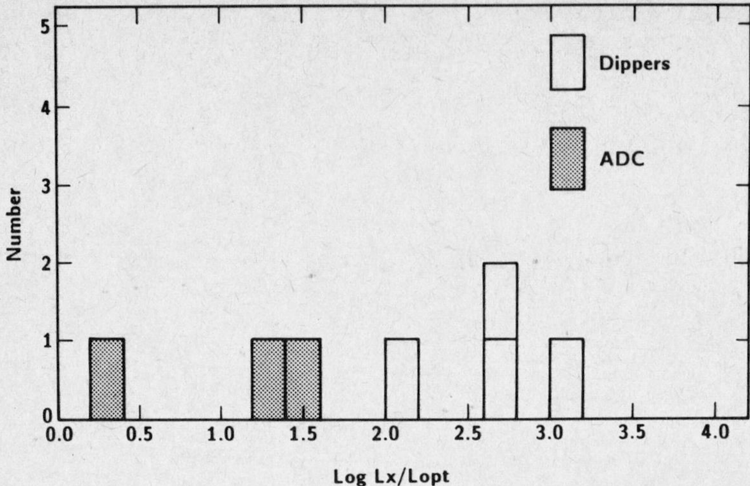

Figure 10 Number distribution of L_x/L_{opt} for three ADC sources and four dipping sources with optical counterparts.

accretion disk which is visible irrespective of whether the central X-ray source is hidden from direct view. The idea that the ADC is optically thin is consistent with theoretical estimates that suggest that it is very difficult to sustain an ADC of the size required in, say, X1822–371 if it is optically thick (Fabian, Guilbert and Ross, 1982). It is also consistent with observations of sources such as Her X-1 (section 3.1.4) and XBT0748–676 (section 3.2.3) where there is direct evidence for an ADC that scatters a small percentage of the source flux. (However, in X1624–490 there may be an ADC that scatters ~25% of the source flux; section 3.2.8). A consequence of the above interpretation is that we expect to detect only the intrinsically brightest LMXRB as purely ADC sources, due to observational selection effects.

6. THE CAUSE OF THE DISK STRUCTURE

In the case of X1822–371 there is clear evidence that the X-ray source is obscured prior to the eclipse by the companion star. This obscuration is centered at about phase 0.8 (measured from eclipse), which is the time when we are looking most directly at the region where the mass-transfer gas stream from the companion joins the disk (Figure 1). This coincidence led to early suggestions that turbulence stemming from the interaction of the gas stream with the disk caused this part of the rim to be thicker than the rest, thus obscuring the X-ray source (e.g. White and Holt 1982; Mason et al. 1980).

There are a number of problems with this simple idea. It is by no means clear that there is enough energy in the gas stream/disk interaction to sustain the large degree of thickening required ($h/r \sim 0.25$ for X1822–371); and, perhaps more importantly, there is clear evidence in a number of dipping systems, particularly XB1916–053 (section 3.2.1) and XBT0748–676 (section 3.2.3), for more than one dip per cycle, and therefore more

than one 'bulge' on the disk. Even in X1822–371, there is evidence for a second 'bulge' at
~ phase 0.2 in addition to the one at phase 0.8 (White et al. 1981; Mason and Córdova
1982b). No convincing argument has been made to account for a second bulge in the
simple gas-stream model.

Other possibilities have been explored. Lubow and Shu (1975) pointed out that
the natural width of the gas stream that overflows the L1 point of the companion star
(of order 10^9 cm) can exceed the thickness of the edge of the disk in the *thin* disk
approximation. Thus part of the gas stream may pass over the top of the disk and
loop around the X-ray source in a ballistic trajectory. Based on these ideas, Walter et al.
(1982), seeking to explain the two dips per cycle observed in XB1916–053, suggested that
the second dip was caused by material in the parts of this ballistic trajectory opposite the
companion star in its orbit, at the turning point where it is furthest away from the X-ray
source. The cause of the primary dip in this model though is not clear; nor is it clear
how any part of the gas stream is able to project high enough out of the orbital plane
to occult the X-ray source in a system whose orbital inclination, based on the absence of
eclipses by the companion, is < 80° (Smale, Mason, and White 1986).

Frank, King, and Lasota (1986) have considered a similar scenario. In their model,
a fraction of the material in the gas stream again misses the top of the disk and falls in
towards the X-ray source on a ballistic trajectory. This material circularises, according
to the prescription of Lubow and Shu (1975), at a radius of about 10^{10} cm from the
X-ray source. This circularisation radius is much smaller than the size of the disk and is
determined by the specific angular momentum of matter flowing through the L1 point.
The result is a ring-like thickening of the disk at this radius. Material in the gas stream
that strikes this ring splashes out of the orbital plane to form a 'bulge', or thickening
in the ring that extends about half way around the disk. This scenario is thus similar
to that sketched in Figure 1, except that the disk structure is much closer to the X-ray
source than the edge of the disk. The advantage of this model over one which has the
structure at the disk edge is to be found in the calculations of Lubow and Shu (1976),
who find that the thickness of the gas stream will decrease very little before it hits the
ring. Consequently the angular size that the stream subtends at the X-ray source is
increased above that which is possible at the outside of the disk, alleviating the problem
of having material at large distances above the orbital plane.

The Frank, King, and Lasota model is, however, not in accord with a number of the
observations. Most strikingly, the radius of the obscuring ring, and the ADC within it,
is predicted to be ~0.15 $R_\odot$, which is much smaller than the empirically derived size of
the ADC in X1822–371 (~0.3 $R_\odot$), and the size of the obscuring region in that source
(~0.6 $R_\odot$). Further, the model predicts dips only in the phase region 0.3–0.8 (i.e. on that
side of the disk away from the companion star), whereas observations of XBT0748–676
(one of the few dipping sources where there is an unambiguous phase reference) show that
dips also occur between phase 0.8 and 1.2, i.e. either side of the eclipse phase (section
3.2.3).

Another possible interpretation of the data is suggested by the work of Sawada,
Matsuda, and Hachisu (1986). These authors have performed two-dimensional hydrodynamic calculations of Roche-lobe overflow in a semi-detached binary system. They find

that two or three (depending on conditions) spiral shaped shocks are formed in the accretion disk due to the gravitational influence of the mass-donating star. It is suggested that gas in the accretion flow loses angular momentum in these shocks, and thus spirals in towards the compact star *without* the need to invoke viscosity. Gas behind the shocks can form subsonic pockets which have an enhanced thickness (Sawada, Matsuda, and Inoue 1986). The location and angular extent of these shocks are very reminiscent of the type of structure that is required to explain the observations, making this model a promising one for future investigation.

7. THE ENERGY SPECTRUM OF THE DIPS

In almost all cases the X-ray spectrum of dipping LMXRB becomes harder during the dips, suggesting photoelectric absorption. The spectral evolution of the dips in three sources, XB1916–053, XBT0748–676, and XB1254–690, has been examined in detail using data from the EXOSAT observatory (Smale, Mason, and White, 1986; Parmar et al. 1986; Courvoisier et al. 1986). It is found in all three cases that the data are *not* adequately fit by a simple model in which a constant underlying source spectrum is modified by a variable column of cold absorbing material. A model which does fit the data is one which consists of two components, both of which have the same slope as the spectrum outside the dips, and only one of which suffers photoelectric absorption. The normalisation of both components is a free parameter, however. In all three sources the spectral component that is not photoelectrically absorbed is most important in the mid-range of source intensities, contributing a much smaller fraction of the total flux in the deepest parts of the dips.

One physical interpretation of such a spectrum is that part of the emission from the source is scattered rather than being absorbed. However, one should be cautious about over-interpreting the fits since the above model is not unique. For example, another model which fits almost as well, has the same number of free parameters, but is less justifiable physically, is one where both the slope and absorption of a single component spectrum are allowed to vary (Smale, Mason, and White 1986). Nevertheless the general statement can be made that the spectrum of these sources during the dips exhibits a low-energy excess above that which would be expected for pure photoelectric absorption of an invariant underlying spectrum.

A number of explanations of this behaviour are possible. Frank, King, and Lasota (1986) have shown that, if the material causing the dips is sufficiently close to the X-ray source, it will be subject to an ionization instability of the type described by Krolik, McKee, and Tarter (1981). This results in a two phase medium in which small, relatively dense and cool clouds are embedded in a hot, lower density medium of much larger volume. The cool clouds would be responsible for the photoelectric absorption, and the hot intercloud gas for the scattering. Alternatively, the absorbing material may be in an intermediate ionisation state in which the low-z elements are highly ionised, but the high-z elements (such as iron) are not (e.g. Ross 1979). The effect of this is that there will be more soft X-rays transmitted than would be expected for absorption by a cold medium. This explanation would be qualitatively consistent with the fact that the 'soft-excess' is least important at the bottom of the dips where the density of the absorbing column is likely to be highest (and the ionisation parameter $\xi = L_x/nr^2$ is lowest). A third explanation, considered by Parmar et al. (1986), is that the soft excess is an artefact of the way in which the data are binned, and nothing to do with the source itself. In any situation where the column density is varying much more rapidly than the time bin over which spectra are integrated, the integrated spectrum will consist of a sum over a range of columns. This can in principle mimic the 'soft X-ray excesses' that are seen. Unfortunately it is difficult to assess how much of an effect this will have on the results without knowledge of the underlying frequency spectrum of variability in the dips.

7.1 Elemental Abundances in the Absorbing Material

The X-ray spectrum during a dip contains information about the abundance of elements in the absorbing material, which means in effect the abundance of elements on the surface of the mass-donating star. Specifically, the ratio of scattered flux to that which is photoelectrically absorbed yields the number of electrons compared to the number of atoms of heavy elements that possess K-shell electrons. Extracting this information from the data on XB1916–053, XBT0748–676, and XB1254–690, is made more difficult by the apparent multi-component nature of the dip spectrum, discussed earlier in this section, and the fact that we do not yet understand the cause of this complexity well enough to model it uniquely. However, as pointed out earlier, in the deepest part of the dips the spectrum can be reasonably well approximated by a single, photoelectrically absorbed, component (i.e. the emission from the second spectral component, or low energy excess, becomes small), so by confining attention to the deepest part of the dips it should be possible to obtain a comparatively model free estimate of the heavy-element abundance. In this way, it is found that the heavy element abundance in XB1916–053 and XB1254–690 is within a factor of two that in the sun, whereas the abundance in XBT0748–676 is between a factor of two and seven times *less* than solar (Smale, Mason, and White, 1986; Courvoisier et al. 1986; Parmar et al. 1986). Strictly, only those atoms with K-shell electrons are measured, so the interpretation of these numbers as an abundance assumes that K-shell ionisation is not important.

It should be noted that the abundance number given above for XB1916–053 conflicts with that derived by White and Swank (1982). This is because the latter authors fitted their data under the assumption that a *single* absorbed spectral component was an adequate representation at all intensity levels, something we now know to be untrue. It is also worth pointing out that the very short orbital period of XB1916–053 probably means that the mass-donating star in that system is severely hydrogen deficient (e.g. Nelson, Rappaport and Joss 1985).

7.2 Energy Independent Dips: X1755–338

The dips in X1755–338 are unusual in that they are shallow and not associated with any detectable change in the hardness of the X-ray spectrum (section 3.2.5). Dips are notoriously variable, and other sources may at times exhibit events that are equally shallow. However, the dips in X1755–338 are consistently so; of the seven dips recorded from the source with EXOSAT in two separate observations, none exceeded a depth of about 30%; and even the shallowest dips in other sources are accompanied by an obvious spectral hardening. It should be emphasised that in all other respects the dips in X1755–338 are indistinguishable from those in other sources. They last for a similar fraction of the cycle, have an irregular morphology, and are highly variable down to timescales of seconds.

A number of possible explanations of the X1755–338 phenomenon have been considered. Frank, King, and Lasota (1986) point out that, if the spectrum of the central source is made up of more than one, spatially separated component, it is possible to arrange the relative fluxes and slopes of the two components such that the overall spectral hardness changes little when the harder component is absorbed. However, this is a rather contrived situation, and while it may be possible to balance the flux in two relatively broad X-ray bands so as to get a constant 'hardness ratio', it is not clear that the overall shape of the spectrum could be kept constant to the tolerance required by the data.

Sztajno and Frank (1984) have suggested that X1755–338 might be an ADC source, and that the dips are due to the partial occultation of the ADC by structure on the disk. In order to explain the rapid variability of the X-ray flux during the dips, they suggested that the height of the disk structure is oscillating rapidly. The problem with this model is that other ADC sources do not behave in this way, and there is no indication of why X1755–338 might be different from, say, X1822–371. Further, the L_x/L_{opt} ratio of X1755–338 is high, which is evidence that we are seeing the central source directly (section 5.2).

The remaining option, perhaps the simplest and most interesting, is that the dips in X1755–338 are caused by scattering in material that does not contain any atoms with K-shell electrons capable of absorbing photoelectrically in the X-ray range. A pleasing feature of such an explanation is that the electron column density required to produce the dips in X1755–338 is very similar to the columns inferred in other dipping sources. Scattering would dominate if the absorbing material were completely ionised; however adopting even the most optimistic assumptions regarding distance and source luminosity, the absorbing material would have to be closer than about 10^{10} cm from the X-ray source for photo-ionisation to be a viable mechanism for stripping the heavy elements (Mason, Parmar, and White 1985), and even at this distance it would be surprising if some cool, dense pockets of gas didn't remain. An alternative is that the abundance of heavy elements in the absorbing gas is very low; A value < 0.002 times the solar abundance relative to electrons is required (White et al. 1984). Note that this value is not subject to the same uncertainties that beset the abundance determinations of sources discussed in section 7.1. There is as yet no observational evidence to contradict this conclusion; in particular there is no detectable iron emission in the X-ray spectrum of X1755-338.

8. ECLIPSING LMXRB: INCLINATIONS AND COMPANION MASSES

The X-ray emission is periodically eclipsed by the mass-donating companion in six of the LMXRB discussed in section 3; the ADC sources X2129+470, X1822−371, and X0921−630; the dipping sources XBT0748−676 and XBT1658−298; and Her X-1. The length of the eclipse can be used in conjunction with the mass ratio of the system to place constraints on the orbital inclination of the binary, provided that the companion star fills its Roche lobe. When the X-ray source is point-like, we have, following Rappaport and Joss (1983):

$$A + B \log q + C \log^2 q \approx [\cos^2 i + \sin^2 i \sin^2 \theta_e]^{\frac{1}{2}}$$

where q is the mass ratio (M_x/M_c), θ_e is the eclipse half-angle, i is the orbital inclination, and A, B, and C are constants $(A = 0.376, B = -0.227, C = -0.028$, if the companion star co-rotates with the orbital motion). If there are no observational constraints on the mass ratio, assumptions must be made about the nature of the two stars in the binary in order to proceed. If the X-ray source is a neutron star (very likely in the case of the bursters) then its mass is probably about 1.4 $M_\odot$. To constrain the mass of the companion, one can adopt a mass-radius relation; for example, the relation for an unevolved main sequence star, $R_c \approx 0.9 M_o$ (e.g. Whyte and Eggleton 1980). In this way we can obtain a probable lower limit to the inclination of XBT0748−676 and XBT1658−298, and a probable upper limit to the companion star mass in these systems (Table 2). An upper limit to the inclination is obtained if it is assumed that the companion still sustains nuclear burning, in which case $M_c \geq 0.085\ M_\odot$.

TABLE 2: Orbital Constraints on Eclipsing LMXRB

Source	Period (hours)	Type[1]	Orbital Inclination (degrees)	Companion Mass ($M_\odot$)	Reference
XBT0748−676	3.8	d	73−83	0.45−0.08	52
X2129+470	5.2	a	80−90	0.65−0.2	85
X1822−371	5.6	a	76−84	0.2−0.35	33
XBT1658−298	7.2	d	72−82	0.9−0.08	10
Her X-1	40.8	p,e	84−90	2.07−2.29	44
X0921−630	216.3	a	> 80	< 1	31

Notes: 1. See key with Table 1.

Similar information can be obtained for the ADC sources, where the eclipse is partial. White and Holt (1982) have modelled the eclipse in X2129+470 under the assumption that the companion is a main sequence star; the values of the orbital inclination and companion mass derived are given in Table 2. In the case of X1822−371 and X0921−630, the orbital velocity of the X-ray source is constrained by optical spectroscopic measurements (Cowley, Crampton and Hutchings, 1982a,b; Mason et al. 1982; 1986), which means that it is possible to determine the mass of the companion star explicitly without assuming a mass-radius relation. In both cases it is found that the companion is undermassive for its size compared to an isolated, un-evolved star. The data on these sources are listed in Table 2, together with the orbital solution for the pulsar system Her X-1, for which considerable observational data exist (e.g. Middleditch and Nelson 1976; Bahcall, Joss, and Avni 1974).

9. CONCLUSION

A considerable body of observational data has been accumulated in a relatively short time on the properties of Low Mass X-ray Binaries. The periodic X-ray variability of these systems can be understood within the framework of a simple empirical model of obscuration by vertical structure in an accretion disk. A satisfactory understanding of how such structure comes about is still lacking, but it is hoped that progress in this area, spurred by the need to understand the considerable phenomenology that is now revealed, will promote an improved knowledge of the physics of accretion disks in general. X-ray spectral data on the dipping sources conveys information about the abundance of elements in the mass-donating star in the binary system, whereas observations of eclipsing systems yield estimates of the masses of the stars in the binary. These types of measurements provide important clues to the evolutionary history of the LMXRB.

Acknowledgements

I am grateful to Len Culhane and Alan Smale for reading the manuscript, and to Simon Rosen for constructing Figure 1. I thank the Royal Society for support.

References

1. Alpar, M. A. 1979, *M.N.R.A.S.*, **189**, 305.

2. Bahcall, J. N., Joss, P.C., and Avni, Y. 1974, *Ap. J.*, **191**, 211.

3. Beuermann, K., and Osborne, J. 1985, *Space Sci. Rev.*, **40**, 117.

4. Branduardi-Raymont, G., Corbet, R. H. D., Parmar, A. N., Murdin, P. G., and Mason, K. O. 1981, *Space Sci. Rev.*, **30**, 279.

5. Branduardi-Raymont, G.. Corbet, R. H. D., Mason, K. O., Parmar, A. N., Murdin, P. G. and White, N. E. 1983, *M.N.R.A.S.*, **205**, 403.

6. Breedon, L. and Watson, M. G. 1986, Contributed Paper at this Workshop.

7. Chester, T. J. 1979, *Ap. J.*, **230**, 107.

8. Chevalier, C. and Ilovaisky, S. A. 1981, *Astr. Ap.*, **94**, L3.

9. Chevalier, C. and Ilovaisky, S. A. 1982, *Astr. Ap.*, **112**, 68.

10. Cominsky, L. and Wood, K. 1984, *Ap. J.*, **283**, 765.

11. Corbet, R. H. D., Thorstensen, J. R., Charles, P. A., Menzies, J. W., Naylor, T., and Smale, A. P. 1986, *M.N.R.A.S.*, **222**, 15p.

12. Córdova, F. A., Mason, K. O., and Kahn, S. M. 1985, *M.N.R.A.S.*, **212**, 447.

13. Courvoisier, T., J.-L., Parmar, A. N., Peacock, A., and Pakull, M. 1986, *Ap. J.*, (in press).

14. Cowley, A. P. and Crampton, D. 1975, *Ap. J.*, **201**, L65.

15. Cowley, A.P., Crampton, D., and Hutchings, J. B. 1979, *Ap. J.*, **231**, 539.

16. Cowley, A.P., Crampton, D., and Hutchings, J. B. 1982a, *Ap. J.*, **255**, 596.

17. Cowley, A.P., Crampton, D., and Hutchings, J. B. 1982b, *Ap. J.*, **256**, 605.

18. Crampton, D., Cowley, A. P., Stauffer, J., and Hutchings, J. B. 1986, *Ap. J.*, **306**, 599.

19. Crosa, L. M. and Boynton, P. E. 1980, *Ap. J.*, **235**, 999.

20. Fabian, A. C., Guilbert, P. W., and Ross, R. R. 1982, *M.N.R.A.S.*, **199**, 1045.

21. Frank, J., King, A. R., and Lasota, J.-P. 1986, *Astr. Ap.*, (in press).

22. Giacconi, R., Gursky, H., Kellogg, E., Levinson, R., Schreier, E., and Tananbaum, H. 1973, *Ap. J.*, **184**, 227.

23. Gottlieb, E. W., Wright, E. L., and Liller, W. 1975, *Ap. J.*, **195**, L33.

24. Hassall, B. J. M., Pringle, J. E., Ward, M. J., Whelan, J. A. J., Mayo, S. K., Echevarria, J., Jones, D. H. P., and Wallis, R. E. 1981, *M.N.R.A.S.*, **197**, 275.

25. Ilovaisky, S. A. and Chevalier, C. 1986, Contributed Paper at this Workshop.

26. Krolik, J. H., McKee, C. F., and Tarter, C. B. 1981, *Ap. J.*, **249**, 422.

27. Lewin, W. H. G. and Joss, P. C. 1983, *in "Accretion Driven Stellar X-ray Sources"*, *ed.* W. H. G. Lewin and E. P. J. van den Heuvel, Cambridge University Press, Cambridge, U.K., *pg.* 41.

28. Lubow, S. H. and Shu, F. H. 1975, *Ap. J.*, **198**, 383.

29. Lubow, S. H. and Shu, F. H. 1976, *Ap. J.*, **207**, L53.

30. Mason, K. O. 1985, *Space Sci. Rev.*, **40**, 99.

31. Mason, K. O., Branduardi-Raymont, G., Córdova, F. A., and Corbet, R. H. D. 1986, *M.N.R.A.S.*, (in press).

32. Mason, K. O. and Córdova, F. A. 1982a, *Ap. J.*, **255**, 603.

33. Mason, K. O. and Córdova, F. A. 1982b, *Ap. J.*, **262**, 253.

34. Mason, K. O., Córdova, F. A., Corbet, R. H. D., and Branduardi-Raymont, G. 1985, *Space Sci. Rev.*, **40**, 225.

35. Mason, K. O., Córdova, F. A., and White, N. E. 1986, *Ap. J.*, (in press).

36. Mason, K. O., Middleditch, J., Nelson, J. E., White, N. E., Seitzer, P., Tuohy, I. R. , and Hunt, L. K. 1980, *Ap. J.*, **242**, L109.

37. Mason, K. O., Murdin, P. G., Tuohy, I. R., Seitzer, P., and Branduardi-Raymont, G. 1982, *M.N.R.A.S.*, **200**, 793.

38. Mason, K. O., Parmar, A. N., and White, N. E. 1985, *M.N.R.A.S.*, **216**, 1033.

39. McClintock, J. E., London, R. A., Bond, H. E., and Grauer, A. D. 1982, *Ap. J.*, **258**, 245.

40. McClintock, J. E., and Remillard, R. A. 1986, *Ap. J.*, **308**, 110.

41. McClintock, J. E., and Remillard, R. A., and Margon, B. 1981, *Ap. J.*, **243**, 900.

42. McHardy, I. M., Pye, J. P., Fairall, A. P., and Menzies, J. W. 1986, *M.N.R.A.S.*, (in press).

43. Middleditch, J., Mason, K. O., Nelson, J. E., and White, N. E. 1981, *Ap. J.*, **244**, 1001.

44. Middleditch, J., and Nelson, J. E. 1976, *Ap. J.*, **208**, 567.

45. Motch, et al. 1984, I.A.U. Circ. # 3951.

46. Motch, C., Chevalier, C., Ilovaisky, S. A., and Pakull, M. W. 1985, *Space Sci. Rev.*, **40**, 239.

47. Motch, C., Pedersen, H., Beuermann, K., Pakull, M. W., and Courvoisier, T. J.-L. 1986, *Ap. J.*, (in press).

48. Mouchet, M., Angebault, L. P., and Ilovaisky, S. A. 1985, I.A.U. Circ. # 4076.

49. Nelson, L. A., Rappaport, S. A. and Joss, P. C. 1985, *Nature*, **316**, 42.

50. Parmar, A. N., Pietsch, W., McKechnie, S., White, N. E., Trumper, J., Voges, W., and Barr, P. 1985a, *Nature*, **313**, 119.

51. Parmar, A. N., White, N. E., Sztajno, M., and Mason, K. O. 1985b, *Space Sci. Rev.*, **40**, 213.

52. Parmar, A. N., White, N. E., Giommi, P., and Gottwald, M. 1986, *Ap. J.*, **308**, 199.

53. Parsignault, D. R., Gursky, H., Kellogg, E. M., Matilsky, T., Murray, S., Schreier, E., Tananbaum, H., Giacconi, R., Brinkman, A. C. 1972, *Nature (Phys. Sci.)*, **239**, 123.

54. Pedersen, H., Cristiani, S., d'Ororico, and Thomsen, B. 1985, I.A.U. Circ. # 4047.

55. Pedersen, H. and Mayer, M. 1985, I.A.U. Circ. # 4039.

56. Pedersen, H., van Paradijs, J., and Lewin, W. H. G. 1981, *Nature*, **294**, 725.

57. Pietsch, W., Steinle, H., Gottwald, M., and Graser, U. 1986, *Astr. Ap.*, **157**, 23.

58. Rappaport, S. A. and Joss, P.C. 1983, *in "Accretion Driven Stellar X-ray Sources"*, *ed.* W. H. G. Lewin and E. P. J. van den Heuvel, Cambridge University Press, Cambridge, U.K., *pg.* 1.

59. Ross, R. R. 1979, *Ap. J.*, **233**, 334.

60. Sanford, P. W. and Hawkins, F. J. 1972, *Nature (Phys. Sci.)*, **239**, 135.

61. Sawada, K., Matsuda, T., and Hachisu, I. 1986, *M.N.R.A.S.*, **221**, 679.

62. Sawada, K., Matsuda, T., and Inoue, M. 1986, Preprint No. KUGD 86-1, Kyoto University.

63. Smale, A. P., Mason, K. O., and White, N. E. 1986, (in preparation).

64. Stella, L., Priedhorsky, W., and White, N. E. 1987, *Ap. J.*, (in press).

65. Swank, J. H., Taam, R. E., and White, N. E. 1984, *Ap. J.*, **277**, 274.

66. Sztajno, M. and Frank, J. 1984, *Astr. Ap.*, **138**, L15.

67. Taam, R. E. 1980, *Ap. J.*, **214**, 358.

68. Tanaka, Y. 1985, *in "X-ray Astronomy '84"*, *ed.* M. Oda and R. Giacconi, Institute of Space and Astronautical Science, Tokyo, *pg.* 125.

69. Tananbaum, H., Gursky, H., Kellogg, E., Levinson, R., Schreier, E., and Giacconi, R. 1972, *Ap. J.*, **174**, L143.

70. Thorstensen, J. R. 1987, *Ap. J.*, (in press).

71. Thorstensen, J. R., Charles, P. A., Bowyer, S., Briel, U. G., Doxsey, R. E., Griffiths, R. E., Schwartz, D. 1979, *Ap. J.*, **233**, L57.

72. Thorstensen, J. R., Charles, P. A., and Tuohy, I. R. 1983, preprint.

73. Ulmer, M. P. et al. 1980, *Ap. J.*, **235**, L159.

74. van Paradijs 1983, *in "Accretion Driven Stellar X-ray Sources"*, *ed.* W. H. G. Lewin and E. P. J. van den Heuvel, Cambridge University Press, Cambridge, U.K., *pg.* 189.

75. van der Klis, M., Jansen, F., van Paradijs, J., and Stollmann, G. 1985a, *Space Sci. Rev.*, **40**, 287.

76. van der Klis, M., Parmar, A. N., van Paradijs, J., Jansen, F., and Lewin, W. H. G. 1985b, I.A.U. Circ. # 4044.

77. Voges, W., Kahabka, H., Ögelman, H., Pietsch, W., and Trümper, J. 1985, *Space Sci. Rev.*, **40**, 339.

78. Vrtilek, S. D. and Halpern, J. P. 1985, *Ap. J.*, **296**, 606.

79. Vrtilek, S. D., Swank, J. H., Kahn, S. M., Kelley, R. L., and Grindlay, J. E. 1986, Contributed Paper at this Workshop.

80. Wade, R. A., Quintana, H., Horne, K., and Marsh, T. R. 1985, *P.A.S.P.*, **97**, 1092.

81. Walter, F. M., Bowyer, S., Mason, K. O., Clarke, J. T., Henry, J. P., Halpern, J., and Grindlay, J. E. 1982, *Ap. J.*, **253**, L67.

82. Warner, B. 1976, *in "IAU Symposium 73, The Structure and Evolution of Close Binary Systems"*, *ed.* P. Eggleton, S. Mitton, and J. Whelan, Dordrecht: Reidel, *pg.* 85.

83. Watson, M.G., Willingale, R., King, A. R., Grindlay, J. E., and Halpern, J. 1985, I.A.U. Circ. # 4051.

84. White, N. E., Becker, R. H., Boldt, E. A., Holt, S. S., Serlemitsos, P. J., and Swank, J. H. 1981, *Ap. J.*, **247**, 994.

85. White, N. E. and Holt, S. S. 1982, *Ap. J.*, **257**, 318.

86. White, N. E. and Mason, K. O. 1985, *Space Sci. Rev.*, **40**, 167.

87. White, N. E., Parmar, A. N., Sztajno, M., Zimmermann, H. U., Mason, K. O., and Kahn, S. M. 1984, *Ap. J.*, **283**, L9.

88. White, N. E. and Swank, J. H. 1982, *Ap. J.*, **253**, L61.

89. Whyte, C. A. and Eggleton, P. 1980, *M.N.R.A.S.*, **190**, 801.

THEORY AND OBSERVATIONS OF TIME-DEPENDENT ACCRETION DISKS

Frank Verbunt
Max Planck Institut für Extraterrestrische Physik
8046 Garching bei München
Federal Republic of Germany

Abstract

In recent years rapid progress has been made both observationally and theoretically in the study of time-dependent accretion-disks, in particular with respect to cataclysmic variables. In the first part of this review I describe the theory of time-dependent disks. In the second part I discuss dwarf nova outbursts and compare the theoretical predictions of two models — the disk instability and the red-star instability — with the observational evidence. It is found that both models have their successes and failures. In general, the optical observations are in better agreement with the disk instability model, the ultraviolet observations with the red-star instability.

1. The theory of time-dependent accretion disks

1.1 Equations

Virtually all the current work on accretion disks follows the approach first suggested by Shakura and Sunyaev (1973), in which the vertical and radial, or local and global, structure equations of the disk are solved separately, and in which the viscosity is assumed to be proportional to the pressure: $\nu = 2\alpha P/3\omega$. In order to simplify the equations, the following assumptions are made:

A. the disk is axisymmetric. This implies that only the gravitational force of the star in the center of the disk is taken into account, and that of the companion star neglected; and also that the Coriolis forces due to the binary revolution are neglected.

B. the disk is geometrically thin. This implies that pressure forces are neglected, and that the velocities in the direction perpendicular to the disk plane are small with respect to the radial velocities $v_z << v_r$.

C. $\alpha \leq 1$. This implies that radial velocities are small with respect to rotational velocities: $v_r << v_\phi$.

With these simplifications, the radial component of the equation of motion reduces to the statement that the rotation velocities are Keplerian, whereas the other equations can be grouped in a global diffusion equation, and local structure equations. Schematically (for more detail see Appendix. Cf. Verbunt 1982):

$$\left.\begin{array}{r}\text{mass} \quad \text{conservation}\\ \text{motion}-\phi\end{array}\right\} \Rightarrow \frac{\partial \Sigma}{\partial t} = \frac{3}{r}\frac{\partial}{\partial r}\left(r^{1/2}\frac{\partial}{\partial r}(\nu\Sigma r^{1/2})\right) \quad (1)$$

$$\text{motion} - \text{r} \quad \Rightarrow \quad v_\phi^2 = \frac{GM}{r} = r^2\omega^2$$

$$\left.\begin{array}{r} \text{motion} - \text{z} \\ \text{energy production} \\ \text{energy loss} \\ \text{state} \\ \text{viscosity} \end{array}\right\}$$

These give local relations between pressure, density, temperature, etc. In particular the product $\nu\Sigma$ can be found from Σ:

$$\nu\Sigma = f(\Sigma) \quad (2)$$

Here Σ is the vertically integrated density (the surface density) of the disk.

In the case of thermal equilibrium the energy loss equals the energy production, and the effective temperature of a disk surface element can be written:

$$\sigma T_{eff}^4 = \frac{9\nu\Sigma\omega^2}{8} \quad (3).$$

In a stationary disk the product $\nu\Sigma$ can be calculated everywhere in the disk from Eq.(1), with the appropriate boundary condition:

$$(\nu\Sigma)_{stat} = \frac{\dot{M}_e}{3\pi}\left(1 - \sqrt{\frac{R}{r}}\right) \quad (4).$$

With Eq.(3) it follows that the radiation emitted by each disk surface element depends only on $\dot{M}_e$, and not on the values of ν or Σ. The reason for this is that the product $\nu\Sigma$ has adjusted itself everywhere in the disk to accommodate a local mass transfer equal to that imposed on the disk from outside, $\dot{M}_e$. Hence the effective temperatures in a stationary disk do not depend on the details of the viscous processes, and can be calculated with some confidence. In order to calculate the density and pressure in the disk, knowledge of the viscosity is necessary, and in fact it turns out that an anomalously high viscosity is needed if the pressure is to remain in accordance with assumption B above. Our understanding of these anomalous effects is rather tenuous, and hence the density and pressure are not well known even in a stationary disk.

1.2 Application to cataclysmic variables

The time-dependent evolution of the disk can be studied as follows. Start with a given density distribution. With Eq.(2) $\nu\Sigma$ can be found. With Eq.(1) the new Σ-distribution after a time Δt then follows. From this $\nu\Sigma$ is found again, etc. The time scale on which disturbations in the accretion disk propagate, depends critically on the viscosity, and hence any study of time-dependent accretion disks is inherently very uncertain. If the relation between surface density and viscosity $f(\Sigma)$ is smooth and single-valued, large changes in the disk can only be the result of large changes in $\dot{M}_e$. If the function $f(\Sigma)$ is multi-valued, however, large changes in the viscosity, and hence in the disk structure, are possible even when $\dot{M}_e$ is constant (Bath & Pringle 1981, 1982). It was noted by Meyer and Meyer-Hofmeister (1981) that use of a realistic opacity-law in the equation for energy loss leads to a multi-valued function $f(\Sigma)$ for an $\alpha = $ constant disk at temperatures $\leq 10^4$ K: in a given density interval the local disk structure can be either hot and mainly ionized at high viscosity or cool and mainly neutral at low viscosity (cf. Eq.(3)).

Calculations of the function $f(\Sigma)$ have been performed by several authors, either analytically or numerically, and applied to the study of time-dependent disks (Faulkner, Lin & Papaloizou 1983, 1985, Cannizzo, Ghosh & Wheeler 1982, Mineshige & Osaki 1983, 1985, Meyer & Meyer-Hofmeister 1984, Smak 1984b). Some of these functions are shown in Figure 1. Each curve consists of three branches: the hot solution (indicated in the left-most curve in Figure 1 as AA'), the cool solution (BB') and a connecting branch (AB). Whereas the hot and cool solutions are thermally stable, the connecting branch is unstable: if the region on this branch gets slightly hotter (cooler), it will move away until the hot (cool) branch is reached. The hot branch is relatively well-defined (given the assumptions), but the cool branch may be unstable to convection, or optically thin. Also the opacity at low temperature is poorly known. Because of this the position of the cool branch is very uncertain, and differs appreciably between different authors.

Consider a disk ring to which the left-most curve in Figure 1 applies, and set the mass transferred towards this ring at a rate $\dot{M}_e$ such that its equilibrium temperature is $T_{eff} = 10^{3.8}$ K. The equilibrium solution for the ring lies on the thermally unstable branch, and the ring cannot

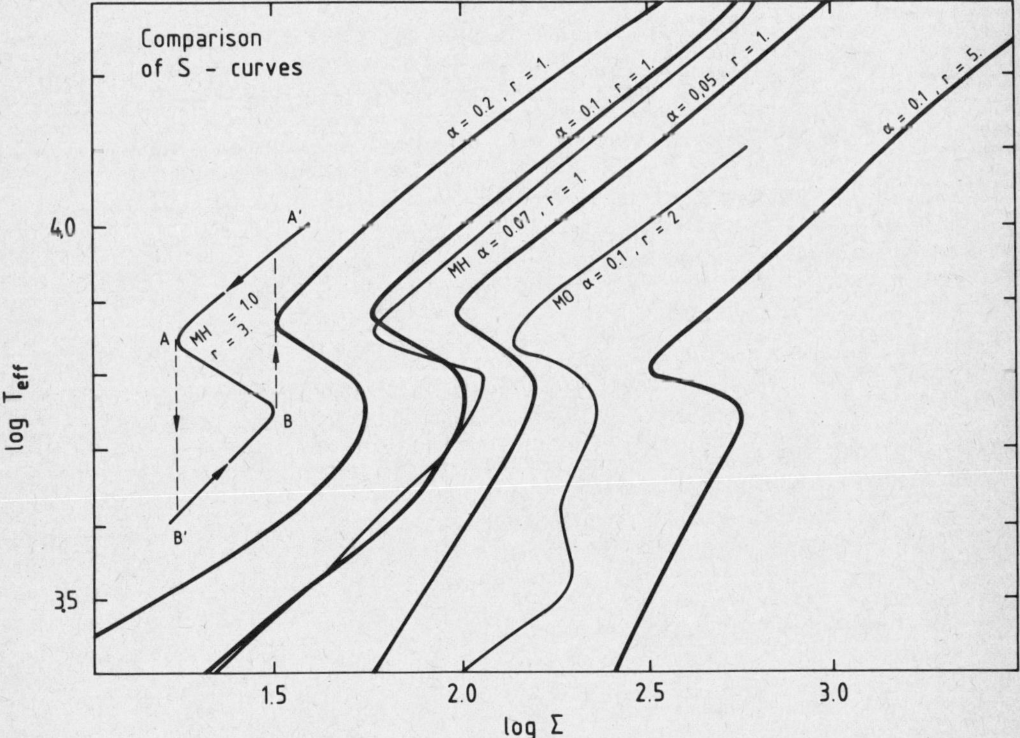

Figure 1. Solutions of the vertical structure equations for an accretion disk around a 1 $M_\odot$ object as obtained by several authors. Each curve is labelled with the values of α and radius r (in units of 10^{10} cm) for which it was calculated. The heavily drawn curves give the relation between the effective temperature T_{eff} - equivalent to the product $\nu\Sigma$ through Eq.(3) - and the surface density Σ as obtained by Smak (1984b), to show the effect of changing α or r. These curves can be compared with those obtained by Meyer & Meyer-Hofmeister (1981), labelled MH, and by Mineshige & Osaki (1983), labelled MO. Indicated on the left-most curve is the cycle that a disk ring with unstable equilibrium solution will undergo.

remain there. Suppose it is at the cool branch. At this branch the viscosity is lower, and the ring is not able to transport $\dot{M}_e$. Hence its density increases on a viscous time scale, and the ring moves up along B'B. When point B is reached, a further increase of Σ forces the disk to jump to the hot branch: it will go out of thermal equilibrium, and increase its temperature on a thermal time scale, until the hot branch is reached. There, the viscosity is too high, and the ring will transport more than $\dot{M}_e$: its density decreases on the viscous time scale, until point A is reached and the ring jumps down to the cool solution on the thermal time scale. To calculate the behaviour of the ring during the jump, one has to solve the structure equations out of thermal equilibrium. To save computing time, one may assume that the viscosity changes with T_{eff} on a thermal time scale t_{th}:

$$\frac{\partial \nu \Sigma}{\partial t} = -\frac{\nu \Sigma - (\nu \Sigma)_{eq}}{T_f t_{th}} \qquad (5)$$

where T_f is an adjustable parameter, and $(\nu \Sigma)_{eq}$ the equilibrium value of $f(\Sigma)$ (Pringle, Verbunt and Wade 1986).

Thus the S-curves for $f(\Sigma)$ lead to a cycle of the ring between a hot bright and a cool dim state. When many rings in a disk undergo such cycles in phase, the disk can be said to undergo outbursts of high $\dot{M}$ and high luminosity between intervals of low $\dot{M}$ and low luminosity. If one ring in a disk changes its viscosity, the neighbouring rings will be affected subsequently: the disturbance spreads. Smak (1984b) discovered that no large outbursts of the disk are found if one uses the S-curves of Figure 1. The reason for this is that the time spent by a ring on either the hot or the cold branch is short. If a ring   jumps from the cool to the hot branch, the disturbance caused by this has not travelled very far before the ring goes back to the cool branch. Hence the different rings of the disk undergo outbursts out of phase, leading to small changes in the overall luminosity. In order to obtain larger outbursts, it is necessary to increase the local instability artificially. This can be done in a variety of ways. A convenient one is to use different values of α for the hot and cool branches, e.g. $\alpha_{hot} = 0.2$ and $\alpha_{cool} = 0.05$. From Figure 1 it can be seen that each cycle then covers a much larger range in density, allowing each ring to remain much longer on the hot (cool) branch as the density drops (builds up). This allows synchronization of the cycles for many rings in the disk and the occurrence of large outbursts.

It is worth noting that the local instability disappears if one assumes $\alpha_{hot} = 0.05$ and $\alpha_{cool} = 0.2$. Thus, whether or not global outbursts occur in the disk due to a multi-valued viscosity depends on the detailed form of the function $f(\Sigma)$, which will remain unknown until a better understanding of the viscous processes is obtained.

The variety of model disk outbursts that can be obtained through variation of the model parameters is very large. If the function $f(\Sigma)$ of Eq.(2) is a smooth, single-valued function, outbursts can only be caused by an instability of the mass-transfer from the red dwarf. Several mechanisms have been proposed, but none of them is sufficiently well understood to predict the form, size and frequency of the enhanced mass-transfer bursts. In the modelling the frequency is therefore set equal to the observed outburst frequency, and the form and size of the bursts in $\dot{M}_e$ are free parameters. If the length of the burst of enhanced mass transfer is short with respect to the viscous transport time in the disk, the shape of the outburst is determined mainly by the viscous time scale of the disk. During the rise the disk deviates strongly from the stationary disk solutions. If the length of the burst of enhanced mass transfer is longer than the viscous time scale in the disk, the shape of the outburst will follow that of the transfer burst. The disk will be close to the stationary solutions throughout.

For the disk instability $\dot{M}_e$ can be set constant. Disk instabilities can start in different regions of the disk (Smak 1984b). At low values of $\dot{M}_e$ a large fraction of the mass fed into the disk at its outer edge will be able to diffuse inwards even in quiescence, and the outbursts will start at the inner edge of the disk, near the central object, where the required densities are lowest (see Figure 1). If the value of $\dot{M}_e$ is chosen somewhat higher, the critical density is reached sooner, and the outbursts occur at shorter intervals. At high values of $\dot{M}_e$ the build-up in the outer region of the disk — where the mass arrives — can be faster than the build-up in the inner region of the disk, which is limited by the viscous transport in quiescence. The outburst then starts at the outer edge. For even higher values of $\dot{M}_e$ no outbursts occur, as the whole disk sits on the hot branch. The values of α_{hot} and α_{cool} determine the length of the outbursts, low viscosities leading to longer outbursts. Thus by careful choice of $\dot{M}_e$, α_{hot} and α_{cool} almost any observed lightcurve can be reproduced. In the disk instability model the disk during the rise is always far from the stationary solution, whether the rise is fast or slow.

In both the transfer and the disk instability models the outbursts end because the hot region starts to decline from the outside, and becomes smaller and smaller until it collapses onto the white dwarf.

1.3 Other applications: low-mass X-ray binaries and active galaxies

Some low-mass X-ray binaries emit the bulk of their X-rays during outbursts of enhanced accretion, that have a rise time of days, decay times of months, and outburst intervals of years. These soft X-ray transients form a similar subgroup of the low-mass X-ray binaries as the dwarf novae of cataclysmic variables (van Paradijs & Verbunt 1984). The outbursts may be caused either by enhanced mass transfer from the companion star or by an accretion disk instability. In contrast to the situation in cataclysmic variables irradiation of the disk by the strong X-ray flux from the accreting neutron star or black hole will probably greatly affect the disk behaviour. For example, if the disk is heated throughout, the disk will always be on the hot branch, and the disk instability discussed in the previous section is suppressed (Meyer & Meyer-Hofmeister 1982).

Another instability may occur in the disk region close to the compact object where radiation pressure is important. The structure of this region is poorly understood, but there may be multiple solutions of the vertical structure at given Σ. Again, it depends on unknown details of the viscosity whether the instability occurs or not, and to some extent the instability is an artefact of the α-prescription (Piran 1978). If the instability exists it may generate outbursts of seconds with intervals of tens of seconds. Our understanding of the physics in the disk regions involved is too limited to make a useful comparison with observations such as those of the rapid burster (Taam & Lin 1984).

It can also be speculated that these inner disk instabilities are connected to the quasi-periodic oscillations in low-mass X-ray binaries. The radial transport timescales of seconds are too long for the quasi-periodic variations in milliseconds. However, the disk instability will lead to changes in disk thickness at each radius on a time scale of the local Keplerian rotation period. This could give rise to variable occultation of the central X-ray source and to sufficiently fast variations in the observed luminosity (see the article of van der Klis in these proceedings).

Accretion disks around massive black holes in the active galactic nuclei may also be unstable to the instabilities discussed above, or they may not. The expected timescales are of the order of the Keplerian period of the unstable disk region if variable occultation due to changes in disk thickness is invoked, or some orders of magnitude longer if the observed changes are changes in the accretion luminosity. The shortest possible timescale of hours is variation in occultation due to an instability in the radiatively dominated region of a disk around a not too massive black hole. Changes in accretion rate due to radial transport due to the instability of Figure 1 can take millions of years in a sufficiently large disk.

2. Comparison of theory and observations
of dwarf nova outbursts

The intense theoretical activity on accretion disks of the last years has its counterpart in a rapid progress of the quantity and quality of observations of cataclysmic variables. It is therefore possible to compare the theoretical predictions of the time-dependent disk calculations with observations of dwarf novae. Below I discuss first the optical measurements of the hot spot luminosity and of the disk radius during outbursts, then the ratio of ultraviolet to optical flux — or equivalently the slope of the ultraviolet-optical spectrum — and the development of the optical lines through the outburst, and finally the ultraviolet and X-ray luminosity during quiescence.

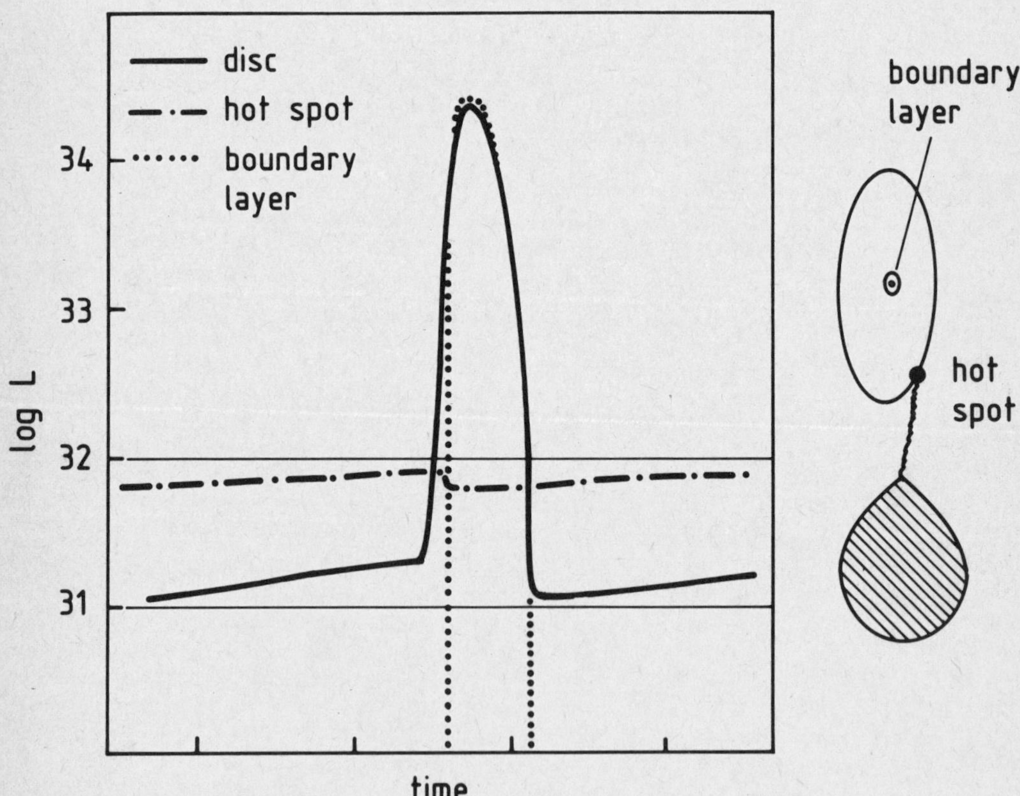

Figure 2. Theoretical lightcurves of a dwarf nova outburst calculated with the disk instability model for VW Hyi. The distance between tick marks on the horizontal axis is 10^6 s. The outburst starts near the outer edge of the disk. Shown are the luminosities of the accretion disk, of the boundary layer between disk and white dwarf, and of the hot spot where the gas stream from the companion star hits the outer edge of the disk. The program used is described in Pringle, Verbunt & Wade 1986. On the right a sketch of a cataclysmic variable.

2.1 Hot spot luminosity

The matter flowing from the companion star hits the disk at its outer edge, causing a hot spot on it. In eclipsing systems this spot is eclipsed separately from the white dwarf (see Figure 2), and from the eclipse depth its luminosity can be derived. Observations of the luminosity of the hot spot during the rise to outburst maximum of VW Hyi (Haefner, Schoembs & Vogt 1979), CN Ori (Schoembs 1982) and OY Car (Vogt 1985) do not show enhancement of the hot spot luminosity in excess of a factor 3. In the mass-transfer model the dwarf nova outburst starts with an increase of the mass flow onto the hot spot by more than an order of magnitude. In the disk instability model the disk is expected to expand at the onset of an outburst (see next paragraph) and this would lead to a moderate decrease in the luminosity of the hot spot, as the mass leaving the inner Lagrangian point has less time to be accelerated before it reaches the disk edge, and impacts at lower velocity (Figure 2).

Thus the observed moderate increase of the hotspot luminosity is not predicted by either the transfer or the disk instability models.

2.2 Disk radius

Figure 3 shows measurements of disk radii for three dwarf novae obtained from eclipse observations of the hot spot. In U Gem and Z Cha there is evidence that the size of the disk is larger after the outburst than before. This is what is expected in the disk-instability model where the outburst starts in the outer disk region. In this model a large fraction of the matter entering the disk is stored in the outer disk region, until this region jumps to the hot branch. Then most of the collected matter starts diffusing inwards, but some of it has to move outwards for angular momentum to be conserved. Thus this model predicts an increase of the disk size during the outburst. After the outburst the low-angular-momentum material arriving at the outer edge of the disk will cause the outer disk radius to shrink to a smaller radius with lower specific angular momentum. A model outburst calculated by Smak (1984b) reproduces this behaviour rather well for U Gem (Figure 3).

In contrast the outburst in the mass-transfer instability model starts with the arrival of a large amount of matter with low angular momentum. This causes the disk to shrink, after which it soon expands again. A problem with the interpretation of changes in the disk radius is that they also seem to occur well away from outbursts, as in the case of OY Car (Figure 3).

Nonetheless, the observations so far are in better agreement with the disk instability model than with the red dwarf instability model.

2.3 Optical lightcurves

The optical lightcurves of dwarf novae show a large variety (Figure 4). Some of these lightcurves by themselves are no good discrimators between different models. The number of parameters in both the red-dwarf instability and the disk instability models is so large that many lightcurve can be modelled equally well in three different ways: a disk instability starting in the disk near the central object, a disk instability starting in the outer region of the disk, and a red-dwarf mass-transfer instability.

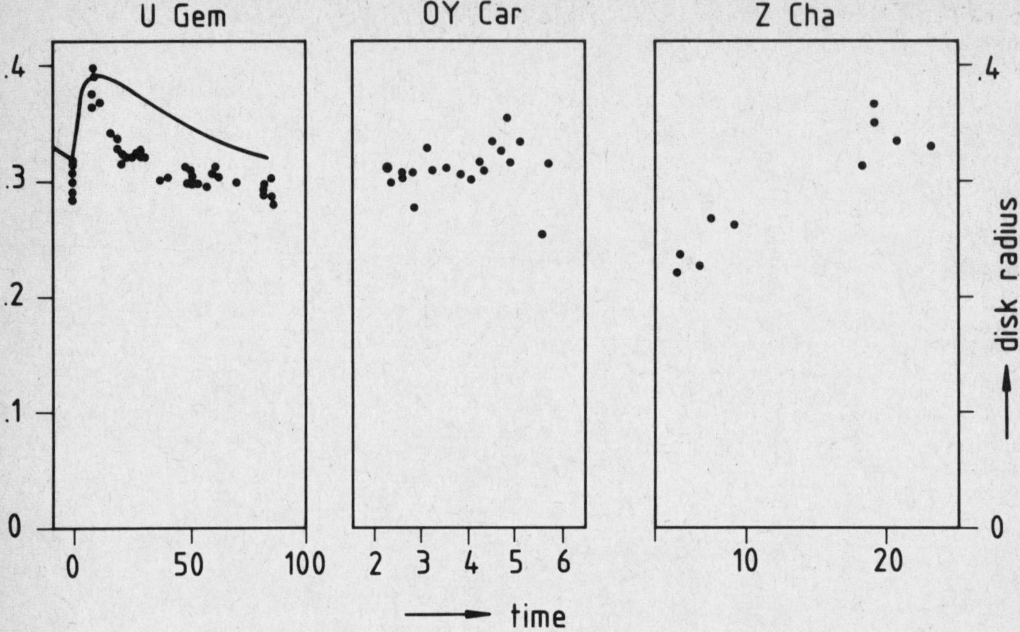

Figure 3. Disk radii (in units of the distance between the two stars) in dwarf novae as obtained from eclipse timings of the hot spot. Time is in units of days for U Gem and Z Cha, and in units of hundred orbital periods (100 × 0.063 days) for OY Car. The measurements are from Smak 1984a (U Gem), Cook 1985a (OY Car) and 1985b (Z Cha). An outburst occurred on day 0 for U Gem, and on day 13 for Z Cha. The drawn line is a disk instability calculation for U Gem by Smak (1984b). For Z Cha see also O'Donoghue (1986).

The transfer instability model has difficulty in modelling fast declines like those of VW Hyi, and tends to produce a slower decline (Bath & Pringle 1981). The disk instability has difficulty in producing flat-topped outbursts like those of SS Cyg, and tends to produce shortlived maxima (e.g. Cannizzo, Wheeler & Polidan 1986). It also has difficulty in producing long quiescent intervals, like those of $\sim 80 - 100$ days observed in U Gem, or — an extreme case — 30 years in WZ Sge!

A large variety in outburst lightcurves is also observed in single systems. Such variety in SS Cyg can be well explained with the transfer model (Bath, Clarke & Mantle 1986). If the mass transfer burst is short, the rise and decline times of the outburst are determined by the viscous time scale in the disk. For very long mass transfer bursts, however, the rise and decline of the outburst will reflect directly the form of the mass transfer burst.

2.4 Ratio of ultraviolet to optical flux

Ultraviolet observations of dwarf novae have been obtained with the Astronomical Satellite of the Netherlands, with the International Ultraviolet Explorer, and with the Voyagers. It has been found for a number of systems that the rise of the ultraviolet flux is delayed with respect to the optical rise. These systems include VW Hyi (Hassall et al. 1983, Schwarzenberg-Czerny et al. 1985), RX And (Pringle and Verbunt 1984), WX Hyi (Hassall, Pringle & Verbunt 1986), all observed with IUE, and SS Cyg (Polidan & Holberg), observed with Voyager. ANS observations of SU UMa show a double rise, each of which shows a spectrum redder than the quiescent spectrum (Wu & Panek 1983). Smak (1984b) and Cannizzo, Wheeler & Polidan (1986) obtain delays of the ultraviolet flux in their disk instability models, in particular for outbursts that start in the outer region of the disk. This may well be due to the use by these authors of black body spectra to describe the spectra emitted by surface elements with a given temperature. This leads to a serious underestimate of the ratio of ultraviolet to optical fluxes at the temperatures around 10^4 K in the beginning of the outburst.

Pringle, Verbunt & Wade (1986) compare the ultraviolet and optical spectra of two well observed systems with both disk instability and red dwarf instability models. They use Kurucz atmosphere model fluxes to describe the spectra emitted by surface elements of the disk. Some of their results are shown in Figure 5. The first system investigated is VW Hyi, which shows optical outbursts lasting 4-6 days with quiescent intervals of 2-3 weeks. Its outbursts start with a fast rise of the optical flux to maximum. During this fast rise the ratio of ultraviolet to optical flux is

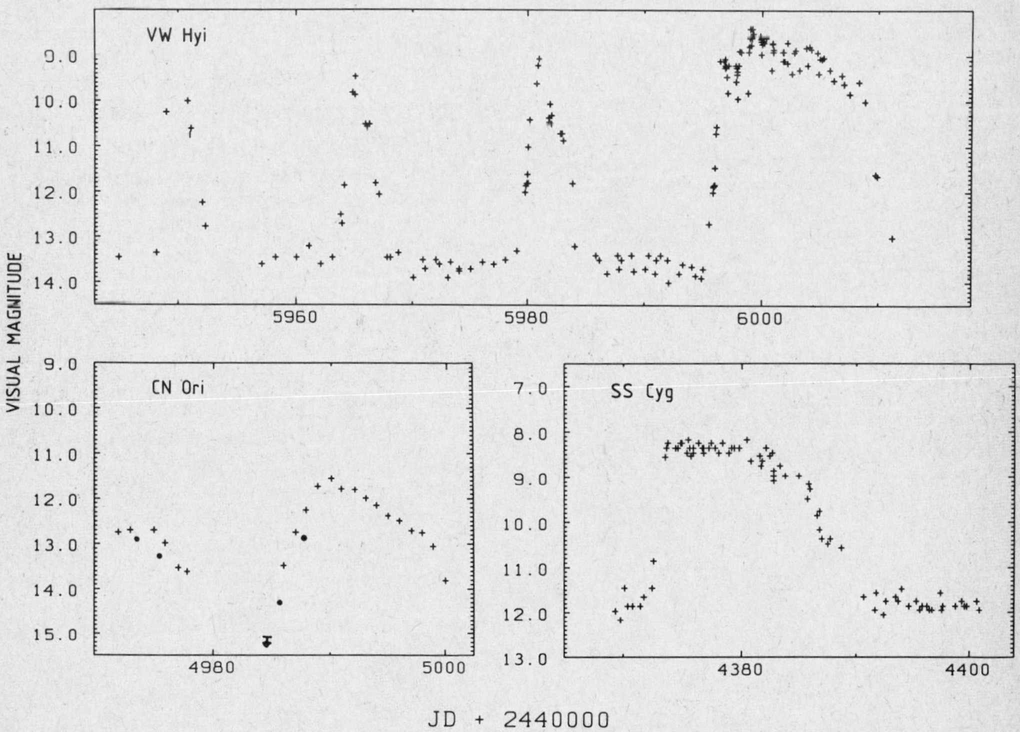

Figure 4. Optical lightcurves of VW Hyi, CN Ori and SS Cyg. The data were obtained by amateur observers of the Variable Star Section of the Royal astronomical Society of New Zealand for VW Hyi, and of the American Association of Variable Star Observers for CN Ori and SS Cyg. The importance of the work of these diligent observers to the research on cataclysmic variables can hardly be overestimated. Some IUE FES magnitudes are also shown for CN Ori.

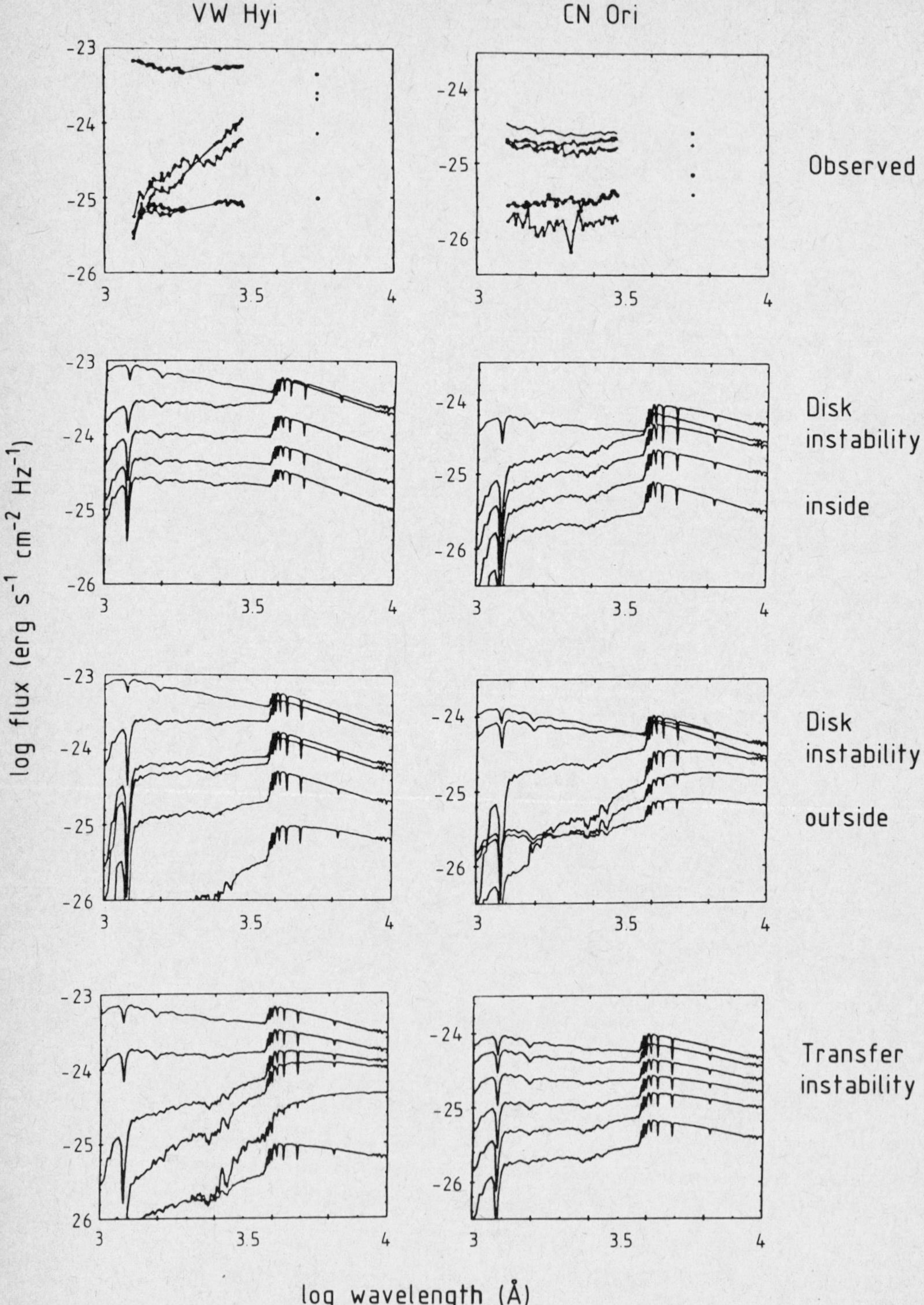

VW Hyi
CN Ori
Observed
Disk
instability
inside
Disk
instability
outside
Transfer
instability
log flux (erg s^{-1} cm^{-2} Hz^{-1})
log wavelength (Å)

Figure 5. Observed and modelled spectra of the rise to outburst maximum for VW Hyi and CN Ori. From top to bottom are shown the ultraviolet - optical spectrum as observed with IUE (including its optical sensor), as calculated for a disk instability starting near the white dwarf, as calculated for a disk instability starting near the outer edge of the disk, and as calculated for a red-star instability. The models were chosen to reproduce both the outburst length and the length of the quiescent interval. The fast rise of VW Hyi with its large ultraviolet lag, and the slow rise of CN Ori with a small ultraviolet lag are both best reproduced by the red-dwarf instability. (Adapted from Pringle, Verbunt & Wade 1986).

very low. During the decline the ultraviolet and optical fluxes drop together. The other system is CN Ori, which showed outbursts lasting 1-2 weeks almost without quiescent intervals during the period of the IUE observations. The optical rise to maximum is slow for CN Ori, and the ultraviolet flux rises together with the optical. The decline spectra are very similar to the rise spectra.

For both systems the optical lightcurves can be fitted with all three types of models enumerated in the previous section. For both systems the ultraviolet behaviour is best described with the transfer instability model.

The problem with the disk instability model for VW Hyi, is that the region of the disk where the model outburst starts — whether in the inner or outer region of the disk — moves up to the hot branch before the outburst has spread very far. The model outbursts therefore start with a ring with high temperature ($> 10^4$ K), which radiates as strongly in the ultraviolet as in the optical wavelength regions. In the transfer instability model for VW Hyi the temperatures rise appreciably only after the increased mass transfer has affected a large part of the disk. Thus the optical flux rises before the ultraviolet flux. In this model the form of the outburst is determined by the viscous time scales in the disk.

In order to explain the rapid succession of outbursts in CN Ori with the disk instability starting in the outer disk region, the start of a new outburst has to precede the completion of the collapse on the white dwarf of the hot region produced by the previous outburst. In this model the ultraviolet flux is therefore still declining when the optical starts to rise again, contrary to observation. If the disk instability starts in the inner region of the disk the ultraviolet and optical fluxes rise together. The ultraviolet to optical flux ratio is smaller during the rise of this model than during the decline, because the temperatures during rise are lower. In the mass transfer instability model the outburst follows the mass transfer burst during the rise, and declines sligthly slower. Throughout the outburst the disk is close to the stationary solution corresponding to the momentary value of $\dot{M}_e$. The spectra during the rise and during decline are therefore very similar. It is interesting to note, that in this model the high temperature region spreads outwards from the center, even though the outburst is caused by an increased mass input at the outer disk edge.

The spectra during the decline are the same for all models: the ultraviolet and optical fluxes drop together. The mass transfer instability models have some difficulty in modelling declines as fast as those of VW Hyi.

The ultraviolet to optical flux ratio is best described with the mass transfer models.

2.5 Optical lines

Optical spectra obtained throughout an outburst of SS Cyg show that the broad emission wings disappear at the onset of the outburst. If the wings are interpreted as originating from the material rotating at highest velocities around the white dwarf, this indicates that the inner region of the disk is affected right at the onset (Clarke, Capel & Bowyer 1984). A disk instability starting in the inner region of the disk can be expected to show such an effect. The mass transfer model for CN Ori outbursts discussed above indicates that the observation may be compatible with a transfer instability model as well, provided that the rise is slow. (The rise of the outburst in question was indeed a slow one.)

2.6 Quiescent ultraviolet and X-ray fluxes

Ultraviolet and X-ray observations of dwarf novae in quiescence show appreciable fluxes in these wavelength regions. The red-dwarf instability model can explain this with moderate values of $\dot{M}_e$ in quiescence. The disk instability models predict very low accretion rates in quiescence for systems with outbursts well separated by quiescent intervals. This is reflected in the low luminosities of the accretion disk and of the boundary layer during quiescence, for example in VW Hyi (Figure 2). The low accretion rates further lead to low temperatures in the disk, and the

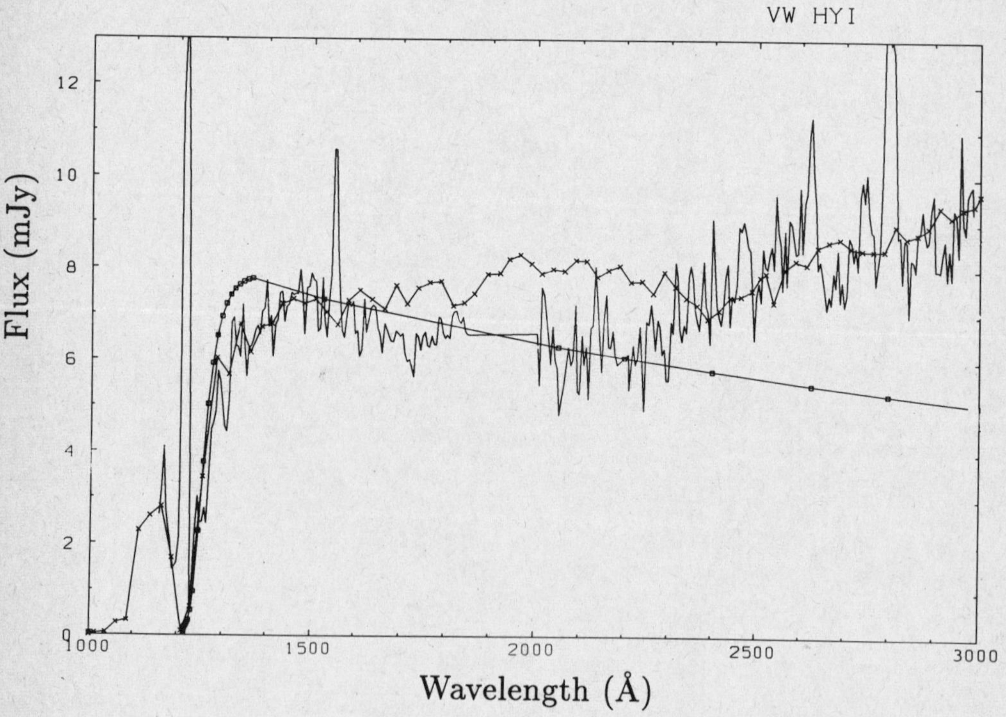

Figure 6. The quiescent ultraviolet spectrum of VW Hyi obtained with IUE on 1980 Dec 28. Also shown are a Wesemael et al. (1980) model spectrum for a white dwarf with log g = 8.0 and T = 20000 K (⊟⊟) and a Kurucz (1979) model spectrum for log g = 4.5 and T = 12000 K (✕✕) The surface area of the model spectra have been chosen to give the observed flux at 1500 Å. The area taken for the Wesemael et al. spectrum is compatible with the white dwarf surface, the Kurucz spectrum needs a larger area, which may correspond to a small disk.

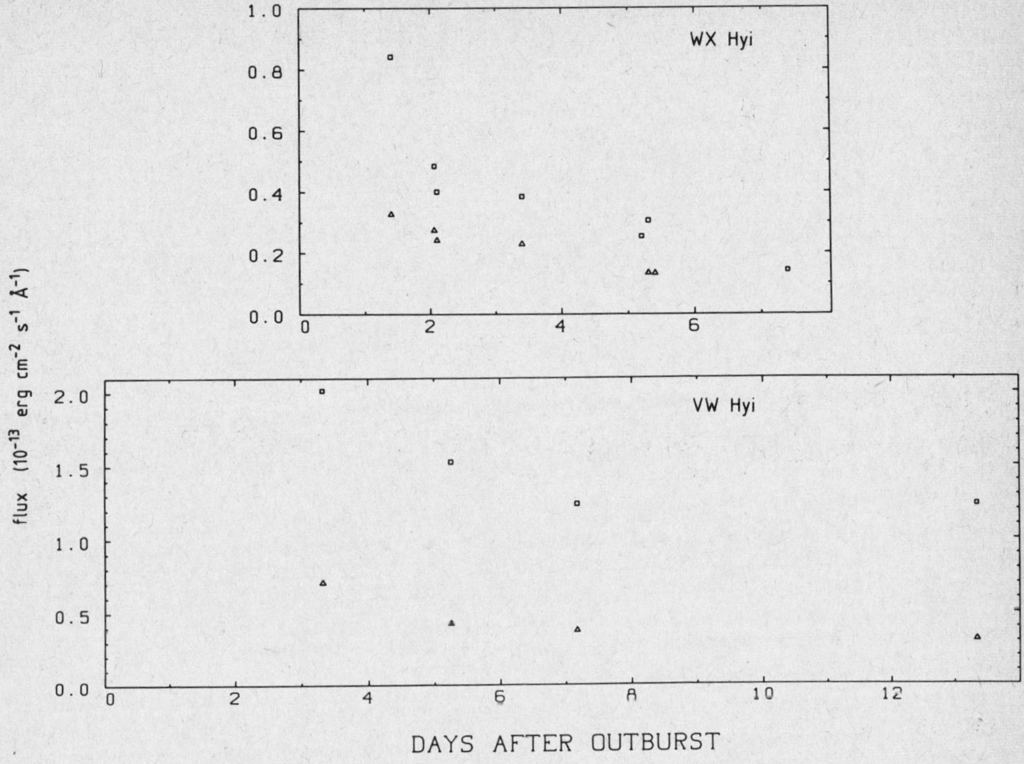

Figure 7. The continued decline of the ultraviolet flux during quiescence. Shown are the fluxes averaged over 80 Å bins around 1460 Å ($\triangle$) and around 2880 Å ($\square$) for data from different outburst intervals of WX Hyi (see Hassall, Verbunt & Pringle 1986) and from a single outburst interval of VW Hyi (see Verbunt et al. 1986).

disk instability model predicts an ultraviolet flux in quiescence orders of magnitude lower than observed. Adherents of the disk instability theory have argued that the white dwarf contributes to the ultraviolet flux in quiescence (e.g. Smak 1984c). The turnover in the flux at wavelengths $\lambda \simeq$ 1400 Å observed in VW Hyi has also been attributed to the white dwarf (Mateo & Szkody 1984). Alternatively, the turnover can be explained as due to a low maximum temperature $T_{eff} < 12000$ K in the disk (Verbunt et al. 1986). In Figure 6 a quiescent ultraviolet spectrum of VW Hyi is compared with a white dwarf model spectrum and with a cool stellar spectrum (as may apply to the disk). The cool disk fits the turnover better, but gives fluxes too high at $\lambda \simeq 2000$ Å. From the ultraviolet alone it is not possible to determine the relative importance of the white dwarf and disk contributions to the ultraviolet flux.

In the dwarf nova Z Cha the distance of the system and the size and temperature of the white dwarf in quiescence have been determined from optical eclipse observations (Horne & Cook 1985; Wood et al. 1986). The white dwarf parameters can be constrained even more if one interprets the broad absorption lines in which the quiescent optical emission lines are embedded as white dwarf atmospheric lines. The ultraviolet flux of the white dwarf can now be predicted and compared to the IUE observations. The result is that most of the flux at wavelengths $\lambda < 2000$ Å may well be due to the white dwarf, and that the disk becomes increasingly important at longer wavelengths (Marsh & Horne 1986). At the moment it is not clear whether the flux contribution of the disk between 2000 and 3000 Å can be accounted for in the disk instability model.

In the disk instability model the disk builds up in density during quiescence, and hence an increase in the disk and boundary luminosities is predicted between well separated outbursts (see Figure 2). This is at variance with the observations, which show a decline of the ultraviolet fluxes during quiescence in WX Hyi and in VW Hyi (Figure 7), also in the flux range where the disk contribution dominates ($\lambda \simeq 3000$ Å). The X-ray (i.e. boundary layer) flux of VW Hyi also declines during quiescence (Van der Woerd & Heise 1986).

Notwithstanding the fact that much of the ultraviolet flux at the shorter wavelengths ($\lambda < 2000$ Å) may be due to the white dwarf, the quiescent ultraviolet and X-ray observations are in better agreement with the transfer instability model.

2.7 Conclusions

The comparison between the theoretical predictions and the observational evidence gives a confused picture, some observations (notably those of the disk radius) supporting the disk instability model, some (the ultraviolet and X-ray observations) supporting the transfer instability model. Theoretically there is no clear preference either: the disk instability models depend on an ad hoc enlargement of the local instability, and there is no detailed theory for the mass transfer instability.

On the other hand, many observations indicate the existence of variations in $\dot{M}_e$. Amongst these are the prolonged low states observed in systems like AM Her, MV Lyr and TT Ari; the superoutbursts in systems like VW Hyi (the last outburst shown in Figure 4 is an example); and the prolonged period of time that systems like Z Cam spend at a luminosity level halfway between outburst maximum and minimum. All of these are generally attributed to changes in the mass tranfer rate from the companion star. Disk modellers have also invoked increased mass transfer to explain the flat-topped bursts like those of SS Cyg. (It is interesting to note that the optical lightcurve of the superoutburst of VW Hyi is rather similar to that of the flat-topped outburst of SS Cyg - cf. van Paradijs 1983.)

Thus the variability of the mass transfer is well attested, whereas the existence of a disk instability still has to be proved.

References

Bath, G.T., Clarke, C.J. & Mantle, V.J. 1986, *Mon. Not. R. astr. Soc.* **221**, 269.

Bath, G.T. & Pringle, J.E. 1981, **194**, 967; 1982, **199**, 267.

Cannizzo, J.K., Ghosh & Wheeler, J.C. 1982 *Astrophys. J. (Letters)* **260**, L83.

Cannizzo, J.K., Wheeler,J.C. & Polidan, R.S. 1986 *Astrophys. J.* **301**, 634.

Clarke, J.T., Capel,D. & Bowyer,S. 1984 *Astrophys. J.* **287**, 845.

Cook, M.C. 1985a, *Mon. Not. R. astr. Soc.* **215**, 211.

Cook, M.C. 1985b, *Mon. Not. R. astr. Soc.* **216**, 219.

Faulkner, J., Lin, D.N.C. & Papaloizou, J. 1983, *Mon. Not. R. astr. Soc.* **205**,359; **205**,487. 1985, **212**, 105.

Haefner, R., Schoembs, R. & Vogt, N. 1979, *Astron. Astrophys.* **77**, 7.

Hassall, B.J.M., Pringle, J.E., Schwarzenberg-Czerny, A., Wade, R.A., Whelan, J.A.J. & Hill, P.W. 1983, *Mon. Not. R. astr. Soc.* **203**, 865.

Hassall, B.J.M., Pringle, J.E. & Verbunt, F. 1985, *Mon. Not. R. astr. Soc.* **216**, 353.

Horne, K.D. & Cook, M.C. 1985, *Mon. Not. R. astr. Soc.* **314**, 307.

Kurucz, R.L. 1979, *Astrophys. J. Suppl. Ser.* **40**, 1.

Marsh, T. & Horne, K.D. 1986 *Mon. Not. R. astr. Soc.* in preparation.

Mateo, M. & Szkody, P. 1984, *Astron. J.* **89**, 863.

Meyer, F. & Meyer-Hofmeister, E. 1981, *Astron. Astrophys.* **104**, L10; 1984, **132**, 143.

Meyer, F. & Meyer-Hofmeister, E. 1982, *Astron. Astrophys.* **106**, 34.

Mineshige, S. & Osaki, Y. 1983, **35**, 377; 1985, **37**, 1.

O'Donoghue, D. 1986, *Mon. Not. R. astr. Soc.* **220**, 23p.

Piran, T. 1978, *Astrophys. J.* **221**, 652.

Polidan, R.S. & Holberg, J.D. 1984, *Nature* **309**, 528.

Pringle, J.E. & Verbunt,F. 1984, in *Fourth European IUE Conference*, eds. E. Rolfe & B. Battrick, ESA SP 218, p.377.

Pringle, J.E. Verbunt, F. & Wade, R.A. 1986, *Mon. Not. R. astr. Soc.* **221**, 169.

Schoembs, R. 1982, *Astron. Astrophys.* **141**, 124.

Schwarzenberg-Czerny, A., Ward, M.J., Hanes, D.A., Jones, D.H.P., Pringle, J.E., Verbunt, F. & Wade, R.A. 1985, *Mon. Not. R. astr. Soc.* **212**, 645.

Shakura, N.I. & Sunyaev,R.A. 1973, *Astron. Astrophys.* **24**, 337.

Smak, J. 1984a, *Acta Astr.* **34**, 93.

Smak, J. 1984b, *Acta Astr.* **34**, 161.

Smak, J. 1984c, *Acta Astr.* **34**, 317.

Taam, R.E. & Lin, D.N.C. 1984, *Astrophys. J.* **287**, 761.

Van der Woerd, H. & Heise, J. 1986, *Mon. Not. R. astr. Soc.* in press.

Van Paradijs, J. 1983, *Astron. Astrophys.* **125**, L16.

Van Paradijs, J. & Verbunt, F. 1984, in *High energy transients in astrophysics*, ed. Woosley, S., American Institute of Physics, New York, p.49.

Verbunt, F. 1982, *Space Sci. Rev.* **32**, 379.

Verbunt, F., Hassall, B.J.M., Pringle, J.E., Warner, B. & Marang, F. 1986, *Mon. Not. R. astr. Soc.* in press.

Vogt, N. 1985, *Astron. Astrophys.* **144**, 124.

Wesemael, F., Auer, L.H., Van Horn, H.M. & Savedoff, M.P. 1980, *Astrophys. J. Suppl. Ser.* **43**, 159.

Wood, J., Horne, K.D., Berriman, G., Wade, R.A., O'Donoghue, D. & Warner, B. 1986, *Mon. Not. R. astr. Soc.* **219**, 629.

Wu, C.C. & Panek, R.J. 1983, *Astrophys. J.* **271**, 754.

Appendix

In this appendix the equations of Section 1.1 are derived. The hydrodynamic equations for a viscous fluid are:

Equation of mass conservation:

$$\frac{\partial \rho}{\partial t} + \nabla.(\rho \mathbf{v}) = 0 \qquad \text{A1}$$

with ρ the density and $\mathbf{v}$ the flow velocity. Equation of motion:

$$\rho(\frac{\partial \mathbf{v}}{\partial t} + (\mathbf{v}.\nabla).\mathbf{v}) = -\rho \nabla \Phi - \nabla P - \nabla.\mathbf{t} \qquad \text{A2}$$

Here Φ is the gravitational potential, P the total pressure and $\mathbf{t}$ the viscous stress tensor. Energy equation:

$$\rho(\frac{\partial \epsilon}{\partial t} + (\mathbf{v}.\nabla)\,\epsilon) = -P\nabla.\mathbf{v} - (\mathbf{t}.\nabla).\mathbf{v} - \nabla.\mathbf{q} \qquad \text{A3}$$

expressing the change of internal energy ϵ of a unit element of the fluid as it moves, in terms of the work done by pressure, of the heat generated by viscous forces, and of the energy loss due to divergence of the energy flow $\mathbf{q}$.

Choose a cilindrical coordinate system r, ϕ, z with the origin in the center of the compact object, and $z = 0$ the orbital plane. The disk is assumed symmetric with respect to $z = 0$. To simplify Eqs. A1 - A3 we use the assumptions stipulated in Section 1.1. Further we integrate over the vertical direction, and define the half thickness h of the disk by

$$\rho h = \int_0^\infty \rho(z)dz$$

Then from A1:

$$\frac{\partial}{\partial t}(\rho h) + \frac{1}{r}\frac{\partial}{\partial r}(r\rho h v_r) = 0 \qquad \text{A4}$$

with v_r the radial component of the velocity. From the r-component of A2:

$$v_\phi^2 = \frac{GM}{r} = r^2\omega^2 \qquad \text{A5}$$

with M the mass of the compact object. From the ϕ-component of A2:

$$r\rho h v_r \frac{\partial}{\partial r}(rv_\phi) = -\frac{\partial}{\partial r}(r^2 h t_{r\phi})$$

We rewrite this using A5 and the definition of the $r\phi$-component of the viscous stress tensor:

$$t_{r\phi} \simeq -\rho\nu(\frac{\partial v_\phi}{\partial r} - \frac{v_\phi}{r}) = \frac{3}{2}\rho\nu\frac{v_\phi}{r}$$

(here ν is the kinematic viscosity coefficient) as

$$-r\rho h v_r = 3r^{\frac{1}{2}}\frac{\partial}{\partial r}(r^{\frac{1}{2}}h\rho\nu) \qquad \text{A6}$$

From the z-component of A2 we find

$$P = \frac{1}{2}\rho h^2\omega^2 \qquad \text{A7}$$

From A3:

$$Q^- = \frac{9}{4}\rho h \nu \omega^2 \qquad \text{A8}$$

where Q^- is the energy flow from a column in the disk with unit surface and height h.

To these equations one must add an equation of state, an energy loss equation and an equation for the viscosity. Many different choices have been made (see summary in Verbunt 1982). For example, the model of Shakura & Sunyaev (1973) has:

$$P = P_{gas} + P_{radiation} = \rho \frac{k}{\mu m_p} T + \frac{1}{3} a T^4 \qquad \text{A9}$$

$$Q^- = \frac{2ac T^4}{3\kappa \rho h} \qquad \text{A10}$$

(i.e. the energy flow from the column is due to radiation diffusion from the $z = 0$ plane to the surface of the disk. κ is the opacity.)

$$\nu = \frac{2\alpha P}{3\omega} \qquad \text{A11}$$

Define $\Sigma = 2\rho h$. Combination of A4 and A6 gives the global diffusion equation Eq. (1), and from the local structure equations A7 to A11 one finds Eq. (2). Eq. (3) is found from A8 by equating the energy loss from a column with unit surface with σT_{eff}^4.

In the stationary case ($\frac{\partial}{\partial t} = 0$) the mass flow through the disk, $-2\pi r \times 2\rho h \times v_r$, is constant — see A4 —, $\dot{M}_e$, say. A6 can then be integrated:

$$(\nu \Sigma)_{stat} = \frac{\dot{M}_e}{3\pi} + const \times r^{-\frac{1}{2}}$$

If the constant is chosen to have zero viscosity at the inner radius $r = R$ of the disk, Eq. (4) follows. The inner radius of the disk can be the radius of the compact object, or the radius of the magnetosphere.

ON THE LONG TERM ACTIVITY OF POP. I BINARY SYSTEMS CONTAINING AN X-RAY PULSAR

L. Stella [1,2,3], N.E. White [1,2], R. Rosner [4]

[1] EXOSAT Observatory, ESOC, Robert Bosch Strasse 5, 61 Darmstadt, FRG.
[2] Affiliated to Astrophysics Division, Space Science Dept. of ESA.
[3] On leave from I.C.R.A., Dept. of Physics 'G. Marconi', University of Rome, Italy.
[4] Harvard-Smithsonian Center for Astrophysics, Cambridge, MA 02138.

Abstract

We review the observed properties of the population I binaries containing X-ray pulsars. We point out that many of the pulsing X-ray transients in OB star systems with pulse periods of 1-10 s, when in outburst, lie close to the critical equilibrium point where the corotation radius equals the magnetospheric radius. This suggests that many of these systems usually are dormant because of the centrifugal drag exerted by the pulsar magnetosphere which inhibits accretion and that a small increase in the mass accretion rate can turn them into very luminous transient X-ray sources. If a system lies close to the equilibrium point and it is in an eccentric orbit, luminous outbursts may occur around the time of periastron passage. A correlation between the X-ray luminosity and the pulsar frequency is found and interpreted as the result of centrifugal inhibition of accretion preventing low luminosity, fast spin period magnetised neutron stars accreting.

1.   Introduction

Most of the optically identified X-ray pulsars are in binary systems containing O or B type stellar companions (see e.g. Rappaport and van den Heuvel 1982; White 1984), whose natural mass-loss is in many cases capable of powering the X-ray source via gravitational capture by a magnetised neutron star. For the OB supergiant X-ray binaries with orbital periods of a few days, it is debatable how much extra material is provided by Roche or quasi-Roche lobe overflow (Conti 1978; Petterson 1978), but for the remainder of the systems with orbital periods of tens or hundreds of days, stellar wind accretion onto a neutron star is the only viable mechanism to produce the observed X-ray emission.

In the simple theory of direct wind accretion onto a magnetised neutron star (Davidson and Ostriker 1973; Lamb, Pethick and Pines 1973) the (Bondi) accretion radius is, for plausible wind parameters and the canonical neutron star parameters, larger than the corotation radius, which is in turn larger than the characteristic dimensions of the magnetospheric boundary surface. The accretion rate is then set by the flow at the accretion radius, and the neutron star magnetosphere must accommodate itself accordingly. The time averaged accretion rate and hence the X-ray luminosity over timescales of $\gtrsim 1$ day in such a model will depend continuously on the stellar wind parameters, since the flow to the magnetosphere is controlled by conditions in the vicinity of the accretion radius.

The long timescales ($\gtrsim 1$ day) variability properties of the <u>persistent</u> X-ray pulsars in OB star binaries consist of (a) irregular variations, presumably caused by corresponding overall changes in the stellar wind parameters and/or (b) periodic variations associated with a moderately eccentric orbit (e $\lesssim$ 0.5; e.g. 4U1223-62; Watson, Warwick and Corbet 1982). In the latter case the wind parameters at the neutron star accretion radius are modulated by the varying separation of the neutron star and the OB star companion.

The variations discussed above are always limited to a dynamic range of < 100 such that the X-ray flux does not usually drop below the detection threshold of the current generation of non-imaging X-ray instruments ($\lesssim 1$ μJy). While these variations are successfully explained by the direct wind accretion model, many Be star systems are associated with X-ray transient outbursts, with a ratio of the maximum to minimum luminosity, $L_X(max)/L_X(min)$, greater than a factor of 100 (Maraschi, Treves and van den Heuvel 1976). For the purpose of our discussion we divide the transient activity into two different classes:

<u>Class I - Periodic Transient Activity</u>. Transient outbursts that recur periodically and have $L_X(max)/L_X(min) \gtrsim 100$. For those systems where a pulse timing orbital solution is available (currently only for the shorter orbital system, with $P_0 < 50$ days) it is found that the neutron star is typically in a moderately eccentric orbit (e $\lesssim$ 0.5) and that the outbursts occur close to the time of periastron passage. This suggests enhanced accretion caused by the closer proximity of the neutron star to the Be star companion. These periodic X-ray outbursts are to be distinguished from those caused simply by variations in the density of the stellar wind around an eccentric orbit. The dynamic range of typically greater than 300 (V0332+53; cf. Stella et al. 1985) cannot be reconciled with the predictions of the direct wind accretion model. Even allowing for the additional modulation induced by the steepest velocity profile considered plausible for a radiatively driven stellar wind (cf. Henrichs 1980) the orbital flux variations in the direct-wind accretion model are limited at most to a factor of $\lesssim 50$.

<u>Class II - Irregular Transient Activity</u>. Transient outbursts that typically last several tens of days and involve an increase in luminosity in excess of a factor of 100-1000. The timing of these outbursts is unrelated to any underlying orbital period. In those cases where the orbit is known and the outburst lasts longer than the orbital period, an orbital modulation at the level predicted by the direct wind accretion model is not always seen (e.g. 4U0115+63; Kriss et al. 1983).

In order to explain the large orbital modulation of the class I transients a number of mechanisms for causing an additional surge in accretion onto the compact object have been considered. Charles et al.(1983) discuss the possibility that the 16 day class I outbursts seen from A0538-66 are caused by the companion star temporarily filling its Roche lobe at periastron passage, causing it to spill extra material onto the neutron star. While intermittent Roche lobe overflow might explain the outbursts of A0538-66 (the orbital eccentricity in this system is not precisely determined), in other systems where class I behaviour is seen and the orbit is well known the companion is far from filling its Roche lobe at periastron. In these cases it has been suggested (e.g. Kelley, Rappaport and Ayasli 1983) that the neutron star orbital plane and the equatorial plane of the companion are misaligned. If the Be star is surrounded by a centrifugally supported disk (many of the these stars rotate close to break up velocity), then an outburst should be seen when the neutron star crosses the disk plane. This model predicts that two outbursts should be seen per orbit, seperated in phase by at least $\Delta\phi \simeq (E - e \sin E)/\pi$ where $E = 2 \arctan\left[(1-e)^{1/2}(1+e)^{-1/2}\right]$. Some confirmation of this geometry has recently come from the determination of the 8.4 day orbit of the 4.2 day recurrent flaring system X1907+097 (Makishima et al. 1984). However, X1907+097 is a persistent system and only one outburst per orbit is observed instead in the class I transient systems. Only by invoking special geometries such as circular orbits, or a disk that has a limited extent relative to the orbital dimensions, can this problem be avoided (Apparao 1985).

Since Be stars are in general known to undergo mass ejection (shell) episodes, this has naturally lead to the suggestion that the class II X-ray outbursts are the result of enhanced accretion as the shell reaches the neutron star. While to zeroth order this provides a plausible explanation, the detailed properties of those outbursts which have been studied in detail are difficult to understand in the context of such a simple picture. During an outburst from 4U0115+63 in 1980 the optical star brightened by 1.7 mag ~ 60 days before the turn on of the X-ray source, far longer than would be expected for the material to reach the neutron star orbit (Kriss et al. 1983). In another case the transient source V0332+53 was observed to display class I X-ray outbursts between November 1983 and January 1984 with a period of 34 days (Stella et al. 1985). However, a factor of 20 brighter outburst in 1973 that lasted ~ 100 days was clearly of Class II and showed no evidence for a 34 day modulation (Terrell and Priedhorsky 1984).

Mediation of the inward flow outside of the neutron star magnetosphere by an accretion disk has been suggested as a reason for the optical/X-ray time delay observed from 4U0115+63 and the failure to detect any strong orbital modulation (Kriss et al. 1983). The presence of an accretion disk during the Class II outburst of V0332+53 in 1973, but not during the Class I outbursts in 1983-84, might also account for the difference in the properties of the 1973 and 1983-84 outbursts. However, it is far from clear that in these stellar wind-driven systems that the stellar wind captured by the neutron star will contain sufficient angular momentum to form a disk that has a lifetime of tens of days (cf. Wang 1981).

The early discussion of the properties of magnetised neutron stars accreting from a stellar wind emphasised the importance of the interaction of the rotating magnetosphere with the inflowing material (Davidson and Ostriker 1973; Lamb, Pethick and Pines 1973). It was pointed out that the specific angular momentum of the inflowing material may cause an accretion disk to form that will apply a torque on the neutron star. The X-ray pulsar rotation frequency will thus increase on relatively short timescales of 100-10,000 years until an equilibrium rotation is reached where the centrifugal force acting on the wind material close to the magnetospheric boundary can prevent accretion onto the neutron star surface. Conversely, if the magnetospheric radius exceeds the corotation radius, as might be expected when the neutron star is young and rotating rapidly, then a braking torque will be applied by the magnetosphere 'propelling' away the inflowing material (Illarionov and Sunyaev 1975). A mass ejection episode from the companion during the X-ray dormant spin-down interval can cause the magnetosphere to be compressed sufficiently for the system to briefly turn on as a bright X-ray pulsar (Fabian 1975; Maraschi, Traversini and Treves 1983).

In this paper we investigate some of the regimes for a magnetised neutron star that interacts with the stellar wind of an OB star and compare the results with the considerable body of observational data that now exists on these systems (cf. Stella, White and Rosner 1986). We emphasise how the transition to the regime in which accretion onto the neutron star is inhibited by the centrifugal drag of the rotating magnetosphere with the inflowing material can cause dramatic changes in X-ray luminosity even for a quite modest change in the stellar wind parameters. This provides the necessary sensitivity of the accretion rate onto the neutron star to the conditions in the vicinity of the accretion radius, which is required to explain the transient activity observed in so many of these systems.

2.   The Observed Properties

In Table 1 we list the known or suspected OB star X-ray binary systems and summarise some key observational properties: their pulse periods, orbital periods,

TABLE 1 [a]

Source	P_S (s)	P_O (d)	$L_x(max)$ (erg/s)	$\dfrac{L_x(max)}{L_x(min)}$	e	Class	Spectral type
(1) A0538-66	0.069	16.7	1.2×10^{39}	$>1.8 \times 10^4$	>0.4	I	B2IIIe
(2) 4U0115+63	3.6	24.3	8×10^{36}	$>2.3 \times 10^4$	0.34	II	Be
(3) V0332+53	4.4	34.2	2×10^{37}	$>1 \times 10^4$	0.31	I+II	Be ?
(4) 2S1553-54	9.3	30.6	1×10^{37}	>30	0.09	II	?
(5) A0535+26	104	111	2×10^{37}	700	0.2-0.4	II+I(?)	09.7 IIIe
(6) GX304-1	272	133	3×10^{35}	>25	?		B2Vne
(7) 4U1145-62	292	188	6×10^{36}	250	?	II+I(?)	B1Vne
(8) X Per	835	580 ?	1×10^{34}	4	?		09.5(III-V)e
(9) LMC X-4 [b]	14	1.4	4×10^{38}	>20	0.0		07(III-V)
(10) A1118-62	405	?	5×10^{36}	>70	?	II(?)	09.5(III-V)e
(11) 1E1145-616	297	?	3×10^{36}	100	?	?	B1I
(12) SMC X-1	0.7	3.9	6×10^{38}	110	7×10^{-4}	?	BOI
(13) Cen X-3	4.8	2.1	8×10^{37}	30	8×10^{-4}		06-8 fp
(14) Vela X-1	283	9.0	6×10^{36}	50	0.09		B0.5Ib
(15) 4U1538-52	529	3.7	4×10^{36}	10	?		BOI
(16) 4U1223-62	699	41	1×10^{37}	30	0.47		B1.5Ia
(17) 4U1700-37	?	3.4	3×10^{36}	40	?		06.5f
(18) A0114+65	?	11.6	2×10^{34}	1	?		B0.5IIIe
(19) γ Cas	?	?	2×10^{33}	2	?		B0.5(II-V)e
(20) SMC X-2	?	?	8×10^{37}	7	?		B1.5Ve
(21) SMC X-3	?	?	2×10^{37}	5	?		09(III-V)e
(22) X1907+097	437	8.4	7×10^{35}	10	0.24		?
(23) EXO 2030+37	41.2	38	4×10^{38}	>5000	0.31	II	?
(24) 1E1048.1-5937	6.4	?	1×10^{36}	>20	?	II(?)	?

[a] The data in this table are taken from Rappaport and Van den Heuvel (1982), Bradt and McClintock (1983), Joss and Rappaport (1984) and Corbet (1984) and references therein, except for V0332+53 (Terrell and Priedhorsky 1984, Stella et al 1985), X1907+097 (Makishima et al. 1984), EXO 2030+375 (Parmar et al. 1985; White et al. 1985) and 1E1048.1-5937 (Seward, Charles and Smale 1986). The data on the optical counterpart of A0535+36 are from Giangrande et al. (1980). The values of L_x (min) for the transient systems correspond to the upper limits to the (undetected) quiescence flux. We have excluded flares on timescales of hours in the evaluation of $L_x(max)/L_x(min)$.

[b] Roche lobe overflow system.

maximum observed X-ray luminosities, $L_X(max)/L_X(min)$, eccentricity, class of transient activity seen and the spectral type of the optical counterpart. The pulse periods range from 69 ms to 835 s, the maximum luminosities from 10^{34} to 10^{39} erg/s and the orbital periods from 1.4 day to greater than several hundred days. Transient activity is exclusively confined to the Be star systems, with many of these systems not detected in quiescence. This reflects the fact that the time averaged mass loss rate from these stars is two to four orders of magnitudes lower than those of the supergiant stars in systems such as Cen X-3. In the case of A0535+26 it is not yet clear whether the source is detected in quiescence, such that the $L_X(max)/L_X(min)$ around the orbit is only known to be > 10 (Priedhorsky and Terrell 1983); we have therefore placed a question mark against the Class I transient activity from the system.

In Table 1 we have also included the other OB star binary systems from which pulsations have not been detected to date. Except for the lack of a large coherent modulation of their X-ray flux, these sources possess the same properties as the Pop. I X-ray pulsar systems. The failure to detect pulsations might reflect the fact that these systems contain a very slowly rotating neutron star $(P > 1000$ s) or that the magnetic dipole axis is coaligned with the rotation axis of the neutron star.

The evolution of the outbursts of V0332+53 is shown in Figure 1. In 1973 the source underwent a giant class II outburst during which the X-ray flux rose linearly to 1.4 Crab in ~ 50 days and then decayed in a similar manner (Terrell and Priedhorsky 1984). In 1983-84 three periodic class I outbursts were observed with EXOSAT (Stella et al. 1985) each with a peak flux of $\lesssim 100$ mCrab. This series of observations resulted in the discovery of 4.4 s X-ray pulsations. Doppler timing allowed the determination of the orbital parameters giving a 34 day period with an eccentricity of 0.31 and the times of periastron passage coincident with the times of the three outbursts. In the decay phase of each outburst the flux decreased abruptly within 3 days to < 0.2 mCrab implying orbital flux variations larger than a factor of ~ 300. An OB star was found within the EXOSAT 10 arcsecond position (cf. Bernacca, Iijima and Stagni 1984 and references therein). Our discussion will focus on explaining the properties of this system.

It is evident from Table 1 that the fast X-ray pulsars $(P_S < 10$ s) display the most luminous outbursts. This is illustrated in Figure 2 where the maximum X-ray luminosity $L_X(max)$ is plotted as a function of the pulsar period P_S. The linear correlation coefficient between the maximum X-ray luminosity and the pulsar frequency is 0.86 (corresponding to a probability of $\sim 10^{-7}$ of obtaining a larger coefficient from a random distribution). There is no clear distinction between the $L_X(max)$-P_S relation for the supergiant (squares) and the Be star systems (circles and diamonds). For the two sub-samples a coefficient of linear correlation between $L_X(max)$ and the pulsar frequency of 0.905 and 0.9997, respectively, is obtained.

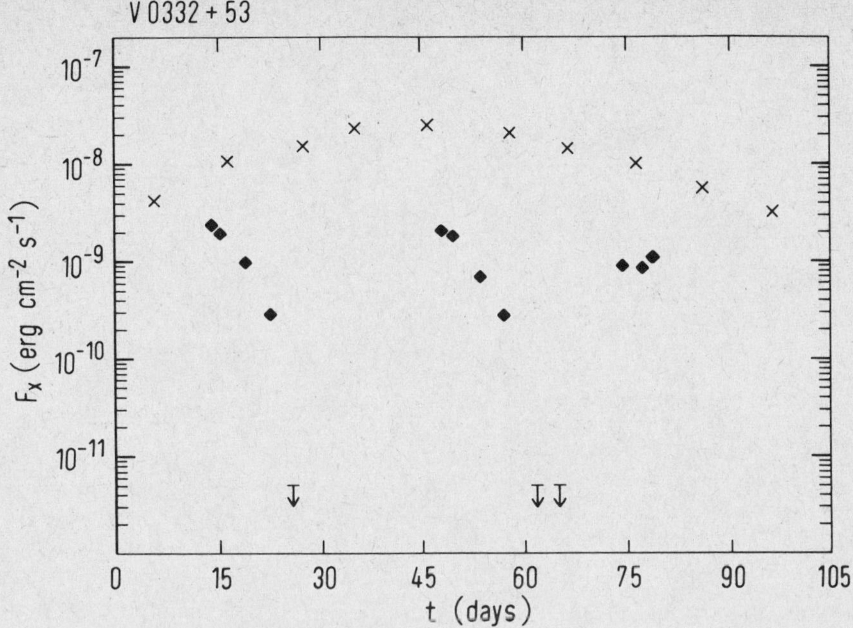

Figure 1:
The evolution of the 1973 giant outburst (crosses) and the 1983-84 periodic outbursts (diamonds) of V0332+53. The origin of the time axis corresponds, respectively, to JD 2,441,827 and JD 2,445,645 and the energy range for the X-ray fluxes is, respectively, 3-12 keV and 1-15 keV (the data are taken from Terrell and Priedhorsky 1984 and Stella et al. 1985). The arrows represent the 3σ upper limits to the source flux during the quiescent phases between the periodic outbursts. The maximum luminosity given in table 1 for 1.5 kpc distance has been obtained by applying a bolometric correction (derived from the EXOSAT X-ray spectra) to the maximum flux observed in 1973 with the Vela 5B satellite.

In Figure 3 the pulse periods are plotted against the corresponding orbital periods. A correlation between the orbital period P_O and the pulsar spin period P_S limited to the Be star and transient systems has been recently noted by Corbet (1984). The pulsar spin period in the supergiant systems does not show any correlation with the orbital period. For a given orbital period the supergiant systems possess systematically longer pulsar spin periods than the Be star systems.

3. Stellar Wind Accretion

In the theory of wind accretion onto a magnetised rotating neutron star three characteristic radii can be defined (Davidson and Ostriker 1973; Lamb, Pethick and Pines 1973):

(i) The accretion radius r_a, which from Bondi and Hoyle (1944) is approximated by:

$$r_a = 4\ GM_X/v_O^2 \tag{1}$$

(M_X is the neutron star mass, v_O the stellar wind velocity). This radius gives the characteristic impact parameter for the wind material to be captured by the neutron star.

(ii) The magnetospheric radius r_m, obtained by equating the magnetic field pressure of the neutron star magnetosphere to the ram pressure of the accreting plasma:

$$B(r_m)^2/8\pi = \rho(r_m)\ v(r_m)^2 \quad , \tag{2}$$

where $B(r_m)$, $\rho(r_m)$ and $v(r_m)$ are respectively the pulsar magnetic field and the inflowing plasma density and velocity. At r_m the neutron star magnetic field begins to dominate the dynamics of the flow.

(iii) The corotation radius r_c , at which the X-ray pulsar corotation velocity is equal to the Keplerian velocity,

$$r_c = (GM_X P_S^2/4\pi^2)^{1/3} \quad , \tag{3}$$

where P_S is the X-ray pulsar spin period.

(a) Direct Wind Accretion.
When $r_a > r_m$ and $r_c > r_m$ the captured wind material flows from the accretion radius down the magnetospheric radius, where it is stopped by a collisionless

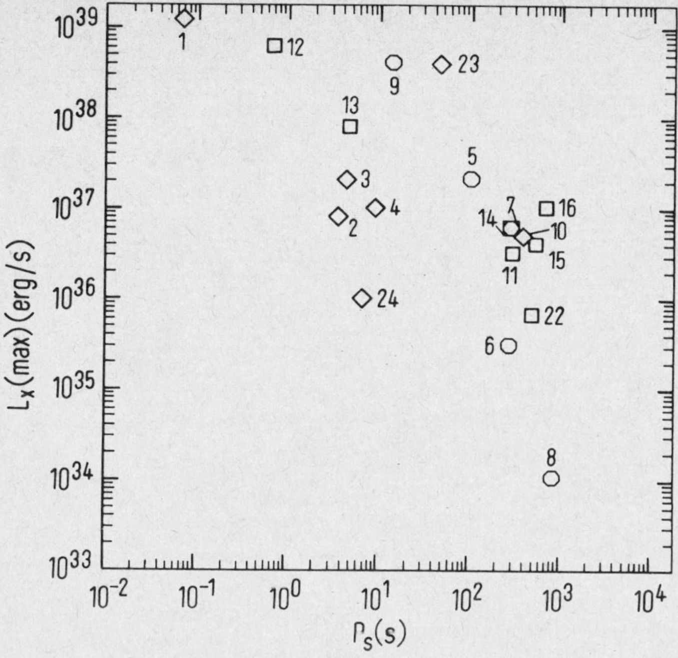

Figure 2:
Maximum X-ray luminosity of the OB star X-ray pulsar systems versus spin period. The diamonds, the circles and the squares represent, respectively, the transients and the persistent Be star systems and the supergiant star systems. The anti-correlation which is clearly apparent in this graph is discussed in the text. Source numbers refer to Table 1.

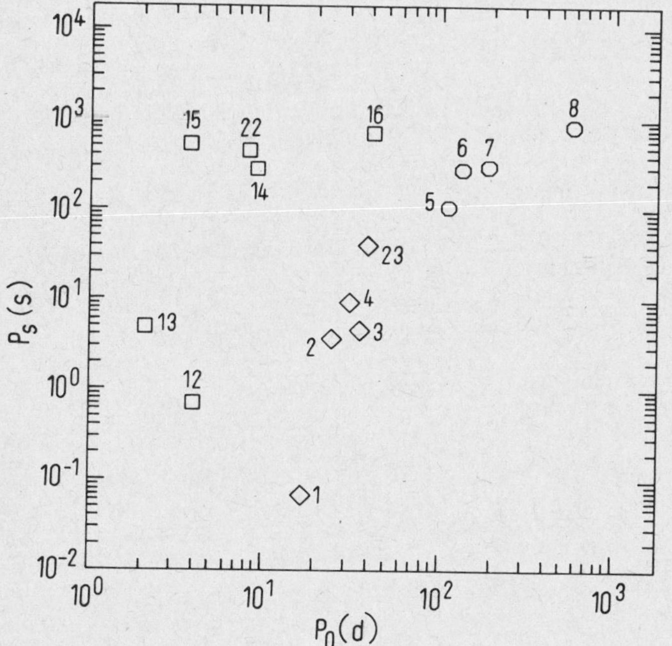

Figure 3:
Spin periods (P_S) of the Pop. I X-ray pulsar systems plotted against the corresponding orbital periods (P_0). A significant correlation between the two periods is found only for the Be star systems sub-sample.

shock. It then penetrates the pulsar magnetosphere presumably in a sporadic fashion by means of a variety of Rayleigh-Taylor and Kelvin-Helmholtz instabilities (cf. Lamb 1984). Although the details of the instabilities which modulate the accretion process are still highly uncertain, this will probably yield time variable emission from the accreting material on dynamical timescales (≤ 1 s).

The time-averaged X-ray luminosity is governed by the stellar wind parameters at r_a. Combining equation (1) with Kepler's third law and assuming all the gravitational potential energy is converted into X-rays gives an expression for the X-ray luminosity L_X of:

$$L_X \simeq 4 \times 10^{35} \left(\frac{R_X}{10^6 \text{cm}}\right)^{-1} \left(\frac{M_X}{1.4 M_\odot}\right)^3 \left(\frac{M_*}{10 M_\odot}\right)^{-2/3} \left(\frac{P_O}{1\text{d}}\right)^{-4/3} \left(\frac{\dot{M}_*}{10^{-8} M_\odot/\text{yr}}\right) \left(\frac{v_O}{10^8 \text{cm/s}}\right)^{-4} \text{ erg/s} \quad (4)$$

where $\dot{M}_*$ is the mass loss rate from the primary, M_* is the mass of the primary star ($M_* \gg M_X$), R_X is the neutron star radius and P_O is the orbital period of the system. Equation (4) uses a circular orbit approximation and neglects the effects of the orbital velocity on the dynamics of accretion, as the former is always much smaller than the asymptotic wind velocity v_O.

(b) <u>Centrifugal Inhibition of Accretion</u>.

If $r_c < r_m < r_a$ the wind material which penetrates through the accretion radius is stopped at the magnetospheric boundary (r_m) and cannot penetrate any further because, as $r_m > r_c$, the drag exerted by the pulsar magnetic field is super-Keplerian. As discussed by Illarionov and Sunyaev (1975) the pulsar magnetospheric boundary assumes a prolate configuration which strongly shocks the wind material by supersonic rotation. This may eject beyond the accretion radius some or all of the material via the 'propeller' mechanism. If the wind material accumulates at the magnetopause more rapidly than the ejection rate, a build up of material outside the magnetosphere may occur (Maraschi, Traversini and Treves 1983). This regime of inhibition of accretion has been discussed in connection with the spin-down evolution of young X-ray pulsars in binary systems (Illarionov and Sunyaev 1975).

By imposing $r_m = r_c$ we obtain a minimum luminosity at which accretion is possible.

$$L_{x,c}(\text{min}) \simeq 2 \times 10^{37} \left(\frac{R_X}{10^6 \text{cm}}\right)^{-1} \left(\frac{M_X}{1.4 M_\odot}\right)^{-2/3} \left(\frac{\mu}{10^{30} \text{Gcm}^3}\right)^2 \left(\frac{P_s}{1\text{s}}\right)^{-7/3} \text{ erg/s} \quad (5)$$

where μ is the magnetic dipole moment of the neutron star. Below this value an X-ray pulsar 'turns off' as a consequence of the centrifugal mechanism discussed above. The minimum luminosity given by equation (5) is reduced by a factor of

~0.5 when applied to disk accretion (Wassermann and Shapiro 1983).

For any of the stellar wind accreting X-ray pulsars to go below $L_{x,c}(min)$ requires a corresponding threshold in the stellar wind parameters. This is given by

$$P_{s,c}(min) \simeq 5.4 \left(\frac{M_X}{1.4M_\odot}\right)^{-11/7} \left(\frac{M_\star}{10M_\odot}\right)^{2/7} \left(\frac{\mu}{10^{30}Gcm^3}\right)^{6/7} \left(\frac{P_0}{1d}\right)^{4/7} \left(\frac{\dot{M}_\star}{10^{-8}M_\odot/yr}\right)^{-3/7} \left(\frac{v_0}{10^8cm/s}\right)^{12/7} \ s \quad (6)$$

where we have used the condition $r_m = r_c$, Kepler's third law in the circular orbit case and a free fall approximation for the flow within r_a. For fixed values of the stellar wind parameters $\dot{M}_\star$ and v_0, and the pulsar magnetic dipole moment μ (the range of possible values of $M_\star$ and M_X plays only a marginal role) equation (6) defines a critical line in the P_s-P_0 plane below which accretion is inhibited by the drag of the pulsar magnetosphere.

The transition between the accreting and the non-accreting regime will not occur abruptly because even when accretion is inhibited close to the equatorial plane of the neutron star, some fraction of the magnetospheric boundary at high latitudes will still be open to the penetration of the stellar wind material as the centrifugal drag is locally sub-Keplerian. The details of this process strongly depend on the geometry of the accretion flow close to the magnetospheric boundary and, in turn, on the distortions in the shape of the magnetospheric boundary induced by the accretion flow. In the case of disk accretion a very sharp on/off transition is likely to be generated. For a spherical accretion flow onto an undistorted dipolar magnetospheric boundary, the flow down the polar cusps will be very small (Elsner 1976). Because the magnetosphere is softest at its equator, the most extended on/off transition is obtained when the dipole axis lies in the equatorial plane of the neutron star. If the mass inflow towards the magnetospheric boundary is reduced by a factor of f with respect to the critical value $\dot{M}_X(min)$ = $L_{x,c}(min) \ R_X/(GM_X)$, the accretion onto the neutron star surface will take place only from a solid angle $\sim (1-\sin\beta)/4\pi$, where $f^{3/7} = \cos\beta \left[(1+3\cos^2\beta)/4\right]^{1/4}$. For f=0.1 the accretion rate and the X-ray luminosity will thus be reduced to 1.2 % of the value given by equation (5). This example illustrates that even in a very conservative case the transition to the centrifugally inhibited regime occurs rather sharply. In the following discussion we will thus assume that the condition $r_m=r_c$ ($< r_a$) corresponds to the onset of the 'propeller' mechanism.

4. The Transient Activity and the $L_x(max)$ versus P_s Relation

In Figure 4 we plot the maximum X-ray luminosity observed from these systems, together with the lines of constant $L_{x,c}(min)$ corresponding to equation (5) for a variety of magnetic dipole moments. The region below the line excludes accretion onto the neutron star and provides an upper limit for μ in each system. This diagram demonstrates that for most of the persistent sources $r_c > r_m$ for $\mu = 10^{30}$ G cm^3. If we were to impose the condition $r_c = r_m$ for all these systems, magnetic field strengths of $\sim 10^{14}$ G are required for many of the longer period pulsars. This in turn would require a strong correlation of the magnetic field strengths with the pulse period which, as discussed in section 6, does not seem plausible. At the other extreme the super-Eddington luminosity and the rapid rotation period of A0538-66 requires a lower than average field strength of $\sim 10^{11}$ G for the source to turn on (see also Skinner et al. 1982). The proximity of many of the transient systems to the $B = 10^{12}$ G ($\mu = 10^{30}$ G cm^3) critical line (the neutron star magnetic field indicated by cyclotron line measurements) suggests that only a small decrease in $\dot{M}_*$ (and/or an increase in v_0) will cause accretion in these systems to be inhibited by magnetospheric centrifugal forces. Conversely a small increase in $\dot{M}_*$ (and/or a decrease in v_0) could turn a dormant source into a luminous X-ray transient.

By using the minimum observed X-ray luminosity before the source 'turns off' as a consequence of the centrifugal inhibition of accretion (and the canonical neutron star masses and radii), equation (5) can provide an estimate of the pulsar magnetic dipole moment (Figure 5). For V0332+53 the minimum luminosity measured just before turn off ($L_x(min) \simeq 7 \times 10^{34}$ (d/1.5 kpc)^2 erg/s; Stella et al. 1985) provides a value for μ of 3.5×10^{29} (d/1.5 kpc) G cm^3. This estimate depends on the uncertainty in the evaluation of the distance to V0332+53. The estimate of 1.5 kpc relies upon the assumption that the primary is a main sequence Be star (cf. Bernacca, Ijima and Stagni 1984 and references therein). Such an evaluation of the pulsar magnetic field, while not direct, is complementary to that obtained from the detection of cyclotron lines in the X-ray spectrum of Her X-1 (Trümper et al. 1978) and 4U0115+63 (Wheaton et al. 1979; White, Swank and Holt 1983)[5]. In view of the uncertainties connected with the measurement of the X-ray luminosity at turn off, the estimate based on equation (5) cannot be more precise than a factor of 2-3. Based on their position in the $L_x(max)$ - P_s plane of Figure 4, the supergiant systems SMC X-1 and Cen X-3 are also likely candidates to undergo a transition to a

---

[5] It is important to realise that equation (5) provides an estimate of the magnetic dipole moment μ of the neutron star, since any higher multipolar component is likely to be negligible at $r \sim r_m$. In the most widely accepted interpretation, the X-ray cyclotron lines provide, instead, a measurement of the magnetic field strength at the neutron star surface, which can include a significant contribution from higher multipole components. The surface magnetic field strengths derived from equation (5) can thus be smaller than those obtained from cyclotron line measurements.

dormant state, as a consequence of only a small reduction in the mass capture rate from the companion star.

The gross characteristics of the $L_x(max)-P_s$ anti-correlation that we have pointed out in §II is naturally explained by the centrifugal barrier mechanism (cf. equation (5)) since only those systems above a particular critical luminosity will be seen (for a given μ). If the magnetic fields were the same and if all these sources were at the critical point $r_c = r_m$ then L_x should be proportional to $P_s^{-7/3}$. The fact that the observed relation of $L_x(max)$ and P_s ($L_x(max) \propto P_s^{\alpha}$ where $\alpha = -0.72 \pm 0.19$) is considerably less steep than $-7/3$ is likely to result both from observational selection effects and the fact that, as noted earlier, $r_c = r_m$. The use of $L_x(max)$ (rather than $L_x(min)$, as equation (5) would require) would cause an observational bias as this requires that all the persistent sources have the same dynamic range of luminosity, which seems unlikely (extensive studies of the minimum X-ray luminosity below which OB Star X-ray pulsar transient systems 'turn off' have still to be carried out). Also, if fast pulsars like A0538-66 (P_s <0.1s) are to become X-ray sources at a luminosity not strongly exceeding the Eddington limit they must either possess low magnetic fields ($B \sim 10^{11}$G) to overcome the centrifugal barrier, or they must slow down to much longer rotation periods. The long period pulsar with periods of several hundred seconds can accrete at much lower luminosities, so that the majority of the low luminosity systems where $r_c \approx r_m$ would have escaped detection with the present generation of X-ray detectors.

5.   The P_s versus P_0 Relation

As discussed above the onset of the centrifugal inhibition of accretion requires a corresponding threshold in the stellar wind parameters. In Figure 6 we plot the critical lines corresponding to equation (6) for a variety of neutron star and stellar wind parameters. The wind parameters for the lines A, A' and B, B' were chosen as to correspond to those typical of a supergiant and Be star, with

$$\gamma = (\dot{M}_\ast/10^{-8}\ M_\odot\ yr^{-1})\ (v_0/10^8\ cm\ s^{-1})^{-4} = 1\ \text{and}\ 100$$

respectively (note that in this description the dependence on these two stellar wind parameters cannot be disentangled). It is interesting to note that most of the persistent sources occupy the upper part of the graph, well above the critical lines A and B corresponding to $\mu = 10^{30}$ G cm^3. As we have noted before, they are unlikely to undergo a transition to a non-accreting quiescence state, unless the magnetic field strengths are substantially higher than 10^{12} G. The transient systems are usually in their dormant state and all occupy the lower part of the P_0-P_s plane, close to the critical lines of Figure 6. As we have discussed for V0332+53, a change in the stellar wind parameters can cause these systems to suddenly 'turn on'. The Be X-ray

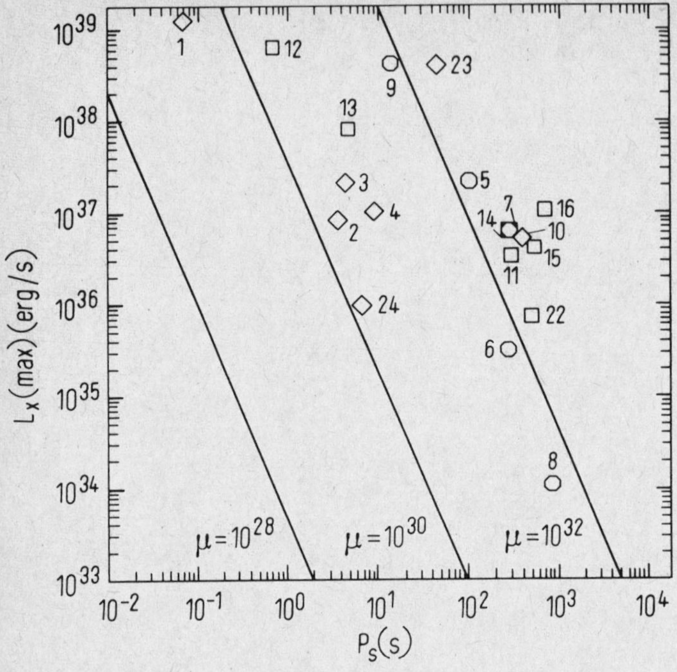

Figure 4:
The critical line below which accretion is inhibited by the centrifugal drag due to the rotating magnetosphere (cf. equation (5)) is given for various values of μ in the $L_X(max)$ - P_S plane. Here and in the following figures we use $M_\star = 10\ M_\odot$, $M_X = 1.4\ M_\odot$ and $R_X = 10^6$ cm.

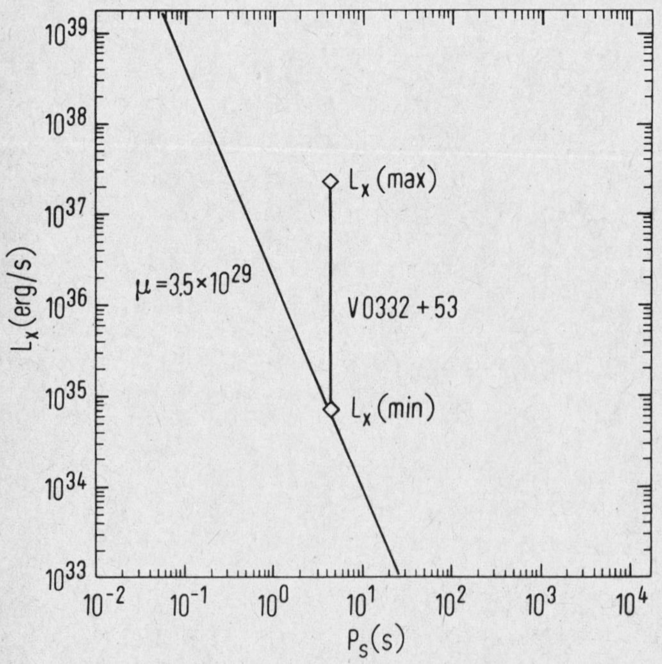

Figure 5:
Evaluation of the pulsar magnetic dipole moment based on equation (5). The dynamic range in the luminosity of V0332+53 for a distance of 1.5 kpc is indicated. $L_X(max)$ refers to the 1973 giant outburst and $L_X(min)$ to the minimum detected X-ray luminosity during the 1983-1984 outbursts.

transient systems may represent the tip of the iceberg of a population of dormant sources, for which an increase in the mass ejection rate from the primary star and/ or a decrease in the wind velocity can cause them to briefly turn into highly luminous X-ray sources.

The effects of an eccentric orbit can be approximated in Figure 6 by indicating the equivalent range of orbital periods $P_0(1-e)^{3/2} \leq P_0(eq) \leq P_0(1+e)^{3/2}$ that corresponds to the circular orbit periods for the range of binary separations of an eccentric orbit. This is shown in Figure 7 for V0332+53; during an orbital cycle the source moves back and forth along the indicated range of equivalent orbital periods. Depending on the values of the primary star wind parameters defined by γ , three different situations can occur. If the critical line defined by equation (6) lies above the range of $P_0(eq)$ the X-ray pulsar will be in its dormant state. An increase of $\dot{M}_\star$ and/or a decrease of v_0 can cause the critical line to move down in the P_0-P_s plane to intersect the segment representing the range of $P_0(eq)$. In this case the X-ray pulsar can become active only in an orbital phase interval centered around periastron. For the remainder of the cycle the system is X-ray dark. A larger increase in γ can move the critical line below the segment of $P_0(eq)$ and the X-ray pulsar will thus be in its active state for all the orbital phases. This provides a very natural explanation for the large orbital modulation observed in the 1983-1984 outburst of V0332+53 (Stella et al. 1985) and the failure to detect a comparably strong orbital modulation during the more luminous 1973 outburst (Terrell and Priedhorsky 1984).

The peculiar behaviour of the transient 4U0115+63 for which the 1980 optical activity preceeded the X-ray outbursts by $\sim$ 60 days, can also be understood in a similar manner. The measured surface magnetic field of B = 10^{12} G (Wheaton et al. 1979; White, Swank and Holt 1983) and the minimum luminosity observed during the 1972 outbursts of $\sim 10^{36}$ erg/s (Forman, Jones and Tananbaum 1976) imply that this system is very close to $r_m = r_c$ during that part of the outburst. The enhanced mass ejection episode from the primary probably occurred when the primary brightened by 1.7 magnitudes, but for the first $\sim$ 60 days was not enough to ensure $r_m < r_c$ at any orbital phase, so that the X-ray pulsar remained inactive. At a certain point in the outburst, the pressure of the inflowing material was sufficient to overcome the centrifugal barrier and the X-ray pulsar turned on. This view is also supported by the fact that, at the end of the 1980 outburst, the X-ray flux decayed more rapidly than the optical flux (Kriss et al. 1980).

6.   Discussion

The issue as to whether or not a young rapidly rotating pulsar can be spun

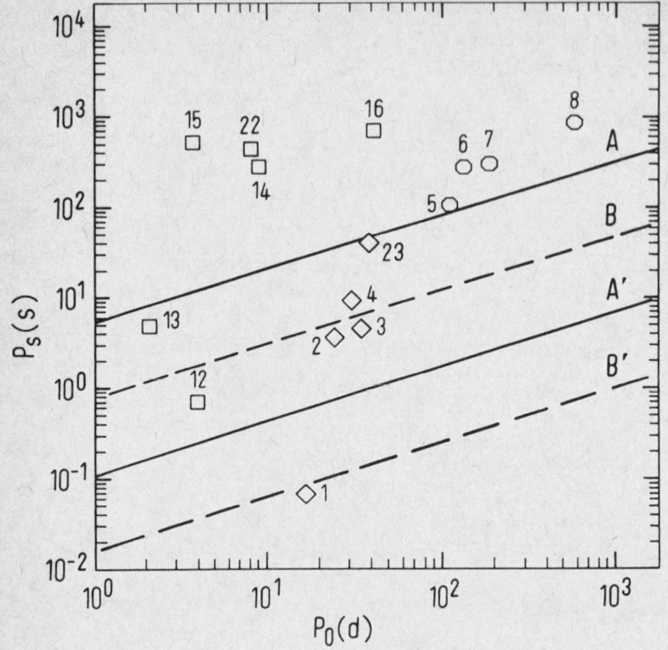

Figure 6:
The critical lines given by equation (6) are shown for selected values of $\dot{M}_*, v_0$ and μ in the spin period (P_S) - orbital period (P_O) plane. The lines A correspond to $\gamma = (\dot{M}_*/10^{-8}\ M_\odot\ yr^{-1}) \times (v_0/10^8\ cm\ s^{-1})^{-4} = 1$ The lines B are obtained by increasing γ by a factor of 100. The magnetic dipole moment is $\mu = 10^{30}\ G\ cm^3$ ($\mu = 10^{28}\ G\ cm^3$) for the lines A and B (A' and B').

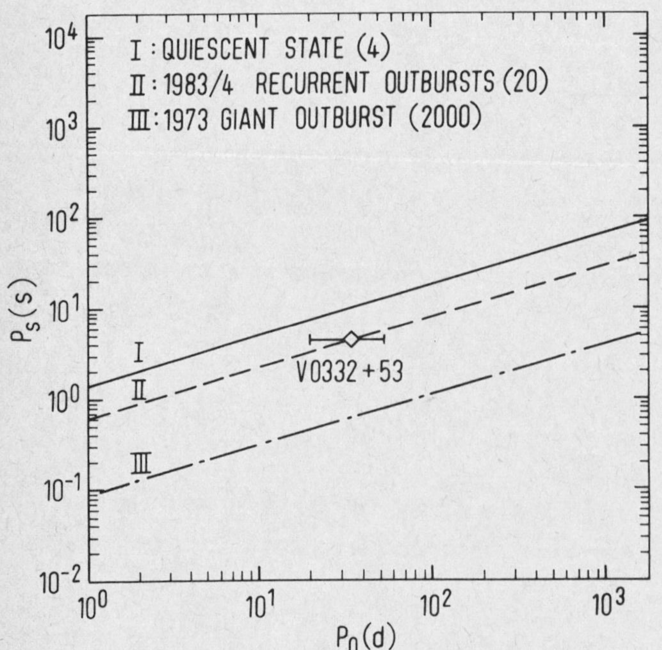

Figure 7:
Position of V0332+53 in the orbital period (P_O) - spin period (P_S) plane. The range of equivalent orbital periods $P_O(eq)$ is indicated. The critical line corresponding to equation (6) is given for three different choices of the stellar wind parameter $\gamma = (\dot{M}_*/10^{-8}\ M_\odot\ yr^{-1}) \times (v_0/10^8\ cm\ s^{-1})^{-4} = 4, 20, 2000$, which are suitable to model, respectively, the quiescent state, the 1983-1984 periodic outbursts, and the 1973 giant outburst. The value of $\mu = 3.5 \times 10^{29}\ G\ cm^3$ is obtained from Figure 5.

down on a timescale short compared to the lifetime of the companion remains contraversial and we will not repeat the discussion here (see Henrichs 1983 and references therein for the current picture). Irrespective of the spin history of the X-ray pulsars, our results have shown that most of the long period systems (P_S longer than several hundred seconds) must lie well away from the equilibrium point where $r_c = r_m$. The assertion by Corbet (1984) that all these systems are close to the equilibrium point requires instead (independent of any assumptions regarding the properties of the stellar wind) a correlation between pulse period and magnetic field strength, with fields as strong as 10^{14} G in the longest period systems (cf. Figure 4). Such field strengths are excessive, but not implausible. However, this would be contrary to the estimate of $\sim 10^{12}$ G obtained from the spin down rate of radio pulsars and contrary to the measured field strengths in Her X-1 and 4U0115+63.

If these long period pulsars are far from the equilibrium point then this requires that either they were born with such periods, or that up until recently the wind parameter γ ($= \dot{M}_{*} v_0^{-4}$) was several orders of magnitude lower and that these systems have yet to be spun up to the new equilibrium period.

The transient X-ray pulsars with periods of a few seconds, when active, all lie close to the equilibrium point if it is assumed that they have magnetic moments close to 10^{30} G cm^3. It is therefore encouraging that in the one case for which measurements of the surface field strength (from the detection of a cyclotron line) are available, namely 4U0115+63, directly shows this assumption to be justified. These pulsars spend less than 10 % of their lifetime in outburst, and during the remaining time they are presumably undergoing 'propeller' spin down. Measurements of the pulse period from one outburst to the next could give some indication of the spin down rates, although because the spin up rates during outburst can be high, it is essential to measure the pulse period at the start of each outburst.

The ratio of the maximum to minimum luminosity of the transient outbursts from the Be star can be strongly amplified by the effects of the centrifugal barrier. If a neutron star lies just below the critical point only a small change in the stellar wind parameters will turn it into a bright X-ray source. We have identified one system, V0332+53, where during an outburst in 1983/84 this critical point straddled the orbit, such that for one part of the orbit the system was in the direct wind accretion regime, and the other part accretion was inhibited by magnetospheric centrifugal forces.

The origin of the correlation between the pulse and orbital periods of the Be star X-ray binaries noted by Corbet (1984) is not clear. The only Be star systems identified to date with orbital periods less than 50 days all have pulse periods of less than 100 s. The spin down timescale for the pulsars in these systems should be somewhat quicker than the longer period systems, however as yet no long period pulsars have been identified in short period binaries. This correlation is suggestive of a connection between the birth period of the X-ray pulsar and the orbital period. However, Savonije and van den Heuvel (1977) have ruled out tidal synchronisation of the helium projenitor (after core contraction), as the timescale required seems to be far too long. Nonetheless, given that the Be stars in these systems should form a homogeneous sample with respect to their age, it is not clear why long orbital periods tend to be associated with long pulse periods.

The 69 ms X-ray pulsar in A0538-66 provides evidence that at least in one case the pulsar was born spinning rapidly. Normally we would not expect to see such systems, however in this case the magnetic dipole moment of the neutron star must be at least a factor of 10 lower than is normally assumed. This allows the centrifugal barrier to be overcome for accretion rates that result in luminosities up to a factor of ten in excess of the Eddington limit. The fact that this is likely to be a young system (the highly eccentric orbit suggests this) and yet has a neutron star with a magnetic field of only $\sim 10^{11}$ G (cf. Skinner et al. 1982) is notable because it either means that neutron stars are born with a range of field strengths, rather than the canonical 10^{12} G that is usually assumed, or that the young neutron star has a very high multipole component that becomes dipolar as the system ages. Variations in the centrifugal barrier in a manner similar to that discussed for V0332+53 provides a very natural explanation for both the high $L_x(\max)/L_x(\min)$ around one orbital cycle and the long quiescent intervals seen from this source, without any need to invoke intervals of Roche lobe overflow at the time of periastron passage (cf. Charles et al. 1983).

7.   Conclusion
____

Our investigation of some of the wind regimes in Pop. I binary systems containing an X-ray pulsar leads to a natural explanation of the transient activity observed in many of these systems in terms of the centrifugal mechanism of inhibition of accretion. In particular we can model the large dynamic range in the X-ray luminosity between the 'on' and the 'off' states and the periodic and/or irregular character of the outbursts. The model is particulary successful in interpreting the bimodal transient activity observed in the sources (like V0332+53) which display both irregular giant outbursts and periodic low-level outbursts close to periastron passage. In this context, the luminosity at which an X-ray pulsar transient 'turns off', if it can be measured by an all sky monitor, would provide a useful estimate of the magnetic dipole moment of the neutron star.

References

Apparao, K.M.V. 1985. Ap.J., 292, 257.

Bernacca, P.L., Iijima, T., and Stagni, R. 1984. Astr. Astrophys. 132, L8

Bondi, H. and Hoyle, F. 1944. M.N.R.A.S., 104, 273.

Bradt, H.V.D., and McClintock, J.E. 1983. Ann.Rev.As.Ap., 21, 13.

Charles, P.A. et al. 1983. M.N.R.A.S., 202, 657.

Conti, P.S. 1978. Astr. Astrophys., 63, 225.

Corbet, R.H.D. 1984. Astr. Astrophys., 141, 91.

Davidson, K., and Ostriker, J.P. 1973. Ap.J., 179, 585.

Elsner, R.F. 1976. Ph.D Thesis, University of Illinois.

Fabian, A.C. 1975, M.N.R.A.S., 173, 161.

Forman, W., Jones, C., Tananbaum, H. 1976. Ap.J.(Letters), 206, L29.

Giangrande, A., Giovannelli, F., Bartolini, C., Guarnieri, A., and Piccioni, A.
    1980. Astr. Astrophys. Suppl., 40, 289.

Henrichs, H.F. 1980, in 'Highlights in Astronomy', Ed. A. Wayman (IAU) Vol.5,
    p.541.

Henrichs, H.F. 1983, in "Accretion Driven Stellar X-ray Sources", Ed. W.H.G. Lewin
    and E.P.J. van den Heuvel, p.393.

Illarionov, A.F., and Sunyaev, R.A. 1975. Astr. Astrophys., 39, 185.

Joss, P.C., and Rappaport, S. 1984. Ann.Rev.As.Ap., 22, 537.

Kelley, R.L., Rappaport, S., Ayasli, S. 1983. Ap.J., 274, 765.

Kriss, G.A., Cominsky, L.R., Remillard, R.A., Williams, G. and Thorstensen, J.R.
    1983. Ap.J., 266, 806.

Lamb, F.K., Pethick, C.J., and Pines, D. 1973. Ap.J., 184, 271.

Lamb, F.K. 1984, in 'High Energy Transients', Ed. S.E. Woosley (New York: AIP),
    p.179.

Makishima, K., Kawai, N., Koyama, K., Shibazaki, N., Nagase, F., Nakagawa, M. 1984.
    Publ. Astron. Soc. Japan, 36, 679.

Maraschi, L., Treves, A., and van den Heuvel, E.P.J. 1976. Nature, 259, 292.

Maraschi, L., Traversini, R., and Treves, A. 1983. M.N.R.A.S., 204,1179

Parmar, A.N., Stella, L., Ferri, P. and White, N.E. 1985. IAU Circ. No. 4066.

Petterson, J.A. 1978. Ap.J., 224, 625.

Priedhorsky, W.C., and Terrell, J., 1983. Nature, 303, 681.

Rappaport, S., and van den Heuvel, E.P.J. 1982. IAU Symposium No. 98 'Be Stars'
    Ed. H. Haschek, H. Groth (Dordrecht, Reidel) p.327.

Savonije, G.J., and van den Heuvel, E.P. 1977. Ap.J., 214, L19.

Seward, F.D., Charles, P.A., and Smale, A.P. 1986. Ap.J., in press.

Skinner, G.K., Bedford, D.K., Elsner, R.F., Leahy, D., Weisskopf, M.C. and
    Grindlay, J.E. 1982. Nature, 297, 568.

Stella, L., White, N.E., Davelaar, J., Parmar, A.N., Blissett, R.J. and van der
    Klis, M. 1985. Ap.J.(Letters), 288, L45.

Stella, L., White, N.E., and Rosner, R. 1986. Ap.J., in press.

Terrell, J., and Priedhorsky, W.C. 1984. Ap.J., 285, L15.

Trümper, J., Pietsch, W., Reppin, C., Voges, W., Staubert, R. and Kendziorra, E. 1978. Ap.J.(Letters), 219, L105.

Wang, Y.M. 1981. Astr. Astrophys., 113, 113.

Wassermann, I., and Shapiro, S.L. 1983. Ap.J., 265, 1063.

Watson, M.G., Warwick, R.S., and Corbet, R.H.D. 1982. M.N.R.A.S., 199, 915.

Wheaton, Wm. A. et al. 1979. Nature, 282, 240.

White, N.E., Swank, J.H., and Holt, S.S. 1983. Ap.J., 279, 711.

White, N.E. 1984. Talk presented at the 1983 NATO Summer School on 'Interacting Binary Stars', in press

White, N.E., Ferri, P., Parmar, A.N., and Stella, L. 1985. IAU Circ. No. 4112.

X-RAY PROPERTIES OF MAGNETIC
CATACLYSMIC VARIABLE SYSTEMS

M.G.Watson

X-ray Astronomy Group, Physics Department,
University of Leicester,
Leicester LE1 7RH, England.

ABSTRACT

The X-ray properties of magnetic cataclysmic variable systems - polars and intermediate polars - are reviewed. Emphasis is placed on the interpretation of various X-ray properties, particularly X-ray spectra and the modulation of the X-ray light curves at the orbital and white dwarf spin periods, in the context of present theoretical ideas of how the X-ray emission components are formed, and the geometry of the accretion flow onto the magnetic polecap of the white dwarf.

1. INTRODUCTION

Cataclysmic variables (CVs) are semi-detached binary systems consisting of a white dwarf which is accreting material lost by a late-type companion star. In a small fraction of CVs, the accreting white dwarf has a surface magnetic field which is large enough to play an important role in determining the system properties, particularly by controlling the flow of accreting material at large distances from the white dwarf surface. Two such classes of CV are presently recognised, each of which contains a dozen objects or less - polar systems, otherwise known as AM Her systems, and intermediate polars, often referred to as DQ Her objects (although there is some uncertainty as to whether DQ Her is actually an intermediate polar). The defining difference between polars and intermediate polars is the degree of synchronisation between the white dwarf and binary system rotation. Polars are essentially completely synchronised (i.e. the white dwarf spin is phase-locked to the orbital rotation). In intermediate polars the white dwarf spin is not synchronised, and the white dwarf typically has a rotation period considerably shorter ($\sim 10 \times$) than the orbital period. In both classes the magnetic field of the white dwarf is strong, measured to be $\sim 10^7$ gauss in polars, and inferred to be of this order in intermediate polars (see contributions by King, and Lamb, in this volume for a discussion of field strengths and related evolutionary questions). The strong field channels the accretion flow onto the magnetic polecap region of the white dwarf in a pseudo-radial geometry, leading to relatively efficient X-ray production. The X-ray emission from magnetic CVs typically has two components : an ultra-soft component which dominates below ~ 1 keV, and a hard component which dominates above ~ 2 keV. Both components are observed in polars, but the ultra-soft component is apparently absent in intermediate polar systems.

In this paper the X-ray properties of these two classes of magnetic CV system are summarised. Several important studies of these systems were carried out prior to the launch of *EXOSAT*, notably with *HEAO-1*, the *Einstein Observatory* and *Ariel 5*. *EXOSAT* observations over the last 3 years have, however, been crucial for the advance of our understanding of these systems in 3 respects : (i) in making observations of a wider range of objects; (ii) in making long, uninterrupted observations spanning many cycles of the spin and/or orbital periods of these systems (by virtue of the long eccentric orbit of the spacecraft); (iii) in obtaining simultaneous X-ray measurements spanning a broad spectral range - particularly important for polars which have two-component X-ray spectra. For some systems the ability to carry out ground-based and IUE observations simultaneous with the *EXOSAT* coverage has also proved enormously valuable. I therefore make no apology in basing most of this review on relatively recent *EXOSAT* observations. The relevant observations were, for the most part, carried out with the Medium Energy (ME) instrument which is sensitive in the $\sim 2 - 20$ keV energy range (for sources at the relatively low flux levels typical of CVs), and the Channel Multiplier Array (CMA) in the focal plane of one of the two Low Energy (LE) imaging telescopes which covers the $\sim 0.05 - 2$ keV band.

2. X-RAY PRODUCTION MECHANISMS

To provide a framework for the observational material, I outline here the basic ideas behind the production of X-rays in a magnetic CV system following e.g. Frank, King & Raine (1985). The magnetic field of the white dwarf in these systems is, as emphasised above, strong enough to control the accretion flow at some radius $R_{mag} \gg R_{wd}$ from the white dwarf. Thus the accretion flow inside R_{mag} will follow the field lines, and this will naturally lead to quasi-radial infall of the accreting material onto a restricted region of the white dwarf surface at the magnetic polecap. The accretion geometry envisaged is shown in Fig.1. It is easy to show that a strong shock must form in the accretion flow close to the white dwarf surface in an accretion column at a height $h \ll R_{wd}$, and that the dominant emission from the shocked gas in the accretion column is optically thin bremsstrahlung at a characteristic temperature $T \sim 10^8$ K - giving a hard X-ray component. Cyclotron losses from the accretion column are also important - this gives optical and UV emission which is highly polarised in polar systems.

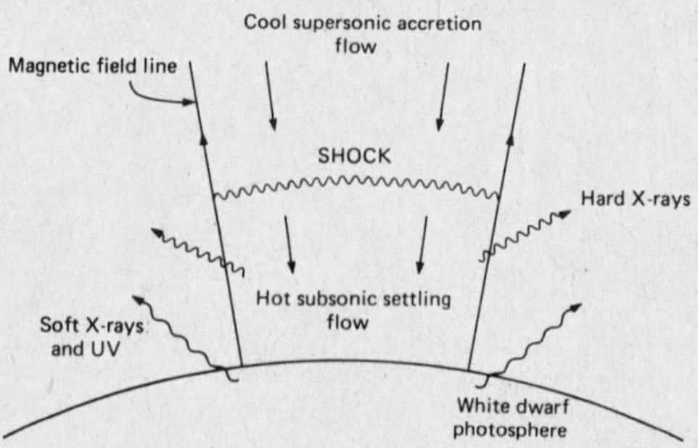

Fig.1 Accretion geometry at the polecap of magnetic white dwarf.

Some fraction of the accretion energy liberated in the accretion column will be transported to the white dwarf surface, either radiatively (heating by the hard X-ray component) or by electron conduction or other bulk transport process. The resultant heating of the white dwarf surface then produces optically thick emission with approximate black-body temperature $\sim few \times 10^5$ K - this is the ultra-soft X-ray component.

The efficiency with which the hard X-ray and soft X-ray components are produced is still the subject of some debate. If, as outlined above, the soft X-ray component is produced only by reprocessing of hard X-rays absorbed at the white dwarf surface, the expected ratio of hard to soft X-rays, L_{hard}/L_{soft} can be written as $(1 + a_X)/(1 - a_X)$ where a_X is the hard X-ray albedo (~ 0.3). (Strictly speaking L_{hard} should include **all** the luminosity generated in the accretion column, i.e. should include the cyclotron component.) This then gives $L_{hard}/L_{soft} \sim 2$. As is discussed in section 4, there is substantial observational evidence for luminosity ratios much smaller than this, implying that the simple radiative accretion column models are not correct, and that energy is transported to the white dwarf surface in some other way. The conflict between the observed ratio and that predicted by radiative column models is often referred to as the 'soft X-ray problem' , although the term 'soft X-ray excess' may be preferable. Another possibility is that the accretion flow may be subject to large-scale inhomogeneities which would invalidate the simple stand-off shock model for the accretion column.

The size of the accretion column, and the region on the white dwarf surface where the soft X-ray component is formed, is likely to be $< R_{wd}$ and indeed $\ll R_{wd}$ in the case of polars. Thus, providing the magnetic and spin axes of the white dwarf are misaligned, the X-ray light curves of magnetic CVs are expected to be strongly modulated at the white dwarf spin period - as is indeed observed.

3. INTERMEDIATE POLARS

3.1 Introduction

The basic properties of the 10 known intermediate polar systems are summarised in Table 1. With the exception of GK Per, all have binary periods $\sim 2 - 5$ hours, and all but EX Hya and SW UMa are above the so-called 'period gap' . *EXOSAT* has measured the X-ray light curves of most of these systems for the first time, and the X-ray periods of all but 2 have been determined from *EXOSAT* observations. The X-ray periods of these systems, which correspond to the white dwarf spin period, range from 350 - 4000 seconds.

3.2 X-ray spectra

The X-ray spectra of intermediate polars above ~ 2 keV are predominantly 'hard' and typically show either bremsstrahlung or power law continua, often with substantial amounts of low energy absorption with $N_H > 10^{22} - 10^{23}$ cm^{-2}. Strong Fe line emission at 6.4 keV, presumably due to fluorescence, is present in several systems (e.g GK Per, Watson *et al.*, 1985a; 3A0729+103, McHardy *et al.*,1986). Fig.2 shows the X-ray spectrum of GK Per. Simple theory predicts that the X-ray emission from the accretion column should have an optically thin bremsstrahlung continuum. This expectation is clearly not met in at least some systems, although it is presently

Table 1 : X-ray properties of intermediate polar systems

Object	P_X(s)	Amp[1] (%)	Dis[2]	P_{orb}(h)	X-ray orb. modn.	References[3]
GK Per	351	60	[EXO]	48.0	?	9, 40
V1223 Sgr (1849-31)	746	15	[EXO]	3.37	poss	22, 24
AO Psc (2252-035)	805	50	[Ein/H]	3.59	yes	44, 30
3A0729+103 (BG CMi)	913	15	[EXO]	3.24	yes	20, 21
SW UMa	954	50[4]	[EXO]	1.36	?	35
H2215-086 (FO Aqr)	1255	70	[EXO]	4.03	?	8, 28, 22
H0542-407	1920	70	[EXO]	6.2:	yes	38
TV Col (0526-328)	1943	20	[EXO]	5.49	yes	43, 33
V426 Oph	3600:	20	[EXO]	6.0:	?	36
EX Hya	4022	10	[Ein]	1.63	yes	17, 2, 10, 19

Notes

(1) Average modulation amplitude at P_X.

(2) Discovery / measurement of the X-ray spin period. [EXO] = EXOSAT, [Ein] = Einstein Observatory, [H] = HEAO-1.

(3) References given relate only to X-ray observations, primarily by EXOSAT plus a few selected earlier X-ray studies.

(4) Amplitude of soft X-ray modulation. The hard X-ray light curve of SW UMa has not yet been determined.

Omitted from this Table are the DQ Her systems (DQ Her, AE Aqr and V533 Her) which have X-ray and spin properties which differ from other intermediate polars (e.g. Warner, 1983), and TT Ari whose intermediate polar status has not yet been confirmed.

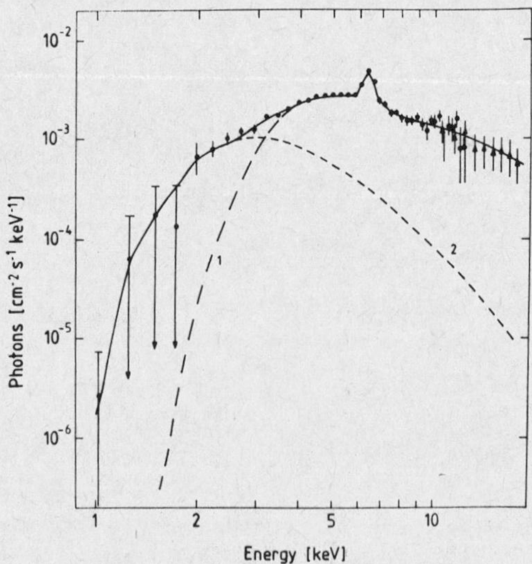

Fig.2 X-ray spectrum of GK Per obtained with the *EXOSAT* ME. The dashed curves indicate the contributions of the two spectral components fitted. Component 1 : power law (+ Fe line) with $\alpha = 0.55$ and $N_H = 1.7 \times 10^{23}$ cm^{-2}; component 2 : thermal bremsstrahlung with $kT = 5$ keV and $N_H = 4.3 \times 10^{22}$ cm^{-2}. See Watson *et al.* (1985a) for further details.

not clear to what extent spectral variability might be responsible for producing the deviations from a simple thermal continuum. The measured column densities are much higher than the expected line-of-sight interstellar values, clearly implying that absorption must occur within the systems. Column densities of the order $10^{22} - 10^{23}$ cm^{-2} are in accord with those expected in simple models of the accretion column (e.g King & Shaviv 1984).

With such high local column densities, there seems to be little possibility of detecting the ultra-soft X-ray component from the white dwarf surface in intermediate polars. Indeed, until recently there were no soft X-ray detections of these systems. Most intermediate polars have now been detected as very weak sources in the *EXOSAT* CMA detector at count rates higher than expected from a simple extrapolation of the spectrum determined at higher energies. This implies that there must be an additional spectral component with emission below 1 keV. This component could be the true ultra-soft X-ray component (see section 2) originating at the white dwarf surface, or might alternatively be a 'softer' or less absorbed hard X-ray component. To complicate the issue there is also evidence for a second, softer spectral component in the hard (i.e. > 2 keV) spectrum of several systems (e.g. Watson *et al.*, 1985a; Osborne *et al.*, 1985; McHardy *et al.*, 1986).

Present observations do not generally provide an unambiguous answer to the nature of the second spectral component (e.g. Osborne *et al.* 1985), because of the lack of spectral resolution in the soft X-ray measurements and the difficulties in separating two components uniquely at higher energies. *EXOSAT* grating observations, which provide good spectral resolution in the soft X-rays band, have been made, however, for EX Hya by Cordova *et al.* (1985). Their results demonstrate that the soft X-ray spectrum of EX Hya is **not** consistent with that expected from the ultra-soft component. They suggest that the CMA detection is due some fraction of the 'hard' component being significantly less absorbed : this may be of some significance for detailed models of the accretion column. The possibility of there being a contribution from a completely separate component is not, however, ruled out by their data.

Further clues to the possible origins of this second spectral component are gained from considering the orbital modulation of the X-ray flux, as is discussed in section 3.4 below.

3.3 X-ray light curves : spin modulation

The hard X-ray light curves of all the intermediate polars are clearly modulated at the spin period of the white dwarf with modulation amplitudes ranging from 10 to 100 %. (At optical wavelengths intermediate polars show strong modulation at either the white dwarf spin period or the beat period between the spin and the orbital periods, or sometimes both, e.g. Warner 1983.). The basic shape of the X-ray light curves at the spin period is rather similar, with a characteristic, slightly asymmetric sinusoidal shape present in each. Two examples are shown in Fig.3. There are some minor deviations from this smooth modulation, for example the dip at maximum in the light curves of both AO Psc and V1223 Sgr (e.g White & Marshall, 1981; Osborne *et al.*, 1985).

Given that the hard X-ray emission is optically thin, it is clear (King and Shaviv, 1984) that the modulation of the hard X-ray light curves of magnetic CV systems can only result from occultations of parts of the accretion column by the white dwarf body as it rotates. The observed smooth shape of the hard X-ray light curves in intermediate polars, together with the fact that

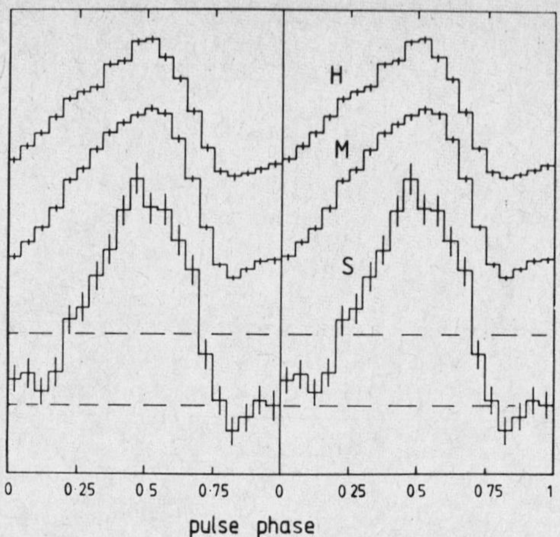

Fig.3(a) X-ray light curves for GK Per obtained with the *EXOSAT* ME folded at the 351-second pulse period in 3 energy ranges - S : 1.4 − 3.1 keV; M : 3.1 − 5.6 keV; H : 5.6 − 10.8 keV. From Watson *et al.* (1985a).

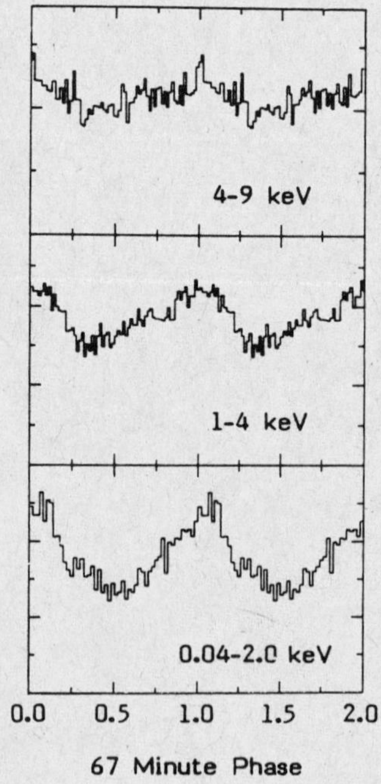

Fig.3(b) X-ray light curves for EX Hya folded at the 67-min. (4022 s) pulse period. The upper panels are *EXOSAT* ME data, the lower panel *EXOSAT* CMA data. Adapted from Mason (1985).

most of these systems do not have extreme modulation amplitudes (Table 1), then demonstrates that the area of the white dwarf surface covered by the accretion column cannot be small since this would result, on average, in much more highly modulated, sharper, light curves. Interpreting present results on the basis of the probabilities of observing any particular sort of light curve presented by King & Shaviv requires that the accretion column occupy a fraction $f > 0.1$ of the white dwarf in intermediate polars. The implications of the large size of the polecap region, and the clear requirement that one polecap dominate the observed hard X-ray emission, are further discussed in Hameury, King and Lasota (1986).

The hard X-ray light curves of most, if not all, intermediate polars show a systematic increase in the modulation depth with decreasing X-ray energy (see Fig.3). For those intermediate polars which have adequate *EXOSAT* CMA observations, the CMA light curves appear to be similar to those measured at higher energies with, as expected from the trend with energy, even higher modulation depths. Such an energy dependence of the modulation depth is expected as a result of the difference in photoelectric absorbing columns along and across the accretion column (King and Shaviv, 1984). The increase in modulation depth with decreasing X-ray energy in some systems does not, however, show the dependence expected if it is due to photoelectric absorption by cold material. This may be due to the fact that the material is partially ionised, or to the presence of a second spectral component which is not modulated at the spin period, thus producing an effective flux offset which can have a large effect on the estimates of modulation depths.

3.4 X-ray light curves : orbital modulation

Apart from the possibility of eclipses, there is no obvious reason to expect any orbital modulation in the X-ray light curves of intermediate polars since the accreting material must lose all knowledge of the orbital orientation inside R_{mag}. It is somewhat surprising, therefore, that half of these systems show a pronounced orbital variation, amounting to a 50 % modulation in the case of 3A0729+103 (see Table 1 and Fig.4). The orbital modulation observed typically consists of a broad, smooth feature covering a substantial range of orbital phase, with the amplitude of the orbital modulation decreasing with energy in at least two systems (3A0729+103 : McHardy *et al.* 1986, EX Hya : Mason, 1985). As well as this broad feature, EX Hya also shows a narrow, partial X-ray eclipse feature (Cordova *et al.*, 1985).

King (1985) and Hameury, King and Lasota (1986) have recently proposed a model which can explain both these orbital effects, and the presence of a second spectral component (see section 3.2). In their model there is a second X-ray emission region where the accretion flow impacts the white dwarf magnetosphere. Shock heating of the impacting material produces temperatures $\sim few$ keV and luminosities $\sim 0.1\,L_{acc}$ in a region located at a radius $R_{mag} \sim 10^{10}$ cm from the white dwarf. The luminosity of this region can be comparable to the hard X-ray luminosity L_X from the accretion column if the latter is produced at low efficiency (i.e. if $L_X \sim \eta L_{acc}$ with $\eta \sim 0.1$), as may indeed be the case (see section 4.1).

The presence of this second emission region can clearly explain the two-component spectra seen in some intermediate polars, and also the observed orbital modulations. Hameury *et al.* suggest three ways of producing an orbital modulation : (i) eclipse of the magnetospheric emission region by the secondary - this can explain the narrow dip in EX Hya; (ii) orbital-phase variable

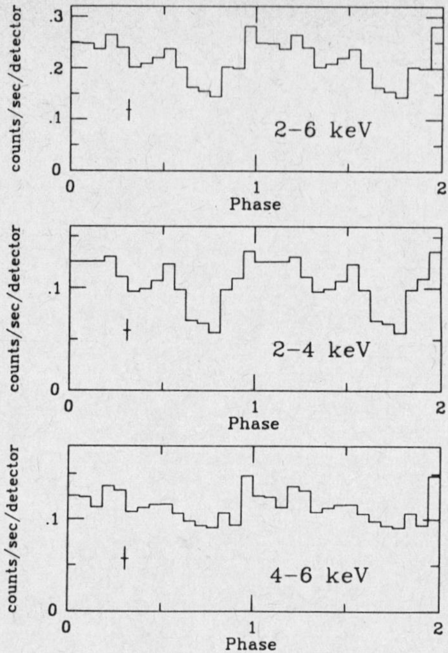

Fig.4 X-ray light curves of 3A0729+103 obtained with the *EXOSAT* ME folded at the 3.24 hour orbital period. From McHardy *et al.* (1986).

absorption of the emission from the magnetospheric emission region by the accretion stream - this can easily produce the broad, energy-dependent modulation discussed above; (iii) the possibility that some fraction of the accreting matter may cross the field lines and hit the white dwarf at a position fixed in the orbital frame, leading to the possibility of occultation by the white dwarf body.

4. POLARS

There are presently eleven polar systems known, nine of which have been observed with *EXOSAT*. *EXOSAT* observations have provided the first detailed X-ray light curves for over half of these objects. Their X-ray properties are listed in Table 2.

4.1 X-ray spectra and the soft X-ray problem

The canonical X-ray spectrum of a polar system has two components : a hard, 'bremsstrahlung' continuum with $kT > 10$ keV and an ultra-soft 'black-body' with $kT \sim 0.03$ keV (e.g. AM Her : Rothschild *et al.*, 1981). *EXOSAT* observations have lead to soft X-ray (CMA) detections of all the polars studied. *EXOSAT* grating observations, and broad-band filter measurements have confirmed this as being a ultra-soft component in several systems (e.g. Heise *et al.*, 1985; Osborne *et al.*, 1986a). The results on the hard X-ray component, however, are somewhat surprising. Of the 9 systems observed by *EXOSAT*, only 6 are detected in the ME instrument, and one of these is a marginal detection.

Table 2 : X-ray properties of polar systems

Object	P_{orb}(h)	F_{HX}[1]	F_{SX}[2]	Comments / references[3]
EF Eri (0311-227)	1.35	5	1	Refs. 27, 45, 41, 5.
E1114+182 (DP Leo)	1.50	< 1	—	Ref.6. Not observed by EXOSAT, F_{HX} limit from HEAO-1 measurement in ref.6.
VV Pup	1.67	< 0.1	1	Refs. 29, 23.
E1405-451 (V834 Cen)	1.69	1	0.7	Refs. 23, 7.
PG1550+191 (MR Ser)	1.89	< 0.2	0.03	Ref. 42.
H0139-68 (BL Hyi)	1.89	< 0.1	0.03	Refs. 1, 4, 34. EXOSAT fluxes quoted here relate to low-state observations only.
CW1103+254 (ST LMi)	1.90	0.5	0.5	Refs. 3, 23.
AN UMa	1.91	< 0.2	0.3	Refs. 37, 23.
AM Her	3.09	6	7	Ref. 13, and references therein.
H0538+608	3.1:	2	—	Ref. 31. Not observed by EXOSAT, F_{HX} from HEAO-1 observations in ref.31.
E2003+225 (QQ Vul)	3.71	0.2	0.5	Ref. 25 and references therein.

Notes

(1) Average hard X-ray flux in the 2 - 10 keV band (ME count/sec $\approx 1.4 \times 10^{-11}$ erg cm^{-2} s^{-1}).

(2) Average soft X-ray flux measured by the EXOSAT CMA with thin Lexan filter (~ 0.04 - 2 keV band, CMA count/sec). *The F_{HX} and F_{SX} values quoted are mainly taken from published light curves and hence are only estimates.*

(3) References given relate only to X-ray observations, primarily by EXOSAT plus a few selected earlier X-ray studies.

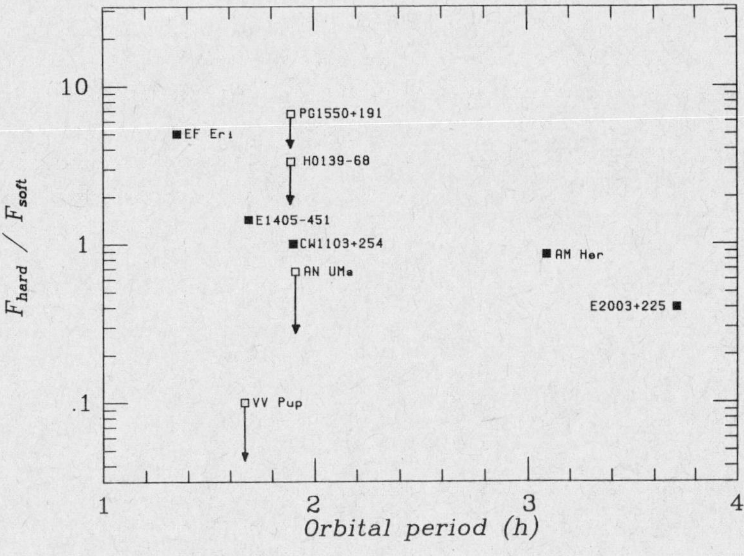

Fig.5 Soft X-ray (CMA) to hard X-ray (ME) count rate ratios as a function of orbital period for 9 polar systems.

These results are particularly relevant to the 'soft X-ray problem' , i.e. the mismatch between observed and expected ratio of soft to hard X-ray luminosities for radiative accretion column models (see section 2). A detailed discussion of the observational data is beyond the scope of this review. Nevertheless it is instructive to take a simple empirical approach, i.e. to compare the observed *EXOSAT* CMA and ME count rates directly. This approach has the advantage that the response of the CMA and ME detectors is such that they essentially measure the ultra-soft and hard X-ray components independently. Fig.5 shows this count rate ratio plotted as a function of orbital period for the 9 polars observed with *EXOSAT* (see Table 2).

The precise conversion of this ratio to the luminosity ratio L_{hard}/L_{soft} depends principally on the temperature of the soft component and the assumed absorbing column density N_H, quantities which are not in general accurately known. For two systems, AM Her and E2003+225, both of which have count rate ratios < 1 in Fig.5, the luminosity ratios are known to be < 1 with some certainty since they are based on *EXOSAT* grating observations of the soft X-ray spectrum (Heise *et al.*, 1985, Osborne *et al.*, 1986a). Thus for the two other systems which have low count rate ratios in Fig.5, AN UMa and VV Pup, it is difficult to avoid the conclusion that both have a soft X-ray excess, providing that their black-body temperatures and column densities are not wildly different to those for AM Her and E2003+225. For the other systems which have higher observed count rate ratios there may be no excess, unless these happen to be systems with rather high column densities or low black-body temperatures in which case the required corrections will push the derived luminosity ratio into the soft X-ray excess regime. (The ratios for CW1103+254 and E1405-451 are already close to 1.)

4.2 X-ray light curves

All polar systems show a strong modulation of both their hard and soft X-ray light curves at the orbital period. Two examples are shown in Fig.6. The hard X-ray light curves, which are only well determined in 5 systems (see Table 2), show a variety of shapes, ranging from the smooth, quasi-sinusoidal modulation seen in EF Eri (reminiscent of intermediate polars) to the more highly modulated curves seen in CW1103+254 and E1405-451 (see Fig.6 & compendium in Mason, 1985). The soft X-ray light curves show a similar variety, but overall appear to be more complex. The most notable feature of the soft X-ray light curves are the narrow, eclipse-like dips which are observed in 4 polars (see Fig.6 and Mason, 1985). Some of this complexity may be an illusion since most, if not all, polars show very pronounced flickering in their soft X-ray light curves which is not averaged out in folded light curves based on observations spanning, typically, only a few orbital cycles.

The hard X-ray light curves of the polar systems can be understood, in general terms, as being due the varying occultation of the emission region in the accretion column as the white dwarf rotates in exactly the same way as proposed for intermediate polars (see section 3.3). From this point of view it is clear that the more strongly modulated hard X-ray light curves of most of the polars observed must argue for significantly smaller polecap areas than found in intermediate polars, in accord with other methods of estimating the polecap area which indicate fractional areas $f \sim 10^{-3}$ to 10^{-5} (e.g. see Liebert & Stockman, 1985). In this picture the soft X-ray light curves should closely resemble the hard X-ray light curves, but be modified by three important effects :

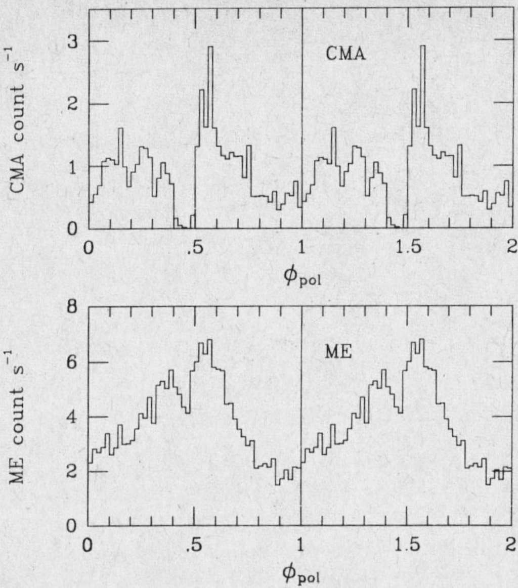

Fig.6(a) Soft X-ray (CMA) and hard X-ray (ME) light curves for EF Eri folded at the 1.35 hour orbital period and shown as a function of linear polarisation phase. (See Watson *et al.*, 1986b for further details).

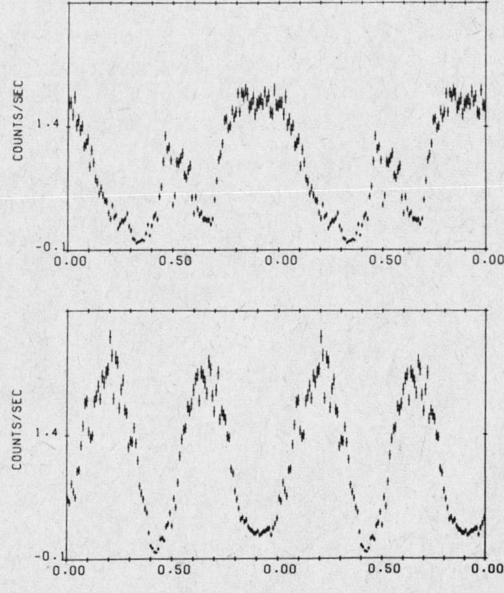

Fig.6(b) Soft X-ray (CMA) light curves for E2003+225 for 2 different epochs folded at the 3.7 hour orbital period and shown as a function of linear polarisation phase. Upper panel : 1985 day 161; lower panel : 1985 day 257. From Osborne *et al.* (1986b).

(i) projection and limb-darkening effects, since the soft X-rays come from an optically thick region; (ii) phase-variable absorption as the line of sight crosses different parts of the accretion column and the matter immediately surrounding it; (iii) large amplitude flickering. It seems likely that the observed soft X-ray light curves may eventually be successfully be interpreted in this way, but at present there is insufficient information, both observationally and theoretically to enable the required modelling to be carried out explicitly.

4.3 Soft X-ray dips

The narrow, phase-stable dips in the soft X-ray light curves of 4 polars presented, until recently, something of a problem (e.g. Mason, 1985). It now seems clear that this phenomenon can be explained by an occultation of the emission region by the accretion stream at large distances from the white dwarf (e.g. Patterson *et al.*, 1984, King and Williams, 1985) leading to photoelectric absorption of the X-rays and free-free absorption of optical and infra-red emission. The fact that these dips only occur in those systems where the active pole is in the hemisphere towards the observer adds substantial weight to this explanation, since in these systems the accretion stream will cross the line of sight to the polecap provided the inclination angle exceeds the colatitude of polecap. In the specific case of EF Eri the dips are observed not only in soft X-rays and as partial, energy-dependent dips in the hard X-ray light curve (Patterson et al., 1981, Watson *et al.*, 1985b), but also in the infra-red and intermittently in V band photometry. A simultaneous observation of one dip in all 4 bands is shown in Fig.7. As King and Williams demonstrate, the observed characteristics of these dips in EF Eri - phase width, depth and implied column density - are in accord with what is expected for reasonable parameters for the accretion stream. The phasing of these absorption dips is an important measurement since it is directly related to the orbital motion, rather than simply the white dwarf rotation, and thus can be used, in principle, to measure deviations from synchronism in polar systems.

4.4 Changes in the light curve

In contrast with intermediate polars whose X-ray light curves seem to be rather stable on timescales of a few years, several polar systems have shown substantial changes in their light curve morphology. The most famous example is AM Her for which the relative phasing soft and hard X-ray light curves was observed to have changed by $\sim 180°$ in *EXOSAT* observations (Heise *et al.*, 1985). Recent results on E2003+225 also show dramatic changes, in this case in the shape of the soft X-ray light curve (see Fig.6 and Osborne *et al.*, 1986b). The anomalous behaviour of AM Her, and perhaps also the changes in E2003+225, are very plausibly related to changes in the accretion geometry in these systems, and in particular to a change in the balance between accretion between the two magnetic poles of the white dwarf.

EF Eri

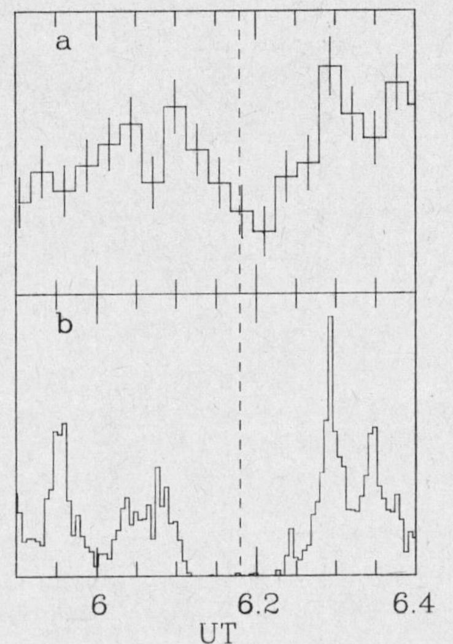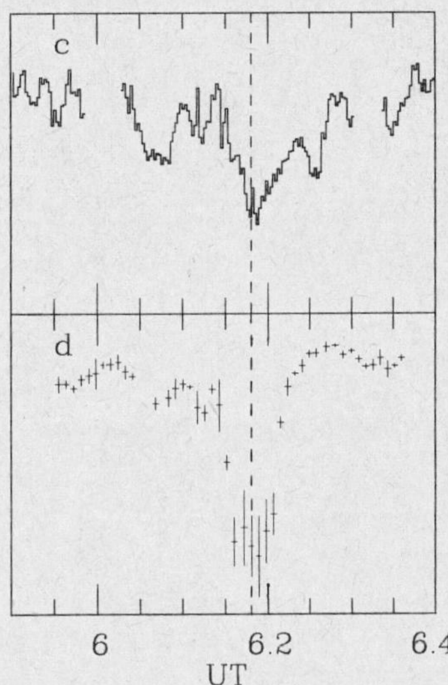

Fig.7 Detailed light curves of a 'dip' in EF Eri observed on 1983 November 29. (a) hard X-rays (ME : 1.5 - 3 keV); (b) soft X-rays (CMA); (c) V-band data; (d) K-band data (C.Motch, private communication, 1986).

5. CONCLUSIONS

Magnetic white dwarf systems offer unique opportunities for the study of a variety of aspects of the accretion onto compact objects. X-ray studies in particular have enabled us to investigate the accretion process in unprecedented detail and have illuminated a range of important considerations in physics of accretion onto magnetised white dwarfs.

EXOSAT observations have played a crucial role in these investigations and it is abundantly clear that further progress in this field is dependant on the development of more detailed theoretical models to explain the wealth of observational material covered in this review.

Acknowledgements

This review has benefitted from many stimulating discussions with colleagues working in this field. I would particularly like to thank Andrew King, Keith Mason and Julian Osborne.

REFERENCES

[1] Agrawal, P.C., Riegler, G.R., & Rao, A.R., 1983. *Nature*, **301**, 318.

[2] Beuermann, K., & Osborne, J., 1985. *Sp. Sci. Rev.*, **40**, 117.

[3] Beuermann, K., & Stella, L., 1985. *Sp. Sci. Rev.*, **40**, 139.

[4] Beuermann, K., Schwope, A., Weissieker, H., Motch, C., 1985. *Sp. Sci. Rev.*, **40**, 135.

[5] Beuermann, K., Stella, L., & Patterson, J., 1986. *Ap.J.*, submitted.

[6] Biermann, P., Schmidt, G.D., Liebert, J., Stockman, H.S., Tapia, S., Kuhr, H., Strittmatter, P.A., West, S., & Lamb, D.Q., 1985. *Ap.J.*, **293**, 303.

[7] Bonnet-Bidaud, J.M., Beuermann, K., Charles, P., Maraschi, L., Motch, C., Mouchet, M., Osborne, J., Tanzi, E., & Treves, A. 1985. In *Recent results on Cataclysmic Variables*, Proc. Bamberg ESA Workshop, p.155.

[8] Cook, M.C., Watson, M.G., & McHardy, I.M., 1984. *Mon. Not. R. astr. Soc.*, **210**, 7P.

[9] Cordova, F.A., & Mason, K.O., 1984. *Mon. Not. R. astr. Soc.*, **206**, 879.

[10] Cordova, F.A., Mason, K.O., & Kahn, S.M., 1985. *Mon. Not. R. astr. Soc.*, **212**, 447.

[11] Frank, J., King, A.R., & Raine, D., 1985. *Accretion Power in Astrophysics*, Ch.6 (Cambridge University Press).

[12] Hameury, J.-M., King, A.R., & Lasota, J.-P., 1986. *Mon. Not. R. astr. Soc.*, **218**, 695.

[13] Heise, J., Brinkmann, A.C., Gronenschild, E., Watson, M.G., King, A.R., Stella, L., & Kieboom, K., 1985. *Astron.Astrophys*, **148**, L14.

[14] King, A.R., & Shaviv, G., 1984. *Mon. Not. R. astr. Soc.*, **211**, 883.

[15] King, A.R., 1985. In *Recent results on Cataclysmic Variables*, Proc. Bamberg ESA Workshop, p.133.

[16] King, A.R., & Williams, G., 1985. *Mon. Not. R. astr. Soc.*, **215**, 1P.

[17] Kruszewski, A., Mewe, R., Heise, J., Chlebowski, T., van Dijk, W., Ballker, R., 1981. *Sp. Sci. Rev.*, **30**, 221.

[18] Liebert, J. & Stockman, H.S., 1985. In *Cataclysmic Variables and Low-mass X-ray Binaries*, p.151 (eds. Patterson & Lamb, Reidel).

[19] Mason, K.O., 1985. *Sp. Sci. Rev.*, **40**, 99.

[20] McHardy, I.M., Pye, J.P., Fairall, A.P., Warner, B., Cropper, S., & Allen, S., 1984. *Mon. Not. R. astr. Soc.*, **210**, 663.

[21] McHardy, I.M., Pye, J.P., Fairall, A.P., & Menzies, J.W., 1986. *Mon. Not. R. astr. Soc.*, submitted.

[22] Osborne, J.P., Mason, K.O., Bonnet-Bidaud, J.M., Beuermann, K., & Rosen, S., 1984a. In *X-ray Astronomy '84*, Proc. Int. Symp. on X-ray Astronomy (Bologna), p.63 (eds. Oda & Giacconi, ISAS).

[23] Osborne, J., Maraschi, L., Beuermann, K., Bonnet-Bidaud, J.M., Charles, P.A., Chiappetti, L., Motch, C., Mouchet, M., Tanzi, E.G., Treves, A., & Mason, K.O., 1984b. In *X-ray Astronomy '84*, Proc. Int. Symp. on X-ray Astronomy (Bologna), p.59 (eds. Oda & Giacconi, ISAS).

[24] Osborne, J.P., Rosen, R., Mason, K.O., & Beuermann, K., 1985. *Sp. Sci. Rev.*, **40**, 143.

[25] Osborne, J., Bonnet-Bidaud, J.M., Bowyer, S., Charles, P.A., Chiapetti, L., Clarke, J.T., Henry, J.P., Hill, G.J., Kahn, S., Maraschi, L., Mukai, K., Treves, A., Vrtilek, S., 1986a. *Mon. Not. R. astr. Soc.*, in press.

[26] Osborne, J., Mukai, K., Beuermann, K., *et al.*, 1986b. Paper presented at the Tenerife Workshop *Physics of Accretion onto Compact Objects*, Tenerife, April 1986.

[27] Patterson, J., Williams, G., & Hiltner, W.A., 1981. *Ap.J.*, **245**, 618.

[28] Patterson, J.P., & Steiner, J.E., 1983. *Ap.J.(Lett.)*, **264**, L61.

[29] Patterson, J., Beuermann, K., Lamb, D.Q., Fabbiano, G., Raymond, J.C., Swank, J., & White, N.E., 1984. *Ap.J.*, **279**, 785.

[30] Pietsch, W., Pakull, M., Tjemkes, S., Voges, W., Kendziorra, E., & van Paradijs, J., 1984. In *X-ray Astronomy '84*, Proc. Int. Symp. on X-ray Astronomy (Bologna), p.67 (eds. Oda & Giacconi, ISAS).

[31] Remillard, R.A., Bradt, H.V., McClintock, J.E., Patterson, J., Roberts, W., Schwartz, D.A., & Tapia, S., 1986. *Ap.J.(Lett.)*, **302**, L11.

[32] Rothschild, R.E., Gruber, D.E., Knight, F.K., Matteson, J.L., Nolan, P.L., Swank, J.H., Holt, S.S., Serlemitsos, P.J., Mason, K.O., & Tuohy, I.H., 1981. *Ap.J.*, **250**, 723.

[33] Schrijver, J., Brinkmann, A.C., van der Woerd, H., Watson, M.G., King, A.R., van Paradijs, J., & van der Klis, M., 1985. *Sp. Sci. Rev.*, **40**, 121.

[34] Schwope, A., & Beuermann, K., 1985. In *Recent results on Cataclysmic Variables*, Proc. Bamberg ESA Workshop, p.173.

[35] Shafter, A., Szkody, P., & Thorstensen, J., 1986. *Ap.J.*, in press.

[36] Szkody, P., 1986. *Ap.J.(Lett.)*, **301**, L29.

[37] Szkody, P., Schmitt, E., Crosa, L., Schommer, R., 1981. *Ap.J.*, **246**, 223.

[38] Tuohy, I.R., Buckley, D.A.H., Remillard, R.A., Bradt, H.V, & Schwartz, D.A., 1986. *Ap.J.*, in press.

[39] Warner, B., 1983. In Proc. IAU Colloq.72 *Cataclysmic Variables and Related Objects*, p.155 (eds.Livio & Shaviv, Reidel).

[40] Watson, M.G., King, A.R., & Osborne, J., 1985a. *Mon. Not. R. astr. Soc.*, **212**, 917.

[41] Watson, M.G., King, A.R., Williams, G., Heise, J., 1985b. In *Recent results on Cataclysmic Variables*, Proc. Bamberg ESA Workshop, p.169.

[42] Watson, M.G., & King, A.R., 1986. In preparation.

[43] Watts, D.J., Greenhill, J.G., Hill, P.W., & Thomas, R.M., 1982. *Mon. Not. R. astr. Soc.*, **200**, 1039.

[44] White, N.E., & Marshall, F.E., 1981. *Ap.J.(Lett.)*, **249**, L25.

[45] White, N.E., 1981. *Ap.J.(Lett.)*, **244**, L85.

EVOLUTION OF MAGNETIC CATACLYSMIC BINARIES

D. Q. Lamb and F. Melia
*Department of Astronomy and Astrophysics
and Enrico Fermi Institute
University of Chicago*

I. INTRODUCTION

More than a third of all cataclysmic variables (CV's) with established binary periods contain strongly magnetic degenerate dwarfs (Ritter 1985). The history of the manner in which the magnetic character of these systems was established suggests that many more CV's—perhaps most—are magnetic. By comparison, only $\approx$ 2% of isolated degenerate dwarfs are magnetic (Angel, Borra, and Landstreet 1981; Schmidt and Liebert 1986). The magnetic CV's divide naturally into two subclasses: the DQ Herculis stars (recently reviewed by Warner 1983, 1985) and the AM Herculis stars (recently reviewed by Liebert and Stockman 1985). Each subclass currently has about a dozen members.

In magnetic CV's, the magnetic field of the degenerate dwarf funnels the accretion flow near the star, producing an accretion shock at the stellar surface, and pulsed optical and X-ray emission. The known DQ Her systems typically have binary orbital periods $P_b \gtrsim 3$ hours. They show evidence of an accretion disk (Warner 1983, 1985) and the degenerate dwarf is believed to have been spun up to a short rotation period P by the accretion torque (see, *e.g.*, Lamb and Patterson 1983; Hameury, King, and Lasota 1986). The optical light from these systems shows little or no optical polarization ($<$ 1 %; Warner 1983, 1985; Penning, Schmidt, and Liebert 1986). The observed modulation of the optical light at P is due partly to emission directly from the degenerate dwarf and partly to reprocessing of the pulsed X-ray emission, while that at the sideband $P_{side} = (P^{-1} - P_b^{-1})^{-1}$ is due to reprocessing of the pulsed X-ray emission from a point at rest in the binary system (see, *e.g.*, Warner 1985). These properties are summarized in Table 1.

In contrast, the known AM Her systems typically have $P_b \lesssim 3$ hours. The degenerate dwarf in them has a magnetic moment μ_1 sufficient to prevent the formation of a disk. Its magnetic moment is also sufficient to couple the degenerate dwarf to the companion star and synchronize its rotation period P with the orbital period P_b, in spite of the accretion torque (see, *e.g.*, the review by Lamb 1985). The optical light from these systems is strongly polarized ($\gtrsim$ 10 %; Liebert and Stockman 1985). These properties are summarized in Table 1.

While the above properties adequately characterize the known DQ Her and AM Her systems, theory indicates that systems can exist which have some properties of each. Therefore, we shall henceforth define DQ Her stars to be magnetic CV's with *asynchronously* rotating degenerate dwarfs and AM Her stars to be those with *synchronously* rotating [$\equiv (P_b - P)/P_b \ll 1$] degenerate dwarfs.

TABLE 1

PROPERTIES OF MAGNETIC CATACLYSMIC VARIABLES

DQ Her stars	AM Her stars
• $P \ll P_b$	• $P = P_b$
• $P_b \gtrsim 3$ hours	• $P_b \lesssim 3$ hours
• Little or no optical polarization (< 1 %)	• Strong optical polarization ($\gtrsim 10$ %)
• Optical modulated at P, P_b, and P_{side}	• Optical, UV, and X-rays modulated at $P = P_b$
• UV(?) and X-rays modulated at P, P_b	
• Accretion disk present	• No accretion disk
• $\mu_1 \sim 10^{31} - 10^{33}$ G cm^3 ($B_1 \sim 10^5 - 10^7$ G)	• $\mu_1 \sim 10^{33} - 10^{34}$ G cm^3 ($B_1 \approx 2 - 3 \times 10^7$ G)

The possibility that DQ Her stars evolve into AM Her stars has recently received considerable attention (Chanmugam and Ray 1984; King, Frank, and Ritter 1985; King 1985; Lamb 1985). According to this hypothesis, DQ Her and AM Her stars are intrinsically similar but are observed at different evolutionary epochs. This picture is attractive for several reasons. First, as a close binary evolves, its orbital period P_b decreases (to a minimum period $P_{b,min}$). Thus, the known DQ Her systems with periods $P_b \gtrsim 3$ hours will eventually have periods $P_b \lesssim 3$ hours, like those of the known AM Her stars. Second, the Roche lobe of the degenerate dwarf shrinks as P_b decreases, and the radial extent of the disk therefore also decreases. At the same time, the Alfvén radius $r_A^{(d)}$ for disk accretion increases because the mass transfer rate decreases. Eventually, these two radii cross, and the disk must disappear. Third, magnetic coupling (of almost any kind) between the magnetic degenerate dwarf and the (magnetic) secondary increases rapidly as the binary separation a and the mass transfer rate decreases with decreasing P_b, so that synchronization becomes more likely.

In this talk, we explore this picture using the physics of magnetic cataclysmic binary evolution. We describe the results of recent calculations (Lamb and Melia 1986), in which the transition between the DQ Her and AM Her states is determined by balancing the accretion torque against the MHD synchronization torque (Lamb, Aly, Cook, and Lamb 1983). Previous workers (Chanmugam and Ray 1984; King, Frank, and Ritter 1985) used a heuristic approach and assumed that synchronization occurs when $r_A^{(0)} \approx a$, where $r_A^{(0)}$ is the Alfvén radius for spherical accretion (Ghosh and Lamb 1979) and a is the orbital separation of the binary.

We identify four regimes: (1) When $\mu_1 \lesssim 10^{31}$ G cm^3, the magnetic field of the degenerate dwarf is unable to funnel the accretion flow; these systems are not DQ Her stars and may show little or no evidence of a magnetic field. (2) Systems with 10^{31} G cm$^3 \lesssim \mu_1 \lesssim 10^{33}$ G cm^3 are always DQ Her stars. (3) Systems with 10^{33} G cm$^3 \lesssim \mu_1 \lesssim 10^{35}$ G cm^3 are DQ Her stars which evolve into AM Her stars, assuming μ_1 is constant throughout their evolution (an assumption which may not be valid). These three regimes agree qualitatively with those of King, Frank, and Ritter (1985). (4) Systems with 10^{35} G cm$^3 \lesssim \mu_1$ are always AM Her stars.

We confirm that AM Her stars do not have accretion disks, while DQ Her stars may or may not have them, depending on the value of μ_1, as pointed out by King (1985) and Hameury, King, and Lasota (1986). The ranges of μ_1 for the known DQ Her and AM Her stars allowed by observation are very large, due partly to uncertainties in the theory of disk and stream accretion (Lamb and Patterson 1983; Hameury, King, and Lasota 1986) and partly to the dependence of the ranges on the mass M_1 of the degenerate dwarf. Nevertheless, it is likely that the μ_1's of the known DQ Her stars (including those with longer rotation periods) are significantly less than those of the known AM Her stars, unless all known DQ Her and AM Her stars have $M_1 \gtrsim 1\ M_\odot$. This differs from the conclusion reached by King (1985) and King, Frank, and Ritter (1985). The criteria for synchronization and for the presence or absence of a disk in DQ Her systems are also uncertain, due to uncertainties in the theory of the MHD synchronization torque and to the dependence of the criteria on M_1. Nevertheless, it appears that the known DQ Her stars will not evolve into AM Her stars, assuming that μ_1 is constant throughout their evolution. This differs from the conclusion reached by Chamugam and Ray (1984), King (1985), and King, Frank, and Ritter (1985).

The fact that the strong magnetic field of the degenerate dwarf dramatically alters the optical, UV, and X-ray appearance of magnetic CV's has come to be appreciated (see the reviews by King 1983, 1985; and Lamb 1983, 1985). That it can also alter the evolution of the binary itself is only now becoming recognized. Spin-up and spin-down of the magnetic degenerate dwarf temporarily speeds up and slows down the binary evolution of DQ Her stars (Ritter 1985; Lamb and Melia 1986). Co-operative magnetic braking may speed up the binary evolution of AM Her stars when the degenerate dwarf has a particularly strong ($B_1 \gtrsim 5 \times 10^7$ G) magnetic field (Schmidt, Stockman, and Grandi 1986). Here we point out that synchronization of the magnetic degenerate dwarf injects angular momentum into the binary, driving the system apart and producing a *synchronization-induced period gap* (Lamb and Melia 1986). This process yields an entirely new way of producing ultra-short period binaries. The secondaries in these binaries are hydrogen-rich rather than hydrogen-depleted or pure helium (*cf.* Nelson, Rappaport, and Joss 1985; Lamb *et al.* 1986).

II. PHYSICS OF MAGNETIC CV EVOLUTION

Figure 1 shows the distribution of binary periods P_b for the magnetic CV's and for all CV's. Several features are evident. First, the two distributions are similar. Second, both have a cutoff at a minimum period $P_{b,min} \approx 80$ minutes. Third, both show the well-known "period gap" between 2 and 3 hours. Fourth, most of the DQ Her stars have $P_b \gtrsim 3$ hours, while most of the AM Her stars have $P_b \lesssim 3$ hours, as already noted.

Figure 2 shows the observed mass loss rate $|\dot{M}_2|$ from the secondary as a function of orbital period for the magnetic CV's (Patterson 1984). The rates shown assume that $|\dot{M}_2|$ is approximately the same as the mass transfer rate into the disk in the case of DQ Her stars and that $|\dot{M}_2|$ is approximately the same as the mass accretion rate $\dot{M}_1$ onto the degenerate dwarf in the case of AM Her stars, which have no disk. Here and in subsequent figures, we use a double-valued axis for the orbital period, so that systems continue to evolve from right to left after minimum period (denoted by the vertical dotted line at $P_b = 1.2$ hours). Data for the DQ Her stars are shown as filled squares and data for the AM Her stars as filled circles; no error bars are shown, but the uncertainties in $\dot{M}_2$ are a factor ≈ 3 (Patterson 1984). From this figure, it is evident that $|\dot{M}_2|$ is smaller and shows less scatter at shorter orbital periods. The distribution of observed $|\dot{M}_2|$'s for the set of all CV's is similar.

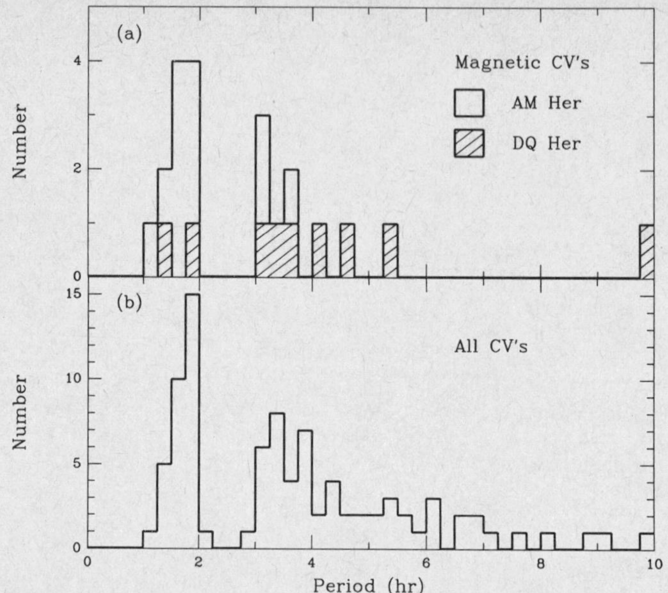

Fig. 1.–Orbital period distributions of CV's. a) Magnetic CV's. The DQ Her stars are shown as shaded boxes. b) All CV's.

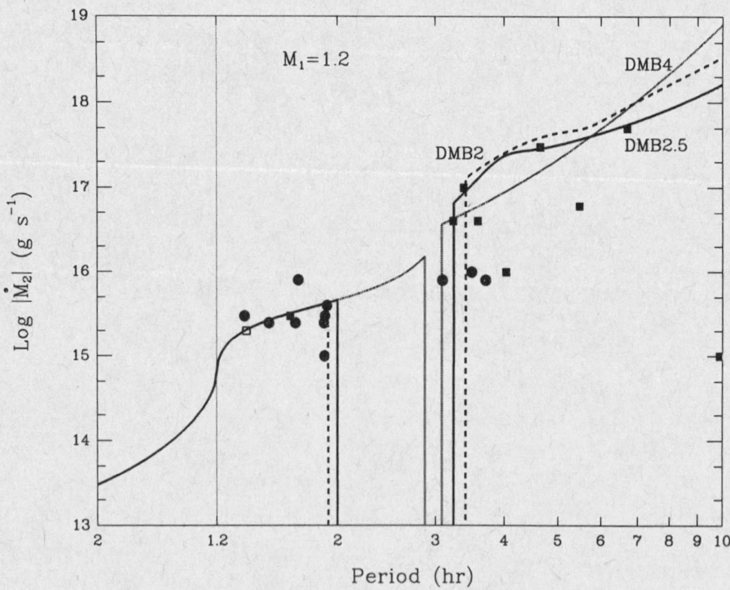

Fig. 2.–Mass transfer rate as a function of orbital period for several disrupted magnetic braking models. DMB2 and DMB2.5 give period gaps similar to that observed, but mass transfer rates longward of the gap which are about a factor of ten larger than those inferred from observation. DMB4 gives a period gap that is too short, but a mass transfer rate that is closer to that inferred from observation. Shortward of the gap, the evolution is driven only by GR and is therefore the same in all three models.

Substantial progress has been made in recent years in understanding the qualitative features of close binary evolution evident in Figures 1 and 2. Stable mass transfer is possible only if the mass ratio $q \equiv M_1/M_2 \lesssim 1$, *i.e.* the less massive star transfers mass to the more massive one. This implies that the secondary stars in CV's have masses $M_2 \lesssim 1.2\ M_\odot$. For binary periods $P_b \gtrsim 10$ hours, such low-mass secondaries must be significantly evolved in order to fill their Roche lobe. It is therefore believed that such binaries are driven by nuclear evolution of the secondary (Whyte and Eggleton 1980; Webbink, Rappaport, and Savonije 1983; Taam 1983b). Assuming conservative mass transfer $(\dot{M}_1 = -\dot{M}_2)$, $\dot{P}_b > 0$ in these systems since $q > 1$. AE Aqr, with $P_b = 9.9$ hours, and GK Per, with $P_b = 46$ hours, are thought to be evolving due to this mechanism (*cf.* Whyte and Eggleton 1980, Patterson 1984).

a) Angular Momentum Loss

Binaries with orbital periods $P_b \lesssim 10$ hours contain secondary stars whose nuclear evolution proceeds too slowly to drive mass transfer at the rates observed. This is consistent with observational evidence that these secondary stars are not significantly evolved (Patterson 1984). These binaries are therefore believed to evolve due to the loss of orbital angular momentum. The rate of change of the angular momentum of the system can be written as

$$\dot{J} = \dot{J}_1 + \dot{J}_2 + \dot{J}_b \, , \tag{1}$$

where $\dot{J}_1$ and $\dot{J}_2$ are the rate of change of the spin angular momenta of the degenerate dwarf and the secondary, respectively, and $\dot{J}_b$ is the rate of change of the orbital angular momentum of the binary.

Gravitational radiation (GR) drives the evolution of the binary by extracting orbital angular momentum, $\dot{J}_{GR} = \dot{J}_b$ (Kraft, Mathews, and Greenstein 1962; Paczyński 1967; Faulkner 1971). This mechanism gives a minimum orbital period of about 1.2 hours, corresponding to the transition of the secondary from the main sequence to the degenerate sequence (Paczyński and Sienkowicz 1981; Rappaport, Joss, and Webbink 1982), and therefore successfully accounts for the observed cutoff seen at $P_b \approx 1.3$ hours (see Figure 1). However, evolution due to GR gives mass transfer rates as large as those observed only for $P_b \lesssim 3$ hours.

Consequently, other mechanisms which could drive binary evolution effectively at intermediate binary periods (3 hours $\lesssim P_b \lesssim 10$ hours) have been investigated. One possibility that has recently received a great deal of attention is magnetic braking (Verbunt and Zwaan 1981; Taam 1983a; Spruit and Ritter 1983; Rappaport, Verbunt, and Joss 1983; Patterson 1984; Verbunt 1984). According to this idea, the secondary star has a strong magnetic field which couples it to its stellar wind and extracts spin angular momentum from the star, $\dot{J}_{MB} = \dot{J}_2$. Since the secondary is largely convective, tidal coupling between the secondary and the binary system is very effective $(P_{tidal} \sim 100$ years; Zahn 1977). This keeps the secondary synchronized, so that the angular momentum is lost from the binary system itself $(\dot{J}_{MB} = \dot{J}_b)$. For large but conceivable values of the companion's magnetic field $(B_2 \sim 300 - 1000$ G) and stellar wind $(\dot{M}_{wind} \sim 10^{-10} M_\odot/yr^{-1})$, magnetic braking can produce mass transfer rates as large as those observed for $P_b \lesssim 10$ hours.

b) The Period Gap

The gap in the orbital period distribution shown in Figure 1 may be the result of very low mass transfer rates (the systems are too faint to be seen), a brief episode of very high mass loss from the secondary at $P_b \approx 3$ hours (the system loses mass too quickly to be seen), or the fact that the systems do not evolve through the gap at all (they avoid it somehow).

Spruit and Ritter (1983) have shown that a gap of the first kind can be produced by an angular momentum loss mechanism $\dot{J}_{add}$ acting in addition to GR, if it is sufficiently strong and ceases suddenly at $P_b \approx 3$ hours. In this picture, a gap occurs because the radius of the secondary star becomes larger than its main sequence value when the star is driven out of thermal equilibrium, *i.e.* $\tau_{EVOL} < \tau_{KH}$, where τ_{KH} is the Kelvin-Helmholz thermal time scale, and τ_{EVOL} $(\approx \tau_{add})$ is the evolutionary time scale. When the additional angular momentum loss mechanism ceases $(\dot{J}_{add} \to 0)$, the secondary star shrinks to its main sequence radius and comes out of contact, ending mass transfer. The system continues to evolve to shorter binary periods and smaller separations, but now on the much longer time scale τ_{GR} $(>> \tau_{add})$. The secondary comes back into contact when the radius of its Roche lobe equals the main sequence radius of the star, and mass transfer is re-established, but at the much lower rate given by evolution due to GR.

Robinson *et al.* (1981) noted that the secondary becomes fully convective at $P_b \approx 3$ hours. Subsequently, Spruit and Ritter (1983), and Rappaport, Verbunt, and Joss (1983) suggested that magnetic braking is disrupted when the secondary becomes fully convective, possibly because the magnetic field of the secondary disappears, and that this produces the period gap in the way just described (with $\dot{J}_{add} = \dot{J}_{MB}$). The latter authors parametrize the magnetic braking torque by

$$N_{MB} = -3.8 \times 10^{-30} f M_2 R_\odot^4 \left(\frac{R_2}{R_\odot} \right)^\eta \Omega_b^3 \text{ dyne cm} , \tag{2}$$

where M_2 and R_2 are the mass and radius of the companion star, respectively, $\Omega_b = 2\pi / P_b$ is the binary orbital frequency, and f is a constant of order unity.

Figure 2 compares the $\dot{M}_2(P_b)$ relation given by disrupted magnetic braking models and the observed $\dot{M}_2$'s for magnetic CV's. The curves labeled DMB2, DMB2.5, and DMB4 correspond to Equation (2) with indices $\eta = 2$, 2.5, and 4, and $f = 1$. This figure shows that disrupted magnetic braking provides an elegant explanation of the qualitative features of CV evolution.

However, more detailed comparison shows model DMB4 gives a period gap that is much too narrow. Models DMB2 and DMB2.5 give period gaps that are about right, but mass loss rates that are much too large. Furthermore, these curves are lower bounds to the mass transfer rate given by magnetic braking, since they assume $M_1 = 1.2\ M_\odot$ (lower mass degenerate dwarfs give rates that are as much as a factor of ten larger). The possibility that non-conservative mass transfer could explain the discrepancy has been noted by Rappaport, Verbunt, and Joss (1983); however, this would require mass loss from the system at $\gtrsim 10$ times the rate of accretion onto the degenerate dwarf, a situation we regard as unlikely.

There is thus an inherent difficulty in the disrupted magnetic braking model for the period gap: The choices of f and η that give period gaps of about the right size give $\dot{M}_2$'s that are much too large, while those that give about the right $\dot{M}_2$'s give period gaps that are much too narrow. Therefore, it would seem that other explanations of the period gap should be pursued.

The discrepancy between the mass transfer rates longward of the period gap predicted by the disrupted magnetic braking model and those observed poses a dire problem when investigating the subtle questions which interest us here, i.e whether the magnetic field of the degenerate dwarf channels the accretion flow, whether an accretion disk is present or absent, and whether or not synchronization occurs. We handle the problem by adopting the following phenomenological model of CV evolution. Longward of the period gap, we take the braking law (3) with $f = 0.15$ and $\eta = 4$, which fits the observed mass transfer rates reasonably well (see Figure 3). At $P_b = 3$ hours, we abruptly reduce the mass of the secondary to $0.18 M_\odot$, a mass which brings the secondary back into contact at $P_b = 2$ hours. Shortward of $P_b = 3$ hours, the evolution is driven by GR alone. This model gives a period gap of the second kind defined earlier; we shall henceforth refer to it as DMB4*.

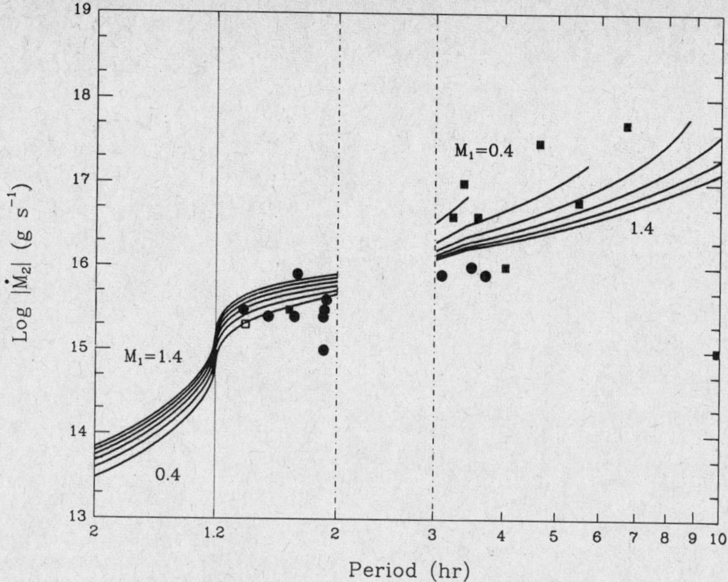

Fig. 3.–Mass transfer rate as a function of orbital period, assuming a rate longward of the period gap equal to that given by the magnetic braking law (2) with $f = 0.15$ and $\eta = 4$, and a rate shortward of the period gap equal to that given by gravitational radiation.

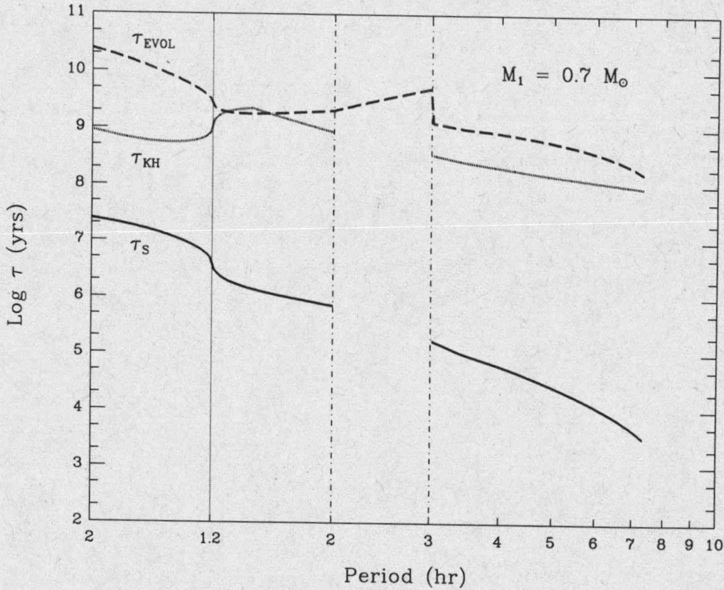

Fig. 4.–The evolutionary time scale τ_{EVOL} of the binary, the thermal time scale τ_{KH} of the secondary star, and the synchronization (or spin-up) time scale τ_S *at synchronization* (they are equal then) as functions of orbital period P_b, assuming $M_1 = 0.7\ M_\odot$ and $\dot{M}_2$ as shown in Figure 3.

Figure 4 shows the evolutionary time scale τ_{EVOL} and the thermal time scale τ_{KH} for model DMB4* as functions of the orbital period P_b. The evolutionary time scale increases at $P_b = 3$ hours, since the magnetic braking torque ceases then and $\tau_{GR} > \tau_{add} = \tau_{MB}$. The jump is not as large as that in models DMB2, DMB2.5 and DMB4, which have $\dot{J}_{MB} \gg \dot{J}_{GR}$. For model DMB4*, $\tau_{KH} \lesssim \tau_{EVOL}$, except between $P_b \approx 1.7$ hours and $P_{b,min}$ (≈ 1.2 hours). The time scale τ_{KH} decreases after $P_{b,min}$ because the secondary becomes increasingly degenerate.

c) Existence of an Accretion Disk

When mass transfer first begins, the trajectory of the stream from the secondary star passes by the degenerate dwarf at a radius ϖ_{min}, swings around, and intersects itself at a somewhat larger radius (Lubow and Shu 1975). The stream then circularizes and forms a ring. The radius of the ring is

$$\varpi_{max} \approx X^2 \, r_{L1} \,, \tag{3}$$

where r_{L1} is the radius of the inner Lagrangian point, and $X \approx 0.3 - 0.4$ is a factor that takes into account the ejection angle of the stream at r_{L1} (Lubow and Shu 1975). Eventually the ring spreads, forming a disk. Figure 5 shows R_1, ϖ_{min}, ϖ_{max}, and r_{L1} (measured from the center of the degenerate dwarf) relative to the binary separation a as a function of P_b, assuming $M_1 = 0.7 M_\odot$.

Whether or not a disk *forms* in a magnetic CV depends on the size of the Alfvén radius $r_A^{(d)}$ for disk accretion relative to ϖ_{min}. If $r_A^{(d)} < \varpi_{min}$, the answer is yes; if $r_A^{(d)} > \varpi_{min}$, what happens depends on a more detailed discussion of the physics involved (*cf.* Hameury, King, and Lasota 1986). Whether or not a disk *persists* in a magnetic CV depends on the size of $r_A^{(d)}$ relative to ϖ_{max}. If $r_A^{(d)} < \varpi_{max}$, the answer is yes; if $r_A^{(d)} > \varpi_{max}$, the answer is no. Thus the presence or absence of a disk depends not only on the present physical parameters of the binary ($P_b, \dot{M}_1, \mu_1$, etc), but also on its history.

Consider what happens if most CV's come into contact at long orbital periods and $\mu_1 \lesssim 10^{33}$ G cm^3. The Roche lobe of the primary is large and typically $r_A^{(d)} < \varpi_{min}$ (see Figure 5). A disk therefore forms. As long as mass transfer continues, the radius ϖ_{min} is irrelevant and the disk persists until $\varpi_{max} < r_A^{(d)}$. [This differs from King (1985) and Hameury, King, and Lasota (1986), who conclude that disks disappear when $r_A^{(d)} > \varpi_{min}$.] Thus we expect disks to exist longward of the period gap in most systems with $\mu_1 \lesssim 10^{33}$ G cm^3.

Assuming that mass transfer ceases when a system enters the period gap, any disk present quickly disappears. When the system leaves the gap and mass transfer resumes, a disk certainly forms if $r_A^{(d)} < \varpi_{min}$ and does not form if $\varpi_{max} < r_A^{(d)}$. When $r_A^{(d)}$ lies in the narrow range $\varpi_{min} < r_A^{(d)} < \varpi_{max}$, the answer is not clear cut; what happens depends on a more detailed consideration of the physics involved (see, e.g., Hameury, King, and Lasota 1986). Such a discussion lies beyond the scope of the present talk. However, we make the following remarks. The Alfvén radius $r_A^{(s)}$ for accretion from a stream is smaller than that for accretion from a disk because the velocity of the matter is similar in both cases ($\approx$ free fall) but the density in the stream is higher. In fact, an estimate of the minimum radius r_{min} at which the accretion stream is disrupted by the magnetic field of the degenerate dwarf gives $r_{min} < \varpi_{min}$ [*cf.* Lamb 1985; Equation (20)].

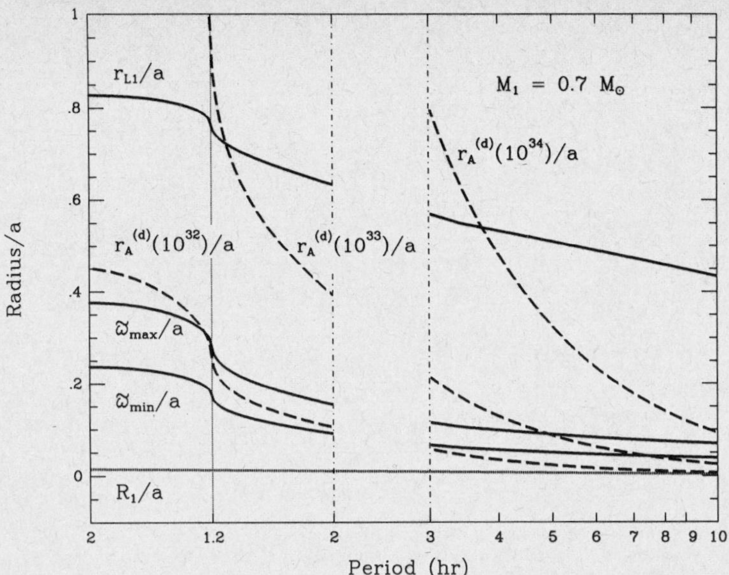

Fig. 5.–Various radii relative to the binary orbital separation a as functions of P_b, assuming $M_1 = 0.7\ M_\odot$ and $\dot{M}_2$ as given in Figure 3. Shown are the radius R_1 of the degenerate dwarf, the radius ϖ_{min} of minimum approach of the stream from the secondary star when mass transfer first begins, the circularization radius ϖ_{max}, the Alfvén radius $r_A^{(d)}$ for disk accretion, and the radius r_{L1} of the inner Lagrangian point (measured from the center of the degenerate dwarf).

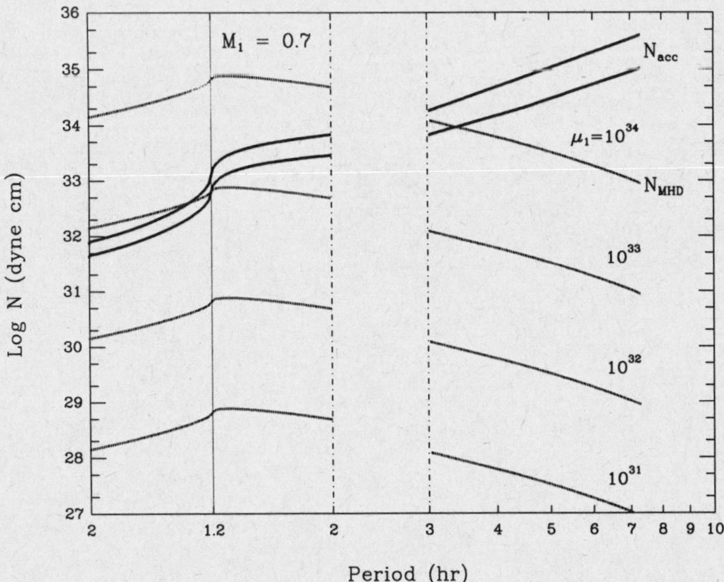

Fig. 6.–Accretion torque N_{acc} and MHD synchronization torque N_{MHD} as functions of P_b, assuming $M_1 = 0.7\ M_\odot$, $\mu_1 = 10^{31} - 10^{34}$ G cm^3, and $\dot{M}_2$ as shown in Figure 3.

The stream will be disrupted by the time it reaches this radius, but its trajectory may not be affected if the stream is Rayleigh-Taylor unstable, as seems likely. Moreover, when a system comes back into contact at $P_b \simeq 3$ hours, a brief stage of very high mass transfer ensues because the mass loss time scale for the envelope of the secondary is shorter than the thermal time scale τ_{KH} for the star as a whole (Webbink 1976, 1985). Thus, for a brief period $r_A^{(s)}$ is much smaller than its subsequent equilibrium value. We therefore expect that the stream will intersect itself and establish a disk in the case when the equilibrium value of $r_A^{(d)}$ lies in the narrow range (factor of ≈ 1.5) between ϖ_{min} and ϖ_{max}. Motivated by these considerations, we assume a disk always forms unless $\varpi_{max} < r_A^{(d)}$.

d) Accretion Torque

The accretion torque arises from the coupling of accreting matter to the magnetic field of the degenerate dwarf. It can be written as

$$N_{acc} \approx n(\omega_s)\,\dot{M}_1\,l_{acc}\,, \tag{4}$$

where $\dot{M}_1$ is the mass accretion rate onto the degenerate dwarf and l_{acc} is the angular momentum per unit mass of the accreting matter. The function $n(\omega_s)$ is a generalization of the dimensionless torque for disk accretion defined by Ghosh and Lamb (1979) and $\omega_s \equiv \Omega/\Omega_K(r_A^{(d)})$ is the fastness parameter, where $\Omega \equiv 2\pi/P$ is the angular frequency of rotation of the degenerate dwarf, and $\Omega_K(r_A^{(d)})$ is the Keplerian angular frequency at the inner edge of the accretion disk. When a disk is present, we use the $n(\omega_s)$ given by Ghosh and Lamb [1979: Equation (10)]; when a disk is absent, we use $n(\omega_s) = \Theta(1 - \omega_s)$, where Θ is the Heaviside function and now $\omega_s = \Omega/\Omega_K(r_A^{(s)})$.

In magnetic CV's, the value of l_{acc} varies as the size of the Alfvén radius for disk accretion changes relative to the sizes of R_1 and ϖ_{max}, where R_1 is the stellar radius of the degenerate dwarf. When $r_A^{(d)} < R_1$,

$$l_{acc} \approx (GM_1 R_1)^{1/2}\,. \tag{5}$$

When $R_1 < r_A^{(d)} < \varpi_{max}$,

$$l_{acc} \approx (GM_1 r_A^{(d)})^{1/2}. \tag{6}$$

In the absence of a disk ($\varpi_{max} < r_A^{(d)}$),

$$l_{acc} \approx (GM_1 \varpi_{max})^{1/2}\,, \tag{7}$$

independent of the radius at which matter couples onto the magnetic field of the degenerate dwarf, since the angular momentum per unit mass of the accreting matter is nearly the same as what it had when it left the secondary at r_{L1} (Coriolis forces modify it, but the change is not large for the component of angular momentum perpendicular to the orbital plane—see Campbell 1986).

The accretion torque is quite insensitive to the presence or absence of a disk:

$$\frac{N_{acc}(r_{L1})}{N_{acc}(R_1)} \lesssim \left(\frac{\varpi_{max}}{R_1}\right)^{1/2} = \left(\frac{10 \times 10^9 \text{cm}}{0.7 \times 10^9 \text{cm}}\right)^{1/2} \approx 4\,. \tag{8}$$

Thus, N_{acc} lies in a narrow band whose width is a factor of ≈ 4 at long orbital periods and a factor of $\lesssim 1.5$ near period minimum. This is illustrated in Figure 6, which shows the accretion torque as a function of binary orbital period for a 0.7 $M_\odot$ primary, assuming μ_1 in the range $10^{31} - 10^{34}$ G cm^3.

e) Synchronization Torque

Several synchronization torques based on magnetic coupling of the degenerate dwarf to the secondary have been proposed (Joss, Katz, and Rappaport 1979; Campbell 1983; Lamb, Aly, Cook, and Lamb 1983; for a critical review, see Lamb 1985). Here we adopt the MHD synchronization torque (Lamb, Aly, Cook, and Lamb 1983), which is given by

$$
N_{MHD} \approx \begin{cases} \alpha\gamma a R_2^2 \left(\dfrac{\mu_1}{a^3}\right)^2 & (\mu_2 = 0) \\ \gamma \dfrac{\mu_1\mu_2}{a^3} & (\mu_2 > \mu_1 > 0) \end{cases} ,
\tag{9}
$$

where the quantity α is the fractional area of the secondary threaded by the magnetic field of the degenerate dwarf (here assumed equal to the cross-sectional area of the convective region of the secondary), and γ is the pitch angle of the degenerate dwarf's dipole magnetic field at the secondary. In the MHD torque picture, synchronization is achieved by the coupling between the degenerate dwarf and the secondary star due to the field lines that thread the secondary. These field lines are not the same as those along which accretion occurs; thus the condition for synchronization is not related to the heuristic condition $r_A^{(0)} = a$.

Synchronization rather requires that two conditions be met (Lamb 1985). First, the synchronization torque must exceed the accretion torque if it is to spin down the degenerate dwarf to synchronous rotation; i.e.

$$
N_{MHD} > N_{acc} .
\tag{10}
$$

Equation (10) tells us that when synchronization occurs, the synchronization and spin-up times scales are equal $[\tau_{syn} = (I_1\Omega_1)/N_{MHD} = (I_1\Omega_1)/N_{acc} = \tau_{spin}]$. Second, the synchronization torque must be able to bring about synchronous rotation faster than the binary is evolving; i.e.

$$
\tau_S < \tau_{EVOL} ,
\tag{11}
$$

where we have set $\tau_S \equiv \tau_{syn} = \tau_{spin}$. Otherwise, the rotation period of the degenerate dwarf will not be able to "catch up" to that of the binary. Figure 4 shows the time scale τ_S at the epoch of synchronization. We see that condition (11) is always well-satisfied, except in the period gap. In the gap, Eq.(11) gives a lower bound on the μ_1 for which the degenerate dwarf synchronizes.

III. RESULTS

We have carried out calculations for a variety of cases. Here we describe the results of calculations which assume (1) disrupted magnetic braking model DMB4*, as defined in §II, (2) conservative mass transfer between novae outbursts (Taam, Flannery, and Faulkner 1980), (3) secondary magnetic moment $\mu_2 = 0$, (4) degenerate dwarf rotation period $P \approx P_{eq}$, and (5) aligned rotation of the degenerate dwarf. Occasionally, we shall refer to the results of our other calculations for comparison.

a) Criteria for Funneling of the Accretion Flow

The magnetic field of the degenerate dwarf funnels the flow of accreting matter near the stellar surface and produces pulsed optical, UV, and X-ray emission if $r_A^{(d)} \gtrsim 3R_1$. We find that this occurs for $\mu_1 \gtrsim 10^{31}$ G cm^3. Systems with smaller values of μ_1 are not DQ Her stars, and may exhibit little or no observational evidence of magnetic fields. These results agree qualitatively with those of King, Frank, and Ritter (1985).

b) Criteria for Existence of an Accretion Disk

Figure 7 shows the values of the magnetic moment μ_1 above which an accretion disk is absent as a function of orbital period P_b. Curves are shown for degenerate dwarf masses $M_1 = 0.4 - 1.4\ M_\odot$. We terminate the curves at long orbital periods P_b when the condition for stable mass transfer is violated; the solid curves correspond to conservative [i.e., $\beta = 1$ in the notation of Rappaport, Joss, and Webbink (1982)] mass transfer, while the dashed extensions correspond to fully non-conservative ($\beta = 0$) mass transfer. The former are relevant to observation, since we assume that the mass transfer is conservative between nova outbursts. Note that *the stability condition constrains the masses of degenerate dwarfs to be high in DQ Her systems with long orbital periods*. The disk criteria curves slope downward toward shorter orbital periods and through minimum period because the decreasing mass transfer rate $|\dot{M}_2|$ (see Figure 3) leads to an increasing Alfvén radius $r_A^{(d)}$ (see Figure 5). Whether a disk is present or absent during the period gap depends on whether the mass transfer rate is very high or very low then.

Figure 8 shows the μ_1 criteria for the existence of an accretion disk *divided by* $|\dot{M}_2|^{1/2}$. The solid and dashed portions of the curves have the same meaning as in Figure 7. Although the required μ_1-values cannot be read directly from this plot, it has the advantage that the curves are universal, in the sense that they are *independent* of the mass transfer rate. They are therefore independent of the evolutionary scenario and of the variations in $\dot{M}_2$ at a given P_b observed among different systems (see Figures 2 and 3).

Uncertainties in the theory of disk accretion lead to an uncertainty of a factor of 2 - 3 in the μ_1 criteria. Figures 7 and 8 show that the P_b at which a disk disappears is very sensitive to the value of μ_1 and M_1; the uncertainty in the μ_1 criteria and in the mass of the degenerate dwarf therefore leads to a much larger uncertainty in P_b.

c) Criteria for Synchronization

Figure 9 shows the magnetic moment μ_1 required for synchronization as a function of binary period P_b, while Figure 10 shows the μ_1 criteria for synchronization *divided by* $|\dot{M}_2|^{1/2}$. Curves are shown for degenerate dwarf masses $M_1 = 0.4 - 1.4\ M_\odot$, assuming $\mu_2 = 0$. When $\mu_2 \neq 0$, the μ_1 criteria become smaller longward of the period gap but are little changed shortward of the gap. The solid and dashed portions of the curves have the same meaning as in Figures 7 and 8. The curves again slope downward toward shorter orbital periods and through $P_{b,min}$ because of the decreasing mass transfer rate $|\dot{M}_2|$. The sharply lower value of $\dot{M}_2$ shortward of the period gap, which arises because the evolution is due only to GR, enhances the probability that systems undergo the transition from DQ Her star to AM Her star at $P_b \approx 2$ hours (Chanmugam and Ray 1984; King, Frank, and Ritter 1985; Lamb 1985).

Comparison of Figures 7 and 9, or 8 and 10, shows that no accretion disk is present when synchronization occurs. Since the accretion torque is then given by Equation (7), the sychronization criteria are independent of uncertainties in the theory of accretion from disks and streams. However, uncertainties in the theory of the MHD synchronization torque again lead to an uncertainty of a factor of 2 - 3 in the μ_1 criteria. Figures 9 and 10 show that the P_b at which the transition from DQ Her star to AM Her star occurs is also very sensitive to μ_1 and M_1. The uncertainty in the μ_1 criteria and in the mass of the degenerate dwarf therefore again leads to a much larger uncertainty in P_b.

Our results show that the μ_1 criterion given by the heuristic condition $r_A^{(0)} = a$ lies only about a factor of 3 lower than that given by the physical argument of balancing the accretion torque and the MHD synchronization torque, even though the two approaches lead to criteria with very different dependences on the physical variables. This agreement, which is comparable to the

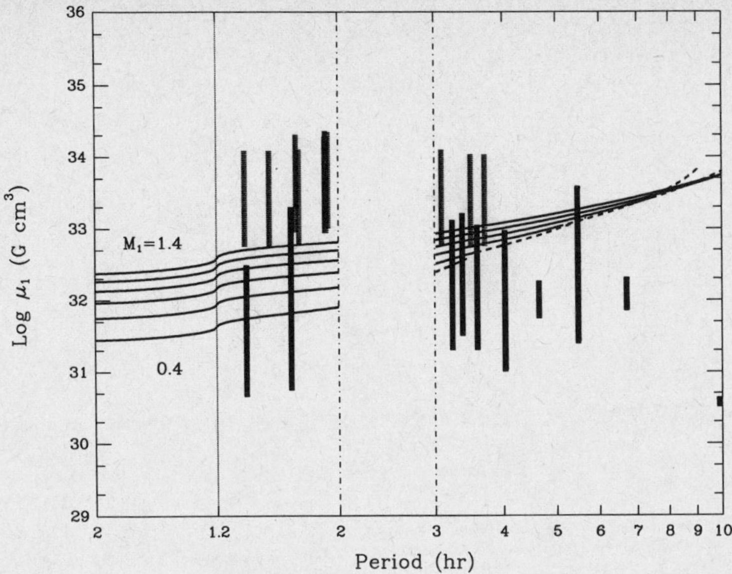

Fig. 7.–Maximum magnetic moment μ_1 for which a disk can exist as a function of P_b, assuming $M_1 = 0.4 - 1.4\ M_\odot$. The ranges of μ_1 for the DQ Her stars inferred from observation are shown as solid vertical bars while those for the AM Her stars are shown as shaded vertical bars.

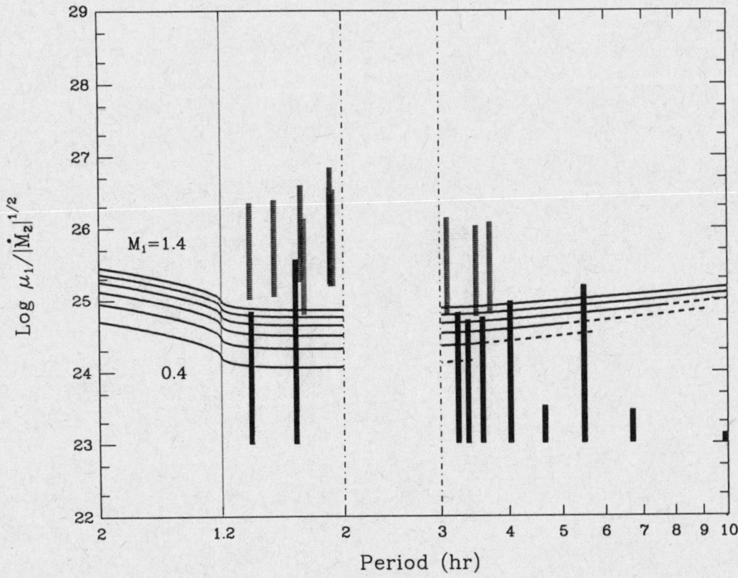

Fig. 8.–Maximum magnetic moment μ_1 for which a disk can exist *divided* by $|\dot{M}_2|^{1/2}$, assuming $M_1 = 0.4 - 1.4\ M_\odot$. The curves are universal in the sense that they are independent of $\dot{M}_2$. The notation for the ranges of μ_1 inferred from observation is the same as in Fig. 7.

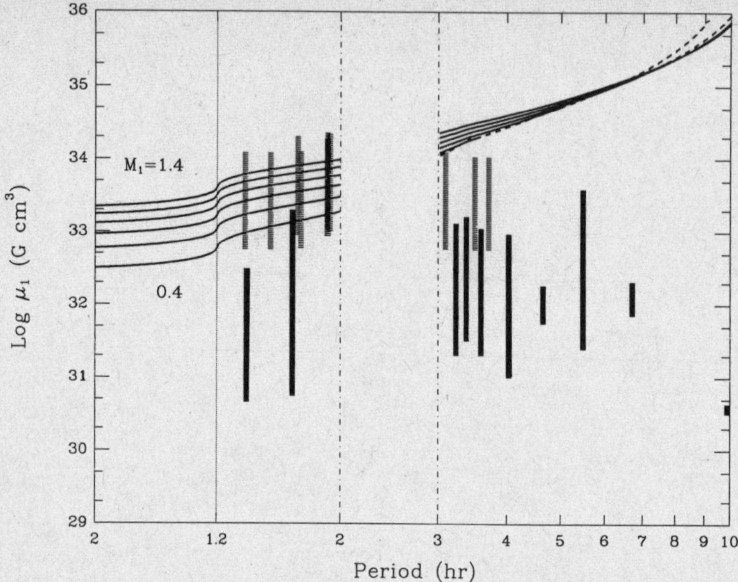

Fig. 9.–Minimum magnetic moment μ_1 required for synchronization as a function of P_b, assuming $M_1 = 0.4 - 1.4\ M_\odot$. The notation for the ranges of μ_1 inferred from observation is the same as in Fig. 7.

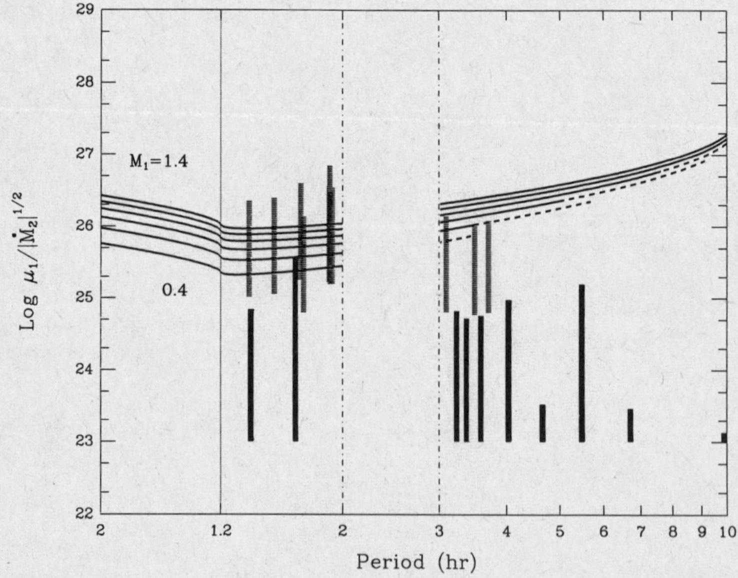

Fig. 10.–Minimum μ_1 required for synchronization *divided* by $|\dot{M}_2|^{1/2}$ as a function of P_b, assuming $M_1 = 0.4 - 1.4\ M_\odot$. These curves are universal in the sense that they are independent of $\dot{M}_2$. The notation for the ranges of μ_1 inferred from observations is the same as in Fig. 7.

uncertainties in the theory, is due partly to coincidence and partly to the fact that the length scales in the problem (e.g., ϖ_{max}, R_2, and a) are all of similar size.

Comparison of Figures 7 and 9 (or 8 and 10) allows us to identify four regimes: (1) In systems which have $\mu_1 \lesssim 10^{31}$ G cm^3, the magnetic field of the degenerate dwarf is unable to funnel the accretion flow near its surface. These systems therefore are not DQ Her stars, and may show little or no evidence of a magnetic field. (2) Systems which lie above this but below the disk criterion curve (10^{31} G cm$^3 \lesssim \mu_1 10^{33}$ G cm^3) are DQ Her stars with disks, while systems which lie above this but below the synchronization criterion curve are DQ Her stars without disks (King 1985; Hameury, King, and Lasota 1986). (3) Systems which lie in the band traversed by the synchronization criterion curve (10^{33} G cm$^3 \lesssim 10^{35}$ G cm^3) are DQ Her stars which evolve into AM Her stars (Chanmugam and Ray 1984; King, Frank, and Ritter 1985; Lamb 1985), assuming μ_1 is constant throughout their evolution (an assumption which may not be valid). (4) Systems which lie above this band are AM Her stars.

d) Synchronization-Induced Period Gap

Synchronization in magnetic CV's leads to a new kind of period gap and gives an entirely new way of producing ultra-short period binaries. A gap arises because the angular momentum J_1 residing in the degenerate dwarf is injected directly into the binary system by the MHD torque when synchronization occurs. This drives the system apart, bringing the secondary out of contact and ending mass transfer. The radius R_2 of the secondary then shrinks back toward its main sequence value $R_{2,MS}$. The secondary is brought back into contact by angular momentum loss due to GR. Since R_2 can be 20 - 50 % larger than $R_{2,MS}$ for some evolutionary scenarios (see, e.g., Verbunt 1984), the resulting period gap can be far larger than the gap $\tau_{GAP}^{(0)}$ which would result from the injection of J_1 alone:

$$\tau_{GAP}^{(0)} \equiv \frac{J_1}{\dot{J}_b} \approx \frac{I_1 \Omega_{eq}}{\dot{J}_b} . \tag{12}$$

It is for this reason that we use the term "synchronization-induced period gap." For the particular evolutionary model discussed here (DMB4*), $\tau_{KH} \approx \tau_{EVOL}$ and the secondary star is only modestly out of thermal equilibrium. Therefore the actual duration τ_{GAP} of the gap is only somewhat greater than $\tau_{GAP}^{(0)}$.

If synchronization occurs during the usual $2 - 3$ hour period gap, as may often happen (see Figures 9, 10 and 13), it only lengthens the duration of the gap. When $P_b \approx 1.5$ hours, the mass of the secondary is approximately $0.086 M_\odot$. As P_b decreases further, the secondary leaves the main sequence (no equilibrium configuration balancing thermal energy generated by nuclear burning and gravitational energy exists). Just after minimum period (≈ 1.2 hours), the secondary reaches the degenerate sequence. If synchronization occurs between these two orbital periods, the secondary shrinks all the way to $R_{2,DS} << R_2$, where $R_{2,DS}$ is its radius on the degenerate sequence, before contact is resumed. This produces a synchronization-induced period gap whose duration can approach the evolutionary time scale τ_{EVOL} for the system. This possibility is realized if $\tau_{KH} < \tau_{EVOL}$; otherwise, the secondary shrinks little, and $\tau_{GAP} \approx \tau_{GAP}^{(0)}$. For the particular evolutionary model discussed here (DMB4*), $\tau_{KH} > \tau_{EVOL}$ from $P_b \approx 1.2 - 1.7$ hours and the duration of the synchronization-induced gap is suppressed in most of this interval except near $P_{b,min}$. These features are shown in Figure 11, which compares τ_{GAP}, τ_{EVOL}, and τ_{KH}, respectively.

These same features are evident in Figure 12, which shows the orbital period $P_{b,begin}$ at the beginning and the orbital period $P_{b,end}$ at the end of the synchronization-induced period gap.

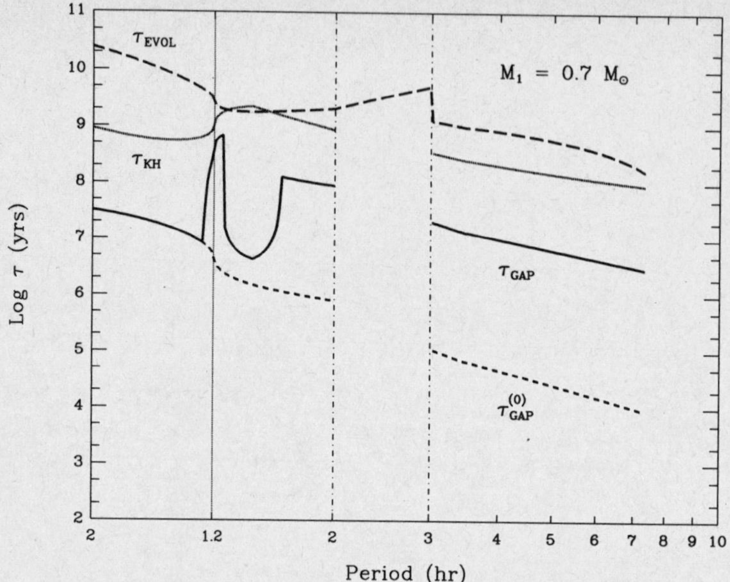

Fig. 11.–Duration τ_{GAP} of the synchronization-induced period gap compared with the evolutionary time scale τ_{EVOL}, the Kelvin-Helmholtz thermal time scale τ_{KH}, and the duration $\tau_{GAP}^{(0)}$ that would result from the injection of J_1 alone [cf. Eq. (12)], for a system containing a 0.7 $M_\odot$ degenerate dwarf.

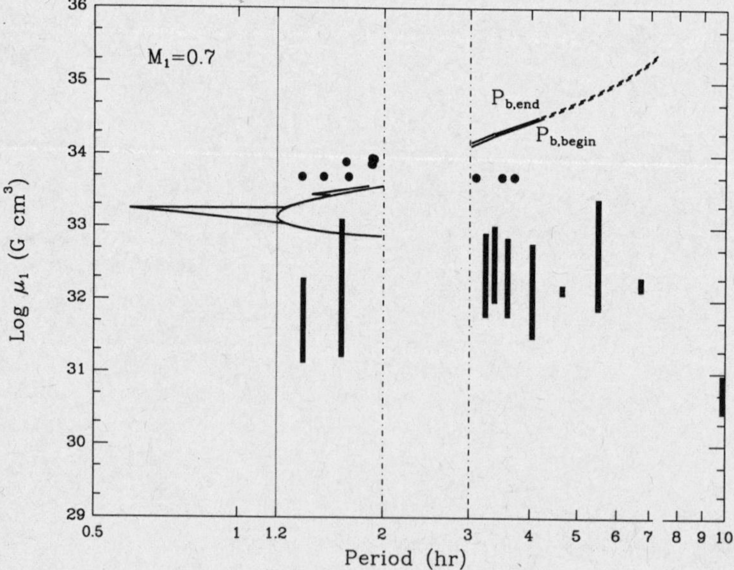

Fig. 12.–Synchronization-induced period gap as a function of magnetic moment μ_1 for a 0.7 $M_\odot$ primary. The rightward curve shows the binary orbital period $P_{b,begin}$ at which the system goes out of contact (this curve is identical to the one for the minimum magnetic moment required for synchronization; see Figure 10), and the leftward curve shows the orbital period $P_{b,end}$ at which the system comes back into contact. The large gap near minimum period arises because the secondary star lies below the main sequence and contracts all the way to the degenerate sequence when synchronization brings it out of contact.

In this figure, we use a single-valued orbital period axis in order to show $P_{b,end}$; systems thus evolve toward the left prior to period minimum and toward the right afterward. The curve of $P_{b,begin}$ is identical to those in Figure 9, but is now shown for a 0.7 $M_\odot$ primary.

Before the period gap and after period minimum, the secondary is very close to thermal equilibrium and $P_{b,end}$ lies close to $P_{b,begin}$. Between $P_b \approx 1.5$ hours and $P_{b,min} \approx 1.2$ hours, the secondary lies off the main sequence and has not yet reached the degenerate sequence. If synchronization occurs during this interval, there is the possibility that $P_{b,end} << P_{b,begin}$. In the particular evolutionary scenario DMB4*, much of this behavior is suppressed because $\tau_{KH} > \tau_{EVOL}$ then, as discussed above. Nevertheless, if $\mu_1 \approx 1 - 4 \times 10^{33}$ G cm^3, synchronization leads to orbital periods as short as $P_b \approx 40$ minutes. This process is an entirely new way of producing ultra-short period binaries. The secondaries in these binaries are hydrogen-rich rather than hydrogen-depleted or pure helium (*cf.* Nelson, Rappaport, and Joss 1985; Lamb *et al.* 1986).

IV. DISCUSSION

a) Magnetic Moments μ_1 from Observation

i) DQ Her stars

The optical spectra of DQ Her stars are dominated by continuum and line emission from the disk, and to date no absorption lines have been detected from the photosphere of the degenerate dwarf in these systems. The optical light shows little or no polarization (Kemp, Swedlund, and Wolstencroft 1974; Penning, Schmidt, and Liebert 1985) and no evidence of emission or absorption at cyclotron harmonics. Thus the magnetic field of the degenerate dwarf must be inferred by indirect means.

Lamb and Patterson (1983) used the spin behavior of the known DQ Her stars to place bounds on the μ_1's of these stars, assuming the existence of a disk. They used the spin-up rate $\dot{P}$ of these stars to estimate μ_1. Unfortunately, measurements of $\dot{P}$ are available for only a few DQ Her stars (although the situation is changing), and we do not use this method here.

Accretion is not expected to occur if the centrifugal force exceeds the gravitational force at the Alfvén radius (Lamb, Pethick, and Pines 1973). This leads to the condition (Ghosh and Lamb 1979)

$$\omega_s \equiv \Omega/\Omega_K(r_A^{(d)}) \lesssim 0.35 . \tag{13}$$

Inequality (13) gives an estimate of μ_1 only if $P \approx P_{eq}$. DQ Her stars are often assumed to satisfy this approximate equality (Chanmugam and Ray 1984; King, Frank, and Ritter 1985; King 1985; Hameury, King, and Lasota 1986) because the spin-up time scale is far shorter than the evolutionary time scale, *i.e.* $\tau_{spin} << \tau_{EVOL}$ (see Figure 3). However, P may be $> P_{eq}$ for the following reason. The accretion torque $N_{acc} \approx N_o$ when $\omega_s << 1$, where $N_o = \dot{M}_1(GM_1R_1)^{1/2}$ is the nominal value due to the matter stresses tending to spin up the star (Lamb, Pethick, and Pines 1973). However, when $\omega_s \approx 1$, the magnetic stresses tending to spin down the star exceed the matter stresses and the accretion torque becomes large and negative, $N_{acc} \approx -20N_o$ (Ghosh and Lamb 1979). Because of the disparity between the magnitudes of N_{acc} above and below the equilibrium spin period $P_{eq} \equiv 2\pi/\Omega_K(r_A^{(d)})$, P can exceed P_{eq} significantly if the accretion rate $\dot{M}_1$ varies appreciably and the variation lasts a time comparable to τ_{spin} (*cf.* Elsner, Ghosh, and Lamb 1980). The larger the variations in $\dot{M}_1$ and the longer they last, the more pronounced the discrepancy can be.

When $P > P_{eq}$, the inequality (13) gives only an *upper bound* on μ_1, as emphasized by Lamb and Patterson (1983). This upper bound is accurate only to within a factor $2 - 3$ because of uncertainties in the theory of disk accretion. Lamb and Patterson (1983) used $\omega_s = 0.35$ (Ghosh and Lamb 1979) and assumed a mass $M_1 = 1\ M_\odot$. These values yield a conservative upper bound $\mu_1^{(max)}$, since lower masses give lower values of $\mu_1^{(max)}$ ($\sim M_1^{5/6}$). Here we adopt $\omega_s = 1$ and assume a mass $M_1 = 1.2\ M_\odot$, which gives a very conservative upper bound. The resulting estimates of $\mu_1^{(max)}$ for the known DQ Her stars range from 3×10^{32} to 2×10^{33} G cm^3, depending on the value of $\dot{M}_2$ and P.

Lamb and Patterson (1983) also derived a *lower bound* on μ_1 using the fact that the DQ Her stars show pulsed optical and X-ray emission, indicating that the magnetic field of the degenerate dwarf is strong enough to channel the flow of accreting matter near the stellar surface. This requires

$$r_A^{(d)} \gtrsim 3R_1 \ . \tag{14}$$

The resulting lower bound $\mu_1^{(min)}$ depends only weakly on the mass of the degenerate dwarf ($\sim M_1^{1/4} R_1^{7/4} \sim M_1^{-1/3}$), but inequality (14) is only a dimensional argument, and is therefore uncertain by a factor of $2 - 3$. The values of $\mu_1^{(min)}$ for the known DQ Her stars range from 3×10^{31} to 2×10^{32} G cm^3, depending on the value of $\dot{M}_2$.

Recently, King (1985) and Hameury, King, and Lasota (1986) have raised the intriguing possibility that disks are absent in the known DQ Her stars with $P_b \lesssim 5$ hours. The observational bounds on μ_1 allow a self-consistent picture to be constructed for either case, given present uncertainties in the theory of accretion from a disk or stream onto a magnetic star. Thus the issue must be settled by other observations. There is, in fact, other observational evidence that accretion disks exist in these systems (see below). However, if no disk is present, $\mu_1^{(max)}$ could be a factor $\lesssim 2$ larger than the value we have adopted (see below).

ii) AM Her stars

The magnetic fields of seven of the twelve known AM Her stars have been determined directly, either by measuring the Zeeman splitting of absorption lines from the photosphere of the degenerate dwarf during a low state or by measuring the frequencies of cyclotron emission features and deducing the frequency of the cyclotron fundamental (Schmidt, Stockman, and Grandi 1986; Schmidt and Liebert 1986). The magnetic fields all lie in the range $\approx 1 - 3 \times 10^7$ G.

Because μ_1 is the natural physical variable in the problems of interest to us here, we convert the AM Her magnetic field determinations to μ_1 determinations. Because observational estimates of the degenerate dwarf mass in AM Her stars are insecure, we take a range $\approx 0.4 - 1.2\ M_\odot$ for each. This leads to magnetic moments of the AM Her stars in the range $\mu_1 \approx 1 \times 10^{33} - 1 \times 10^{34}$ G cm^3.

b) Comparison of Magnetic Moments

i) DQ Her and AM Her stars

Figures 13 and 14 compare the distribution of magnetic moments μ_1 of the 12 known AM Her stars and of the 10 known DQ Her stars, assuming $0.7\ M_\odot \lesssim M_1 \lesssim 0.9\ M_\odot$.

Figure 13 assumes that disks exist in the DQ Her systems, and uses the $\mu_1^{(min)}$ and $\mu_1^{(max)}$ derived from inequalities (13) and (14) with $r_A^{(d)}$. According to the figure, the μ_1 distributions of the DQ Her and AM Her stars differ significantly. The centroid of the DQ Her distribution lies at $\sim 10^{32}$ G cm^3, while the centroid of the AM Her distribution lies at $\sim 4 \times 10^{33}$ G cm^3. There is no overlap in the distributions.

Figure 14 assumes that no disks exist in the DQ Her systems (with the exceptions of AE Aqr, DQ Her, and V533 Her, for which no self-consistent solution without a disk exists), and

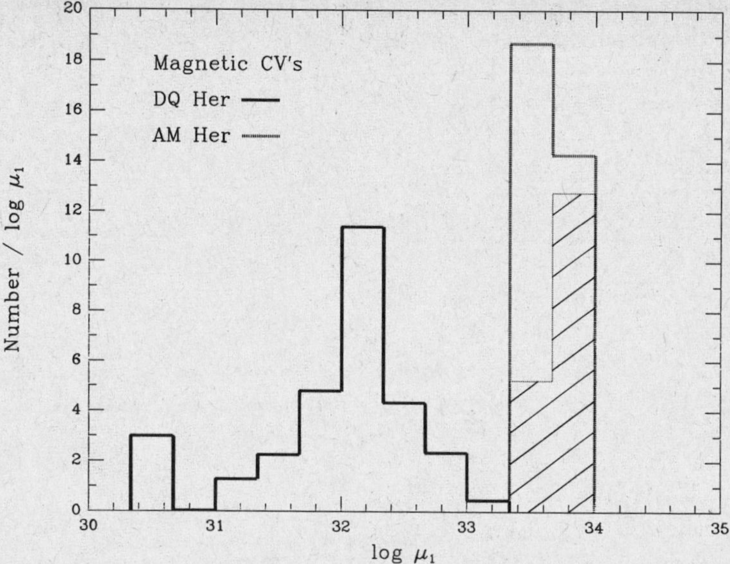

Fig. 13.–Distribution of the magnetic moments μ_1 of the 12 known AM Her stars and of 10 of the known DQ Her stars, assuming disks exist in the latter. The vertical axis gives the number per unit $\log \mu_1$, so that the total number of binaries within a given interval of $\log \mu_1$ is simply the area within that interval. The AM Her stars with directly measured magnetic field values are shown shaded.

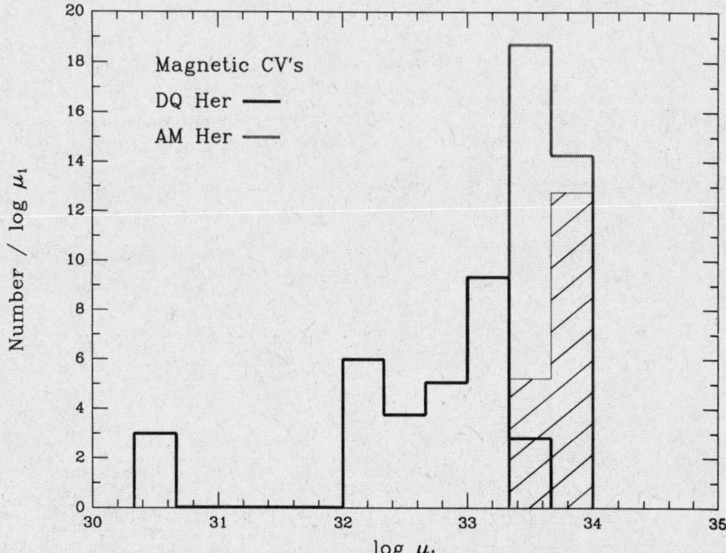

Fig. 14.–Distribution of the magnetic moments μ_1 of the 12 known AM Her stars and of 10 of the known DQ Her stars, assuming no disks exist in the latter (except for AE Aqr, DQ Her, and V533 Her, for which no self-consistent solutions without a disk exist). The AM Her stars with directly measured magnetic field values are shown shaded. The units are the same as in Figure 13.

uses the bounds derived from inequalities (13) and (14) with $r_A^{(s)}$ [Lamb 1985, Equation (20); see also Hameury, King, and Lasota (1986)]. According to the figure, the μ_1 distributions of the two subclasses still differ. The centroid of the DQ Her distribution lies at $\sim 6 \times 10^{32}$ G cm^3, while the centroid of the AM Her distribution is the same as before. There is hardly any overlap in the distributions, but the DQ Her distribution now nestles up against the AM Her distribution.

Since $\mu_1^{(min)} \sim M_1^{-1/3}$ and $\mu_1^{(max)} \sim M_1^{5/6}$, the range in μ_1 for individual DQ Her stars is determined solely by the highest assumed mass. The higher the mass, the broader the range in μ_1. Conversely, the higher the mass assumed for the AM Her stars, the lower the μ_1, since $\mu_1 \sim R_1^3 \sim M_1^{-1}$. Thus the two distributions are more disjoint for smaller values of M_1, and less so for larger values. We find that the two distributions overlap significantly only if $M_1 \gtrsim 1.2\ M_\odot$, assuming disks exist in the DQ Her systems, and only if $M_1 \gtrsim 1.0\ M_\odot$, assuming no disks exist (with the exceptions of AE Aqr, DQ Her, and V533 Her). Thus it is likely that the μ_1's of the known DQ Her stars (including those with longer rotation periods) are significantly less than those of the known AM Her stars, unless all members of the two subclasses have $M_1 \gtrsim 1\ M_\odot$.

This differs from the conclusion reached by King (1985), and Hameury, King, and Lasota (1986) who suggest that the magnetic moments of the AM Her and the DQ Her stars are the same. Their view is motivated by the fact that most (8 of 10) of the DQ Her stars lie longward of the period gap, while most (9 of 12) of the AM Her stars lie shortward of the gap. They note that many more DQ Her stars would be expected shortward of the gap, and conclude that the known AM Her stars have come from systems like the known DQ Her stars.

However, the period distributions of the DQ Her and AM Her stars may be distorted by observational selection effects. For example, the number of AM Her stars shortward of the gap may be enhanced by the high interest in them and by the ease with which their classification can be confirmed by optical polarimetry. As a second example, DQ Her stars shortward of the period gap have rotation periods similar to their orbital periods; such periods are more difficult to detect and confirm. There may also be theoretical reasons for the disparate observed distributions of the two subclasses. For example, longward of the period gap, only low mass $(M_1 \lesssim 0.4 - 0.6\ M_\odot)$ degenerate dwarfs are more likely to synchronize, even if the secondary has a strong $(B_2 \sim 10^2 - 10^3$ G) magnetic field (Lamb and Melia 1986). Since stable mass transfer occurs in such systems only for $P_b \lesssim 3 - 3\ 1/2$ hours (see Figures 7 $-$ 10), very few AM Her stars may exist longward of the period gap. As another example, the μ_1's of magnetic CV's may increase with time if nova explosions progressively expose an internal fossil field (Chanmugam and Gabriel 1972, Lamb 1974), leading to more AM Her stars at short orbital periods.

ii) Magnetic CV's and isolated magnetic degenerate dwarfs

About 2 % of isolated degenerate dwarfs have magnetic fields $B_1 \gtrsim 10^6$ G (Angel, Borra, and Landstreet 1981; Schmidt and Liebert 1986). The number of such stars per logarithmic interval in magnetic field strength is consistent with a uniform distribution from $B_1 \approx 10^6 - 5 \times 10^9$ G, with ≈ 0.5 % per decade (Angel, Borra, and Landstreet 1981; Schmidt and Liebert 1986). There is *no* evidence that the distribution is bimodal, as has been claimed (King 1985).

The difference between the frequency of occurrence of strongly magnetic degenerate dwarfs in CV's and in the field is striking. This may indicate that something about the formation and evolution of degenerate dwarfs in close binaries is favorable to the creation of strong surface magnetic fields. Important differences in the distribution of magnetic field values in the two groups may exist, but unknown observational selection effects and low statistics prevent a meaningful comparison at present.

c) Existence of Accretion Disks

Figures 7 and 8 compare the magnetic moments μ_1 of the magnetic CV's with the criteria for the existence of an accretion disk in the system. The DQ Her stars are shown as solid bars and the AM Her stars as shaded bars. Figure 8, which is independent of the evolutionary model and variations in $\dot{M}_2$ among the individual stars, shows that the three AM Her stars longward of the period gap are expected to have no accretion disks if $M_1 \lesssim 0.9\ M_\odot$, while those shortward of the gap are expected to have no disks for any mass. These results are in good agreement with observations, which indicate that these systems have no accretion disks.

Figure 8 does not allow us to decide whether or not the known DQ Her stars have disks because of the large uncertainties in the actual values of μ_1 for these stars. However, there is strong observational evidence that accretion disks exist in these systems, which includes the following (Warner 1983, 1985 and references therein; Penning 1985):

- A power-law spectrum in the optical and UV light from the system like that expected for emission from a disk.
- Doubled emission lines which reflect the circular velocity of material in the outer disk where the lines originate, and which do not cross back and forth in wavelength with period P_b, as would be expected for accreting matter streaming toward the degenerate dwarf primary (as is seen, *e.g.* in the AM Her stars and in SS433).
- Broad emission line wings, with velocities frequently extending up to 2000 km s^{-1}, reflecting the circular velocities in the inner portion of the disk.
- Broad emission lines phased with pulse phase, due to light from reprocessing of pulsed X-ray emission by the inner edge and surface of the disk.
- Dwarf nova outbursts in EX Hya and SW UMa, the two DQ Her stars with orbital periods below the period gap, implying (within the framework of the disk instability model) disks in these systems.

The existence of disks in these systems indicates that $P << P_{eq}$ in many of them, and that the observed values of $|\dot{M}_2|$ and $\dot{P}$ do not reflect their long-term evolutionary values, as other evidence also indicates. It also implies that the μ_1's of the DQ Her stars are significantly less than the upper bounds derived from their spin behavior. This increases the likelihood that the DQ Her stars have smaller μ_1's than do the AM Her stars (compare Figures 13 and 14).

Recently, Tuohy *et al.*(1986) discovered a cataclysmic variable, H0542-41, which appears to show X-ray emission at the sideband period $P_{side} = (P^{-1} - P_b^{-1})^{-1} \approx 2050$ seconds as well as at $P \approx 1920$ and $P_b \approx 6.5$ hours. Lamb and Mason (1986) have suggested that this system may be a DQ Her star without an accretion disk, and that the modulation of the X-ray emission at P_{side} may be due to magnetospheric gating of the accretion stream. If this picture is verified, the system may teach us a great deal about the appearance of DQ Her stars without accretion disks.

d) Synchronization

Figures 9 and 10 compare the magnetic moments μ_1 of the magnetic CV's with the criteria for synchronization. Again, the DQ Her stars are shown as solid bars and the AM Her stars as shaded bars. Two of the three AM Her stars longward of the gap are not expected to be synchronized and the third (AM Her) is expected to be synchronized only if the degenerate dwarf mass $M \lesssim 0.4\ M_\odot$, assuming $\mu_2 = 0$. However, all three systems may be synchronized if $\mu_2 \sim 10^{33} - 10^{34}$ ($B_2 \sim 100 - 1000$ G) (Lamb and Melia 1986). The AM Her stars with orbital periods shortward of the gap are expected to be synchronized if they have moderate masses $M_1 \lesssim 0.7\ M_\odot$. Observations indicate that several of these systems are synchronously rotating, or nearly so; thus the criterion for synchronization is consistent with the behavior of the AM Her stars. The requirement $\mu_2 = 0$ for synchronous rotation longward of the period gap indicates that the secondary possess a significant magnetic field, as other evidence also indicates (see, e.g, Chanmugam and Dulk 1982).

Figures 9 and 10 show that, even taking into account $\mu_2 \neq 0$ and the theoretical uncertainty the synchronizatin criteria, it appears that the known DQ Her stars will not evolve into AM Her stars. This differs from the conclusion of Chanmugam and Ray (1984) and King, Frank, and Ritter (1986).

V. CONCLUSIONS

Our conclusions are as follows. We have identified four regimes: (1) When $\mu_1 \lesssim 10^{31}$ G cm^3, the magnetic field of the degenerate dwarf is unable to funnel the accretion flow; these systems are not DQ Her stars and may show little or no evidence of a magnetic field. (2) Systems with 10^{31} G cm$^3 \lesssim \mu_1 10^{33}$ G cm^3 are always DQ Her stars. (3) Systems with 10^{33} G cm$^3 \lesssim \mu_1 \lesssim 10^{35}$ G cm^3 are DQ Her stars which evolve into AM Her stars, assuming μ_1 is constant throughout their evolution (an assumption which may not be valid). These three regimes agree qualitatively with those of King, Frank, and Ritter (1985). (4) Systems with 10^{35} G cm$^3 \lesssim \mu_1$ are always AM Her stars.

We have confirmed that AM Her stars do not have accretion disks, while DQ Her stars may or may not have them, depending on the value of μ_1, as pointed out by King (1985) and Hameury, King, and Lasota (1986). The ranges of μ_1 for the known DQ Her and AM Her stars allowed by observation are very large, due to uncertainties in the theory of disk and stream accretion, and to the dependence of the ranges on the mass M_1 of the degenerate dwarf. Nevertheless, it is likely that the μ_1's of the known DQ Her stars (including those with longer rotation periods) are significantly less than those of the known AM Her stars, unless all known DQ Her and AM Her stars have $M_1 \gtrsim 1 \, M_\odot$. This differs from the conclusion reached by King (1985) and King, Frank, and Ritter (1985). The criteria for synchronization and for the presence or absence of a disk in DQ Her systems are also uncertain, due to uncertainties in the theory of the MHD synchronization torque and the dependence of the criteria on M_1. Nevertheless, it appears that the known DQ Her stars will not evolve into AM Her stars, assuming that μ_1 is constant throughout their evolution. This differs from the conclusion reached by Chamugam and Ray (1984), King (1985), and King, Frank, and Ritter (1985).

The fact that the strong magnetic field of the degenerate dwarf dramatically alters the optical, UV, and X-ray appearance of magnetic CV's has come to be appreciated. That it can also alter the evolution of the binary itself is only now becoming recognized. Spin-up and spin-down of the magnetic degenerate dwarf temporarily speeds up and slows down the binary evolution of DQ Her stars (Ritter 1985; Lamb and Melia 1986). Co-operative magnetic braking may speed up the binary evolution of AM Her stars when the degenerate dwarf has a particularly strong $(B_1 \gtrsim 5 \times 10^7$ G) magnetic field (Schmidt, Stockman, and Grandi 1986). Here we have pointed out that synchronization of the magnetic degenerate dwarf injects angular momentum into the binary, driving the system apart and producing a *synchronization-induced period gap* (Lamb and Melia 1986). This process yields an entirely new way of producing ultra-short period binaries. The secondaries in these binaries are hydrogen-rich rather than hydrogen-depleted or pure helium.

We thank Fred Lamb for valuable discussions of the physics of the accretion flow in the magnetospheres of compact objects, and Saul Rappaport, Lorne Nelson, Frank Verbunt, and Ron Webbink for illuminating discussions of the physics of close binary evolution. This research was supported in part by NASA Grants NAGW-830, NAG8520, and NAG8563.

REFERENCES

Angel, J. R. P., Borra, E. F., and Landstreet, J. D. 1981, *Ap. J.*, **45**, 457.

Campbell, C. G. 1983, *M.N.R.A.S.*, **205**, 1031.

Campbell, C. G. 1986, submitted to *M.N.R.A.S.*.

Chanmugam, G., and Dulk, G. A. 1982, *Ap. J. (Letters)*, **255**, L107.

Chanmugam, G., and Gabriel, M. 1972, *Astr. Ap.*, **16**, 149.

Chanmugam, G., and Ray, A. 1984, *Ap. J.*, **285**, 252.

Elsner, R. F., Ghosh, P., and Lamb, F. K. 1980, *Ap. J. (Letters)*, **241**, L155.

Faulkner, J. 1971, *Ap. J. (Letters)*, **170**, L99.

Faulkner, J., Flannery, B, and Warner, B. 1972, *Ap. J. (Letters)*, **175**, L79.

Ghosh, P., and Lamb, F. K. 1979, *Ap. J.*, **232**, 259.

Hameury, J.-M., King, A. R., and Lasota, J. P. 1986, *M.N.R.A.S.*, **218**, 695.

Joss, P. C., Katz, J. I., and Rappaport, S. A. 1979, *Ap. J.*, **230**, 176.

Kemp, J. C., Swedlund, J. B., and Wolstencroft, R. D. 1974, *Ap. J. (Letters)*, **193**, L15.

King, A. R. 1983, in *Cataclysmic Variables and Related Objects*, IAU Colloquium No. 72, ed. M. Livio, and G. Shaviv (Reidel: Dordrecht), p. 181.

King, A. R. 1985, in *Recent Results on Cataclysmic Variables*, Proceedings of ESA Workshop, Bamberg, Germany, 17-19 April 1985, ed. W. R. Burke (ESA SP-236), p.133.

King, A. R., Frank, J., and Ritter, H. 1985, *M.N.R.A.S.*, **213**, 181.

Kraft, R. P., Mathews, J., and Greenstein, J. L. 1962, *Ap. J.*, **136**, 312.

Lamb, D. Q. 1974, *Ap. J. (Letters)*, **192**, L129.

Lamb, D. Q. 1979, in *Compact Galactic X-Ray Sources*, ed. F. K. Lamb and D. Pines (Department of Physics, University of Illinois: Urbana, Illinois), p. 27.

Lamb, D. Q. 1981, in *X-Ray Astronomy in the 1980's*, ed. S. Holt (NASA TM-83848), p. 37.

Lamb, D. Q. 1983, in *Cataclysmic Variables and Related Objects*, IAU Colloquium No. 72, ed. M. Livio and G. Shaviv (Reidel: Dordrecht), p. 299.

Lamb, D. Q. 1985, in *Cataclysmic Variables and Low-Mass X-Ray Sources*, ed. D. Q. Lamb and J. Patterson (Reidel: Dordrecht), p. 179.

Lamb, D. Q., and Mason, K. O. 1986, in preparation.

Lamb, D. Q., Mason, K. O., Cropper, M., and Patterson, J. 1986, submitted to *M.N.R.A.S.*

Lamb, D. Q., and Melia, F. 1986, submitted to *Ap. J.*

Lamb, D. Q., and Patterson, J. 1983, in *Cataclysmic Variables and Related Objects*, IAU Colloquium No. 72, ed. M. Livio and G. Shaviv (Reidel: Dordrecht), p. 229.

Lamb, F. K., Pethick, C. J., and Pines, D. 1973, *Ap. J.*, **184**, 271.

Lamb, F. K., Aly, J.-J., Cook, M., and Lamb, D. Q. 1983, *Ap. J. (Letters)*, **274**, L71.

Liebert, J., and Stockman, H. S. 1985, in *Cataclysmic Variables and Low-Mass X-Ray Sources*, ed. D. Q. Lamb and J. Patterson (Reidel: Dordrecht), p. 151.

Nelson, L., Rappaport, S., and Joss, P. C. 1985, submitted to *Ap. J.*.

Paczyński, B. 1967, *Acta Astra.*. **17**, 287.

Paczyński, B. 1981, *Acta Astr.*, **31**, 1.

Paczyński, B., and Sienkiewicz, R. 1981, *Ap. J. (Letters)*, **248**, L27.

Patterson, J. 1984, *Ap. J. Suppl.*, **54**, 443-493.

Penning, W. R. 1985, *Ap. J.*, **289**, 300.

Penning, W. R., Schmidt, G. D., and Liebert, J. 1986, *Ap. J.*, in press.

Rappaport, S., Joss, P. C., and Webbink, R. F. 1982, *Ap. J.*, **254**, 616.

Rappaport, S., Verbunt, F., and Joss, P. C. 1983, *Ap. J.*, **275**, 713.

Robinson, E. L., Barker, E. S., Cochran, A. L., Cochran, W. D., and Nather, R. E. 1981, *Ap. J.*, **251**, 611.

Ritter, H. 1985, *Catalogue of Catacysmic Binaries, Low-Mass X-Ray Binaries and Related Objects*, 3rd ed., *Astr. Ap. Suppl.*, **57**, 385.

Ritter, H. 1985, *Astr. Ap.*, **148**, 207.

Schmidt, G. D., and Liebert, J. 1986, in *Cataclysmic Variables*, Proceedings IAU Coll. No. 93, ed. H. Drechsel, Y. Kondo and J. Rahe (Reidel: Dordrecht).

Schmidt, G. D., Stockman, H. S., and Grandi, S. 1986, *Ap. J.*, **300**, 804.

Spruit, H. C., and Ritter, H. 1983, *Astr. Ap.*, **124**, 267.

Taam, R. E. 1983a, *Ap. J.*, **268**, 361.

Taam, R. E. 1983b, *Ap. J.*, **270**, 694.

Taam, R. E., Flannery, B., and Faulkner, J. 1980, *Ap. J.*, **239**, 1017.

Tuohy, I. R., Buckley, D. A. H., Remillard, R. A., Bradt, H. V., and Schwartz, D. A. 1986, submitted to *Ap. J.*.

Verbunt, F. 1984, *M.N.R.A.S.*, **209**, 277.

Verbunt, F., and Zwaan, C. 1981, *Astr. Ap.*, **100**, L7.

Warner, B. 1983, in *Cataclysmic Variables and Related Objects*, IAU Colloquium No. 72, ed. M. Livio and G. Shaviv (Reidel: Dordrecht), p. 155.

Warner, B. 1985, in *Cataclysmic Variables and Low-Mass X-Ray Binaries*, ed. D. Q. Lamb and J. Patterson (Reidel: Dordrecht), p. 269.

Webbink, R. F. 1976, *Ap. J. Suppl.*, **32**, 583.

Webbink, R. F. 1985, in *Interacting Binary Stars*, ed. J. E. Pringle and R. A. Wade (Cambridge Univ. Press: Cambridge), p. 39.

Webbink, R. F., Rappaport, S. A., and Savonije, G. J. 1983, *Ap. J.*, **270**, 678.

Whyte, C. A., and Eggleton, P. P. 1980, *M.N.R.A.S.*, **190**, 801.

Zahn, J. P. 1977, *Astr. Ap.*, **57**, 383.

<u>THE EVOLUTION OF MAGNETIC CATACLYSMIC VARIABLES</u>

A.R. King
Astronomy Department
University of Leicester
Leicester LE1 7RH, UK

<u>Summary</u>

The evolution of magnetic cataclysmic variables is reviewed. A simple interpretation of the period distribution of these systems is found possible if their magnetic moments lie in the narrow range $10^{33} - 10^{34}$ Gcm^3: systems at long orbital periods are likely to be asynchronous (intermediate polars) and those at short periods to be synchronous (AM Hers). The origins of this field distribution, the inhibition of accretion disc formation, and the synchronization processes are discussed.

1. Introduction

Cataclysmic variables (CV's) are close binary systems (periods $P \sim$ few hours) in which a white dwarf accretes from a low-mass companion star. If the white dwarf possesses a sufficiently strong magnetic field, the accretion flow will be channelled on to restricted regions of its surface near the magnetic poles, causing much of the accretion luminosity, particularly that in X-rays, to be modulated at the spin period P_{spin} of the white dwarf. Observation naturally divides the ~ 20 known magnetic CV's into two groups, depending on whether $P_{spin} = P$ or $P_{spin} < P$. The synchronously spinning ($P_{spin} = P$) systems are known as *AM Herculis stars*, or 'polars'; direct observational evidence (see below) shows that their polar field strengths are all in the range $\sim (1 - 3) \times 10^7$G, while for the non-synchronous ($P_{spin} < P$) systems (*intermediate polars*, or DQ Her stars) such direct estimates are much harder to come by.

The central question in discussing the evolution of these systems is the distribution of field strengths in the intermediate polars. Two distinct suggestions have been put forward.

(i) *The Weak Field Hypothesis* (Lamb and Patterson, 1983):
the intermediate polars are supposed to have fields systematically weaker (i.e. ~ 10^5, 10^6G) than the AM Hers; the two classes of objects are entirely distinct.

(ii) *The Strong Field Hypothesis* (Chanmugam and Ray, 1984; King, Frank and Ritter, 1985):
the distribution of field strengths in the intermediate polars is supposed to be the same (i.e. ~ a few x 10^7G) as in the AM Hers; magnetic CV's are a homogeneous class, appearing as intermediate polars at long periods and becoming AM Her systems as the orbital period decreases.

In this article I will review the arguments for (i) or (ii) above. The strong field hypothesis (ii) allows a simple picture of the orbital evolution of magnetic CV's leading to a natural explanation for the observed period distribution. By contrast the weak field hypothesis (i) faces severe difficulties in explaining this distribution, requiring the intervention of unknown physical processes and/or hitherto unsuspected selection effects. I suggest reasons why the white dwarf field distribution should have the bimodal character (~ 10^7G or $\leqslant 10^5$G) required by the strong field hypothesis.

In the remainder of the article, (Sections 5 and 6), I shall adopt the strong field hypothesis, and review its consequences for the accretion flows in the intermediate polars, in particular for disc formation, and the synchronization of the white dwarf spin to the orbit in the AM Hers. Throughout, I shall try to identify areas in which further observational and theoretical study may be fruitful.

2. Magnetic CV's and Orbital Evolution

A CV system will appear as 'magnetic' if the field is strong enough to influence the accretion process. A rough measure of this influence is the magnetospheric radius

$$r_\mu \simeq 2.7 \times 10^{10} \; \mu_{33}^{4/7} \; \dot{M}_{16}^{-2/7} \; M_1^{-1/7} \; cm \tag{1}$$

obtained by balancing magnetic and material stresses for assumed spherical symmetry (here μ_{33} and $\dot{M}_{16}$ are the white dwarf magnetic moment (BR_*^3) and the accretion rate, measured in units of 10^{33} Gcm^3 and 10^{16} gs^{-1} and $M_1 \sim 1$ the white dwarf mass in solar masses). This radius can be compared with the white dwarf radius $R_* \sim 10^9$cm and the binary separation

$$a \simeq 3.5 \times 10^{10} \ P_{hr}^{2/3} \ M^{1/3} \ cm \tag{2}$$

(P_{hr} = P/hours, $M = M_1 + M_2 \sim 1$ = total mass of the binary in solar masses). Clearly 'magnetic' CV's need to have $r_\mu > R_*$, so that the minimum magnetic moment is (with $R_9 = R_*/10^9$ cm)

$$\mu_{33} \ (min) \simeq 3 \times 10^{-3} \ \dot{M}_{16}^{1/2} \ M_1^{1/4} \ R_9^{7/4} \tag{3}$$

Intuitively, we should expect the AM Her systems to have a stronger magnetic interaction than given by this minimum value; as their white dwarfs are spinning synchronously, there must be an interaction with the companion star. This suggests the requirement

$$r_\mu \gtrsim a \qquad \text{for AM Hers,} \tag{4a}$$

leading to a minimum magnetic moment

$$\mu_{33} \ (AM) \simeq 0.5 \ \dot{M}_{16}^{1/2} \ P_{hr}^{7/6} \ M^{7/12} \ M_1^{1/4} \tag{4b}$$

Of course (4a,b) is a rather crude criterion: in particular the assumption of spherical symmetry is unsatisfactory, and one should aim for a criterion for torque balance, rather than the disruption of an accretion flow as implied here. In fact, (see eq. 17) we shall find that detailed considerations of torque balance give criteria almost identical to (4a). Further, one should note that observational estimates of μ for AM Her systems give values ($\sim 10^{33} - 10^{34}$ Gcm^3) very close to μ(AM) as implied by (4a,b): accordingly, we may adopt (4a,b) as a working definition of an AM Her system.

The relevance of binary evolution to the discussion of the strong and weak field hypotheses stated above is now clear: as explained in Section 3 below, during the lifetime of a CV both P and $\dot{M}$ decrease, while μ can be taken as constant (Chanmugam and Gabriel, 1972; Fontaine, Thomas and van Horn, 1973). Thus a system with $\mu_{33} = 2$ would be an intermediate polar at long periods ($P_{hr} \sim 5$, $\dot{M}_{16} \sim 10$) and an AM Her system at short periods ($P_{hr} \sim 1.5$, $\dot{M}_{16} \sim 1$). This realization lies at the heart of the strong field hypothesis. To make quantitative estimates we need to know how $\dot{M}$ and P change with time, that is, we must discuss the evolution of CV's.

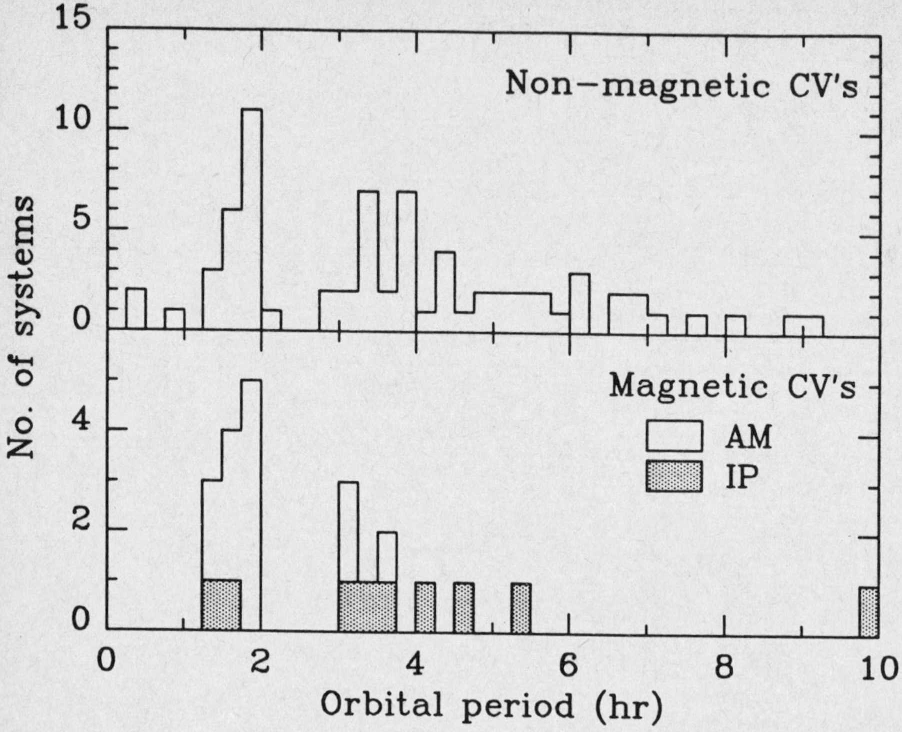

Fig. 1   The distribution of CV's with orbital period.

3. Secular Evolution of CV's

3.1 The General Picture

It is by now well known (see e.g. Savonije, 1983, for a review) that mass transfer in CV's with $P_{hr} \leqslant 10$ must be driven by processes which reduce the orbital angular momentum J and hence also the binary separation a and period P.** Consequently the distribution of CV's with orbital period (Fig. 1) must reflect this orbital (or 'secular') evolution. A framework for understanding this process is now beginning to emerge (see Ritter, 1986, for a review) whose main points may be summarized as follows (see Fig. 2).

(i) At periods $P_{hr} \gtrsim 3$, the evolution is probably driven ($\dot{J} < 0$) by magnetic stellar wind braking of the companion star, (Verbunt and Zwaan, 1981) whose rotation is locked to the orbit by tides. The transfer rates are typically

$$\dot{M}_{16} \sim 10(P_{hr}/3)^{5/3} \tag{5}$$

so that the mass loss timescale $t_{\dot{M}} = M_2/\dot{M} \sim 10^8 y$ for the companion is shorter than its thermal timescale $t_{KH} = GM_2^2/R_2 L_2 \sim 10^9 y$ (M_2, R_2, L_2 are the companion's mass, radius and luminosity). As a result this star is driven progressively out of thermal equilibrium and R_2 is larger than the main-sequence value appropriate to M_2.

(ii) At $P_{hr} \simeq 3$ the driving mechanism becomes inefficient, possibly because the companion becomes fully convective; $\dot{M}$ drops, so that now $t_{\dot{M}} \gg t_{KH}$ and the companion star contracts towards its main sequence radius, detaching from the Roche lobe and reducing $\dot{M}$ to zero (Rappaport, Verbunt and Joss, 1983; Spruit and Ritter, 1983). The star remains detached until angular momentum loss by gravitational radiation has shrunk the orbit sufficiently for the Roche lobe to close down on it again; this occurs at $P_{hr} \simeq 2$.

** The single well-established exception to this amongst CV's is GK Per (ironically an intermediate polar), which has a 48 hr period. Its mass transfer must be driven by the nuclear evolution of the companion in an *expanding* orbit. See Watson, King and Osborne (1985) for a detailed discussion of its evolutionary status.

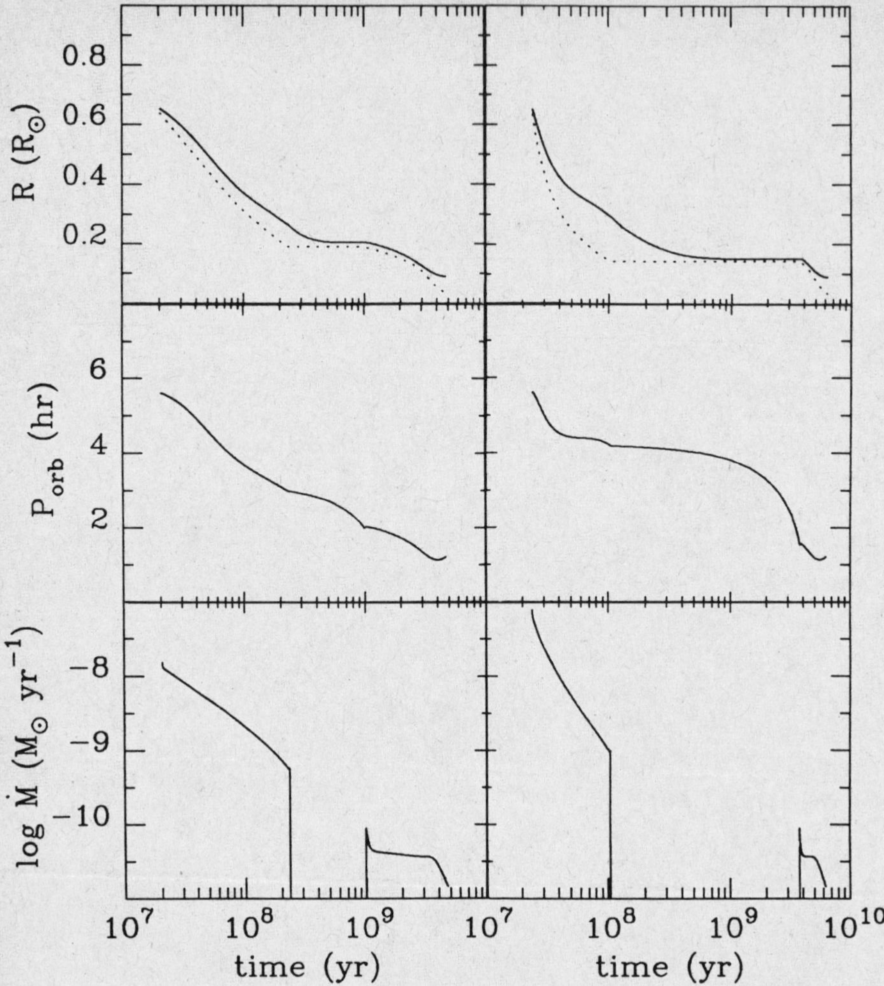

Fig. 2   The evolution of CV's with normal magnetic braking (*left-hand panels*) and enhanced magnetic braking as suggested by Schmidt, Stockman and Grandi (1986) (*right-hand panels*).   The dotted curves in the upper panels show the radius of a main sequence star with a mass given by the instantaneous value of M_2.

(iii) For $P_{hr} \lesssim 2$ gravitational radiation drives the mass transfer at rates $\dot{M}_{16} \sim 1$. At $P_{hr} \simeq 1.3$ the companion has such a low mass that once again $t_{\dot{M}} \ll t_{KH}$; further mass transfer now causes this star to expand adiabatically and P to increase. The star becomes degenerate, mass transfer drops to very low levels and ultimately stops: $P \sim 80$ minutes is thus the minimum orbital period for a CV (Paczynski and Sienkiewicz, 1981; Rappaport, Joss and Webbink, 1982).

This evolution accounts well for the period histogram of CV's, in particular the minimum period $\sim$ 80 minutes and the 'period gap' between 2-3 hours (more precisely 2.1 to 2.9 hours). The latter feature is essentially independent of the nature of the driving mechanism for $P_{hr} \gtrsim 3$, provided only that this can provide transfer rates of the order (5), which are in any case probably required by observation. It is very important to realize however, that only the value of $\dot{M}$ *averaged over epochs* $\gtrsim 10^6$ *yr* (the thermal readjustment time of the companion star) are important for evolution: 'instantaneous' values (e.g. any 'observed' value) for $\dot{M}$ for an individual system can deviate widely from this average and may be a misleading guide to the evolutionary state. A clear observational priority in this area is the detection of more of the secondary stars using red-sensitive CCD spectrographs. This should allow us to study the departures of these stars from the main sequence and ultimately check this overall picture.

3.2 *Deviations from the General Picture: Enhanced Magnetic Braking?*

As pointed out by Ritter (1985), the extreme sharpness of the edges of the period gap in the histogram (Fig. 1) imposes very stringent constraints on the secular evolution above the gap: even small changes (factors $\sim$ 2) in $\dot{M}$ at periods $\gtrsim 3$ hr lead to a significantly different gap. From Fig. 1 we see that the period histogram of magnetic CV's, taken as a homogeneous group, is essentially identical to that of non-magnetic CV's, strongly suggesting that their evolution is identical. In particular the period gap is very well defined for the AM Her systems: 4 out of the first 6 CV's below the gap ($\lesssim$ 2.1 hr) are AM Hers, and AM Her itself is only the third system above the gap ($\gtrsim$ 2.9 hr). This creates severe difficulties for any scheme in which the AM Hers evolve in a systematically different way from other CV's (see Hameury *et al.*, 1986a,b). For example, Schmidt, Stockman and Grandi (1986) propose that mass transfer rates in AM Her systems are 2 - 5 times higher than in other CV's at periods above the gap, because of enhanced magnetic braking involving the *white dwarf's* field. If this idea is incorporated into the evolutionary scheme outlined at the beginning of this Section, one finds the secondary star becomes fully convective, and the system thus detached, at periods 3.5 - 4.2 hours (for $\dot{M}$ increased 2 or 5 times respectively), emerging from the gap only at 1.7 - 1.5 hours (see Fig. 2). This is in flat contradiction to the observed identity of the gap for magnetic and non-magnetic systems. One might try to save this idea by postulating that the enhanced mass transfer in the AM Hers always switches off at an orbital period of 3 hr, irrespective of the internal structure of the secondary, because, for example, the secondary's stellar wind somehow knows about the orbital period. This idea, too, fails, as increasing $\dot{M}$ even by a factor 2 for periods down to 3 hr pushes the secondary star so far out of thermal equilibrium that the period gap extends down to 80 minutes, i.e. to the minimum period, thus excluding all AM Her systems below the gap (see Hameury, King, Lasota and Ritter, 1986a). Similarly, if one were to postulate instead *lower* transfer rates for AM Her systems for periods $\gtrsim$ 3 hr other unacceptable consequences would follow; the secondary stars would not deviate as strongly from thermal equilibrium at P_{hr} = 3, and would have a narrower period gap (or no gap at all). We should then have AM Her systems transferring mass via gravitational radiation at periods in the CV gap: these would have similar luminosities to those below 2 hr, contrary to observation. We must conclude that, if CV's evolve across the period gap in a detached state, the evolution of all of them, magnetic or non-magnetic, is identical.

4. Evolution of Magnetic CV's: Fieldstrengths

4.1 *The Strong and Weak Field Hypotheses and the Period Distribution of Magnetic CV's*

The distribution of magnetic CV's shown in Fig. 1 reveals a second striking result: the AM Her systems are clustered at short periods ($P_{hr} \leq$ 3.8; all but three systems below the period gap) while the intermediate polars are clustered at long periods (above the gap). The strong field hypothesis (Chanmugam and Ray, 1984; King, Frank and Ritter, 1985) gives a natural explanation for this: *all* magnetic CV's have magnetic moments $\mu \sim 10^{33} - 10^{34}$ Gcm^3, and evolve from intermediate polars to AM Hers as their periods decrease, preferentially near the period gap. Fig. 3 shows this in detail, using the evolutionary calculations of Hameury *et al.* (1986a) to determine $\dot{M}$ in the criterion (4b).

The weak field hypothesis (Lamb and Patterson, 1983) is incompatible with a straightforward interpretation of the period histogram for magnetic CV's. We have seen in Section 3 that the secular evolution of magnetic CV's cannot differ systematically from that of non-magnetic ones. Thus, on the weak field hypothesis

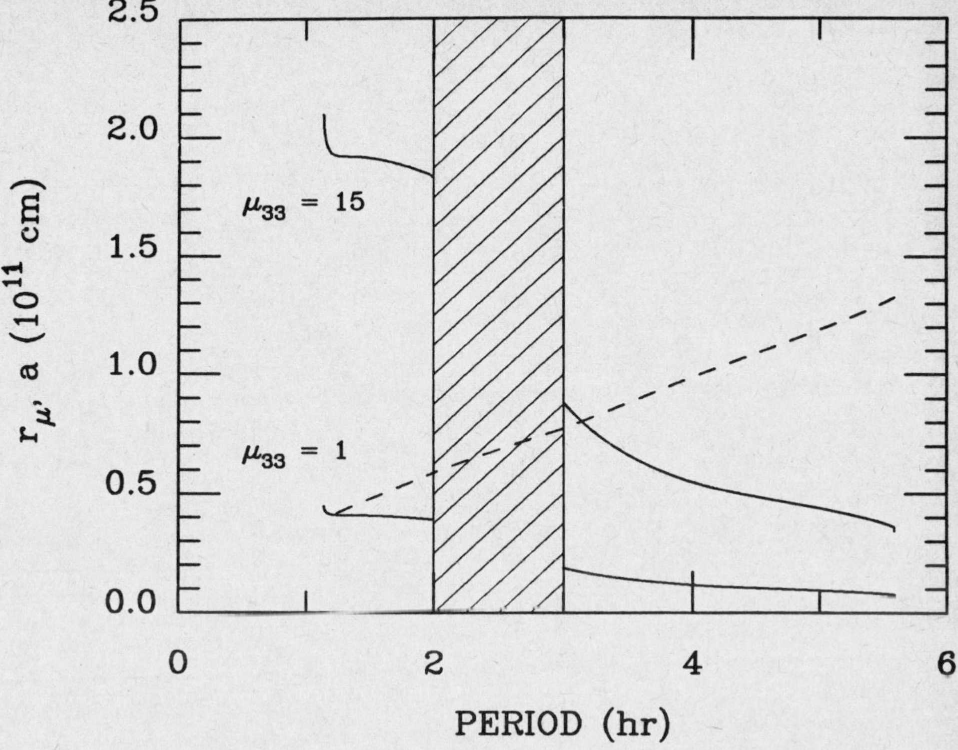

Fig. 3　Magnetospheric radius r_μ (solid line) compared with the orbital separation a (dashed line) for two values of the magnetic moment μ. For periods below the gap, r_μ is usually larger than the orbital separation, and the system is locked; above the gap, the opposite is true. There are however possible exceptions: for relatively low values of μ, a system can be an intermediate polar below the gap, while if $\mu \sim 10^{34}$ Gcm^3, the system can be locked at a period slightly above the gap, thus explaining systems such as EX Hya and AM Her respectively.

that the AM Hers and intermediate polars are distinct classes ($\mu \gtrsim 10^{33}$ Gcm^3 for AM Hers, $\mu < 10^{33}$ Gcm^3 for intermediate polars), the period histogram for each class *separately* ought to resemble that of non-magnetic systems. This is clearly not the case for the AM Her systems. For the intermediate polars the low number of known systems makes a similar conclusion less firm, but this very sparseness is in itself worrying for the weak field hypothesis: much larger numbers should be found, as from (3,4b) systems with $10^{30} \lesssim \mu \lesssim 10^{33}$ Gcm^3 should be intermediate polars at all orbital periods (King, Frank and Ritter, 1985).

4.2 *Distortions of the Period Distribution: Selection Effects?*

The weak field hypothesis thus fails a direct comparison with the observed period distribution of magnetic CV's. There are two ways in which one might nonetheless try to save it: one might postulate either that selection effects distort the observed period distribution, or that unknown physical effects would bring the predicted distribution into agreement with observation. The requirements on selection effects are rather stringent: they must select both against long period AM Hers and short period intermediate polars, but in such a way that the combined distribution contrives to resemble that of non-magnetic CV's very closely. There are no candidates for such effects at present.

Affecting the predicted distribution by new physical processes seems also rather unpromising: we have seen in Section 3 that postulating systematic differences in mass transfer rates and hence in the secular evolution of magnetic systems results in unacceptable consequences for the predicted period distributions. Thus one can only attempt to affect the magnetic field distributions. For example, one might postulate that nova explosions remove the outer layers of the white dwarf and uncover stronger internal fields, so that older systems (i.e. those below the gap) had stronger magnetic moments. Such a mechanism has actually been suggested (King, 1985a) to account for the anomalously *weak* field inferred in DQ Herculis (Nova Her 1934), which differs from the intermediate polars (Warner, 1983) in not being an X-ray source and in having a very much shorter spin period (71s compared with typical values 10-20 min). This magnetic field might be the result of uncovering a previously completely buried field (see 4.3 below). If successful, the application of such an idea to account for the lack of intermediate polars at short periods would amount to saying that the strong field hypothesis holds at least for systems below the gap. However, it is not easy to see why, if the mechanism works at all, it should only do so below the gap.

4.3 The Origin of the Fieldstrength Distribution

In the absence of convincing cases either for selection effects or new physical processes, one must conclude that the strong field hypothesis is to be preferred, i.e. that essentially *all* magnetic CV's have magnetic moments in the narrow range $10^{33} \lesssim \mu \lesssim 10^{34}$ Gcm^3. Possible reasons for this distribution have been suggested (King, 1985b). The lower limit is just what is expected from field-burying because of the white dwarf rotation (Moss, 1979, 1984): for $\mu \lesssim 10^{33}$ Gcm^3, the consequent circulation reduces the surface field severely. The upper limit, which is below fieldstrength values actually measured in some single white dwarfs, may indicate (Liebert and Stockman, 1985; King, 1985b) that enhanced angular momentum losses, perhaps on the lines suggested by Schmidt, Stockman and Grandi (1986) *do* occur, but only for systems with $\mu \gtrsim 10^{34}$ Gcm^3, i.e. larger than any observed value. As we have seen in Section 3, the evolution in such cases is so catastrophic that the systems effectively disappear, being detached for very long epochs.

5. Disc Formation

5.1 Conditions for Disc formation in Close Binaries

Accretion discs are one of the most basic and important features of non-magnetic CV's (see e.g. Frank, King and Raine, 1985 for an introduction, Pringle 1981 for a review). It is natural to ask if they are present in magnetic systems also. The fundamental reason for disc formation in any binary system is that the accreting matter has too much specific angular momentum to land directly on the mass-gaining star. If the latter is non-magnetic, and has radius R_*, and the free accretion stream, if left to itself, would follow a trajectory whose distance of closest approach to the star is R_{min}, a disc will form if

$$R_{min} > R_* . \tag{5}$$

Values of R_{min} for given binary parameters are given by Lubow and Shu (1975) and Flannery (1975) (in fact a reasonable estimate results from treating the accretion stream as a test particle trajectory in a near-parabolic orbit, e.g. Frank, King and Raine, 1985). For non-magnetic CV's one always finds that (5) is satisfied. By contrast, some Algol systems have sufficiently extended mass-gaining stars that (5) is not satisfied and the gas stream hits the star directly.

5.2 Disc Formation in Magnetic CV's

For magnetic CV's one must replace R_* in (5) by the effective size R_{mag} of the white dwarf's magnetosphere, defined as the position where the motion of the gas stream is no longer a free orbit but is strongly influenced by the magnetic field. Thus we require

$$R_{min} > R_{mag} \cdot \tag{6}$$

To evaluate (6) we need an estimate of R_{mag}. This is a global problem in which one must estimate the conditions in the gas stream and the large-scale distortion of the white dwarf magnetic field. Hameury, King and Lasota (1986c) find a value

$$R_{mag} \approx 0.37 \; r_\mu \tag{7}$$

for R_{mag}, with r_μ the spherical magnetospheric radius given by (1). Using estimates of $\dot{M}$ as a function of orbital period, and combining (6) and (7) with the estimates for R_{min}, now leads to the conclusion that $R_{min} < R_{mag}$ for orbital periods below about 5 hours, while $R_{min} > R_{mag}$ at longer periods. Thus for the latter systems there is no reason why discs should not form in the usual way. However, most known magnetic systems have periods below 5 hr, and here the conclusion is less clear.

First, one should note that $R_{min} > R_{mag}$ is a *sufficient*, but possibly not a *necessary* condition for disc formation. Thus even in a system with $R_{min} < R_{mag}$, the accreting matter might nevertheless be able to cross fieldines enough to interact with itself and build up a disc. To investigate this, one must study the growth times of the various magnetohydrodynamical instabilities which try to couple the impacting plasma at R_{mag} to the fieldlines. The nonlinear development of these instabilities is not particularly well understood, so that a firm decision is impossible at present. Hameury, King and Lasota (1986c) tentatively conclude that discs probably do not form in most cases for $P < 5$ hr, and that in the intermediate polars the accreting matter impacts at R_{mag} and is chanelled on to relatively large fractions of (~ 0.1) of the white dwarf surface.

5.3 Observational Evidence

There is some observational support for the conclusion that discs may not form in intermediate polars with orbital periods $\lesssim 5$ hr. First, the impact region at R_{mag} should be a medium-energy X-ray source in its own right, plausibly accounting for orbital X-ray modulations observed in the intermediate polars. (These are much harder to understand if a disc forms, as the X-ray emitting material must then lose all knowledge of the orbital phase.) Second, the predicted large accreting fractions f of the white dwarf surface agree with deductions from the quasi-sinusoidal nature of the X-ray light curves of the intermediate polars (King and Shaviv, 1984). Other possibly observable effects, such as 'amplitude modulation' of the X-rays, are also predicted in the paper by Hameury, King and Lasota (1986c) and should be checked for. Note that the presence of doubled emission lines in the optical is only evidence for circulating optically thin material, rather than an optically thick disc; such material is likely to be present whether or not a disc forms (Hameury, King and Lasota, 1986c).

5.4 Disc Formation and the Orbital Period

We have seen that disc formation is probably hindered for orbital periods less than about 5 hr, and that magnetic systems seem preferentially to synchronize (i.e. become AM Hers) at 3-4 hours or less. Thus no AM Her system should have a disc, in agreement with observation. In fact, the maximum orbital periods P_{disc} and P_{AM} for intermediate polars without discs and AM Her systems respectively should be roughly in the ratio 2:1, according to the simple criteria (4a, 6). From (7) we get

$$\frac{R_{mag}}{R_{min}} = \frac{0.37}{f(q)} \cdot \frac{r_\mu}{a}$$

where $f(q)$ is a function of the binary mass ratio, $q = M_2/M_1$ which can be evaluated from the tables of Lubow and Shu (1975). For $0.1 \leqslant q \leqslant 1$ the ratio $0.37/f(q)$ varies only between ~ 1 and ~ 3, so that we can write

$$\frac{R_{mag}}{R_{min}} \sim 2\,\frac{r_\mu}{a} \tag{8}$$

For periods above the gap, magnetic braking suggests $\dot{M} \propto P^{5/3}$ (see eq. 5). Combining this with (1) and (8) shows that

$$\frac{R_{mag}}{R_{min}} \sim \frac{r_\mu}{a} \propto \frac{P^{-10/21}\mu^{4/7}}{P^{2/3}} = P^{-8/7}\mu^{4/7} \quad . \tag{9}$$

For μ about the same for *all* magnetic CV's, as required by the strong field hypothesis, one thus has

$$\frac{R_{mag}}{R_{min}} \sim \frac{2r_\mu}{a} \propto P^{-8/7} \tag{10}$$

Hence as P decreases during the orbital evolution, a magnetic CV first loses its disc (at $P = P_{disc}$) when $R_{mag}/R_{min} = 1$, and then becomes an AM Her system (at $P = P_{AM}$) when $r_\mu/a \simeq 1$. This requires

$$P_{AM} \sim (2)^{-8/7} P_{disc} \sim 0.5 P_{disc} \tag{11}$$

Note that the precise numerical coefficient in (11) is sensitive to our criterion (4a) for becoming an AM Her system. However, the qualitative result ($P_{AM} < P_{disc}$) is unlikely to be affected by improvements in this criterion.

Finally, it should be noted that it is incorrect, although sometimes asserted, that the formation of a disc at any one epoch guarantees its presence at all later epochs. Any disc will disappear if the mass transfer from the companion is interrupted for a time exceeding its viscous transport time (weeks-months); upon the resumption of mass transfer the conditions on R_{mag} and R_{min} discussed in Section 5.2 above must be applied to the incoming gas stream once again. The fact that magnetic CV's are observed to have quite lengthy ($\sim$ months) low states, in which mass transfer rates drop by at least an order of magnitude, suggests that it may be quite possible for the same object to possess a disc at some epochs and not at others; the remarks above concerning the presence of discs at certain orbital periods only apply *on average* over epochs long compared with the thermal time of the secondary star (cf. the remark at the end of Section 3.1).

6. The White Dwarf Spin: Synchronization

6.1 *The Equilibrium Period*

Accretion torques on magnetic compact stars have been discussed for many years, and the simpler results are by now well known. If the accreting matter has a 'lever arm' R_ℓ (which we might expect to be $\sim R_{mag}$) it is straightforward to show that the star is likely to spin up, on a timescale short compared with the binary evolution time, to a spin period P_{spin} close to the centrifugal limit:

$$P_{centr} = 2\pi \left[\frac{R_\ell^3}{GM_1 M_\odot} \right]^{\frac{1}{2}} . \tag{12}$$

We may parametrize R_ℓ as Φr_μ, where r_μ is given by (1) and $\Phi \sim 1$ is dimensionless.

Then, in principle, a comparison of the observed spin periods with P_{centr} should lead to estimates of μ (e.g. Lamb and Patterson, 1983). However, as Hameury, King and Lasota (1986c) show, uncertainties in the dimensionless factor Φ, and also in how close to P_{centr} the 'equilibrium' spin period actually is, introduce uncertainties of at least a factor 6 in these estimates: hence this method cannot be used to discriminate between the strong and weak field hypotheses. By definition, at spin periods close to 'equilibrium' the timescale $|P_{spin}/\dot{P}_{spin}|$ for changes in the white dwarf rotation rate ought on average to be close to the orbital evolution time $t_{\dot{M}} > 10^8$ yr (see Ritter, 1985b for a discussion, and Section 6.2). Thus, measurements of spin period derivatives $\dot{P}_{spin}$ implying much shorter timescales (e.g. van Amerongen et $al.$, 1985) probably reflect 'short-term' (i.e. < 10^6 yr!) fluctuations in the mass transfer rates.

6.2 Synchronization

According to the strong field hypothesis, the white dwarf spin in an intermediate polar must ultimately synchronize to the orbital rotation as the system becomes an AM Her object. The centrifugal limit (12) makes this process quite plausible, at least in outline. From (12) and Kepler's law one finds

$$\frac{P_{centr}}{P} = \left[\frac{\Phi r_\mu}{a} \right]^{3/2} \left[\frac{M_1 + M_2}{M_1} \right]^{1/3} \tag{13}$$

Since $P_{spin} \leqslant P_{centr}$, one sees that P_{spin} must approach P as r_μ/a tends to unity (King, Frank and Ritter, 1985). Note that this actually implies secular spin*down* of the white dwarf: for $P > 3$ hr, magnetic braking implies $r_\mu/a \propto P^{-8/7}$ (cf. eq. 9 above), so that $P_{centr} \propto P^{5/7}$ from (13), and P_{centr} thus decreases on the evolutionary timescale $t_{\dot{M}} \sim |P/\dot{P}|$. Thus material torques will already make P_{spin} and P quite close by the time $r_\mu \sim a$: once mass transfer switches off as the system passes through the period gap, 'propellor' spindown in the secondary's stellar wind is likely to bring about synchronism (King, Frank and Ritter, 1985).

Although this picture seems promising, a number of areas remain uncertain. First, there are three AM Her systems on the long side of the period gap. For these systems one probably cannot appeal to an extended epoch of low or zero mass transfer in order to have brought about synchronism as envisaged above. Thus while the latter is certainly helpful in reaching synchronism it does not appear to be indispensable. As a related point, even when synchronism is attained, it must be maintained in the face of accretion torques. We must therefore look at the process in more detail: in doing so we can hope to improve on the simple criterion $r_\mu \sim a$

used in Section 2.

Many synchronizing and locking torques have been suggested: all of them can be written in the general form:

$$G_{synch} = B_1 \mu f \qquad (14)$$

Here $B_1 \sim \mu_1/a^3$ is the white dwarf's magnetic fieldstrength near the secondary, μ is the effective magnetic moment of this star and f a dimensionless function. If the secondary has an *intrinsic* (e.g. dynamo-maintained) magnetic moment μ_2, we have $\mu = \mu_2$, and $f \sim 1$ is just a function of the angles between the two moments (Campbell, 1985). This magnetostatic torque averages to zero over a synodic rotation of the white dwarf, so it is only possibly important for locking rather than synchronizing. For an *induced* magnetic moment we can write $\mu = \mu_1 (R_2/a)^\epsilon$. This describes the torques discussed by Campbell (1983) and Joss, Katz and Rappaport (1979) (both $\epsilon = 3$), Lamb *et al.* (1983) ($\epsilon = 2$) and Chanmugam and Dulk (1983) ($\epsilon = 1$), where f is variously a function of quantities such as the secondary's turbulent resistivity, angles between field components, the white dwarf's electrical resistance, and the synodic angular velocity ϵ. (In Chanmugam and Dulk (1983) f involves R_2 linearly as well.) The competing torque is the accretion torque, which for a near-synchronous white dwarf spin must be taken as

$$G_{acc} = \dot{M}b^2\Omega \sim \dot{M}a^2\Omega \qquad (15)$$

with $\Omega = 2\pi/P = [G(M_1+M_2)M_\odot/a^3]^{1/2}$ and $b \simeq a - R_2 \sim a$ the distance from the inner Lagrange point to the white dwarf. Synchronization or locking can occur if $G_{synch} \geqslant G_{acc}$, i.e.

$$\frac{\mu_1 \mu f}{a^3} \geqslant \dot{M} a^2 \Omega \qquad . \qquad (16)$$

We can compare this with the stress-balance condition defining r_μ (cf. eq. 1), which may be written as

$$\frac{\mu_1^2}{r_\mu^6} \sim \frac{\dot{M}v}{r_\mu^2} \qquad (17)$$

where $v = (2GM_1M_\odot/r_\mu)^{1/2}$ is the free-fall velocity at r_μ. Comparing (16) and (17) and using taking $M_2 \ll M_1$ gives

$$\frac{r_\mu}{a} \gtrsim \left[\frac{\mu_1}{\mu f}\right]^{2/7} \tag{18}$$

as the condition for synchronism or locking. The right-hand side is of order unity for $\mu_1 \sim 10^{33}$ Gcm^3 and μf given by any of the torques proposed in the literature. Thus we recover our crude criterion (4a) for AM Her behaviour. A detailed study (Hameury, King, Lasota and Ritter, 1986b) shows that the locking mechanism for the AM Her systems must probably use the *intrinsic* magnetic moment $\mu = \mu_2$ of the secondary. This agrees with the very tight observational limits on any asynchronous rotation (e.g. Biermann *et al.*, 1985). This result has two important consequences.

First, the secondaries in magnetic CV's, and thus presumably in *all* CV's, must retain at least some magnetic field even at periods below the gap: surface fields $B_2 \gtrsim 50G$ are sufficient to lock the white dwarf against the accretion torques. Hence whatever the mechanism which causes the abrupt drop in $\dot{M}$ at the upper edge of the period gap, it cannot be the complete loss of the secondary's field. Second, some kind of propellor-type spindown mechanism is probably required to bring the three AM Her systems with $P > 3$ hr into synchronism: once this is achieved, the intrinsic magnetostatic torque can keep them locked.

7. Conclusions

The evidence presented in this review suggests that a credible and consistent picture of the evolution of magnetic CV's may be emerging. It is already clear that their mass transfer rates cannot differ systematically from those of non-magnetic CV's. The strong field hypothesis, that *all* magnetic CV's have magnetic moments $10^{33} \lesssim \mu \lesssim 10^{34}$ Gcm^3, leads to a simple interpretation of the period histogram: magnetic CV's are intermediate polars at long periods and become AM Hers at periods close to the period gap. Study of the mechanism by which synchronism is maintained in these systems may reveal more about the secondary stars in all CV's.

Lack of space prevents discussion of further topics of importance, such as the low polarizations observed for most intermediate polars (see Hameury, King and Lasota, 1986a). In particular the soft X-ray excess (e.g. King, 1983; King, 1985c) is still with us (see the article by M.G. Watson in this volume) despite premature reports of its disappearance (Cordova and Howarth, 1986). This is important, for the mass transfer rates deduced on the assumption that no excess exists are probably too low to fit easily into the evolutionary scheme outlined in

Section 3.1 above.

Acknowledgements

I thank Drs. J.M. Hameury, J.P. Lasota, H. Ritter and M.G. Watson, as well as many of the workshop participants for valuable discusions.

References

Biermann, P., Schmidt, G.D., Liebert, J., Stockman, H.S., Tapia, S., Kühr, H., Strittmatter, S., West, S. and Lamb, D.Q., 1985. *Ap. J.*, <u>293</u>, 303.

Campbell, C.G., 1983. <u>MNRAS</u>, <u>205</u>, 1031.

Campbell, C.G., 1985. <u>MNRAS</u>, <u>211</u>, 69.

Chanmugam, G. and Dulk, G.A., 1983. In: Proc. IAU Colloq. 72, <u>Cataclysmic Variables and Related Objects</u>, p.233 (eds. Livio and Shaviv, Reidel).

Chanmugam, G. and Gabriel, M., 1972. <u>Astr. Ap.</u>, <u>16</u>, 149.

Chanmugam, G. and Ray, A., 1984. <u>Ap. J.</u>, <u>285</u>, 252.

Cordova, F.A. and Howarth, I.D., 1986, to appear in <u>Scientific Accomplishments of the IUE</u> (eds. Kondo, Y, *et al.*, Reidel).

Flannery, G., 1975. <u>MNRAS</u>, <u>170</u>, 325.

Fontaine, G., Thomas, J.H. and van Horn, H.M., 1973. <u>Ap. J.</u>, <u>184</u>, 911.

Frank, J., King, A.R. and Raine, D.J., 1985. <u>Accretion Power in Astrophysics</u>, Cambridge University Press.

Hameury, J.-M., King, A.R., Lasota, J.-P. and Ritter, H., 1986a. Proceedings of IAU Colloq. 93, Bamberg 1986, in press.

Hameury, J.-M., King, A.R., Lasota, J.-P. and Ritter, H., 1986b, in preparation.

Hameury, J.-M., King, A.R. and Lasota, J.-P., 1986c. <u>MNRAS</u>, <u>218</u>, 695.

Joss, P.C., Katz, J.I. and Rappaport, S.A., 1979. <u>Ap. J.</u>, <u>230</u>, 176.

King, A.R., 1985a. <u>Nature</u>, <u>313</u>, 221.

King, A.R., 1985b. <u>MNRAS</u>, <u>217</u>, 23P.

King, A.R., 1985c. In: <u>Recent Results on Cataclysmic Variables, Proc. Bamberg ESA Workshop</u>, ESA SP-236.

King, A.R., Frank, J. and Ritter, H., 1985. <u>MNRAS</u>, <u>213</u>, 181.

King, A.R. and Shaviv, G., 1984. <u>MNRAS</u>, <u>211</u>, 883.

Lamb, D.Q. and Patterson, J., 1983.   In: Proc. IAU Colloq. 72, <u>Cataclysmic</u>
    <u>Variables and Related Objects</u>, p.229 (eds. Livio and Shaviv, Reidel).

Lamb, F.K., Aly, J.J., Cook, M.C. and Lamb, D.Q., 1983. <u>Ap. J. Lett.</u>, <u>274</u>, L71.

Lubow, S.H. and Shu, F.H., 1975. <u>Ap. J.</u>, <u>198</u>, 383.

Paczynski, B. and Sienkiewicz, R., <u>Ap. J. Lett.</u>, <u>248</u>, L27.

Pringle, J.E., 1981. <u>Ann. Rev. Astr. Ap.</u>, <u>19</u>, 137.

Rappaport, S.A., Joss, P.C. and Webbink, R.F., 1983. <u>Ap. J.</u>, <u>254</u>, 616.

Rappaport, S.A., Verbunt, F. and Joss, P.C., 1983. <u>Ap. J.</u>, <u>275</u>, 713.

Ritter, H., 1985a. <u>Astr. Ap.</u>, <u>145</u>, 227.

Ritter, H., 1985b. <u>Astr. Ap.</u>, <u>148</u>, 207.

Ritter, H., 1986.   In: <u>The Evolution of Galactic X-ray Binaries</u>, p.271 (eds.
    Trümper, Lewin and Brinkmann, Reidel NATO ASI Series).

Savonije, G.J., 1983. Ch. 9 of <u>Accretion-Driven Stellar X-ray Sources</u>, eds. Lewin
    and van den Heuvel, Cambridge University Press.

Schmidt, G.D., Stockman, H.S. and Grandi, S.A., 1986. <u>Ap. J.</u>, <u>300</u>, 804.

Spruit, H.C. and Ritter, H., 1983. <u>Astr. Ap.</u>, <u>124</u>, 267.

van Amerongen, S., Kraakman, H., Damen, F., Tjemkes, S. and van Paradijs, J., 1985.
    <u>MNRAS</u>, <u>215</u>, 45P.

Verbunt, F. and Zwaan, C., 1981. <u>Astr. Ap.</u>, <u>100</u>, L7.

Warner, B., 1983.   In: Proc. IAU Colloq. 72, <u>Cataclysmic Variables and Related</u>
    <u>Objects</u>, p.155 (eds. Livio and Shaviv, Reidel).

Watson, M.G., King, A.R. and Osborne, J., 1985. <u>MNRAS</u>, <u>212</u>, 917.

QUASI-PERIODIC OSCILLATIONS IN

LOW-MASS X-RAY BINARIES

M. van der Klis
Space Science Department of ESA
ESTEC, postbus 299
2200 AG Noordwijk
The Netherlands

1. Introduction

Everybody knows what periodic oscillations are, but for quasi-periodic oscillations (QPO) the situation is not so clear. In this review I shall call every phenomenon 'QPO' that causes a broad bump (local maximum) in the observed power spectrum of a signal (time series). The advantage of this definition is that the power spectrum of the signal only has to be considered to decide whether QPO are present. The disadvantage is that by ignoring half of the available information (the phases of the Fourier components) we may be including many different phenomena into the same class.

There are many ways to produce a broad bump in the power spectrum. A signal containing finite wave trains of n oscillation cycles will have a power spectrum that contains a bump of relative width $\Delta\nu/\nu \sim 1/n$, somewhat depending on the envelope of the wave train. ($\Delta\nu$ is bump width, ν its centroid frequency). An important subclass of wave trains are the oscillating shots, wave trains which have a non-zero (usually positive) time integral. Signals containing constant-frequency oscillations with phase jumps, frequency-modulated oscillations or recurrent flares with recurrence times selected from a certain finite range can also have a broad bump in their power spectra. If the length of the observation is not sufficient to resolve them, a number of closely-spaced coherent oscillations can also show up as a broad bump in the power spectrum and thus be classified as QPO.

For the detection of QPO it is not useful to go to frequency resolutions which much exceed the width of the expected bump: contrary to the case for periodic oscillations the power in QPO gets spread out over more and more bins in the power spectrum as the frequency resolution increases and finally tends to get lost in the noise. This is illustrated in Figure 1, showing a high- and a low-resolution power spectrum of the same data.
A typical QPO power spectrum (Figure 1b) can be divided in three sections: a horizontal part towards the high-frequency end of the spectrum where counting statistics dominate the intensity variations,

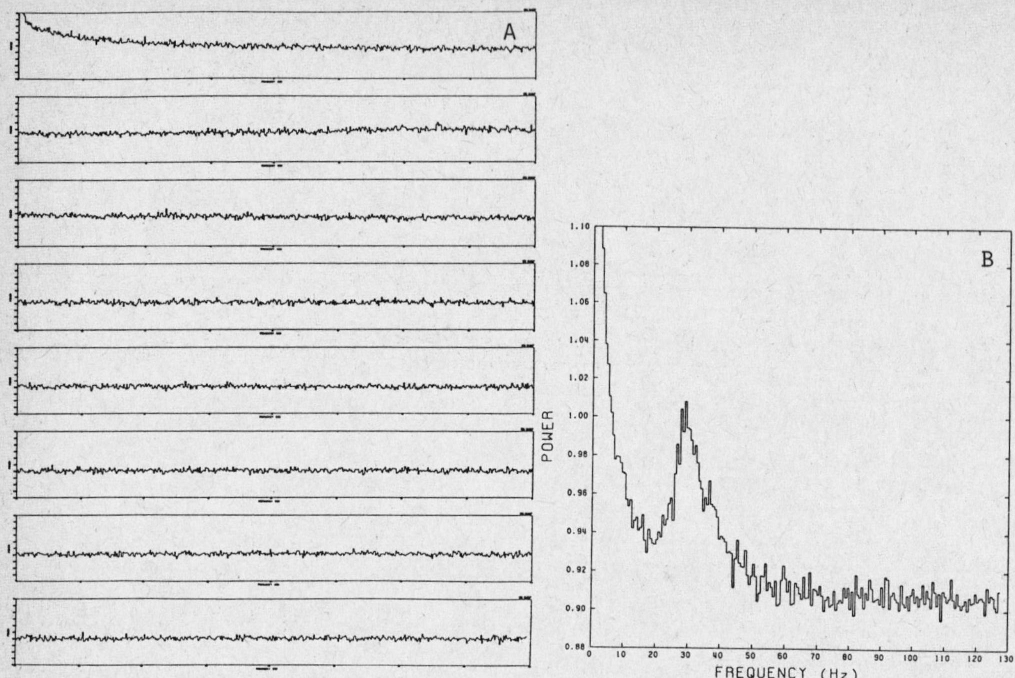

Figure 1. A high- and a low-resolution representation of the same power spectrum of GX 5-1.

the part dominated by the broad QPO peak, and a section that gradually rises towards lower frequencies below the QPO peak which is dominated by 'low-frequency noise'. This 'LFN' is a common phenomenon of noisy time series. It consists of stochastic variations (here occurring in addition to the QPO) of which the slowest have the largest amplitudes.

Table 1 contains a list of low-mass X-ray binaries in which QPO in the X-ray flux were reported. Low-frequency QPO (periods of 50-1000 seconds) in the optical and X-ray flux have been reported since the early days of X-ray astronomy (e.g. Gribbin et al. 1970, Joss et al. 1978, Stella et al. 1984, van der Klis and Jansen 1985, Matsuoka 1985). In many cases these QPO are reported to persist for relatively few cycles (always less than 100) and in some cases they might be an accidental aspect of random shot noise (see Terrell 1972). Only those sources were included in Table 1 for which the authors discuss this possibility.

The sources are listed in Table 1 in order of decreasing QPO frequency. The maximum X-ray luminosity as given by Bradt and McClintock (1983) and the maximum observed persistence of the QPO are also given. The first seven sources listed show the high-frequency (5-50 Hz) low-coherent ($\Delta\nu$ / ν ~ 1) strong (amplitudes typically resulting in a 5% rms flux variation) persistent ($>10^5$ cycles) QPO recently discovered

Table 1 - 'QPO' sources

Source	ν_{QPO} (Hz)	Log L_x[1]	Persistence (log(cycles))	Remarks[1]	References
Cyg X-2 (2142+380)	5,28-45	38.1	6	PB	1,2,3,4
GX 5-1 (1758-250)	5?,20-36	38.6 [2]	6	PB	5,6,7,8
4U 1820-30 (1820-303)	(0.02), 15-35	37.9	(1),6	PB,Bu,GC	9,10,11
Sco X-1 (1617-155)	6-10-20	37.9 [3]	5	PB	12,13,14 15,16
GX 349+2 (1702-363)	5,11?	38.5 [2]	5	PB	17,18,11
GX 3+1 (1744-265)	8.6	38.2 [2]	5	PB,Bu	19
GX 17+2 (1813-140)	7.2	38.4 [2]	5	PB,Bu?	20,11
Rap. Burster (1730-333)	1-5	37.0	3	Bu,GC	21,22,11
Cir X-1 (1516-569)	1.4	38.9	?	Bu?,Tr,RV	23
Terzan 2 (1724-307)	0.092?	36.7	3	Bu,GC	24,25
Cyg X-3 (2030+407)	0.001- 0.02	38.0	1	PB	26
4U 1626-67 (1627-673)	0.001	36.8 [4]	2	pulsar	27,28

[1] Bradt and McClintock (1983)
PB = persistent bright   Tr = transient
Bu = X-ray burst source   RV = rapidly variable
GC = in globular cluster

[2] at 10 kpc

[3] at 1.2 kpc (White, Peacock and Taylor 1985)

[4] ref. 27

with EXOSAT in GX 5-1 (van der Klis et al. 1985a,b) and subsequently in Sco X-1 (Middleditch and Priedhorsky 1985a,b), Cyg X-2 (Hasinger et al. 1985a, 1986a), GX 349+2 (Lewin et al. 1985), GX 17+2 (Stella et al. 1985a), 4U 1820-30 (Stella et al. 1985c) and GX 3+1 (Lewin et al. 1986). These sources form the main subject of this review. They are all very luminous (> 10^{38} erg/s) and classified as 'persistently bright'

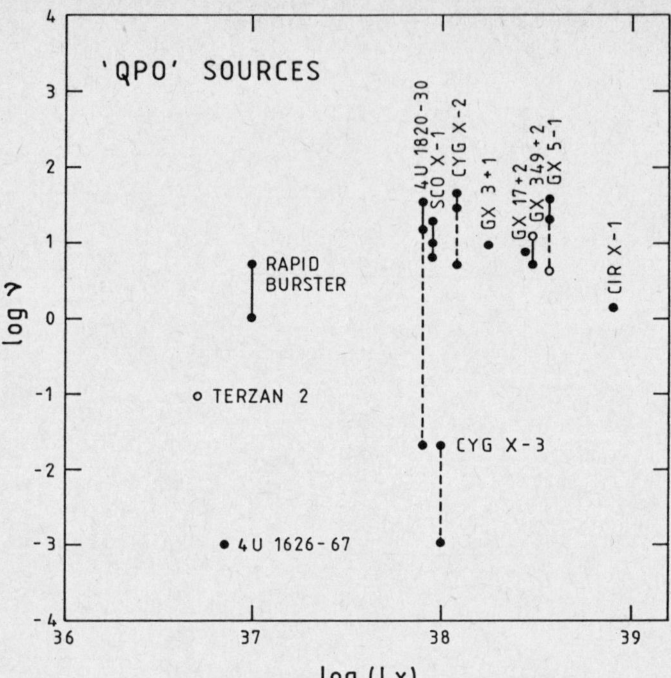

Figure 2. QPO frequency vs. 'Bradt and McClintock (1983)' luminosity.
Data from Table 1.

(Bradt and McClintock 1983). They form a clear cluster in a QPO
frequency vs. X-ray luminosity diagram (Figure 2). The QPO in the rapid
burster (1730-333) which were discovered with the Hakucho satellite by
Tawara et al. (1982) have slightly different properties: lower
frequency range and a persistence of 10^3 cycles at most (see Stella
1985). The results on Cir X-1 (Tennant 1986), which may be a massive
system, and Terzan 2 (Belli et al. 1986) are very recent and will not
be discussed here. The data listed in Table 1 do not allow to determine
whether all observed QPO could be the manifestation of a single
mechanism, but it seems evident that the sources which show rapid QPO
with frequencies above 5 Hz stand out as a group, with the rapid
burster as something of a borderline case.

Five of the seven high-frequency QPO sources are bright galactic bulge
sources which in spite of being the brightest X-ray sources in the
galaxy notoriously lack diagnostic properties: they don't show X-ray
bursts, periodic dips or pulsations and have no optical
identifications. The QPO, which from their high frequencies probably
originate from within approximately 10^2 km from the central source, are
one of the very few available probes into the structure of these
systems.
The remaining two sources, Sco X-1 and Cyg X-2, are low-mass
X-ray binaries which are known to be accreting matter from an evolved
low-mass companion. Except for their position relative to us and the

galactic center, they may be very similar to bright galactic bulge sources.

The following description of the observational data of the QPO in these sources concentrates on three main subjects which have been of particular importance for the development of QPO theories:
1)   the relations between X-ray intensity and QPO frequency,
2)   the connection between QPO and LFN, and
3)   the correlated bimodal X-ray spectral and QPO/LFN behaviour observed in some sources.

This paper will follow a roughly historical line, first discussing the QPO/LFN properties of GX 5-1 and Cyg X-2 in their 'horizontal branch' spectral state, then moving on to the bimodal properties of Sco X-1 and finally comparing this to the bimodal behaviour of GX 5-1 and Cyg X-2. I shall be indicating at some points what is the significance of the discussed observational data for QPO models; for a full review of QPO models see Lewin (these proceedings). We shall be mainly concerned with the three best-studied QPO sources: GX 5-1, Sco X-1 and Cyg X-2.

2. GX 5-1 and Cyg X-2 on the 'horizontal branch'

a) Frequency-intensity relations

Both GX 5-1 and Cyg X-2 show, in their horizontal branch spectral state (see Section 4) a strong positive correlation between X-ray intensity and QPO frequency (van der Klis et al. 1985b, Hasinger et al. 1986a). This is easily seen in Figure 3 where a series of power spectra constructed from data at different intensity levels is given for each source. Note how the QPO peak not only moves to higher frequencies but also widens for increasing intensity. In both cases the intensity-frequency relation is consistent with a straight line (Figure 4); the logarithmic derivative of the relation is ~ 2 for GX 5-1 and ~ 1.7 for Cyg X-2.

From the beginning this positive correlation suggested that the QPO frequency might have something to do with the Keplerian rotation frequency of matter in an accretion disk just outside a magnetosphere. To explain the steepness of the observed relation Alpar and Shaham (1985a,b) proposed that the QPO might arise as a beat between the Keplerian rotation of matter in the accretion disk and the spin of a magnetised neutron star. The derived spin frequency would then be of the order of 10 ms and the neutron star surface field 10^9-10^{10} Gauss.

b) Low-frequency noise (LFN)

As is clearly visible in Figure 3, both sources show LFN of similar strength as the QPO. The shape of this LFN is not consistent with a power law; an exponential function usually provides an acceptable fit (van der Klis et al. 1985b, Hasinger et al. 1986a). In the case of

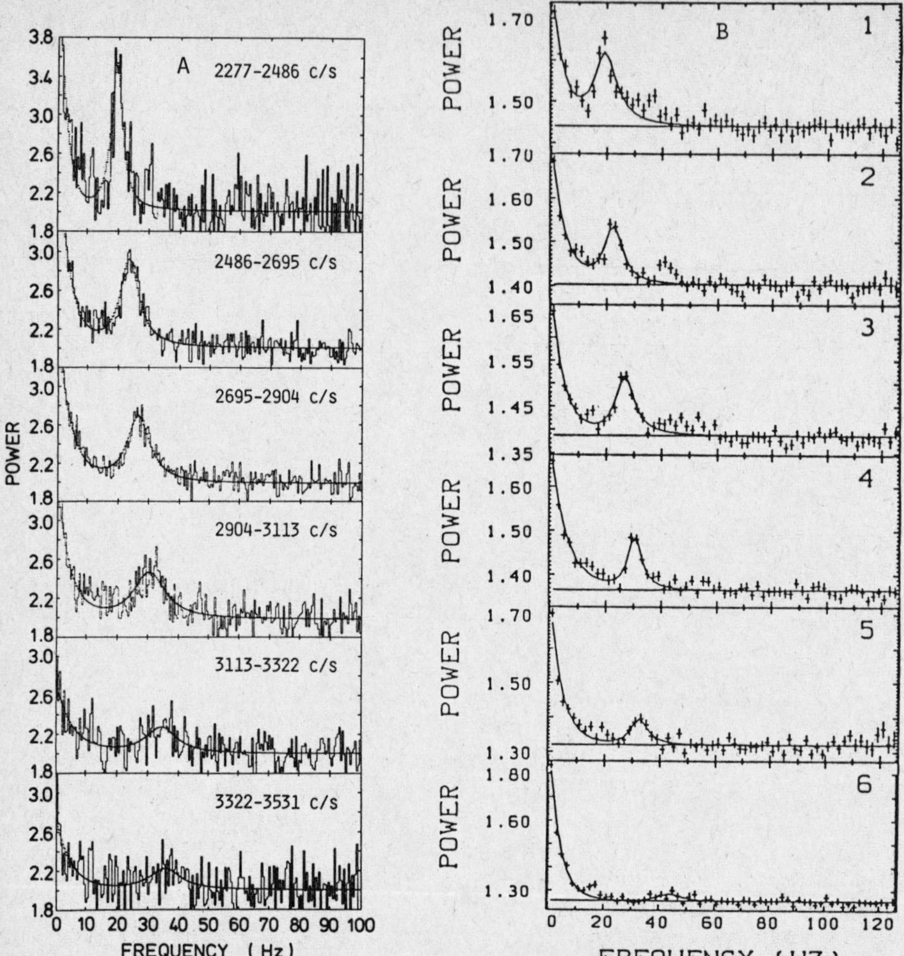

Figure 3. Intensity-selected power spectra of GX 5-1 (a, van der Klis et al. 1985b) and Cyg X-2 (b, Hasinger et al. 1986b) on the horizontal branch.

GX 5-1 the power in the LFN (measured down to 0.5 Hz) and in the QPO are nearly equal, and depend in the same way on source intensity, decreasing from ~ 6% to ~4% rms flux variation (Figure 5a). In Cyg X-2, the intensity dependence is different, but the power in the QPO and in the LFN (also measured down to 0.5 Hz) are still roughly the same (Figure 5b).

These data prompted the development of a version of the beat-frequency model in which the accreting matter contains random clumps and the X-ray signal consists of oscillating shots (each shot is the accretion of one clump, the oscillation is due to a magnetic gating mechanism operating at the beat frequency; Lamb et al. 1985). A prediction of this model is that, somewhat dependent on the assumed QPO wave form, the power in the LFN is at least about equal to the power in the QPO.

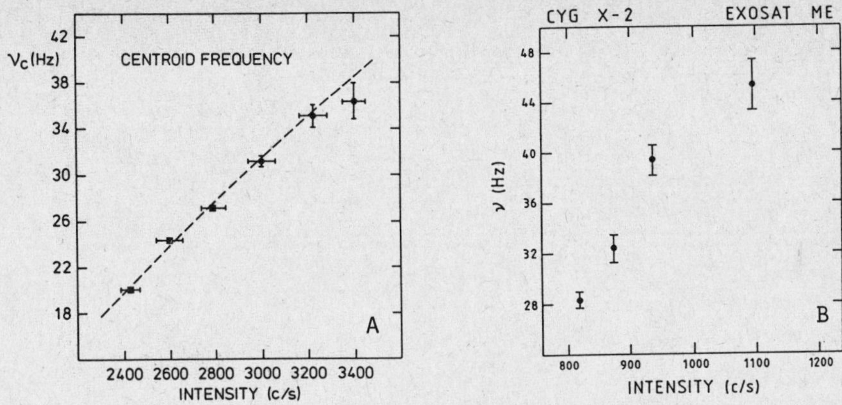

Figure 4. QPO frequency - X-ray intensity relation as observed in GX 5-1 (a, van der Klis et al. 1985b) and Cyg X-2 (b, after Hasinger 1986a) on the horizontal branch.

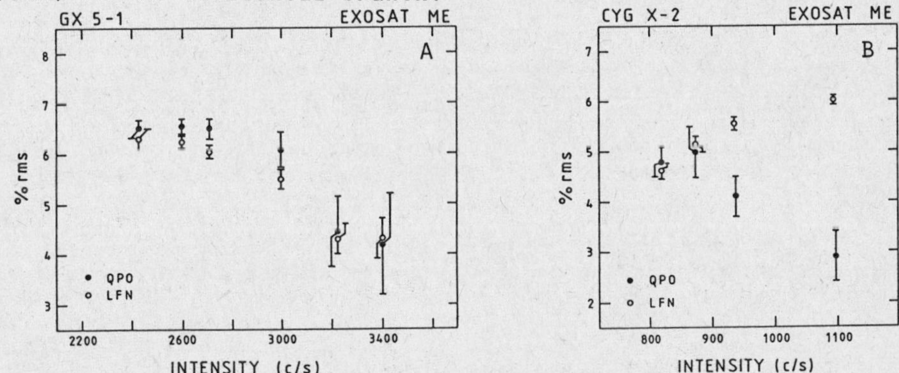

Figure 5. QPO (dots) and LFN (circles) strength as a function of X-ray intensity as observed in GX 5-1 (a, after van der Klis et al. 1985b) and Cyg X-2 (b, after Hasinger et al. 1986a).

3. Sco X-1

a) Bimodal behaviour

An X-ray light curve of Sco X-1 shows a succession of active states containingflaring episodes and quiescent states, which alternate on a time scale of hours to days (Figure 6b). During an active state, intensity dips with a duration of 10-30 min. sometimes occur where the intensity drops to the quiescent level.

The QPO in Sco X-1 were first discovered in the quiescent state (Middleditch and Priedhorsky 1985a,b). Later QPO were also found in the dips in between flaring episodes (van der Klis et al. 1985d, 1986a; Priedhorsky et al. 1986). Figure 6a shows both types of QPO in a time vs. frequency diagram where the power in the intensity variations at any given time and frequency is gray-scale coded (darker for higher power). In the active state the QPO centroid frequency varies between

10 and 20 Hz; in quiescence it remains near 6 Hz. A gradual transition
between the two states is visible.

The frequency-intensity relation of the QPO in the two states is
different: in the dips in the active state a strong positive
correlation very similar to that in GX 5-1 and Cyg X-2 is observed (van
der Klis et al. 1985d), while in quiescence a weak anti-correlation is
seen (Middleditch and Priedhorsky 1985b). Active-state QPO have also
been seen just after the end of the last flaring episode of an active
state, when the source starts a transition into quiescence (van der
Klis et al. 1986a). QPO which show rapid variations (time scales of
minutes) in frequency between 6 and 20 Hz with no single relation to
intensity have been seen on several occasions ('intermediate' state,
van der Klis et al. 1986a; Pollock et al. 1986). The time history of
the QPO in the intensity-frequency diagram can in these cases be loop-
like (both clock- and counter-clock-wise loops have been seen).

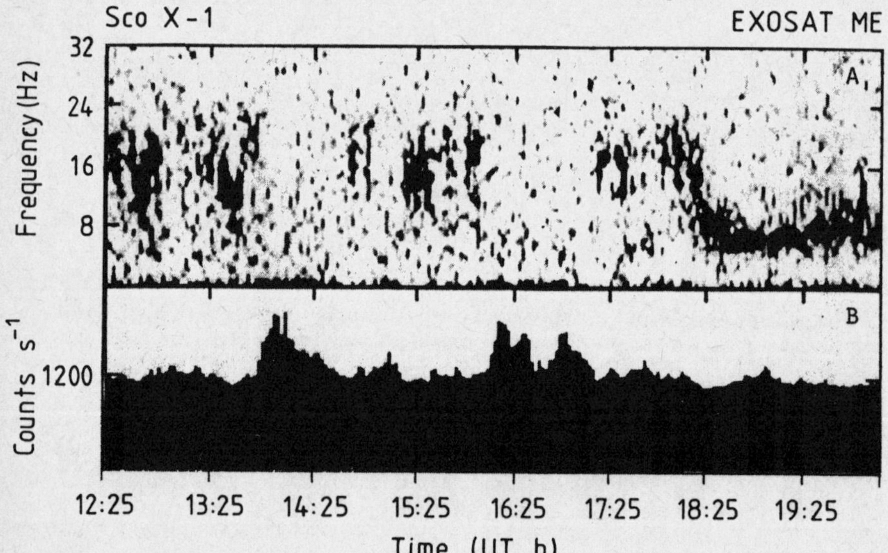

Figure 6. Dynamic power spectrum (a) and X-ray light curve (b) of
Sco X-1 (Priedhorsky et al. 1986).

Figure 7 shows a sample of the frequency-intensity relations observed
in active, 'intermediate' and quiescent states.

When the source switches from one intensity state to the other, its
X-ray spectral characteristics also change: during the flares the
spectrum is hardest, but <u>for the same source intensity</u> the quiescent
spectrum is harder than the spectrum in the active state (White et al.
1976). Figure 8 shows this hardness ratio vs. intensity behaviour as
observed with EXOSAT (Priedhorsky et al. 1986); the quiescent and
active branches are clearly seen. There is a gradual increase of QPO
frequency from 6 to 20 Hz as we move from the quiescent branch via the
branching point into the active branch.

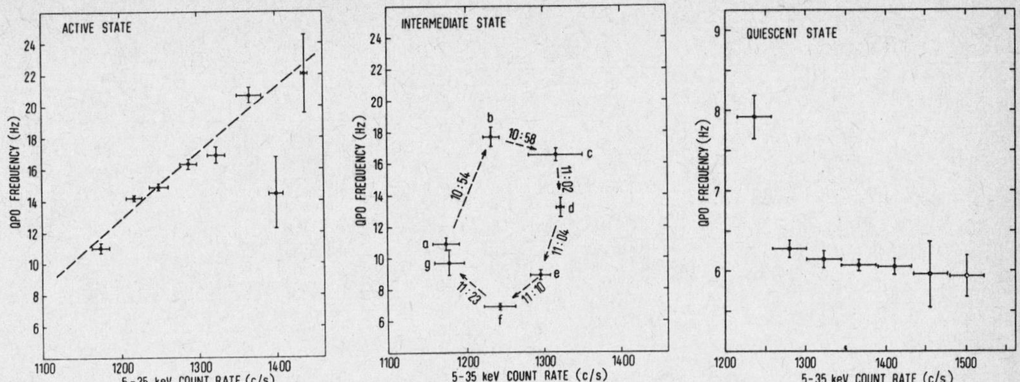

Figure 7. QPO frequency - X-ray intensity relations as observed in Sco X-1 (van der Klis et al. 1986a).

Decomposition of the X-ray spectrum into two components, a blackbody component believed to originate at the neutron star surface and a component with a shape similar to that expected from unsaturized comptonization of soft photons shows that the flaring is dominated by increases in the blackbody component (White et al. 1985; but see also Mitsuda et al. 1985; Figure 9). Contrary to the case of the relations between QPO frequency and total source intensity (Figure 7), there seems to be one simple relation between frequency and blackbody luminosity (van der Klis et al. 1986a; Figure 10).

The fact that different intensity-frequency relations apply in Sco X-1 at different times makes the magnetospheric hypothesis less compelling. The basic prediction of magnetospheric QPO models is a correlation between accretion rate and QPO frequency. To explain the complexity of the observed intensity-frequency relations in Sco X-1 within this framework, an additional hypothesis is required about the relation between accretion rate and X-ray intensity (and -spectrum), and this

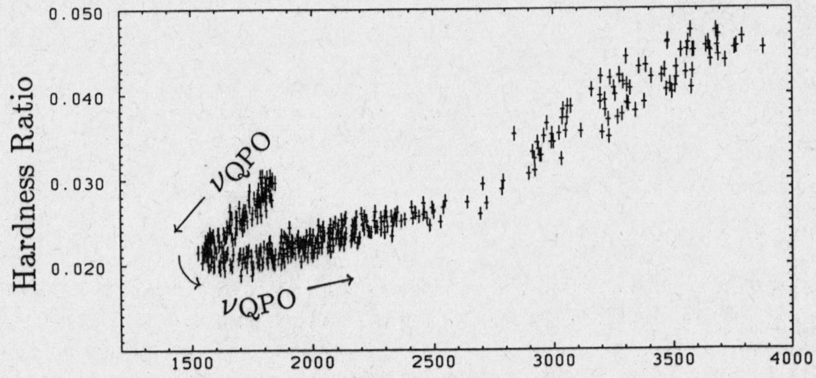

Figure 8. Hardness ratio - intensity diagram of Sco X-1. The continuous increase of QPO frequency along the two-branched curve is indicated (Priedhorsky et al. 1986).

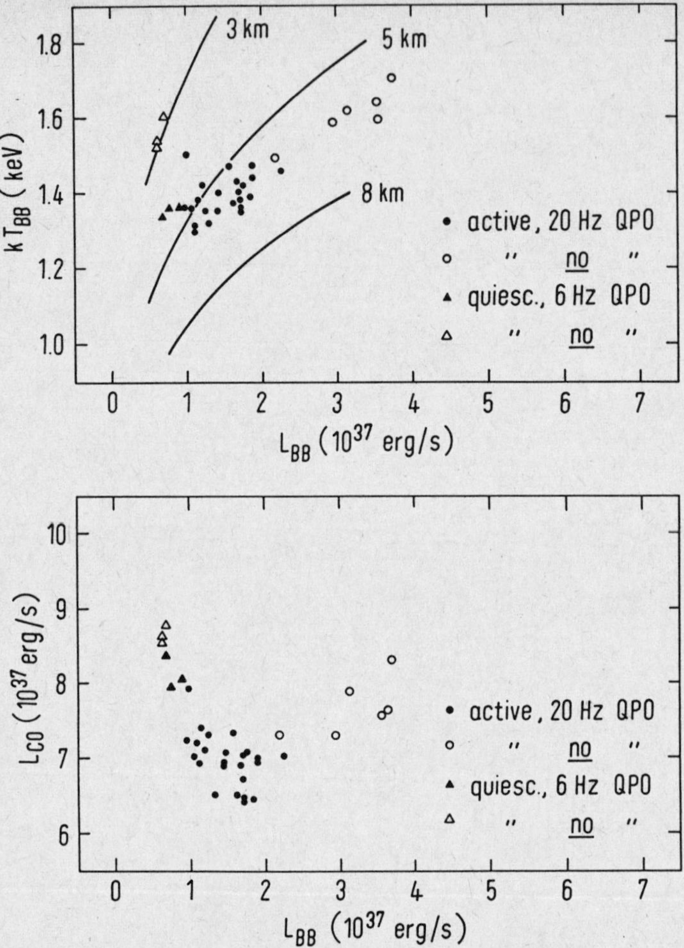

Figure 9. Spectral properties of Sco X-1. Upper frame: blackbody temperature vs. blackbody luminosity. Lower frame: luminosity of comptonized component vs. luminosity of blackbody component. The presence or absence of QPO is indicated (van der Klis et al. 1986a).

additional hypothesis could be chosen in such a way that <u>other</u> than magnetospheric models can be accommodated. Priedhorsky (1986) has proposed that rotation energy fed back from the neutron star into the disk can cause extra luminosity generation in the disk and thereby complicate the accretion rate - X-ray intensity relation. An alternative possibility is that the intensity variations are strongly affected by geometric effects. Van der Klis et al. (1986a) have discussed a model in which the QPO are generated by oscillations in a thick inner accretion disk which partially obscures the X-ray source and in which both QPO frequency and X-ray intensity are determined by the detailed geometry of the obscuring inner disk (which may in turn depend an accretion rate). Both models can probably explain the observed simple relation between blackbody luminosity and QPO frequency.

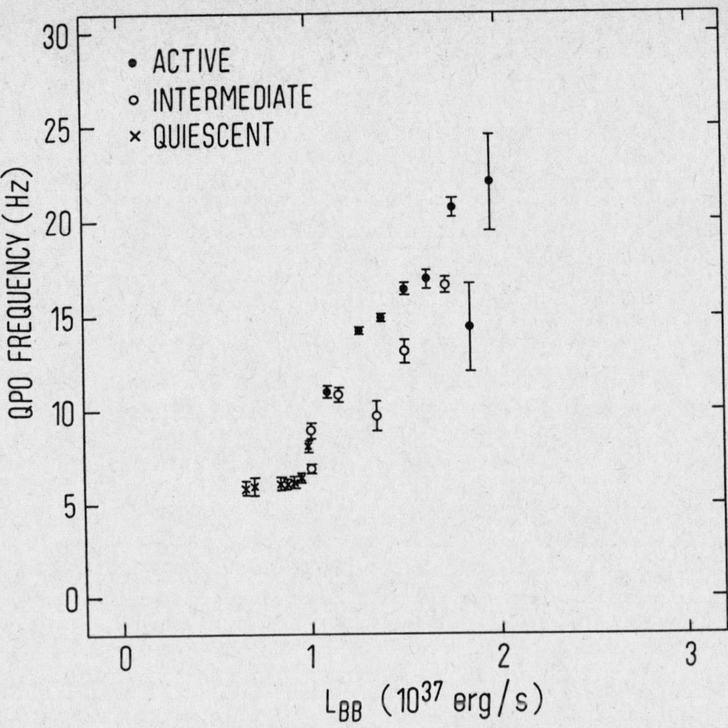

Figure 10. QPO frequency vs. blackbody luminosity in Sco X-1. These are the same data as in Figure 7 (van der Klis et al. 1986a).

b) Low-frequency noise

The prediction of the 'random clump accretion beat-frequency model' that QPO and LFN should have approximately the same power is strongly violated by Sco X-1 (van der Klis et al. 1985d; and even more strongly by the rapid burster, Stella 1985). Figure 11 shows a typical QPO power spectrum of Sco X-1. Some LFN is present, but its power is much

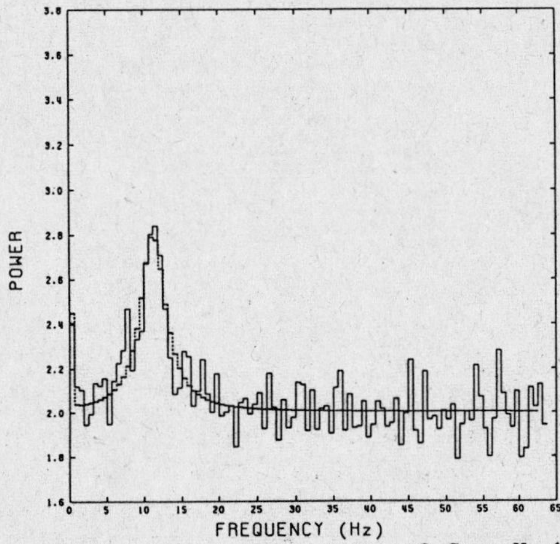

Figure 11. A typical QPO power spectrum of Sco X-1. Note the weakness of the LFN as compared to Figure 3.

less than that of the QPO (usually less than 10% of it down to 1/64 Hz, van der Klis et al. 1986a). Contrary to what is the case for GX 5-1 and Cyg X-2, in Sco X-1 the LFN is found to fit well to a power-law shape. Figure 12 shows how the relative strength of LFN and QPO depend on source intensity: when the source becomes brighter, the QPO disappear but the LFN gets stronger.

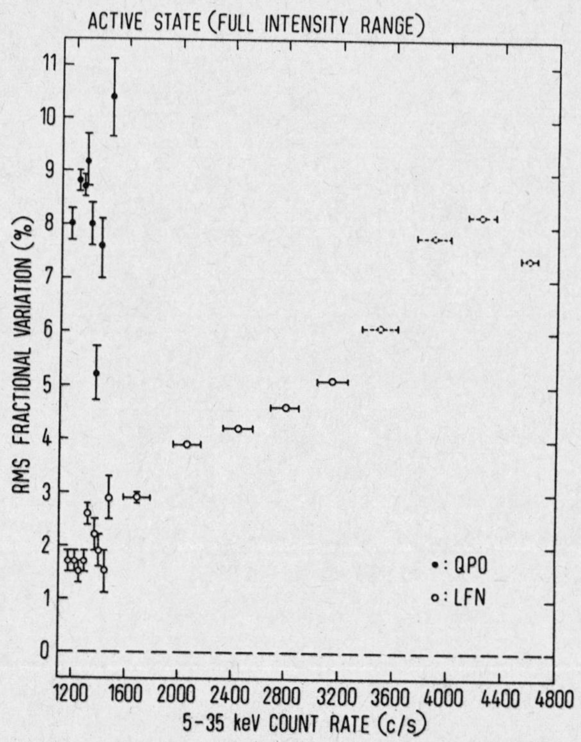

Figure 12. QPO (dots) and LFN (circles) strength as a function of X-ray intensity as observed in Sco X-1 in the active state. The QPO disappears when the source flares (van der Klis et al. 1986a).

To explain the lack of LFN within the beat-frequency model it is necessary to postulate ad-hoc mechanisms to 'line-up' some of the accreting clumps, so that the phase of the QPO can be kept from one oscillating shot to the next (Lamb et al. 1985) or to produce an 'ever-lasting' clump in the Keplerian flow (Shaham 1986). By contrast, in an obscuration model for the QPO it depends on the details of the obscuration geometry (such as inclination) whether the QPO signal consists of oscillating shots (and thus shows QPO-related LFN) or of wave trains with a zero time integral, so that no QPO-related LFN is produced (van der Klis et al. 1986a).

4. <u>Bimodal behaviour of GX 5-1 and Cyg X-2</u>

a) QPO

Both Cyg X-2 (Branduardi et al. 1980) and GX 5-1 (Shibazaki and Mitsuda 1983) show two-branched spectral behaviour somewhat reminiscent of, but not identical to that of Sco X-1. Hardness ratio vs. intensity diagrams as observed with EXOSAT are given in Figures 13 and 14. In the case of GX 5-1 (Figure 13), the pattern is very reproducible. There are two branches, a 'horizontal' branch (Shibazaki and Mitsuda 1983), where the X-ray spectrum slightly softens for increasing source intensity, and a 'normal' branch which shows a positive correlation of intensity and hardness. (This second type of variability is sometimes called 'Sco X-1

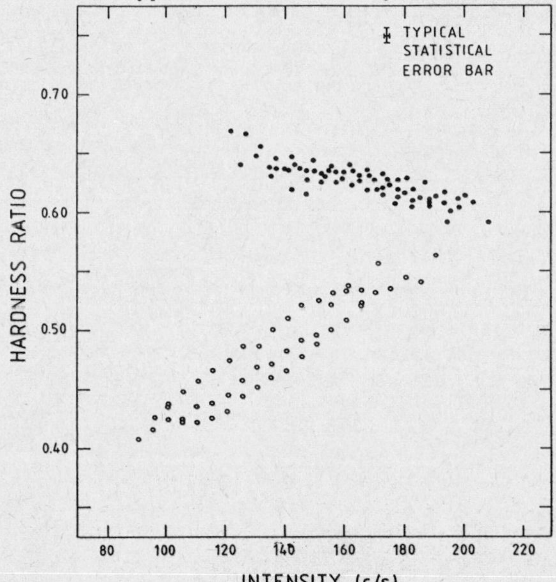

Figure 13. Hardness ratio - intensity diagram of GX 5-1. Dots indicate the presence of strong intensity-dependent QPO, circles a flat power spectrum (see Figure 15). Small systematic offsets between circles are consistent with being instrumental (van der Klis et al. 1986b).

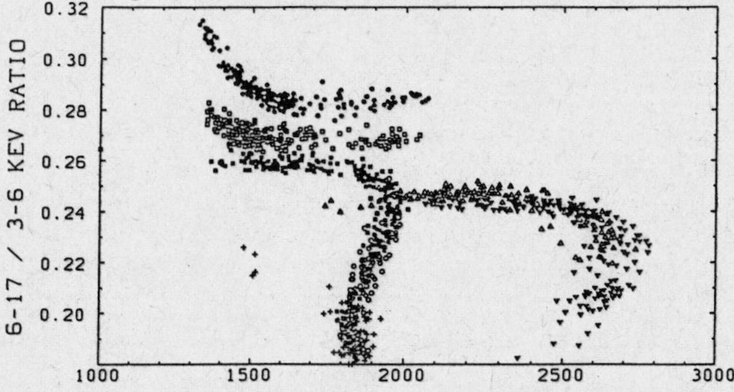

Figure 14. Hardness ratio - intensity diagram of Cyg X-2. Different symbols refer to different observations (Hasinger et al. 1986b).

like', but in the light of the above this is clearly an ambiguous expression.) In Cyg X-2 the same two types of branches can be distinguished (Figure 14), but they can occur in different regions of the hardness ratio - intensity plane (Hasinger et al. 1986b). In both cases, the strong 20-50 Hz intensity-dependent QPO and associated LFN described in Section 2 are only seen in the horizontal branch (van der Klis et al. 1985d, Hasinger et al. 1985b). Figure 15 shows representative power spectra in the two spectral states of GX 5-1; the

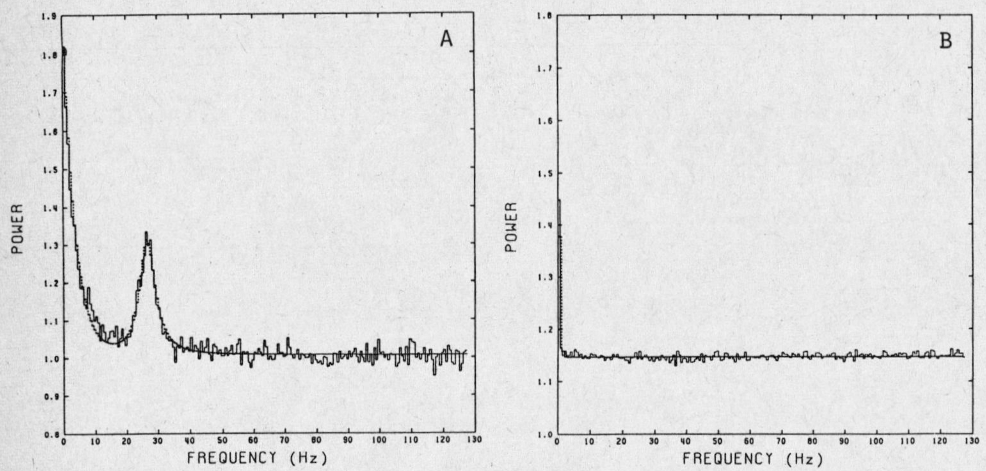

Figure 15. Typical horizontal branch (a) and normal branch (b) power spectra of GX 5-1 (van der Klis et al. 1986b).

normal branch power spectrum is consistent with being flat at frequencies above 1 Hz, while the horizontal branch spectrum shows strong QPO/LFN. In Cyg X-2, ~5.6 Hz QPO whose frequency does not depend perceptibly on intensity are seen in parts of the normal branch (Hasinger et al. 1986b; Figure 16). There is an indication that ~5 Hz QPO may also occur in GX 5-1 on the normal branch (van der Klis et al. 1986b; Figure 17), but the feature is too weak to be detectable independently of the Cyg X-2 result.

Although GX 5-1 and Cyg X-2 show, just as Sco X-1, two branched spectral behaviour with strongly intensity-dependent relatively high-frequency QPO occuring exclusively on one branch and slower (5-6 Hz), much less intensity-dependent QPO on the other, there are also differences. The most important of these differences is the fact that in Sco X-1 QPO frequency gradually increases when the source evolves from the quiescent branch into the active branch, while in GX 5-1 and Cyg X-2 there is a discontinuity in QPO frequency between the two spectral states. While an analogy between the GX 5-1 and Cyg X-2 horizontal- and normal-branch behaviour seems likely, active and quiescent state in Sco X-1 may be something different. This makes classification of the QPO observed in other sources, for example into 'high- and low-frequency' QPO (Stella 1985) uncertain.

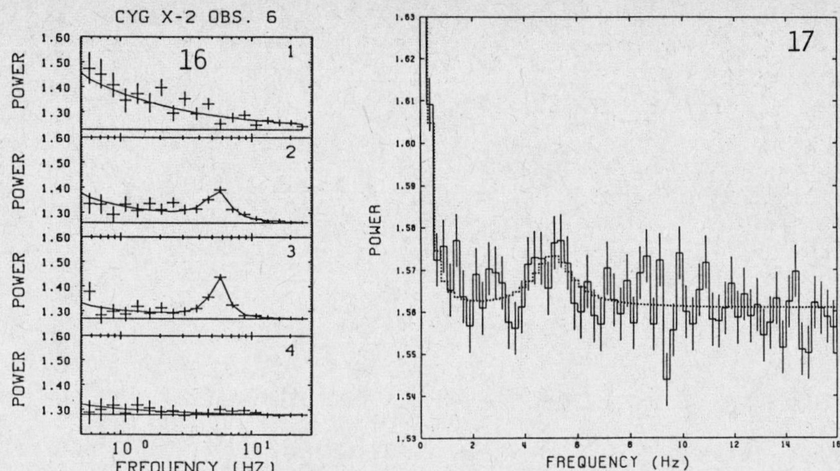

Figure 16. Intensity-selected normal branch power spectra of Cyg X-2 showing 5.6 Hz QPO (Hasinger et al. 1986b).

Figure 17. Summed power spectrum of all normal branch EXOSAT data on GX 5-1 showing possible QPO near 5 Hz. (van der Klis et al. 1986b).

b) Low-frequency noise and <u>very</u> low-frequency noise

Exploration of the power spectrum of GX 5-1 down to frequencies of 10^{-3} Hz (van der Klis et al. 1986b) shows the presence, in both spectral states, of a power-law shaped componenent of the LFN which dominates the power spectrum below 0.5 to 0.1 Hz (Figure 18). This 'very low-frequency noise' (VLFN), which corresponds to the variations of the X-ray flux on time scales of seconds to hours which have long been known to characterize the variability of bright galactic bulge sources (Forman et al. 1976, Parsignault and Grindlay 1978, Ponman 1982), has

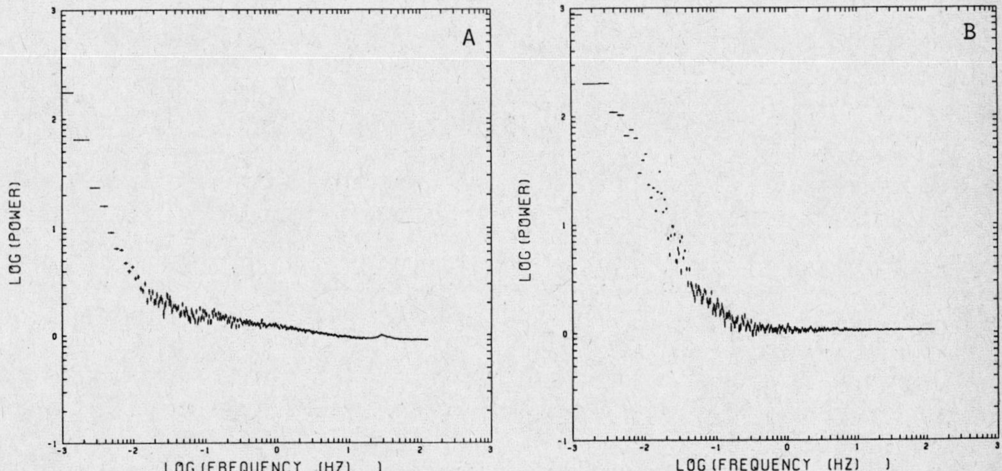

Figure 18. Low-frequency noise down to 10^{-3} Hz in GX 5-1 in the horizontal (a) and normal (b) branch. Steep, power-law shaped, 'very low-frequency noise' is detected in both cases below 0.5-0.1 Hz. QPO and associated LFN up to 20 Hz are only seen on the horizontal branch (van der Klis et al. 1986b).

quite different properties from the QPO-related LFN. Figure 19 shows
its dependence (in the horizontal branch spectral state) on X-ray
intensity as compared with that of QPO and related LFN. While QPO and
LFN decrease in relative strength, VLFN remains approximately constant
as a function of X-ray intensity. In Figure 20 the spectral hardness of
the three timing components is compared. QPO and LFN have an X-ray
spectrum harder than the average flux while VLFN (on the horizontal

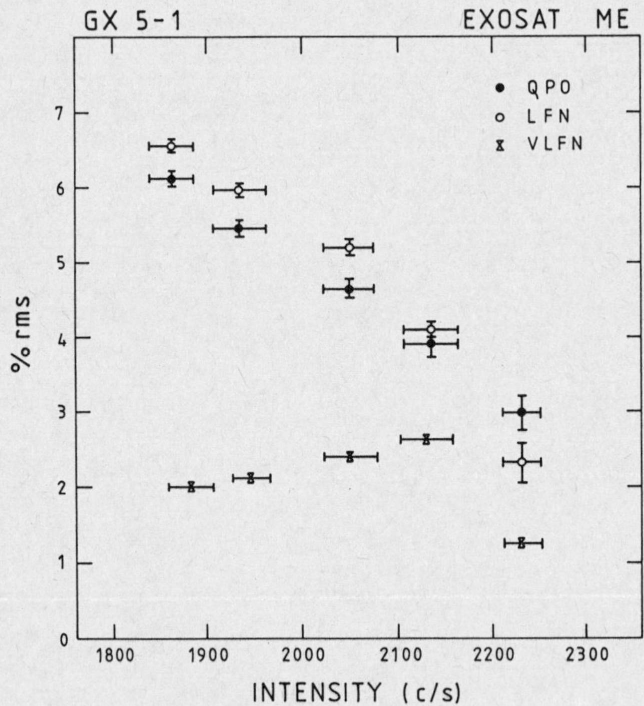

Figure 19. Intensity dependence of QPO, LFN and VLFN in GX 5-1 on the
horizontal branch (van der Klis et al. 1986b).

branch) has a flat relative spectrum. These differences suggest that
while the LFN is related to the QPO, the VLFN is not. The properties of
the VLFN (low frequency, power-law shape, different intensity
dependence from QPO) are very reminiscent of those of the LFN observed
in Sco X-1. If the LFN in Sco X-1 is indeed unrelated to the QPO, then
this would increase the difficulty to account for the observed power
spectrum of this source in oscillating shot models such as the random
clump accretion beat-frequency model.

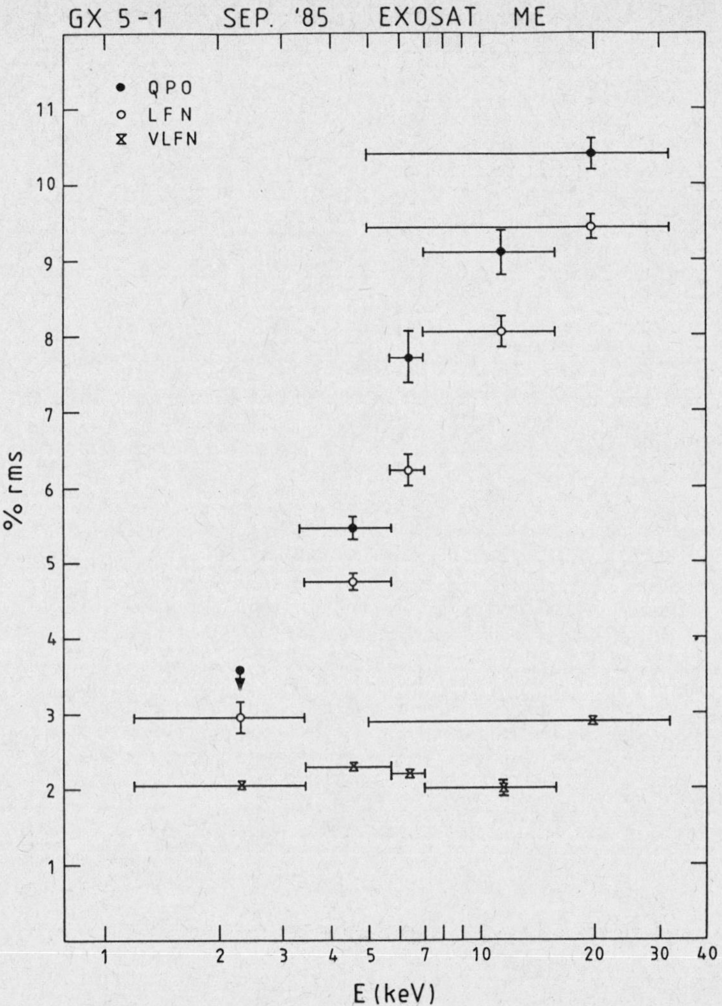

Figure 20. Energy dependence of QPO, LFN and VLFN in GX 5-1 on the horizontal branch (Jansen et al. 1986).

5. <u>Summary</u>

Strong, persistent, high-frequency QPO have been discovered with EXOSAT in seven luminous, persistently bright low-mass X-ray binaries. The detailed properties of these QPO, in particular their dependence on source intensity, are diverse. In the three best-studied sources, GX 5-1, SCO X-1 and Cyg X-2 the known bimodal X-ray spectral behaviour is in all cases found to be strictly correlated to bimodal QPO behaviour. However, while the bimodal QPO properties of GX 5-1 and Cyg X-2 are very similar, those of Sco X-1 are qualitatively different.

A large variety is observed in the relations between source intensity and QPO frequency. If QPO frequency is simply related to accretion rate, as is the prediction of most proposed QPO models, then a matching large variety of relations between accretion rate and X-ray intensity will be required - a strong departure from the traditionally assumed strict proportionality.

QPO-related LFN is only detected in part of the QPO sources. This presents problems for oscillating-shot models such as the random clump accretion beat-frequency model, and could be more easily accounted for in an obscuration model.

Acknowledgements

Stimulating conversations with many participants of the Workshop on 'Accretion onto Compact Objects', April 1986, Tenerife, Spain, the Workshop on 'HE and VHE Behaviour of Accreting X-ray Sources', May 1986, Vulcano, Italy and the 125th IAU Symposium on 'The Origin and Evolution of Neutron Stars', May 1986, Nanjing, China are gratefully acknowledged. I thank Guenther Hasinger and Bill Priedhorsky for permission to reproduce diagrams of their work before publication.

References to Table

1. Hasinger et al. 1985a; 2. Hasinger et al. 1986a; 3. Norris and Wood 1985; 4. Hasinger et al. 1985b; 5. Van der Klis et al. 1985a; 6. Van der Klis et al. 1985b; 7. Van der Klis et al. 1985c; 8. Van der Klis, these proc.; 9. Stella et al. 1985c; 10. Stella 1985; 11. Stella et al. 1984; 12. Middleditch and Priedhorsky 1985a; 13. Van der Klis et al. 1985d; 14. Middleditch and Priedhorsky 1985b; 15. Van der Klis et al. 1986a; 16. Priedhorsky et al. 1986; 17. Lewin et al. 1985; 18. Cooke et al. 1985; 19. Lewin et al. 1986; 20. Stella et al. 1985a; 21. Tawara et al. 1982; 22. Stella et al. 1985b; 23. Tennant 1986; 24. Belli et al. 1986; 25. Morini, priv. comm. 1986; 26. Van der Klis and Jansen 1985; 27. Joss et al. 1978; 28. Li et al. 1980.

References

- Alpar, M.A. and Shaham, J., 1985a, IAU Circ. 4046.
- Alpar, M.A. and Shaham, J., 1985b, Nature 316, 239.
- Belli, B.M., d'Antona, F., Molteni, D. and Morini, M., 1986, IAU Circ. 4174.
- Bradt, H.V.D. and McClintock, J.E., 1983, Ann. Rev. Astron. Ap. 21.
- Branduardi, G., Kylafis, N.D., Lamb, D.Q. and Mason, K.O., 1980, Ap. J. 235, L153.
- Cooke, B.A., Stella, L. and Ponman, T., 1985, IAU Circ. 4116.
- Forman, W., Jones, C. and Tanaubaum, H., 1976, Ap. J. 208, 849.
- Gribbin, J.B., Feldman, P.A. and Plagemann, S.H., 1970, Nature 225, 1123.
- Hasinger, G., Langmeier, A., Sztajno, M. and White, N., 1985a, IAU Circ. 4070.
- Hasinger, G., Langmeier, A., Sztajno, M., Pietsch, W. and Gottwald, M., 1985b, IAU Circ. 4153.
- Hasinger, G., Langmeier, A., Sztajno, M., Truemper, J., Lewin, W.H.G. and White, N.E., 1986a, Nature 319, 469.

- Hasinger, G., Langmeier, A., Sztajno, M. and Pietsch, W., 1986b, in prep.
- Jansen, F.A. et al., 1986, in prep.
- Joss, P.C., Avni, Y. and Rappaport, S., 1978, Ap. J. 221, 645.
- Lamb, F.K., Shibazaki, N., Alpar, M.A. and Shaham, J., 1985, Nature 317, 681.
- Lewin, W.H.G., van Paradijs, J., Jansen, F., van der Klis, M., Sztajno, M. and Truemper, J.E., 1985, IAU Circ. 4101.
- Lewin, W.H.G., van Paradijs, J., Hasinger, G., Penninx, W.H., van der Klis, M., Jansen, F., Langmeier, A., Sztajno, M. and Truemper, J., 1986, IAU Circ. 4170.
- Li, F.K., Joss, P.C., McClintock, J.E., Rappaport, S. and Wright, E.L., 1980, Ap. J. 240, 628.
- Matsuoka, M., 1985 in 'Cataclysmic Variables and Low-Mass X-ray Binaries', D.Q. Lamb and J. Patterson (eds.), D. Reidel, p.139.
- Middleditch, J. and Priedhorsky, W., 1985a, IAU Circ. 4060.
- Middleditch, J. and Priedhorsky, W.C., 1985b, preprint LA-UR-85-3649 to appear in Ap. J.
- Mitsuda, K. et al., 1985, Publ. Astron. Soc. Japan 36, 741.
- Norris, J.P. and Wood, K.S., 1985, IAU Circ. 4087.
- Parsignault, D.R. and Grindlay, J.E., 1978, Ap. J. 225, 970.
- Pollock, A.M.T., Carswell, R.F. and Ponman, T.J., 1986, poster presented at Workshop on 'Accretion onto Compact Objects', Tenerife, April 1986.
- Ponman, T., 1982, MNRAS 201, 769.
- Priedhorsky, W., Hasinger, G., Lewin, W.H.G., Middleditch, J., Parmar, A., Stella, L. and White, N., 1986, preprint, to appear in Ap. J. (Lett.).
- Priedhorsky, W., 1986, preprint, to appear in Ap. J. (Lett.).
- Shibazaki, N. and Mitsuda, K., 1983, ISAS RN 234, 63.
- Shaham, J., 1986, paper presented at IAU Symp. 125, Nanjing, China, May 1986.
- Stella, L., Kahn, S.M. and Grindlay, J.E., 1984, Ap. J. 282, 713.
- Stella, L., Parmar, A.N. and White, N.E., 1985a, IAU Circ. 4102.
- Stella, L., Parmar, A.N., White, N.E., Lewin, W.H.G. and van Paradijs, J., 1985b, IAU Circ. 4110.
- Stella, L., White, N.E. and Priedhorsky, W., 1985c, IAU Circ. 4117.
- Stella, L., 1985, EXOSAT prep. 25.
- Tawara, Y., Hayakawa, S., Kunieda, H., Makino, F. and Nagase, F., 1982, Nature 299, 38.
- Terrell, N.J., 1972, Ap. J. 174, L35.
- Tennant, A., talk presented at Workshop on 'Accretion onto Compact Objects', Tenerife, April 1986.
- Van der Klis, M. and Jansen, F.A., 1985, Nature 313, 768.
- Van der Klis, M., Jansen, F., van Paradijs, J., Lewin, W.H.G., Truemper, J. and Sztajno, M., 1985a, IAU Circ. 4043.
- Van der Klis, M., Jansen, F., van Paradijs, J., Lewin, W.H.G., van den Heuvel, E.P.J., Truemper, J.E. and Sztajno, M., 1985b, Nature 316, 225.
- Van der Klis, M., Jansen, F., van Paradijs, J., Lewin, W.H.G., Truemper, J. and Sztajno, M., 1985c, IAU Circ. 4140.
- Van der Klis, M., Jansen, F., White, N., Stella, L. and Peacock, A., 1985d, IAU Circ. 4068.
- Van der Klis, M., Stella, L., White, N., Jansen, F. and Parmar, A.N., 1986a, preprint, submitted to Ap. J.
- Van der Klis, M. et al. 1986b, in prep.
- White, N.E., Mason, K.O., Sanford, P.W., Ilovaisky, S.A. and Chevalier, C., 1976, MNRAS 176, 91.
- White, N.E., Peacock, A. and Taylor, B.G., 1985, Ap. J. 296, 475.

Quasi-Periodic Oscillations

Models vs. Observations - a Brief Review

Walter H. G. Lewin

Massachusetts Institute of Technology

Cambridge, MA 02139, USA.

Abstract. A brief review is given of the models that have recently been proposed to explain the high-frequency quasi-periodic oscillations observed in several bright low-mass X-Ray binaries. We do not yet know what causes the oscillations, not even whether they are magnetospheric in origin. However, some proposed ideas could well be relevant to the various rather complex aspects of the oscillations. It is likely that more than one mechanism is at work.

1. Models vs. Observations - a Brief Review.

1.1. QPO in Cataclysmic Variables

Quasi-periodic oscillations (QPO) are commonly observed in the optical flux (in a few cases also in the X rays) of dwarf novae in outburst (dwarf novae are accreting white dwarfs; see e.g., Robinson and Nather 1978; Patterson 1981; Cordova and Mason 1983). The timescale of these oscillations ranges from ~10-10^3 sec (frequency ~1 mHz - 0.1 Hz), and the coherence ranges from a few oscillations up to ~10^5 oscillations. Optical QPO with very different coherence can be observed simultaneously at two frequencies. The strength of the optical oscillations is typically less than 1%. Many models have been proposed to explain these oscillations. Not one alone can explain the whole range of complex phenomena; it is almost certain that more than one mechanism is at work.

It is not surprising that the recent models on high-frequency (larger than ~1 Hz) QPO observed in the bright low-mass X-ray binaries (LMXB) reflect some earlier ideas for optical QPO in the Cataclysmic Variables. In scaling the geometry of an accreting white dwarf to that of an accreting neutron star, it is not too difficult to imagine a frequency increase by a factor of order $\sim 10^{2\text{-}3}$.

1.2. Long-Period QPO in LMXB

Before I discuss some of the current models for the high-frequency QPO in the bright LMXB, I want to mention that long-period QPO of $\sim 10\text{-}10^3$ sec have also been observed in several X-ray binaries. The QPO spectrum can be softer as well as harder than the mean source spectrum. The fraction of the modulations in the flux (thus the strength of the QPO) can be enormous (typically ~50%). I list here those cases that I am aware of in sequence of increasing frequency from ~1 mHz to ~2 Hz (1626-67 Joss, Avni and Rappaport 1978; Li et al. 1980; Cyg X-3 Van der Klis and Jansen 1985; GX 349+2 Matsuoka 1985; Her X-1 Voges et al. 1985; 1820-30 Stella, Kahn and Grindlay 1984; GX 339-4 Maejima et al. 1984). Perhaps the 1.5-h oscillations observed by Langmeier et al. (1985) in GX 17+2 are also quasi-periodic; this would then extend the range of long-period QPO in LMXB up to periods of $\sim 5 \times 10^3$ sec. I suspect that these long-period QPO have a different origin than the short-period QPO. However, it is unclear as yet where the dividing line is (perhaps somewhere between 0.01 and 1 Hz).

1.3. QPO Models

The QPO models that I will now discuss have been proposed to explain some of the recent observations of high-frequency QPO in LMXB (Alpar and Shaham 1985a,b; Berman and Stollman 1985; Lamb et al. 1985; Lamb 1986; Hameury, King and Lasota 1986; Boyle, Fabian and Guilbert 1986; Morfill and Truemper 1986a,b).

It is interesting to look at the evolution of some models in historical perspective. The excitement and activity started with the discovery of the intensity-dependent QPO in GX 5-1 in early 1985 (Van der Klis et al. 1985a,b). The centroid frequency of the QPO,

ν, was linearly related to the observed source intensity, I, ($\nu \simeq 1.9 \times 10^{-2} I$ -25 Hz) over the observed range of I from about 2400–3400 cts/sec (1–18 keV). No one paid much attention to this linear relation then, and not now, and even though we mention the linearity in our paper, we do not give it quantitatively (Van der Klis et al. 1985b). Our data can also be fit by a power-law dependence ($\nu \simeq 6.9 \times 10^{-6} I^{1.9}$ Hz). The observed exponent (which I will designate α) led Alpar and Shaham to the idea that QPO could be the result of a beat phenomenon, as earlier suggested by Warner (1983) for optical QPO in cataclysmic variables. Another striking relation was present between the strength of the QPO and that of the low-frequency noise (LFN); the two went "hand in hand" (Van der Klis et al. 1985b).

1.3.1. The Bath Model.

Before I expand on the beat frequency idea, I want to remind you of a model suggested 13 years ago by Geoffrey Bath to explain optical QPO in cataclysmic variables (Bath 1973). He suggested that the QPO frequency was that of the Kepler frequency of matter orbiting a magnetized white dwarf at the magnetopause (inner edge of the accretion disk). This idea can be adapted to magnetized accreting neutron stars in an effort to explain the observed high-frequency QPO in X rays; I will refer to this hereafter as "the Bath model". The radius of the inner edge of the accretion disk, r, depends on the mass and radius of the neutron star, on the magnetic dipole field strength, B, at the surface of the neutron star, and on the mass transfer rate $\dot{M}$ at that radius (see Lamb, Pethick and Pines 1973). Combining the relation between r and $\dot{M}$ with Kepler's law, and assuming that the Kepler frequency, ν_K, equals the QPO frequency, ν, (Bath model) one can easily find that

$$\nu \propto \dot{M}^{3/7} \qquad (1).$$

If we now make the assumption that $\dot{M}$ depends linearly on the observed broad-band X-ray intensity, I, we find that the observed QPO frequency, ν, should be proportional to I to the power 3/7, thus:

$$\nu \propto I^{3/7} \qquad (2).$$

This, however, was not observed for GX 5-1; α was ~2, and not 3/7 (Van der Klis et al. 1985a,b).

1.3.2. The Beat Frequency Model .

In the beat frequency model (Alpar and Shaham 1985a,b), the observed QPO frequency is the difference between the Kepler frequency at the magnetopause, ν_K, and the rotation frequency, ν_n, of the neutron star

$$\nu = \nu_K - \nu_n \qquad (3).$$

If again the assumption is made that the mass transfer rate, $\dot{M}$, at the magnetopause depends linearly on the observed broad-band X-ray intensity, one finds that

$$\alpha = 3\nu_K/7\nu \qquad (4).$$

Combining eqs. 3 and 4 leads to

$$\nu_n = \nu(7\alpha/3 - 1) \qquad (5).$$

If, as an example, we take an approximate average value for the observed QPO frequency, ν, in GX 5-1 of 30 Hz, then, with $\alpha \approx 2$ (as observed), $\nu_K \approx 140$ Hz (eq. 4), and $\nu_n \approx 110$ Hz (eq. 5). Assuming a mass for the neutron star of 1.4 $M_\odot$, with Kepler's law we can then also find the radius of the inner edge of the accretion disk (here ≈ 55 km). If, in addition, one assumes a radius for the neutron star (e.g., 10 km), and one estimates the mass transfer rate at the magnetopause (this can be done from an estimate of the distance to the source, and from the observed broad-band X-ray intensity), one also finds the magnetic dipole field strength, B, at the surface of the neutron star, using the equations given by Lamb, Pethick and Pines (1973). For GX 5-1, B $\approx 6 \times 10^9$ G.

As we just saw, in the case of GX 5-1, the beat frequency model (Alpar and Shaham 1985a,b), leads to the rotation period of the neutron star, and to its magnetic dipole field strength. These results were very pleasing as they fitted well into an existing evolutionary scenario in which the neutron stars in bright LMXB are spun up by accretion and form the progenitors of msec binary radio pulsars (see e.g., Smarr and Blandford 1976; Van den Heuvel 1981; Radhakrishnan 1981; Srinivasan and Van den Heuvel 1982; Alpar et

al. 1982; Radhakrishnan and Srinivasan 1982; Webbink, Rappaport and Savonije 1983;
Taam 1983; Joss and Rappaport 1983; Paczynski 1983; Savonije 1983).

Problems? If the beat frequency idea is correct, it is puzzling that we do not observe in our
GX 5-1 data ~110 Hz **coherent** X-ray pulsations (however, see Lamb et al. 1985, and Lamb
1986). The absence of coherent pulsations may not be the only problem. A few months
after QPO were found in GX 5-1, they were also discovered in Sco X-1 (Middleditch and
Priedhorsky 1985, 1986; see also Priedhorsky et al. 1986; Van der Klis et al. 1986). Here
the observed QPO frequency was **anti-correlated** with the observed source intensity
($\nu \propto I^{-0.6}$). Thus, α was negative. As long as we make the above assumption that $\dot{M}$
is linearly proportional to the observed X-ray intensity, α can not be negative. This can
easily be seen as follows. Any increase in I will then be associated with an increase in
$\dot{M}$. This leads to a smaller radius of the magnetopause, and thus to an increase in the Kepler
frequency. If the Kepler frequency goes up, the QPO frequency will also go up (see eq. 3).

Introduction of the β parameter. Lamb et al. (1985) do not assume that the <u>observed</u> X-ray
intensity is linearly proportional to $\dot{M}$, but rather

$$I \propto \dot{M}^{\beta} \tag{6}$$

Here β can have all values between $-\infty$ and $+\infty$, as can easily be seen. Suppose there is no
change in $\dot{M}$, but there is, e.g., a minute decrease in the observed X-ray intensity as the
result of a small increase in absorption along the line of sight; β would then be $-\infty$. If the
absorption were to decrease, β would become $+\infty$; β can have all values in between. For
"n-fold symmetry" (Lamb et al. 1985; see also below), eq. 3 becomes

$$\nu = n(\nu_K - \nu_n) \qquad [n=1,2,3,......] \tag{7}$$

With the introduction of β we then find for the beat frequency model

$$\alpha = 3n\nu_K/7\beta\nu \tag{8}$$

For $n=1$, and $\nu_K=\nu$ (Bath model), we find

$$\alpha\,\beta = 3/7 \qquad\qquad (9).$$

Combining eqs. 7 and 8, the beat frequency model leads to

$$n\nu_n = \nu(7\alpha\beta/3 - 1] \qquad\qquad (10).$$

The rotation frequency of the neutron star plays no role in the Bath model; the latter is mathematically equivalent to the beat frequency model with $n\nu_n=0$, and $\alpha\,\beta=3/7$ (see eq. 10).

With the introduction of the β parameter (with $\beta \neq 1$) the beat frequency model can obtain any value for the rotation frequency of the neutron star. This can best be illustrated with an example. Priedhorsky et al. (1986) found in Sco X-1 a "6-Hz QPO branch" which was **anti-correlated** with the source intensity ($\alpha \simeq -0.6$), and a "15-20 Hz branch" which was correlated with the source intensity ($\alpha \simeq +3$) (see also Van der Klis et al. 1986). Their results are shown in the figure below. With the observed values for ν and α, (using eq. 10), we can now find values for β to obtain any neutron star rotation rate. For instance for a rotation frequency $n\nu_n=1000$ Hz, β would have to be ~+9 in the 15-20 Hz branch, and ~-120 in the 6-Hz branch. For a rotation rate $(n\nu_n)$ of 100 Hz, the corresponding values for β would have to be ~+1, and ~ -13, respectively. The rotation rate can also be zero (this is mathematically equivalent to the Bath model). Then, the values for β would have to be ~+0.14, and ~ - 0.7, respectively (here, $\alpha\,\beta=3/7$). Near the apex of the "curve" (see figure below), $\alpha = \pm\infty$ (observational fact). If at the same time $\beta=0$ (i.e., the mass transfer rate, $\dot{M}$, changes, but no simultaneous source intensity change is observed), the product $\alpha\,\beta$ becomes undetermined, and this is consistent with any neutron star rotation frequency.

In this context it is interesting to mention that Priedhorsky et al. 1986 (see also Priedhorsky 1986) suggest that in "moving along" the "curve" of the figure below in a clockwise direction, there is a continuous monotonic increase in the mass transfer rate which, however, is not reflected in a continuous increase in the **observed** broad-band X-ray intensity. With increasing mass transfer rate, the **observed** intensity at first decreases (6-Hz branch), and then increases (15-20 Hz branch). If this suggestion is correct, it would mean that at the apex, where the two branches meet, $\beta=0$.

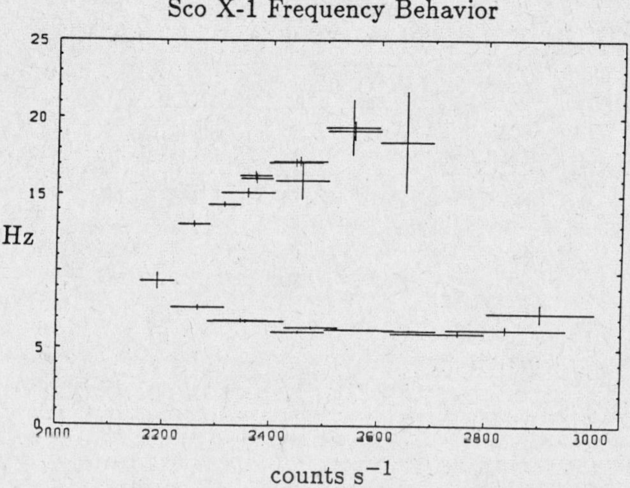

Figure 1: Centroid QPO frequency, ν (in Hz), vs. the 8-20 keV X-ray intensity, I (in cts/sec), observed for Sco X-1 with EXOSAT. In the upper QPO branch (15-20 Hz) the frequency and source intensity are correlated; $\nu \propto I^{+3}$. In the lower branch (~6 Hz), the QPO frequency is anti-correlated with I, here $\nu \propto I^{-0.6}$. This figure is from Priedhorsky et al. 1986.

In my opinion, the main reason for the early rejection of the Bath model in favor of the beat frequency model is no longer valid. With the introduction of $\beta(\neq 1)$, which, of course, is entirely reasonable, the product $\alpha\beta$ counts, and no longer α alone. In the Bath model $\alpha\beta=3/7$, and any observed value of α is allowed (thus also $\alpha \simeq 2$, as observed for GX 5-1, as long as β is then $\simeq 0.21$). In the absence of knowledge of β (β can vary on a timescale of minutes), we are at a loss (however, see Priedhorsky 1986). **Obviously, this does not mean that the beat frequency model is incorrect.** It is too early to decide (see also "note added" below).

A specific beat frequency scenario has been proposed by Lamb, Shibazaki, Alpar and Shaham (1985) (see also Van der Klis et al. 1985b; Berman and Stollman 1985; Lamb 1986). They suggest that the beat phenomenon is the result of blob formation at the inner edge of the accretion disk. The matter in the blobs is gradually stripped off by interaction with the magnetic field which is anchored into a spinning neutron star. The matter will then reach the surface of the neutron star in a quasi-periodic fashion with a frequency equal to the difference between the (variable) Kepler frequency of the blobs and the (fixed) rotation frequency of the neutron star (see eq. 3) [or n times that for n-fold symmetry (see eq. 7)]. While a blob is "milked" in a piecemeal fashion the X-ray flux oscillates (producing QPO), and the mean X-ray flux is temporarily increased. The latter gives rise to low-frequency noise (LFN). The model can be represented by oscillating "shots". The oscillatory part of the shots determines the QPO characteristics; the overall envelopes of the shots determine the LFN characteristics. The blobs are produced at random times, and with a random equatorial azimuthal distribution. For a sufficiently large number of blobs, the "crests" of the oscillations, produced by one blob, will always coincide (or nearly so) with the "valleys" of those produced by others, and both the QPO and the associated LFN will disappear.

In this scenario, LFN is **a logical consequence** of the QPO, and one would expect to observe strong LFN whenever strong QPO are observed. This was the case in GX 5-1 where the QPO strength and the LFN strength went "hand in hand" (see above). However, in later observations of e.g., Sco X-1 and the Rapid Burster this was not the case (Stella et al. 1986; Stella 1986; Lewin 1986; Van der Klis 1986; Van der Klis et al. 1986). Fred Lamb kindly pointed out to me (see also Lamb 1986, and Shaham 1986) that one can adjust the mathematics of the "shots" so that the blobs oscillate more or less in unison (crests from one blob support crests from other blobs). This could lead to strong QPO in the absence (or near absence) of LFN. It remains to be seen, however, whether this mathematical adjustment is physically meaningful.

1.3.3. Other Models. I will now discuss some other models. They will not get as much attention as the model proposed by Alpar and Shaham (1985a,b). This does not mean that I favor their model. It merely reflects that it has been around longer and evolved (Lamb et al. 1985; Lamb 1986) as the observations revealed a more and more complex behavior and richness in the QPO, and LFN characteristics. I thought it was worthwhile to discuss some aspects of this evolution.

Morfill & Truemper (1986a) suggested a magnetospheric beat frequency model in a heavily obscured binary. The supersonic stirring of the disk by an inclined magnetic field causes shock waves which interact with disk inhomogeneities (plasmoids) which are in Keplerian rotation. The plasmoids are hotter than their environment. This results in a quasi-periodic signal at the beat frequency given by eq. 3. Here the QPO signal does not come from the surface of the neutron star but from the plasmoids in the magnetopause. Thus the available energy for the QPO is dictated by the radius of the magnetopause. If the latter were e.g. ten times that of the neutron star, the maximum available energy in the plasmoids (to produce QPO) would be ~10% of the total X-ray flux (for larger radii it would be lower, for smaller radii larger). The QPO mechanism could not be 100% efficient (see Lamb 1986). "Self luminous" models like this one, can therefore perhaps explain QPO with a modest strength, but it is hard to see how they could explain the high percentages observed in e.g., GX 5-1 (up to ~6%), Sco X-1 (up to ~8% at high energy X rays), and the Rapid Burster (up to ~30%) (Tawara et al. 1982, and above references).

In a later version, Morfill and Truemper (1986b) point out that one may expect the formation of large vortices in the disk flow (near the magnetopause), and that vortex shedding (perhaps associated with a characteristic Strouhal frequency) is likely to occur. They show that under certain conditions, and with a constant Strouhal number of ~1.3, the QPO frequency could be about twice the Kepler frequency ($\sim 2\nu_K$). They also mention that other phenomena such as frequency halving or doubling may add further complexity. As an extension on their idea of plasmoid formation, they point out that the observed QPO frequencies could perhaps also be due to the single Keplerian frequency of these plasmoids and/or to a beat frequency as discussed in the above paragraph.

These are all versions of "self-luminous" models, and my comments regarding the rather limited available energy for the QPO should also be valid here. The authors, however, allow for the possibility that the QPO are the result of shadowing of the central source by hot cocoon gas, in which case the limitation in available energy is not a problem (see sect. 1.4).

In their closing comments, Morfill and Truemper (1986b) also suggest that the interaction of the plasmoids with the shocks (the latter are revolving with the neutron star rotation frequency) may bring accretion disc material into co-rotation, and thus making accretion along the field lines onto the polar caps feasible. If that happens, the QPO emission comes from the neutron star surface with a beat frequency (see eqs. 3 and 7); this scenario has similarities with that proposed by Lamb et al. (1985) (see also Lamb 1986).

Hameury, King and Lasota (1985) proposed a scenario which does not require the presence of a magnetopause. The QPO are the result of hot spots rotating in the boundary layer where the accreting matter settles onto the surface of a slowly rotating (with period of order ~1 sec) neutron star. They suggest that transient magnetic fields are generated due to turbulent dynamo action, and that the local hot spots are heated by conduction due to these magnetic fields. QPO are then the result of repeated occultations of the hot spots when they rotate out of our line of sight to the "back side" of the neutron star. The low-frequency noise results from the lifetime of the hot spots, and from those spots that are never occulted due to their high latitude. The authors predict that the QPO spectrum is that of a blackbody with a radius less than that of the neutron star. There is some growing evidence that the QPO spectra are indeed blackbodies for Cyg X-2 (Hasinger, 1986), and Sco X-1 (Van der Klis et al. 1986). A blackbody spectrum, however, is not unique to this model.

The "available" frequencies in this model range all the way from ~1 Hz (at the bottom of the boundary layer, rotating with the neutron star frequency) to ~10^3 Hz (at the very top of the boundary layer, rotating with the Keplerian frequency). If this scenario were operating, it is puzzling why the observed QPO frequencies vary only in such a limited range (e.g., in GX 5-1 by a factor of ~2), and why the width of the peaks in the power spectra of most QPO sources are so narrow (typically ~10% to ~30%). Since here the LFN is a logical consequence of the QPO, it may also be difficult to explain those cases where LFN is nearly absent in the presence of strong QPO (sect. 1.3.2). It is also unclear whether the hot spots could contain such a high fraction of the available gravitational potential energy to explain the observed high strength of QPO (typically ~5% rms variation, but up to ~30% in the Rapid Burster).

A particular hot spot would have to appear and disappear (occultation) at least three or four times to produce the observed QPO. The QPO with a fundamental frequency of ~0.44 Hz in the Rapid Burster (Stella et al. 1986; Stella 1986; Lewin 1986) would thus require a lifetime of the hot spots of at least ~10 sec. According to Jean Paul Lasota (private communication), the lifetime of one hot spot is probably less than a few seconds. Thus, if this is so, the very low-frequency QPO could not be explained with this "sun spot" scenario. It seems quite possible, however, that more than one mechanism is responsible for the various "kinds" of QPO observed (see also sect. 1.2, and Hasinger 1986), and it would therefore be somewhat naive to expect that one model could explain the enormous complexity in QPO behavior.

Boyle, Fabian and Guilbert (1986) proposed that the QPO are produced by a hot ($>10^7$ ^{o}K) accretion disk corona. X rays produced at the central source (the neutron star) scatter off oscillating disturbances in this corona near the inner disk. The oscillations are in a direction perpendicular to the disk plane (disk oscillations); they have a frequency approximately equal to the local Keplerian frequency. Thus, QPO frequencies of ~100 Hz and ~1 Hz, would correspond to an effective scatter radius of ~ 8×10^4 m (~8 stellar radii) and ~1.6×10^6 m (160 stellar radii), respectively. No magnetic field, and no neutron star rotation are required.

The X rays scatter off the $\tau \simeq 1$ surface. Since the observed peaks (QPO) in the power spectra have widths of typically ~10 to ~20%, a very restricted range of radii of this scatter surface is required. The spread in radii of the effectively contributing part of this surface (as seen from Earth) should be no larger than ~30%. Thus this scatter surface must be very steep and well localized. It is not sufficient that the disk oscillations cause this surface to move closer to, or farther away from, the neutron star (by ~30%) as this would not modulate the scattered X rays sufficiently to produce the observed oscillations in the

X-ray signal. The scatter surface should more or less behave as a rising and falling "wall" (with the frequency of the disk oscillations) in order to obtain a strong modulation in those X rays that are scattered. The authors predict that those systems seen at low inclinations are favored (no QPO should be seen from eclipsed systems).

The fraction of X rays that scatter off the accretion disk coronae in 1822-37 and 2129+47 is no more than ~10% (Keith Mason and Nick White, private communication). If this number is typical for LMXB, it is very hard to see how the disk oscillations could produce the high strength of QPO (typically ~5% rms variation, but up to ~30% in the Rapid Burster). If the modulation efficiency (of scattered X rays only) were as high as ~10%, one would expect to observe a modulation in the total X-ray signal (scattered and non-scattered) of only ~1%. If the central source were obscured, the situation would be different, and the observed modulation could then be much higher since only the scattered X-rays are seen. This, however, is not the case for most (if not all) sources (they are very bright) in which QPO has been detected.

The authors appreciate the problems and suggest that non-linear effects might help to localize the $\tau \simeq 1$ surface. Such non-linear effects are unexplored. The physics is not understood, and so far it appears to be only an interesting mathematical exercise until more theoretical work is done.

1.4. Occultation Models

Some of the ideas as put forward by Morfill and Truemper (1986a,b) may work if the plasmoids are not self luminous (see above) but occult the central source. Similarly, the disk oscillations (Boyle, Fabian and Guilbert 1986) could produce a high degree of modulation if the neutron star is obscured by the oscillating disturbances. Van der Klis et al. (1986) show that certain occultation models can explain the absence of LFN in the presence of strong QPO. Occultation models do not suffer from a lack of available energy for the QPO. However, they probably require that the systems that exhibit QPO are seen at a rather large inclination, and it is puzzling why that would be the case for so many bright low-mass X-ray binaries.

2. Concluding Remarks

As of today (April 22, 1986) we do not yet know what causes the QPO, not even whether they are magnetospheric in origin (see note added below). It appears that none of the proposed scenarios can alone explain the richness and complexity in the high-frequency quasi-periodic X-ray oscillations now observed in about eight bright low-mass X-ray binaries (above references; Stella 1985; Lewin and Van Paradijs 1986; Van der Klis 1986, and references therein). However, some of the proposed ideas could well be relevant to the various rather complex aspects of the QPO. In any case it is likely that more than one mechanism is responsible for the enormous variety in the QPO.

In the light of our ignorance, I would like to finish with a humorous quote from Harlow Shapley which was brought to my attention by Ed Chupp (1984).

> A hypothesis or theory is clear, decisive and positive but it is believed by no one but the man who created it. Experimental findings, on the other hand, are messy, inexact things which are believed by everyone except the man who did the work.
>
> Harold Shapley

Acknowledgments.
During the past 18 months I have had numerous discussions with many colleagues and friends about QPO. Their valuable contributions have become an essential and integral part of my thinking. I particularly want to thank Jan van Paradijs, Michiel van der Klis, Guenther Hasinger, Luigi Stella, Ed van den Heuvel and Fred Lamb. I am grateful for a

generous award from the Alexander von Humboldt Stiftung, and for a John Simon Guggenheim fellowship. My work was also supported by NASA grant NAG8-571.

Note added after my talk.

Guenther Hasinger (1986) has recently shown that in the case of Cyg X-2 the rapid variability in high-energy photons lags that of the low-energy photons by several msec; the time lag is smaller when the QPO frequency is higher. He suggests that Comptonization is responsible for this time lag. For a delay of ~3 msec, and a Compton optical depth of ~5, the size of the scattering cloud would be ~170 km. At that radius the Kepler frequency of matter orbiting the neutron star is ~20 Hz (as observed in Cyg X-2). When the Compton scattering cloud moves closer to the neutron star the QPO frequency would become larger, and the time lag smaller. Guenther's new, and very interesting findings are in general support of the Bath model which explains the QPO frequencies in terms of the Keplerian frequencies of orbiting matter at the magnetopause; this model provides no information on the rotation frequency of the neutron star (see text).

References

Alpar, M.A., Cheng, A.F. et al., Nature, **300**, 729, 1982.

Alpar, M.A. and Shaham, J., IAU Circ. No. 4046, 1985a.

Alpar, M.A. and Shaham, J., Nature, **316**, 239, 1985b.

Bath, G.J., Nature Phys. Sci. **246**, 84, 1973.

Berman, N. and Stollman, G., Astron. Astrop. **154**, L23, 1985.

Boyle, C.B., Fabian, A.C. and Guilbert, P.W., Nature, **319**, 648, 1986.

Chupp, E.L., Ann. Rev. Astron. Astrophys., **22**, 359, 1984.

Cordova, F.A. and Mason, K.O., in: "Accretion Driven Stellar X-ray Sources", eds. W.H.G. Lewin and E.P.J. v.d. Heuvel (Cambridge University Press) page 147, 1983.

Hameury, J.M., King, A.R. and Lasota, J.P., Nature, **317**, 597, 1985.

Hasinger, G., Talk presented at the IAU Symp. No. 125 on "The Origin and Evolution of Neutron Stars", Nanjing, China, May 1986.

Joss, P.C., Avni, Y. and Rappaport R., Ap. J., **221**, 645, 1978.

Joss, P.C. and Rappaport, S.A., Nature, **304**, 419, 1983.

Lamb, F.K., in: "The Evolution of Galactic X-Ray Binaries", eds. J. Truemper, W.H.G.
Lewin   and W. Brinkmann, NATO ASI Series C: Math. & Phys. Sci., **167**, 151,
1986 (Reidel, Dordrecht).

Lamb, F.K., Pethick, C.J. and Pines, D., Ap. J., **184**, 271, 1973.

Lamb, F.K., Shibazaki, N., Shaham, J. and Alpar, M.A., Nature, **317**, 681, 1985.

Langmeier, A. et al., in: "The Evolution of Galactic X-Ray Binaries", eds. J. Truemper,
W.H.G. Lewin and W. Brinkmann, NATO ASI Series C: Math. & Phys. Sci.,
167, 253, 1986 (Reidel, Dordrecht).

Lewin, W.H.G., Talk presented at the IAU Symp. No. 125 on "The Origin and Evolution
of Neutron Stars", Nanjing, China, May 1986.

Lewin, W.H.G. and Van Paradijs J., Comments Astrophys., **11**, No. **3**, 127, 1986.

Li, F.K., Joss, P.C., McClintock, J.E., et al., Ap. J., **240**, 628, 1980.

Maejima, Y., Makishima, K., Matsuoka, M., Ogawara, Y., Oda, M. et al., Ap J. **285**,
714, 1984.

Matsuoka, M., in: "Cataclysmic Variables and Low-Mass X-Ray Binaries",
eds. D.Q.Lamb, and J. Patterson (Reidel, Dordrecht), page 139, 1985.

Middleditch, J. and Priedhorsky, J., IAU Circ. No. 4060, 1985.

Middleditch, J. and Priedhorsky, J., Ap. J., **306**, 230, 1986.

Morfill, G.E. and Truemper, J., in: "The Evolution of Galactic X-Ray Binaries", eds.
J. Truemper, W.H.G. Lewin and W. Brinkmann, NATO ASI Series C: Math. &
Phys. Sci., **167**, 173, 1986a (Reidel, Dordrecht).

Morfill, G.E. and Truemper, J., submitted to Nature, 1986b.

Paczynski, B., Nature, **304**, 421, 1983.

Patterson, J., Ap. J. Suppl., **45**, 517, 1981.

Priedhorsky, W., Ap. J. (Lett), **306**, L97, 1986.

Priedhorsky, W., Hasinger, G., Lewin, W.H.G., Middleditch, J., Parmar, A., Stella, L. and
White, N.E., Ap. J. (Lett), **306**, L91, 1986.

Radhakrishnan, V., IAU Asian-Pacific Regional Meeting, Bandung, ed. B. Hydyat, 1981.

Radhakrishnan, V. and Srinivasan, G.M., Current Sci., **51**, 1096, 1982.

Robinson, E.L. and Nather, R.E., Ap J. Suppl. **39**, 461, 1979.

Savonije, G.J., Nature, **304**, 422, 1983.

Shaham, J., Talk presented at the IAU Symp. No. 125 on "The Origin and Evolution of Neutron Stars", Nanjing, China, May 1986.

Smarr, L.L. and Blandford, R.D., Ap. J., **207**, 574, 1976.

Srinivasan, G. and Van den Heuvel, E.P.J., Astron. Astrop., **108**, 143, 1982.

Stella, L., Proceedings of the "Mediterranean School in Plasma Astrophysics", Colymbari, Crete, Greece, Sept-Oct 1985.

Stella, L., Talk presented at the ESA meeting on: "The Physics of Accretion onto Compact Objects", Tenerife, Spain, April, 1986.

Stella, L., Kahn, S.M. and Grindlay, J.E., Ap. J., **282**, 713, 1984.

Stella L., Parmar, A., White, N.E., Lewin, W.H.G., and Van Paradijs, J., IAU Circ. No. 4110, 1985.

Stella, L., Haberl, F., Parmar, A., White, N., Lewin, W., and Van Paradijs, J., to be submitted to Ap. J. 1986.

Taam, R.E., Ap. J., **270**, 694, 1983.

Tawara, Y., Hayakawa, S., Kunieda, H., Makino, F. and Nagase, F., Nature, **299**, 38, 1982.

Van den Heuvel, E.P.J., IAU Symp No. 95, eds. W. Sieber and R. Wielebinsky, page 379, 1981 (Reidel).

Van der Klis, M., This volume, and Talk presented at the IAU Symp. No. 125 on "The Origin and Evolution of Neutron Stars", Nanjing, China, May 1986.

Van der Klis, M., and Jansen, F.A., Nature, **313**, 768, 1985.

Van der Klis, M., Jansen, F., Van Paradijs, J., Lewin, W.H.G., Truemper, J. and Sztajno, M., IAU Circ. No. 4043, 1985a.

Van der Klis, M., Jansen, F., Van Paradijs, J., Lewin, W.H.G., Truemper, J., Van den Heuvel, E.P.J. and Sztajno, M., Nature, **316**, 225, 1985b.

Van der Klis, M., Stella, L., White, N., Jansen, F. and Parmar, A.N., submitted to Ap. J., 1986.

Voges W., et al., in: "X-Ray Astronomy in the EXOSAT Era", ed. A. Peacock (Reidel, Dordrecht), page 347, 1985; also Space Sci. Rev. **40**, 1985.

Warner, B., in: "Cataclysmic Variables and Related Objects", eds. M. Livio and G. Shaviv, page 155, 1983 (Reidel, Dordrecht).

Webbink, R.F., Rappaport, S.A. and Savonije, G.J., Ap. J., **270**, 678, 1983.

THE X-RAY VARIABILITY OF ACTIVE GALACTIC NUCLEI

R.S.Warwick

X-ray Astronomy Group, Physics Department

University of Leicester

Leicester LE1 7RH, England.

ABSTRACT

Recent results relating to the X-ray variability exhibited by active galactic nuclei are reviewed. *EXOSAT* observations have established, contrary to earlier indications, that X-ray variability on a timescale of an hour or less is a relatively common feature of such sources. It is argued that the minimum variability timescale may be a direct indicator of the mass of the black hole in an active galactic nucleus. In some active galaxies the central black hole mass may be $\lesssim 10^7$ $M_\odot$.

1. INTRODUCTION

The current orthodoxy is that active galactic nuclei (AGN) are powered by the release of gravitational potential energy as matter is accreted onto a central massive black hole (e.g. Rees 1984). In many AGN the bulk of the energy thus liberated appears in the X-ray and soft gamma-ray regions of the spectrum (e.g. Rothschild *et al.* 1983; Bassini & Dean 1983). In comparison to the situation for stellar mass objects (see F. Verbunt, these proceedings), details of the accretion process in AGN are poorly understood. It is established, however, that most of the high energy luminosity will be generated close to the central black hole (i.e. within about 10 Schwarzschild radii, R_s; e.g. Shapiro, Lightman & Eardley 1976). The observation of rapid large amplitude X-ray variability confirms that the X-ray emitting regions are compact and, presumably, closely coupled to the primary energy source. X-ray spectral and variability measurements may therefore provide one of the most direct methods of probing the innermost regions of an active galactic nucleus.

In this review the status of X-ray variability studies of AGN is discussed with emphasis on recent results from *EXOSAT*. Attention is necessarily focussed on X-ray bright objects which comprise mainly Seyfert 1 and Seyfert 2 (alias narrow emission line) galaxies plus a small number of quasars; BL Lac objects are not discussed here (see M. Urry, these proceedings).

Present measurements cover timescales ranging from ~30 seconds up to ~15 years (i.e. about seven decades). The former limit is set by the signal-to-noise ratio obtainable on a typical X-ray bright AGN with the present generation of instruments (collimated and imaging), and the latter by the paucity of X-ray measurements prior to *Uhuru*, *Ariel 5* and other satellite observations. The *EXOSAT* programme has added significantly to the available data base. In particular the unique capability of *EXOSAT* to perform long uninterrupted observations (of up to 2×10^5 s) has had a major impact on our understanding of AGN variability. Unfortunately the data are still generally inadequate to properly characterise the variability in terms of power spectra or autocorrelation functions. However a characteristic minimum timescale for significant variations can often be assigned; the two folding timescale defined as $t_v = (dlog_2 S/dt)^{-1}$ is used in the present review.

2. TIMESCALES OF X-RAY VARIABILITY IN AGN

The timescales which may be relevant for different values of the mass of the central black hole are illustrated in Fig.1. Neglecting any effects due to relativistic beaming, the shortest conceivable timescale is given by the light travel time across the black hole, i.e. $t_1 = 10M_6$ s, where M_6 is the black hole mass in units of 10^6 $M_\odot$. However, in an accretion-driven source it is reasonable to presume that the shortest timescale on which substantial variations in luminosity could occur is the dynamical timescale, which may be taken as the orbital timescale close to the black-hole, i.e. $t_d = 1000M_6$ s (at $R=5R_s$). Whether it is correct to use the dynamical timescale to infer AGN black hole masses from X-ray variability measurements has, of course, to be justified. Many processes may tend to smear out or distort intrinsic rapid variations (e.g. scattering in a disc corona or reprocessing in a pair dominated atmosphere; see A.Fabian, these proceedings). A comparison with the galactic X-ray source Cyg X-1, for which there is long standing evidence for the presence of an accreting black hole (see J. McClintock, these proceedings), is appropriate here. Although the fast (several millisecond) variability in Cyg X-1 fits fairly well with the dynamical timescale of a 16 $M_\odot$ object, most of the power appears at timescales in excess of 1 s (e.g. Nolan *et al.* 1981). The processes which give rise to these long timescales and indeed to the form of the overall power spectrum of Cyg X-1, are not well understood. Furthermore, it should be noted that a simple scaling up of the properties of Cyg X-1 implies implausibly small black hole masses for AGN which show large amplitude variability on timescales of 10^5 s and less (see Fig.1).

In the following sections recent results are discussed in the context of long-term (weeks, months and years), medium-term (days) and short-term (hours and less) variability. This division into three timescale regimes is, of course, entirely arbitrary but ties in reasonably well with developments in this field.

2.1 Long-Term Variability

It is well-established that AGN exhibit substantial variations in X-ray luminosity on timescales of weeks or longer (e.g. Marshall, Warwick & Pounds 1981). Recent *EXOSAT* observations have demonstrated that extremely large amplitude variations occur in some sources (e.g. Cen A, Molteni & Morini, this meeting; NGC 5548, Branduardi-Raymont & Bell-Burnell, this meeting; NGC 526A and Mkn 335, Pounds 1985). The eventual full analysis of the *EXOSAT* data base should provide fairly definitive statistics on the X-ray variability of AGN on timescales of months. In particular it will be interesting to check whether the degree of X-ray variability is different for radio-loud and radio quiet objects and whether there is any (inverse) correlation with X-ray luminosity (*cf* Mushotzky 1984).

EXOSAT monitoring of the archetypal X-ray bright Seyfert 1 galaxy, NGC 4151, has proved particular interesting. The X-ray light curve of NGC 4151 measured over a two year period is shown in Fig.2. The dramatic variations apparent in the *EXOSAT* medium energy (ME) band contrast with the essentially stable soft X-ray flux measured in the *EXOSAT* low-energy telescope through the thin lexan filter (LE). This immediately suggests that there are two physically separate X-ray emitting components in the nucleus of NGC 4151. The soft non-varying

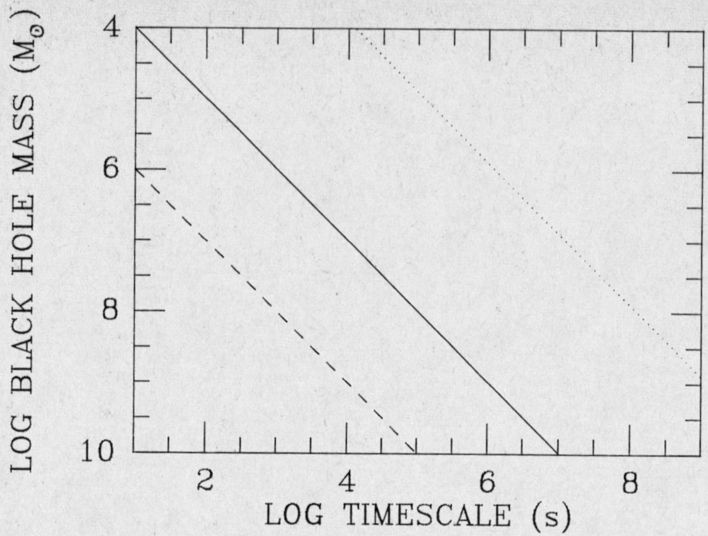

Fig.1 Variability timescales for accretion onto supermassive black holes. Dashed line: the light travel time, t_1, across the black hole. Solid curve: the dynamical timescale, t_d, at R=5R$_s$. Dotted curve: the timescale corresponding (after scaling) to the break in the power spectrum of Cyg X-1 (see Nolan *et al.*, 1981).

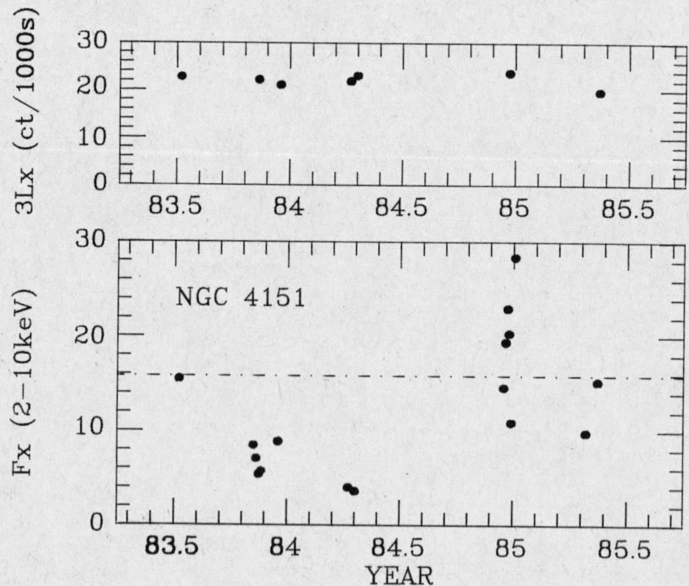

Fig.2 The X-ray light curve of NGC 4151 from *EXOSAT* monitoring. Upper panel: 0.05 - 1 keV band (LE). Lower panel: 2 - 10 keV band (ME).

component may originate in a hot intercloud medium in the narrow-line region (Pounds *et al.* 1986a; Perola *et al.* 1986) whereas the varying hard X-ray flux, which in the *EXOSAT* observations correlates well with the ultraviolet continuum, must be produced close to the nucleus (see M.Penston, these proceedings).

At the present time there is only limited evidence that spectral changes accompany variations in the X-ray luminosity of AGN. Perola *et al.* (1986) have recently found a correlation between X-ray spectral index and X-ray flux in NGC 4151 in the sense that the spectrum softens as the source brightens (see also Baity *et al.* 1984). A similar effect was reported by Halpern (1985) in 3C120 from Einstein MPC observations. *EXOSAT* observations of NGC 4051 by Lawrence *et al.* (1985) show that this spectral variation with intensity can also operate on relatively short timescales. Lawrence *et al.* argue that such a correlation may be explained by thermal Comptonisation models in which a constant temperature plasma scatters a varying soft photon flux. An alternative interpretation is that there are two separate spectral components present which vary in such a way as to produce the observed effect. This latter hypothesis seems less plausible in the case of NGC 4151.

AGN X-ray spectra can also exhibit variations on timescales of months or longer as a result of changes in the broad line region surrounding the active nucleus (e.g. Culhane 1984). Variations in the line-of-sight gas column density can, for example, be inferred from changes in the low-energy absorption spectra. Here two recent results are briefly described.

In the case of NGC 4151 previous measurements (Barr *et al.* 1977) have indicated that the gas column density can vary by up to 10^{23} cm^{-2} over an interval of about a year. In contrast *EXOSAT* observations show no evidence for changes in the low-energy absorption in this source but do confirm that the column is not simply that due to a uniform cool absorbing medium (*cf* Holt *et al.* 1980). The *EXOSAT* spectra are consistent with a model in which the absorber consists of a number of discrete clouds with individual column densities of ~3 x 10^{22} cm^{-2} but with an additional absorbing component (of ~10^{22} cm^{-2}) present to prevent the leakage of X-rays below ~1keV (Pounds *et al.* 1986a; Perola *et al.* 1986). The latter may be located outside the broad-line region, perhaps coincident with the region where the prominent ultra-violet absorption lines are formed.

An *EXOSAT* monitoring programme on the moderate luminosity (Lx ~ 2 x 10^{43} erg s^{-1}) Seyfert 1 galaxy ESO 103-G35 has been more successful in demonstrating that significant changes can occur in the effective line-of-sight gas column of an active galaxy. The first measurements of the X-ray spectrum of this source were performed by *EXOSAT* in 1983 and 1984 and revealed an extremely high column of ~1.8 x 10^{23} cm^{-2} (Pounds 1985). In subsequent observations during 1985 the column density decrease by roughly a factor of 2 in less than three months. This is very comparable to the column change of 1.3 x 10^{23} cm^{-2} in less than 4 months inferred from a comparison of *HEAO 1* and *Ariel 5* observations of NGC 3783 (Mushotzky *et al.* 1980; Hayes *et al.* 1981). The variability timescale of the X-ray continuum in ESO 103-G35, as determined from the *EXOSAT* observations, is ~30 days. This implies a dimension for the X-ray source of ~4 x 10^{15} cm (assuming a mass based on the dynamical timescale, t_d=30 days and a source size of 5R$_s$). The low energy spectral variations can then be modelled as due to an extended absorbing cloud with a transverse velocity of 6000 km s^{-1} (*cf* the width of H$_{alpha}$ in the optical spectrum, Phillips *et al.* 1979) gradually moving out of the line-of-sight and thereby reducing the effective column density to the X-ray source. Since particle densities in broad line region clouds are typically 10^9 to 10^{10} cm^{-3} (e.g. Kwan & Krolik 1981), the thickness of the absorbing cloud must be ~10^{13} to 10^{14} cm indicating a sheet or pancake cloud geometry (Mathews, 1982).

2.2 Medium-Term Variability

X-ray variability on a timescale of days was established in a number of AGN by *ARIEL 5* (Marshall *et al.* 1981). The *HEAO 1/A2* experiment observed variability in 5 out 17 observations of narrow emission line galaxies but with the exception of NGC 6814 and NGC 4151 failed to detect large amplitude day-to-day variability in a sample of 13 Seyfert 1 galaxies (Mushotzky 1984). Einstein observations of a sample of quasars found 4 out of about 30 objects to be variable on a one day timescale (Zamorani *et al.* 1984). Clearly the duty cycle for large amplitude variations in the X-ray luminosity of AGN on a timescale of days is probably quite low (perhaps about 10%) but the data are very sparse. Unfortunately this timescale range is not particularly well covered by *EXOSAT* monitoring programmes.

NGC 4151 again provides an exception to the rule in that in this source the X-ray flux variations appear to be almost continuous. Lawrence (1980) modelled this behaviour as due to flare-like events in the frame-work of a shot noise model. Fig.3 shows the ME light curve of NGC 4151 from an *EXOSAT* observation in May 1985. The timescale for the source flux to halve during the first part of the observation is ~50000 s. Very similar drift rates are apparent in observations reported by Perola *et al.* (1986). From the *EXOSAT* data there is no compelling evidence for any differences in the rise and decay timescales contrary to the assumption of asymmetric sawtooth flares by Lawrence (1980). If the observed variations occur on the dynamical timescale of the system, then a black hole mass of ~5 x 10^7 $M_\odot$ is implied for NGC 4151. In comparison Ulrich *et al.* (1984a,b) obtained a value of ~10^9 $M_\odot$ from a study of the variability and time-lags exhibited by a number of emission lines in the ultra-violet spectrum of NGC 4151.

2.3 Short-Term Variability

Prior to *EXOSAT* the most extensive data base on short term X-ray variability was provided by the *HEAO 1/A2* pointed-phase observations. From an analysis of these data Tennant and Mushotzky (1983) found only one source to be variable on a timescale less than three hours out of a sample of 54 observations of 38 objects and concluded that *"large amplitude short-term variations are not a characteristic of the X-ray emission from active galaxies"*. The one highly variable source in the *HEAO 1* sample was the Seyfert 1 galaxy NGC 6814 which displayed erratic variations on timescales down to ~100 s (Tennant *et al.* 1981; but see also Beall *et al.* 1986). A further good example of rapid variability in a Seyfert galaxy is provided by an Einstein IPC observation of NGC 4051 (Marshall *et al.* 1983) in which the soft X-ray flux doubled in ~2000 s. Tennant & Mushotzky (1983) noted that NGC 6814 and NGC 4051 are both relatively low luminosity AGN ($L_x < 10^{43}$ erg s^{-1}) and speculated that low luminosity sources have a greater probability of being variable than high luminosity objects.

EXOSAT observations have added considerably to our understanding of the variability of AGN on timescales of the order of an hour. The first incontrovertible result was obtained in an uninterrupted 8.5 hour observation of NGC 4051 in which large amplitude variability on a timescale of an hour was detected in both the *EXOSAT* ME and LE

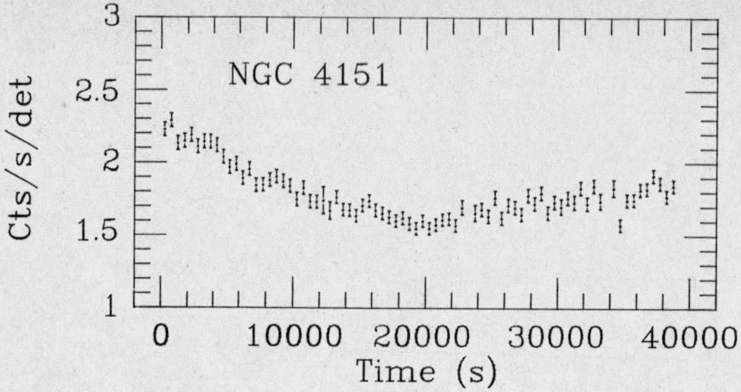

Fig.3 The X-ray (ME) light curve of NGC 4151 from an *EXOSAT* observation in 1985, May 15.

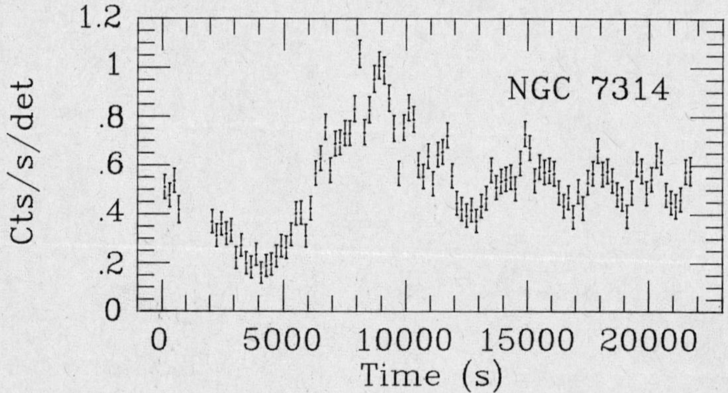

Fig.4 Short term variability in NGC 7314. The light curve has a decaying oscillatory form.

instruments (Lawrence *et al.* 1985). A systematic softening of the spectrum as the source brightened was also seen (see section 2.1). Significant variations on a similar timescale have also been observed by *EXOSAT* in NGC 6814 (Branduardi-Raymont, private communication).

More recent *EXOSAT* observations indicate that rapid variability is by no means as rare in AGN as suggested by the *HEAO 1* results and neither is it restricted solely to low luminosity objects.The spectral survey of Seyfert 1 and narrow emission line galaxies, which is in progress at Leicester (see Pounds 1985), provides a useful comparative sample. In a total exposure of 150-200 hours on about 20 sources rapid variability is seen in 5 sources (see Table 1) with X-ray luminosities ranging from 0.4 to 80 x 10^{43} erg s^{-1} (2-10 keV). In three of these sources, NGC 7314 (see Fig.4), MCG-6-30-15 (Pounds, Turner & Warwick 1986) and Mkn 335 (Pounds *et al.* 1986b) the variability can be crudely described as quasi-sinusoidal, whereas in Akn 120 and III Zw2 the variability arises from individual flaring events. In the case of MCG-6-30-15 analysis shows that the ME and LE light curves are well correlated with no measurable time-delay between variations in the two wavebands (Pounds, Turner & Warwick 1986); this is also the case for NGC 4051 (Lawrence *et al.* 1985).

Short timescale variability has also been reported in *EXOSAT* observations of NGC 3031 (Barr *et al.* 1985), NGC 3516 (G.Reichert *et al.,* this meeting), Cen A (Molteni & Morini, this meeting), NGC 4593 (Barr *et al.* 1986), Mkn 766 (Barr & Giommi, this meeting) and NGC 5506 (I. McHardy, private communication).

At first sight the results from *HEAO 1* and *EXOSAT* would appear to lead to almost diametrically opposite conclusions concerning the incidence of rapid variability in AGN. The problem may, however, be no more than one of sensitivity; in terms of the σ_I/I criterion only one of the five sources in Table 1 would definitely have been detected as a rapidly variable object by *HEAO 1* (see Tennant & Mushotzky 1983). Other *EXOSAT* facilities such as the long uninterrupted observations, continuous monitoring of the background and simultaneous observations by the ME and LE are also advantageous for the detection of rapid variability in relatively faint sources such as AGN.

Interestingly three of the sources for which rapid variability is now established have X-ray spectra which show an excess flux below ˜1 keV (MCG-6-30-15, Pounds, Turner & Warwick 1986; Mkn 335, Pounds *et al.*

1986b; NGC 4051, Lawrence *et al.* 1985). This flux excess probably arises from an additional steep spectral component which, it has been argued, may be the signature of thermal emission from the hot inner region of an accretion disc (Arnaud *et al.* 1985; Pounds *et al.* 1986b). Pounds *et al.* find that in the case of Mkn 335 a thin accretion disc will be disrupted by the radiation pressure unless the central black hole mass is $\lesssim 10^7$ M$_\odot$, a value which is very comparable to that inferred from the variability timescale.

Up to present time there have been *no reports of extremely rapid variability in EXOSAT observations of AGN* (i.e. on timescales of minutes).

3. IMPLICATIONS OF RAPID VARIABILITY IN AGN

From the above discussion it is evident that AGN exhibit a wide range of amplitudes and timescales of X-ray variability. Although in some observations there is the suggestion of quasi-periodicity (see, for example, Fig.4) in general the light curves can be described as chaotic. One parameter which is readily extracted from the data is the minimum timescale for significant flux variations. In a recent paper Barr & Mushotzky (1986; see also P.Barr, these proceedings) have attempted to establish a correlation between the minimum variability timescale and the X-ray luminosity of AGN. Fig.5 shows their diagram (excluding BL Lac objects) with the addition of five points corresponding to the sources in Table 1. It is probably incorrect to argue that this figure demonstrates a correlation given that many sources which have not exhibited obvious variability on medium and short timescales are excluded. However, as noted by Barr & Mushotzky, the diagram does suggest that in AGN the compactness parameter, l, ($l=L_X\sigma_T/Rm_ec^3$), generally does not exceed a value of between 10 to 100 (taking $R=ct_v$, and a bolometric correction ~10). This is just the range in which the sources become dominated by the production of electron-positron pairs provided their high energy spectra remain relatively flat out to ~1 MeV (e.g. Guilbert,Fabian & Rees 1983). Processes occurring in an electron-positron atmosphere may give rise to the canonical form of the hard X-ray spectra and also significantly influence the variability exhibited by AGN (see A.Fabian, these proceedings and references therein).

An alternative interpretation of Fig.5 is possible if it is assumed that the minimum timescale of variability exhibited by an active

TABLE 1 Five sources exhibiting short term variability.

	L_x	$\Delta I/I$	σ_I/I	t_v
NGC 7314	0.4	1.7	0.33	2000
MCG-6-30-15	0.9	0.4	0.13	2150
Mkn 335	4	0.6	0.18	6200
Akn 120	18	0.3	0.08	2250
III Zw2	80	1.3	0.25	2000

Notes

(1) L_x is the 2-10 keV X-ray luminosity in units of 10^{43} erg s^{-1}.

(2) $\Delta I/I$ is the ratio of the peak-to-peak flux to the mean flux.

(3) σ_I/I is the rms to mean flux ratio (for 100 s time bins).

(4) t_v is the estimated flux doubling timescale in s.

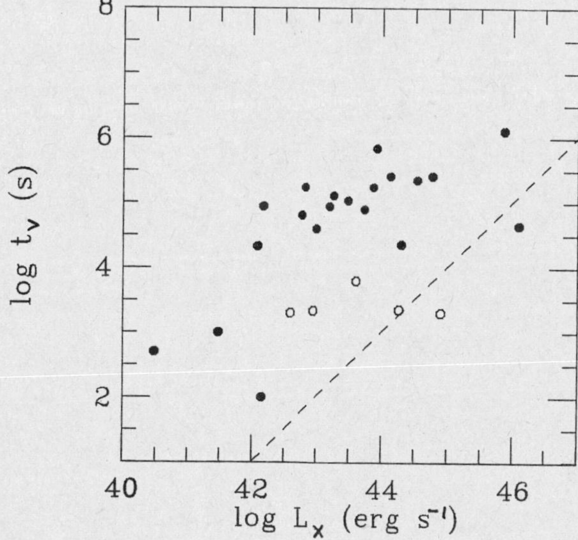

Fig.5 The variability timescale versus X-ray (2-10 keV) luminosity for known variable sources. Filled circles: data from Barr & Mushotzky (1986). Open circles: data from Table 1. The dashed line corresponds to sources radiating at the Eddington luminosity (assuming $t_v = t_d = 1000M_6$).

galaxy is predominantly governed by a single parameter, namely the central black hole mass. Pair processes, for example, will not be important if in general the high energy spectra of AGN steepen well below 1 MeV. Provided the heating and cooling processes operate sufficiently rapidly, much of the observed variability may reflect the system responding on the dynamical timescale to accretion rate fluctuations. Interpreting the measured variability timescales as dynamical timescales implies black hole masses ranging from 10^6 to 10^9 $M_\odot$. The Eddington and efficiency limits are not violated except in the case of 3C273 (Marshall *et al.* 1981) and III Zw2 (see Fig.5) both of which are radio-loud and may be influenced by relativistic beaming effects. Support for this interpretation of the X-ray variability measurements is provided by the recent work of Wandel and Mushotzky (1986). These authors find a highly significant correlation (with some reservations concerning selection effects) for Seyfert 1 galaxies between the X-ray variability timescale and a dynamical mass deduced from the properties of the narrow [OIII] line. Their correlation does not ,however, extend to narrow emission line (Seyfert 2) galaxies.

The implications of X-ray variability measurements in AGN are discussed in more detail by Fabian (these proceedings).

4. FUTURE PROSPECTS

Up to the present time the proper characterisation of X-ray variability of AGN has been impossible because of the lack of sufficiently extended and adequately sampled light-curves. In the latter part of the *EXOSAT* programme a number of very long (i.e. ~ 2 x 10^5 s) continuous observations were performed on sources found to be variable on short timescales in earlier observations (e.g. Mkn 766, Barr & Giommi, this meeting; MCG-6-30-15, Pounds, this meeting). When fully analysed these *"long-look"* observations should provide a much more complete picture of the short term variability exhibited by AGN. Fig.6 shows the ME light curves obtained for two sources as part of this programme. In the case of the Seyfert 2 galaxy, NGC 5506, the variability can be described as a continuous flickering with a peak-to-peak amplitude of about 25 per cent. Interestingly there appear to be no quiescent episodes. The most rapid fluctuations occur on a timescale of ~2000 s. In comparison the light curve of the Seyfert 1 galaxy, MCG-6-30-15, is dominated by a sawtooth type behaviour, namely a slow upward drift over about half a day and then a

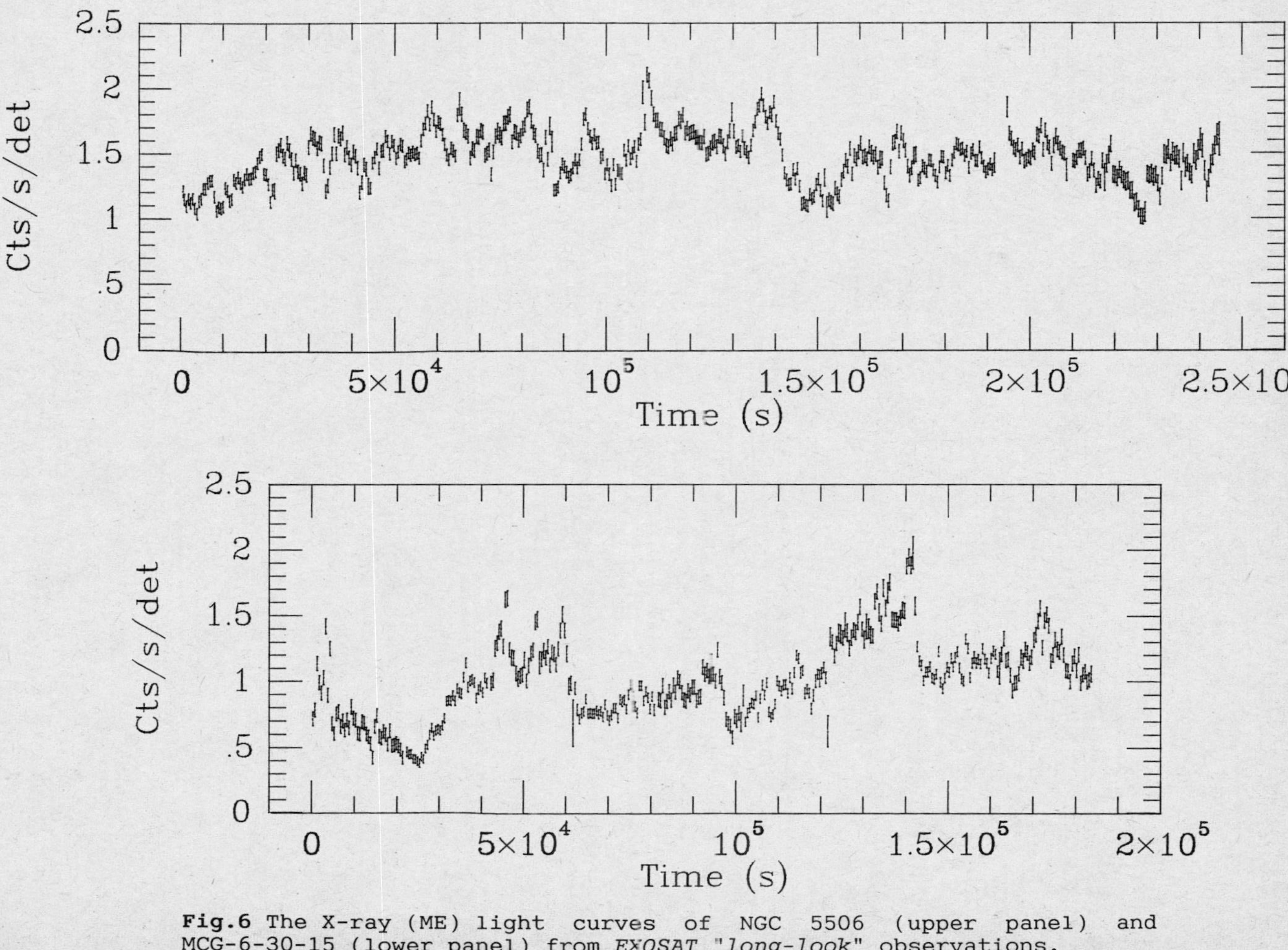

Fig.6 The X-ray (ME) light curves of NGC 5506 (upper panel) and MCG-6-30-15 (lower panel) from *EXOSAT "long-look"* observations.

relatively rapid decline. Occasionally short term flaring, of a form comparable to that seen in the earlier *EXOSAT* observation is superimposed upon this underlying activity.

The detailed analysis of these *"long-look"* observations remains a task for the future. A final comment is needed, however, to put these new riches in perspective. A signal-to-noise ratio of 5 is obtained with the ME detectors on Cyg X-1 after about 50 msec and on a typical X-ray bright AGN after ~60 sec. Purely in terms of the quality of the time series record, a 48 hour observation of an AGN is therefore equivalent to only 2 minutes of data on Cyg X-1. The need for more data and hence new X-ray missions is therefore assured!

ACKNOWLEDGEMENTS

The author would like to thank Ken Pounds, Tracy Turner and Ian McHardy for the use of results prior to publication.

REFERENCES

Arnaud,K.A., *et al.*, 1985. *Mon.Not.R.Astr.Soc.*, **217**,105.
Baity,W.A., *et al.*, 1984. *Astrophys.J.*, **279**,555.
Barr,P. & Mushotzky,R.F., 1986. *Nature*, **320**,421.
Barr,P., *et al.*, 1977. *Mon.Not.R.astr.Soc.*, **181**,43P.
Barr,P. *et al.*, 1986. *Proc. IAU Symp.No.119, in press.*
Bassini,L. & Dean,A., 1983. *Astron.Astrophys.*, **122**,83.
Beall,J.H., *et al.*, 1986. *Astrophys.J., in press.*
Culhane,L., 1984. *Physica Scripta*, **T7**,134.
Guilbert,P.W., Fabian,A.C & Rees,M.,J., 1984. *Mon.Not.R.astr.Soc.*, **205**,593.
Halpern,J.P., 1985. *Astrophys.J.*, **290**,130.
Hayes,M.J.C., *et al.*, 1981. *Space Sci. Rev.*, **30**,39.
Holt,S.S., *et al.*, 1980. *Astrophys.J.Lett.*, **241**,L13.
Kwan,J. & Krolik,J.H., 1981. *Astrophys.J.*, **250**,478.
Lawrence,A., 1980. *Mon.Not.R.astr.Soc.*, **192**,83.
Lawrence,A., *et al.*, 1985. *Mon.Not.R.astr.Soc.*, **217**,685.
Marshall,F., *et al.*, 1983. *Astrophys.J.Lett.*, **269**,L31.
Marshall,N., Warwick,R.S. & Pounds,K.A., 1981. *Mon.Not.R.astr.Soc.*, **194**,987.
Mathews,W.G., 1982. *Astrophys.J.*, **252**,39.
Mushotzky,R.F., *et al.*, 1980. *Astrophys.J.*, **235**,377.
Mushotzky,R.F., 1984. *Adv. in Space Res.*, **3**,157.
Nolan,P.,L., *et al.*, 1981. *Astrophys.J.*, **246**,494.
Perola,G.C., *et al.*, 1986, *Astrophys.J., in press.*
Phillips,M.M., *et al.*, 1979. *Astron.Astrophys.*, **76**,L14.
Pounds,K.A., 1985. *Proc. Japan/US Symposium on Galactic and Extragalactic X-ray Sources*, eds Tanaka,Y. & Lewis,W., ISAS, Tokyo.
Pounds,K.A., Turner,T.J. & Warwick,R.S., 1986. *Mon.Not.R.astr.Soc.*, **221**,7P.
Pounds,K.A., *et al.*, 1986a. *Mon.Not.R.astr.Soc.*, **218**,685.
Pounds,K.A., *et al.*, 1986b. *Mon.Not.R.astr.Soc., in press.*
Rees,M.J., 1984. *Ann.Rev.Astr.Astrophys.*, **22**,471.
Rothschild,R.E., *et al.*, 1983. *Astrophys.J.*, **269**,423.
Shapiro,S.L., Lightman,A.P. & Eardley,D.M., 1976. *Astrophys.J.*, **204**,187.
Tennant,A.F., *et al.*, 1981. *Astrophys.J.*, **251**,15.
Tennant,A.F. & Mushotzky,R.F., 1983. *Astrophys.J.*, **264**,92.
Ulrich,M-H., *et al.*, 1984a. *Mon.Not.R.astr.Soc.*, **206**,221.
Ulrich,M-H., *et al.*, 1984b. *Mon.Not.R.astr.Soc.*, **209**,479.
Wandel,A. & Mushotzky,R.F., 1986. *Astrophys.J., in press.*
Zamorani,G., *et al.*, 1984.*Astrophys.J.*, **278**,28.

BLACK HOLES IN X-RAY BINARIES

Jeffrey E. McClintock
Harvard-Smithsonian Center for Astrophysics

ABSTRACT

The best available evidence that black holes exist comes from dynamical studies of three X-ray binaries, Cyg X-1, LMC X-3, and A0620-00. The probable masses of the compact objects in these systems are $\sim 10\ M_\odot$. Evidence is presented that the masses of Cyg X-1 and A0620-00 are not less than $3.4\ M_\odot$ (for $d \geq 2$ kpc) and $3.2\ M_\odot$, respectively. It is believed that a neutron star with a mass above about $3\ M_\odot$ cannot exist within the framework of general relativity because causality would be violated; the star's great rigidity would cause the speed of sound to exceed the speed of light. It therefore appears highly probable that these objects are black holes.

I. INTRODUCTION

In 1964, seven years before Cygnus X-1 appeared in the limelight as a black-hole candidate, Zeldovich first suggested that one might detect X-rays from black holes in binary systems--black holes that are in a close enough orbit around a companion star to raise tides and capture material. Accreted gas would form a disk, swirling downwards into the potential well surrounding the black hole. Viscous dissipation would heat the gas to temperatures of $10^{8}\ ^{0}K$ so that its bremsstrahlung emission would be in the X-ray band. The X-rays would display rapid, irregular flickering (excerpted from Rees 1980).

I will present a summary case that three X-ray binaries contain black holes, as Zeldovich foresaw more than two decades ago. They are Cygnus X-1, LMC X-3, and A0620-00. In brief, the case for each binary rests on the following three pieces of evidence: (1) There exists a highly luminous ($\sim 10^{38}$ ergs s^{-1}) and variable X-ray source which is powered by gravitational infall onto a compact object. Our detailed knowledge of more than 40 X-ray binaries that pulse or burst, and therefore manifestly contain neutron stars, establishes beyond reasonable doubt that these three binaries contain compact objects that are at least as dense and massive as a neutron star. Many reviews have been published that compellingly make this point (e.g., Joss and Rappaport 1984), and I will not discuss it further. (2) The second piece of evidence is also observational. The observed velocity of the optical star, the orbital period of the binary system, and a few particulars (such as the absence of an X-ray eclipse) imply that the mass of the compact object exceeds three solar masses. (3) The final argument rests on theory. According to theorists, a neutron star of three or more solar masses cannot exist; it would be crushed by its self-gravity to form a black hole. As I will make clear later, this is a robust conclusion. It does not depend at all on the details of the equation of state of nuclear matter.

So, the observational inference that black holes exist rests on the claims of observers that they have found compact objects more massive than three solar masses, and the claims of theorists that such objects are too massive to be neutron stars.

Unfortunately, this evidence is indirect or circumstantial; it is not conclusive. I do believe, though, that it is compelling. I want to stress that the arguments presented here are based on firm dynamical evidence. The weighing of stars in binaries is a foundation piece of 20th century astronomy and is a technique we fully understand.

It is therefore fair to say that the evidence that these three X-ray binaries contain black holes is far stronger than the evidence that has been put forward for the presence of black holes in several other binaries. It is also far stronger than the evidence that active galactic nuclei are powered by supermassive black holes. There the evidence is very suggestive, even seductive, but the case remains weak. Indeed, the observations of active galactic nuclei can be interpreted in terms of a black-hole model, but the interpretation is not at all unique, and plausible alternative models exist (e.g., Flaser and Morrison 1976; Ginzburg and Ozernoy 1977; Bailey and Clube 1978; Duncan and Wheeler 1980; Burbidge 1985).

II.   THE NATURE OF INDIRECT EVIDENCE

I want to put the problem of assessing the indirect evidence that these three binaries contain black holes in the context of science as a human endeavor. As Martin Rees (1980) points out, there is a serious problem with indirect evidence if one is attempting to establish the existence of a remarkable class of objects: "It is impossible to construct a watertight case, since ad hoc and contrived models that do not involve a black hole can always be devised. One's assessment of the odds obviously depends on whether one regards a black hole as inherently absurd, or as a plausible endpoint for stellar evolution."

A nice illustration of Rees's point is Eddington's reaction to Chandrasekhar's work on degenerate matter. Eddington (1935) realized almost immediately that Chandrasekhar's theory implied that the formation of black holes would be the inevitable fate of massive stars. In 1935 Eddington said, "The star apparently has to go on radiating and radiating and contracting and contracting until, I suppose, it gets down to a few kilometers radius when gravity becomes strong enough to hold the radiation and the star can at last find peace." But then he went on to say, "I felt driven to the conclusion that this was almost a <u>reductio</u> <u>ad</u> <u>absurdum</u> of the relativistic degeneracy formula. Various accidents may intervene to save the star, but I want more protection than that. I think that there should be a law of Nature to prevent the star from behaving in this absurd way." And Eddington never did accept Chandrasakhar's result despite the fact that Eddington was one of the first people to appreciate the general theory of relativity (see Shapiro and Teukolsky 1983).

Like Eddington, we all let our emotions enter; we shade the facts as we struggle to reach conclusions. And, of course, that is why the indirect evidence for black holes must be intensely scrutinized.

And now I want to distinguish between indirect evidence and direct evidence by sketching the history of neutron stars and black holes in astronomy. The evidence for the existence of neutron stars is now indisputably direct. Their existence was first suggested by Baade and Zwicky (1934), and their theory was developed shortly thereafter by Oppenheimer and Volkoff (1939). In the early 1960s, neutron stars were invoked to account for the recently discovered quasars and X-ray stars; several possible signatures of neutron stars were suggested (e.g., thermal X-ray emission). The first clearcut evidence, however, was the serendipitous discovery of radio pulsars by Hewish, Bell, et al. (1968). Today it is plain that neutron stars are as real as solar-type stars; consider the following examples: (1) The Crab Pulsar, a core remnant of an historical supernova, has maintained an X-ray bright nebula for centuries by the continuous injection of flywheel-powered electrons; (2) The Crab and Vela timing glitches are thought to be caused by starquakes, and the post-glitch "healing" behavior provides evidence for a superfluid neutron core; (3) the "Millisecond Pulsar" (PSR 1937+21) rotates perilously close to breakup speed, and a thousand times too rapidly to be a white dwarf; and (4) The radii and magnetic field strengths of compact X-ray sources (deduced from observations of bursts, spin-up rates, cyclotron lines, etc.) are about 10 km and 10^{12} G, respectively; surely these objects are neutron stars (e.g., Joss and Rappaport 1984).

The core of the direct argument that neutron stars are physical objects was already contained in Gold's 1968 paper in <u>Nature</u> in which he predicted the spin-down of the pulsars. In essence, he said the following: if it has feathers and it quacks, and it looks like a duck, then it is a duck.

Interestingly, in the case of neutron stars we have gone beyond the question of their existence. They are now tools used to probe the interstellar medium, and more. For example, the orbital decay of the Binary Pulsar PSR 1913+16 provides compelling evidence that gravitational waves exist (Taylor and Weisberg 1982); other pulsars are being used to search for a cosmological background of gravitational radiation (Romani and Taylor 1983). Pulsars may even reinstate astronomy as an applied science; some of them are at least comparable in stability to the best man-made atomic clocks (Davis et al. 1985).

In the case of black holes, the situation is entirely different. Although the concept of a black hole is a venerable one (Michell 1784; Laplace 1795), and their theory, based on general relativity, has been developed in detail (e.g., Schwarzschild 1916; Oppenheimer and Snyder 1939; Kerr 1963; Wheeler 1966; Hawking 1974), nevertheless, we cannot, as yet, point to a single object that is indisputably a black hole. This is in part because stellar black holes are difficult to distinguish from weakly magnetized neutron stars. Consider a 10 $M_\odot$ maximally-rotating black hole (e.g., Bardeen 1970; Thorne 1974): its radius (R ~ 15 km) and gravitational potential well (GM/Rc^2 ~ 0.3) are very nearly the same as that of a 1.4 $M_\odot$ neutron star. Moreover, the period of the last stable orbit is the same for the black hole and the neutron star, namely, 0.5 ms. Given this latter fact (and the early observations of submillisecond variability of radio pulsars), it is surprising that fast X-ray flickering has been widely held to be a bona fide signature of a black hole since the early 1970s. Stella et al. (1985) recently refuted this conventional wisdom; they discovered that the pulsating neutron-star binary V0332+52 mimics the rapid flickering behavior of Cyg X-1.

Recently, White and Marshall (1984) and White et al. (1984) suggested another indirect signature of a black hole--a soft X-ray spectrum (i.e., a spectrum that is rich in low-energy photons). Unfortunately, it is reasonable to expect that a neutron star is also capable of forging this signature. Therefore, although both the temporal and spectral signatures can serve as signposts in the search for black holes, they cannot establish the existence of a black hole.

A second reason that black holes remain elusive is their absence of "feathers" or "hair" (e.g., Carter 1979). That is, the only properties that can in principle be observed outside the surface of the black hole (the event horizon) are mass, angular momentum, and charge. For example, there cannot be an off-axis magnetic field anchored in the hole (e.g., Ruffini 1973), and obviously there is no material surface. Presumably, therefore, periodic pulsations, thermonuclear bursts, quasi-periodic oscillations, and other phenomena that signal the presence of a neutron star are absent. This near absence of evidence provides a set of useful but indirect tests for a black hole. We are, however, left with the following question: how can we unmistakably recognize the presence of a black hole?

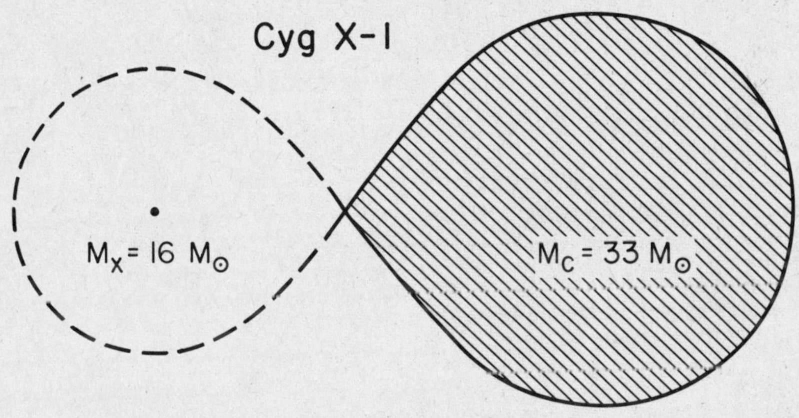

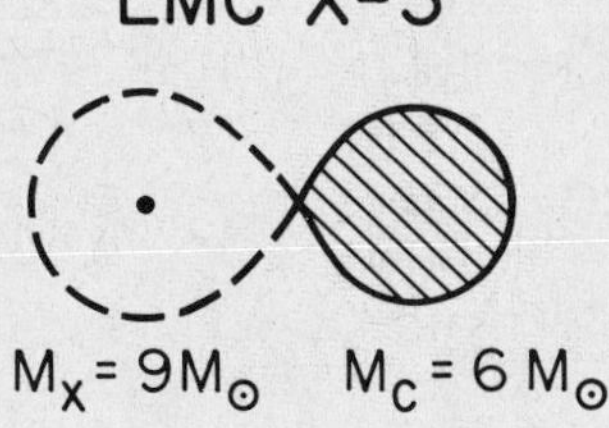

Figure 1. Cartoon comparison of three black-hole binaries. The optical companions (shaded regions) are shown filling their Roche lobes. (Cyg X-1, Gies and Bolton 1986; LMC X-3, Cowley et al. 1983; A0620-00, sketched for i = 40°, McClintock and Remillard 1986). LMC X-1 has also been suggested as a black-hole candidate; however, the dynamical evidence is less compelling, and the orbital period and the optical identification are somewhat uncertain (Hutchings et al. 1983).

Of course, we can hope that black holes, like neutron stars, will reveal their presence in an unexpected fashion. We can also undertake a search for their predicted gravitational effects, which include the following: (1) A distinctive "chirp" of gravitational radiation emitted just prior to the coalescence of two $\sim 10\ M_\odot$ black holes (cf. Clark et al. 1979) (2) Spectral lines that have suffered extreme redshift (Goldman 1981). (3) A rotation of the plane of X-ray polarization due to the Lense-Thirring effect (Conners et al. 1980), (4) A binary companion with an anomalously high orbital velocity.

III. THREE BLACK-HOLE BINARIES

At present we do not have the capability to search for the first three effects mentioned above. The case for black holes rests on the discovery (via the fourth effect) of massive compact objects in the three X-ray binaries that I will now discuss. Some of the nominal characteristics of the host binaries are summarized in Table 1 (see the caption to Fig. 1 for references). The X-ray luminosities are the maximum that have been observed. The visual magnitude of A0620-00 corresponds to a state of X-ray quiescence (see below). For all three systems, the orbits are circular and X-ray eclipses are not observed.

TABLE 1
PROPERTIES OF THREE BLACK-HOLE BINARIES

	Cyg X-1	LMC X-3	A0620-00
L_x (ergs s^{-1})	2×10^{37}	3×10^{38}	1×10^{38}
MK type	09.7Iab	B3V	K5V
d (kpc)	2.5	55	1
m_v	9	17	18
V_c sin i (km s^{-1})	76±1	235±11	457±8
P (days)	5.6	1.7	0.32
$f\,(M/M_\odot)$	0.25±0.01	2.3±0.3	3.18±0.16

The most important parameter, the bottom line of both this talk and Table 1, is the optical mass function:

$$f(M) \equiv (M_x \sin i)^3/(M_x + M_c)^2 = \frac{P\,(V_c \sin i)^3}{2\pi G},$$

where M_x and M_c are the masses of the X-ray source and the companion star, respectively, i is the orbital inclination angle, P is the orbital period, and $V_c \sin i$ is the projected velocity semiamplitude of the optical companion. The term on the right is an observable; it depends only on the measured values of P and $V_c \sin i$. The mass function is, therefore, a cubic equation for M_x, with M_c and i as parameters. The important point to note is that the value of the mass function is, in fact, the least possible value that M_x can take on. This rock-bottom limit on M_x corresponds to a system with a zero-mass companion ($M_c = 0$) viewed at the maximum inclination angle ($i = 90^{\circ}$).

Thus, based on the value of the mass function alone, we can say that A0620-00 is a strong black-hole candidate, $M_x \geq 3.18 \pm 0.16\ M_\odot$. On the other hand, in the case of Cyg X-1 we must argue in addition that the companion star is massive and/or the system is viewed at low inclination. As I discuss at the end of this lecture, the holy grail in the search for black holes (via dynamical observations of X-ray binaries) is a system with a mass function that is plainly 5 $M_\odot$ or greater.

A schematic sketch to scale of models for Cyg X-1, LMC X-3, and A0620-00 are shown in Figure 1. The size of the mass-losing star largely determines the size of the system. The masses of the compact sources are all about 10 $M_\odot$ for the models shown.

a) <u>Cygnus X-1</u>

It was 15 years ago that Cyg X-1 was identified as a premier black-hole candidate. For a summary of the early X-ray, optical, and radio observations see Tananbaum et al. (1972) and Bolton (1972). For an introduction to the extensive literature on the mass of Cyg X-1, see Bahcall (1978) and Gies and Bolton (1982, 1986).

If the optical companion (HDE 226868) is a normal O9.7 supergiant, as it appears to be, then the likely mass of Cyg X-1 is about 16 $M_\odot$ (see Figure 1). Moreover, the mass of the compact X-ray source cannot be less than 7 $M_\odot$ (Gies and Bolton 1986), a conclusion that is supported by the results of an analysis of the optical light variations (Avni and Bahcall 1975). These strong arguments have not provided conclusive evidence that Cyg X-1 is a black hole, and for good reason. The masses of OB companion stars in other X-ray binaries are generally low for their spectral type by a factor of two or so (Hutchings 1979; Rappaport and Joss 1983). Moreover, there exists the possibility that a companion star in a close binary may be <u>severely</u> undermassive for its spectral type (Paczynski 1973; van den Heuvel 1983). A well-known example of this phenomenon is the hot subdwarf HZ22 (Trimble et al. 1973; Young and Wentworth 1982).

Because of these lingering uncertainties, the following argument due to Paczynski (1974) is particularly important. Paczynski (see also Bahcall 1978, and references therein) derived a <u>firm lower limit on M_x that does not depend on the structure of either star or on the evolutionary status of the system.</u> It is based on the absence of an X-ray eclipse, the value of the mass function and the dereddened spectral flux of the companion star. The limit depends on the uncertain distance, d:

$$ M_x \geq 3.4 \ M_\odot \left(\frac{d}{2 \ \text{kpc}} \right)^2. $$

The argument now turns on the question of the distance to Cyg X-1. It is not sufficient to show that the distance is greater than 1.5 kpc ($M_x > 1.9 \ M_\odot$); we require that it exceed 2 kpc ($M_x > 3.4 \ M_\odot$).

Two groups have estimated a lower bound on the distance to Cyg X-1. They determined the run of reddening with distance for ~100 nearby field stars (Margon et al. 1973; Bregman et al. 1973); the distance to Cyg X-1 was thereby inferred from its known reddening (Treves et al. 1980, and references therein). The results clearly favor a distance that is greater than 2 kpc but, unfortunately, they do not rule out the possibility of a lesser distance for the following reasons. First, there are significant uncertainties associated with all reddening-distance measurements (e.g., patchiness in the distribution of dust and the possibility that the source is shrouded in dust and intrinsically reddened). Second, the two reddening studies cited above have serious shortcomings (e.g., neither study included more than three stars beyond 2 kpc and both studies relied heavily on ancient HD spectral classifications). A concluding statement in Bregman et al. (1973) summarizes the situation and indicates the possibility that the distance could be less than 2 kpc: "We conclude that HDE 226868 is not closer than 1 kpc, and that its minimum distance is about 2.5 kpc."

It appears that the best way to strengthen the black-hole candidacy of Cyg X-1 is to adopt Paczynski's approach and to shore up the lower bound on the distance. Additional optical data on the field stars is clearly called for (e.g., MK-quality spectra and a determination of the run of linear polarization with distance). Also, Cyg X-1 is a radio source and it may be possible to set useful distance limits by detecting 21-cm absorption features or by measuring the annual parallax with the VLBI network (cf. Bartel et al. 1986).

In conclusion, we note that alternative models of Cyg X-1 have been proposed that do not invoke a black hole (e.g., see Bahcall 1978; Kundt 1979). Triple-star models have been among the favored contenders; however, the predicted effects of these models have not been detected (Abt et al. 1977; Shafter et al. 1980), and they now appear to be only marginally viable.

b. <u>LMC X-3</u>

Twelve years passed before the second black-hole candidate, LMC X-3 was discovered (Cowley et al. 1983). The new candidate is superior to Cyg X-1 in two respects. First, its systemic velocity establishes that it is a member of the Large Magellanic Cloud and, therefore, its distance is known (d = 55 kpc). Second, the large value of its mass function already exceeds many theoretical mass limits for neutron stars.

Cowley et al. (1983) concluded that the mass of LMC X-3 exceeds 7 $M_\odot$ based on the companion's spectral type (B3V) and the absence of an X-ray eclipse. Using the known distance to LMC X-3, Paczynski (1983) repeated the argument he had given earlier for Cyg X-1 and obtained a mass limit for LMC X-3 that is independent of spectral type: $M_x > 6\ M_\odot$. Thus, the observations of Cowley et al. and the analysis of Paczynski appear to offer nearly conclusive proof that LMC X-3 is a black hole.

Paczynski's argument, however, rests on the assumption that the companion star is the sole source of the optical flux. Although this is true for Cyg X-1, which is a supergiant, it is not true for LMC X-3. Van der Klis et al. (1983) have observed secular (1974-1981) light variations with a maximum excursion of 0.5 magnitudes, which they suggest are probably due to the presence of an accretion disk. The light variations and, in addition, the possibility that the radial velocity data may be affected by large systematic errors have lead Mazeh et al. (1986) to conclude that the mass of LMC X-3 may be as small as about 2.5 $M_\odot$. For this model, the mass and radius of the optical companion are only about 0.7 $M_\odot$ and 2.5 $R_\odot$, respectively. Mazeh et al. argue that such a remnant star could plausibly be formed during the evolution of the binary. They conclude that a "heavy neutron-star" model for LMC X-3 cannot be fully excluded.

Much can be done to rebut the arguments of Mazeh et al. For example, the putative emission-line contamination that may affect the absorption-line velocities can be searched for in high-resolution spectra. Continued monitoring of the long-term light variations at several wavelengths is clearly called for. Ultraviolet observations with the Space Telescope may help to define the nature of the non-stellar light source.

c) <u>A0620-00</u>

Ronald Remillard and I recently identified a third black-hole candidate, A0620-00; the details are published elsewhere (McClintock and Remillard 1986). Here I will summarize the salient points, present new radial velocity data, and pose a puzzle.

A0620-00 is an X-ray nova. For two months in the fall of 1975 it was the brightest celestial X-ray source and was identified optically with a blue star of 12th magnitude. Fifteen months later the star returned to its pre-outburst brightness ($V \approx 18.3$). For the past decade the star has remained at this quiescent level and no X-ray emission has been detected from the system. The quiescent optical spectrum consists of two nearly equal components: A K4V to K7V stellar spectrum and an emission-line component, which is generally attributed to an accretion disk (Oke 1977; Whelan et al. 1977: Murdin et al. 1980).

We determined the orbital period (P=7.75 hours) from photometric data obtained at the McGraw-Hill Observatory during 1981-1985. At KPNO in January 1985, we used the Mayall 4m telescope and the IIDS spectrometer to measure radial velocities. We found that the velocity semiamplitude of the dwarf companion is 457 ± 8 km s^{-1} (Fig. 2; open circles), and the corresponding value of the mass function is f(M) = 3.18 ± 0.16 $M_\odot$. A preliminary analysis of data that we obtained in December 1985 at the MMT Observatory confirms the large amplitude of the velocity curve (Fig. 2; filled circles).

An absolute 3σ lower limit to the mass of the X-ray source (M_x) of 2.70 $M_\odot$ is given by the value of the mass function itself, f(M) = 3.18 ± 0.16 $M_\odot$. This unphysical limit corresponds to a companion star with zero mass.

A firm lower limit on the mass of the compact object can be obtained as follows: If the companion star is on the zero-age main sequence, then its mid-K spectral type implies that its mass is between about 0.5 $M_\odot$ and 0.8 $M_\odot$. Because some X-ray binaries are known to be undermassive for their spectral type (e.g., see Cowley and Crampton 1975; Mason et al. 1982), we make the conservative assumption that the mass of the companion is only 0.25 $M_\odot$, which is the measured mass of the ~K2IV secondary in the old nova GK Per (Watson et al. 1985). For M_c = 0.25 $M_\odot$, i = 85^0 (absence of X-ray eclipses), and a K velocity which is 3 standard deviations less than the best-fit value (Table 3), we find M_x = 3.20 $M_\odot$. <u>Therefore, a firm 3σ lower limit to the mass of the compact object is 3.20 $M_\odot$</u>.

A more realistic lower bound on M_x can be set by using the preferred spectral type and mass of the companion (K5V and 0.7 $M_\odot$) and the absence of X-ray eclipses. In this case we find $M_x > 4.0$ $M_\odot$ (3σ). In addition, if the companion fills its Roche lobe, and we invoke the constraint on the inclination angle imposed by the amplitude of the ellipsoidal variations (i < 50°; see below), we find $M_x > 7.3 M_\odot$ (3σ).

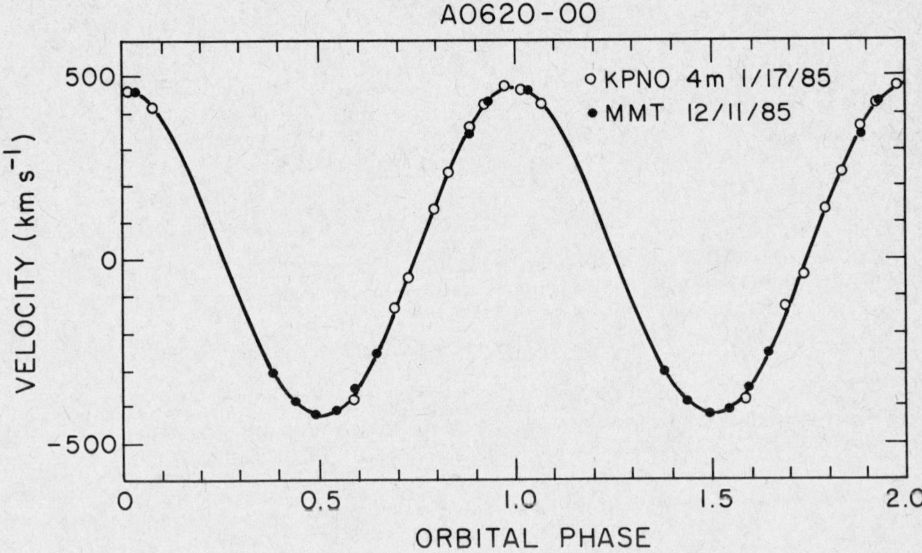

Figure 2. Radial velocities of the K-dwarf companion of A0620-00. The smooth curve is a fit to a circular orbit; the rms of the residual differences between the curve and the data is ≈10 km s⁻¹.

There are four systematic effects that may modify the velocity amplitude and reduce the limiting value of M_x: X-ray heating, tidal distortion, non-synchronous rotation, and spectral contamination by emission lines. In the case of A0620-00, it appears that these effects are not significant (see McClintock and Remillard 1986). For example, the possible contamination of the absorption lines by emission lines, which is one of the most perfidious effects, is not a problem. The center of an absorption line cannot shift by more than a fraction of its ≈1.5Å width, which in turn is only a small fraction of the observed Doppler modulation (16Å or 900 km s⁻¹).

The most obvious and direct approach to firming up the black-hole candidacy of A0620-00 is to establish an upper bound on the orbital inclination angle through studies of the optical light variations, which are primarily due to the ellipsoidal effect. As mentioned above, an approximate analysis indicates that if the K dwarf fills its Roche lobe during quiescence, then the inclination of the orbit is less than 50° and the mass of the black hole exceeds 7.3 $M_\odot$ (3σ; McClintock and Remillard 1986). In future work, which will be based on higher quality data and a Roche-lobe model, it may be possible to determine the values of both the orbital inclination and the degree of filling of the Roche lobe independently.

I will close the discussion of A0620-00 by digressing to remark on a puzzling property of the quiescent state: the presence of a luminous accretion disk and the absence of X-ray emission. As Oke (1977) discovered, about half of the quiescent optical flux is due to a second light source, which is believed to be an accretion disk. The intense, broad, and variable Balmer emission lines, which are always present, strongly support the hypothesis that the second light source is, in fact, an accretion disk (McClintock and Remillard 1986, and references therein).

Now, as Frank Verbunt discussed at this conference, the light from an accretion disk (which is not externally heated) is due to viscous heating. Viscous dissipation will also cause some of the gas to wend its way inward toward the central object. The accretion disk in A0620-00 is as luminous as the K dwarf, $(M_v \approx 7)$, and is, therefore, as luminous as a typical disk in a quiescent cataclysmic variable (Warner 1976). By analogy, one expects a substantial mass-flow rate onto the compact object in A0620-00: $\dot{M} \sim 10^{-11}\ M_\odot$ yr, and, therefore, an X-ray luminosity in quiescence of $\sim 10^{35}$ ergs s^{-1} (for an efficiency of $\sim 10\%$).

Thus, X-ray emission is expected in quiescence; however, none is observed. For example, in 1979 the X-ray luminosity was $< 10^{32}$ ergs s^{-1}, according to results obtained with the Einstein Observatory (Long et al. 1981). Several other X-ray astronomy missions have been active since the end of 1977, and numerous observations of the A0620-00 region have been made. Even the briefest observation would have detected a source at the predicted intensity, ~ 40 mCrab ($L_x \sim 10^{35}$ ergs s^{-1} at d = 1 kpc). The absence of X-rays and the presence of a bright accretion disk is a paradox. Several people at the conference suggested possible solutions: e.g., X-ray shielding of Earth via a bloated accretion disk, a triple-star model, an accretion-disk dam, and special properties of an accreting black hole.

IV. UPPER BOUND TO THE MASS OF A NEUTRON STAR

The maximum mass of a neutron star is a sensitive function of the unknown equation of state for nuclear matter. As indicated in Figure 3, the most realistic and sophisticated models of nonrotating neutron stars imply a maximum mass of 2.0 $M_\odot$, and even the stiffest equation of state predicts a mass of only 2.7 $M_\odot$. Moreover, calculations indicate that rotation cannot increase the maximum mass by more than about 20% (see Shapiro and Teukolsky 1983, and

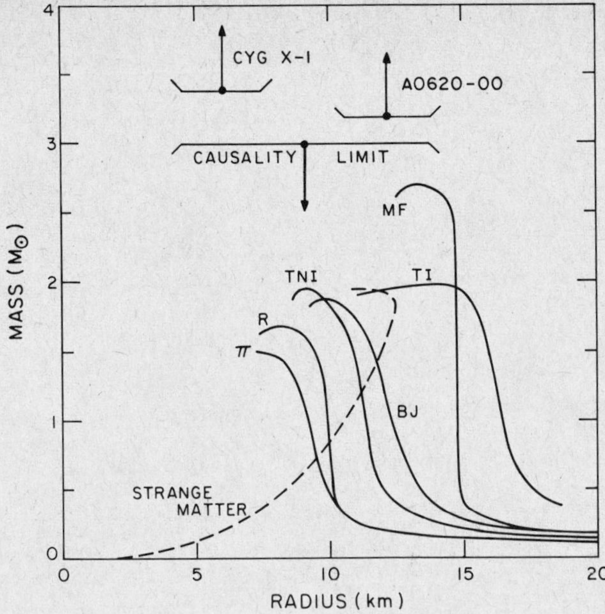

Figure 3. Gravitational mass versus radius for various models of nuclear matter (Arnett and Bowers 1977) and for strange (quark) matter (Alcock 1985). The causality limit (see text) is believed to be a firm upper bound on the mass of a neutron star. The masses of both Cyg X-1 (for d $\geq$ 2 kpc) and A0620-00 exceed this limit, which implies that they are black holes.

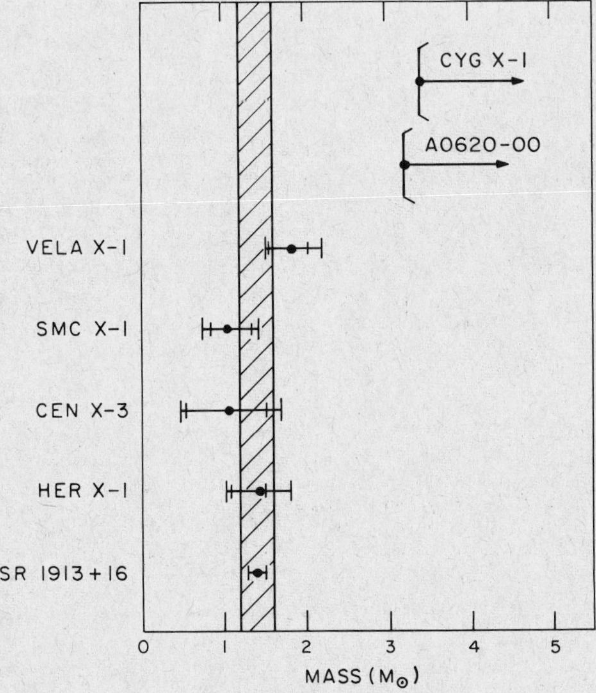

Figure 4. The five best-determined neutron-star masses (after Joss and Rappaport 1984). Four of the masses are derived from observations of binary X-ray pulsars; PSR 1913+16 is a binary radio pulsar. The most probable value for the mass of each neutron star is indicated by the filled circles. For the X-ray binaries, an inner set of error limits is also shown which corresponds to less conservative assumptions: the optical star fills its critical Roche lobe and its rotation period is phase locked to the orbital period. Cyg X-1 (for d $\geq$ 2 kpc) and A0620-00 plainly stand out in the crowd.

references therein).

Fortunately, there is a firm bound on the mass of a neutron star that does not depend on the details of the unknown equation of state at high densities. The key assumptions are the following: general relativity is the correct theory of gravity and causality is obeyed (i.e., the speed of sound is less than the speed of light). These minimal assumptions lead to a strong conclusion: <u>a firm upper bound to the mass of a nonrotating neutron star is about 3 $M_\odot$</u> (Rhoades and Ruffini 1974; Chitre and Hartle 1976).

As shown in Figure 3, the masses of Cyg X-1 (d $\geq$ 2 kpc; see Section IIIa) and A0620-00 exceed this causal limit, which implies that they cannot be neutron stars, and that they are therefore black holes.

The measured neutron-star masses for five systems are shown in Figure 4. The masses are all consistent, within the uncertainties, with a mass range of 1.4±0.2 $M_\odot$ (shaded region). This is the range of masses that might be expected if neutron stars are formed in the collapse of degenerate cores of highly evolved stars (see Joss and Rappaport 1984, and references therein); the masses of Cyg X-1 and A0620-00 greatly exceed this mass range. Furthermore, Cyg X-1 and A0620-00 are plainly more massive than any of the neutron stars that have well-determined masses.

V.   CONCLUSION

For more than a decade, Cyg X-1 has been the solitary example of a black-hole candidate with firm supporting evidence. During the past few years, the situation has improved with the addition of two new candidates, LMC X-3 and A0620-00. Some specific suggestions for further, directed studies of these three systems have been discussed in Section III. In addition, broad-ranging and exploratory observations may prove to be valuable; they may uncover some unexpected and possibly unique characteristic of a black hole.

It is important to make new observations; it is equally important to improve the theory of massive neutron stars. The limiting masses of the three candidates are perilously close to the causality limit. Further definition of this limit, which is not universally accepted (e.g., Caporaso and Brecher 1979), and further work on the equation of state of nuclear matter would probably serve to strengthen the case that black holes exist.

It would be a major advance to discover an X-ray binary with an optical mass function of 5 $M_\odot$ or more. Such a system will not be easy to find. We require an X-ray binary that contains a low-mass ($<$ 1 $M_\odot$) secondary. It must be in a quiescent (X-ray off) state so that the absorption lines of the late-type star are observable, and it must be viewed at a favorably high angle of inclination. Presumably the X-ray source would not pulsate or burst, since a black hole can do neither. Finally, radial velocity data can only be obtained if the companion is brighter than $V \approx$ 21.

These are severe requirements. Apart from the value of its mass function, A0620-00 meets these requirements. It is, however, an especially favorable case for two reasons: it is nearby (d $\approx$ 1 kpc) and it has a luminous secondary (K5V). Its dereddened magnitude is $V_o \approx$ 17. A more typical low-mass X-ray binary has an M-dwarf secondary ($P_{orb} \sim$ 5 hours; White 1985), and is located at a distance $\sim$ 8 kpc. Neglecting extinction, it is $\sim$ 6 magnitudes fainter than A0620-00. Thus, the unreddened magnitude of a typical system is $V_o \approx$ 23; moreover, the extinction is usually sizable (e.g., Bradt and McClintock 1983). It is not possible now, or in the forseeable future, to measure the radial velocities of a star this faint.

Consider, as an example, 4U1755-33 which has a very soft X-ray spectrum and is therefore a probable black hole (White and Marshall 1984).   The absorption-line spectrum of the companion star is overwhelmed by the intense optical continuum produced by X-ray heated gas. Even if the persistent X-ray source were to turn off, the $\sim$ M$\emptyset$V secondary (P = 4.4 hours) would almost certainly be too faint to be observed. Consider also the X-ray novae Cen X-4 and Aql X-1. These systems have quiescent states and observable absorption-line spectra; however, they contain X-ray burst sources, which are almost certainly neutron stars.

In short, known sources do not appear to be especially promising candidates. It is reasonable to hope, however, that new and nearby X-ray transients can be discovered. Unfortunately, we currently lack the appropriate instrument for this search, namely, an all-sky X-ray monitor. Such an instrument would, among other things, allow us to search out and locate X-ray transient sources and to monitor the high and low states of known sources. Only a modest instrument is required. This much-needed capability has been available in the past on an occasional basis. It should now be established as a service facility and operated continuously for decades.

ACKNOWLEDGEMENTS

I am grateful to several people for comments and advice, including Yoram Avni, Ronald Remillard and Harvey Tananbaum. I thank

Donna Irwin for her help in preparing the manuscript. Financial support for my participation in the workshop was provided by the Research Opportunities Fund of the Smithsonian Institution and by the European Space Agency.

REFERENCES

Abt, H.A., Hintzen, P., and Levy, S.G. 1977, Ap.J. **213**, 815.

Alcock, C. 1985, Astronomy colloquium, Harvard-Smithsonian Center for Astrophysics, and private communication.

Arnett, W.D. and Bowers, R.L. 1977, Ap.J.Suppl. **33**, 415.

Avni, Y. and Bahcall, J.N. 1975, Ap.J. **197**, 675.

Baade, W. and Zwicky, F. 1934, Phys.Rev. **45**, 138.

Bahcall, J.N. 1978, Ann.Rev.Astron.Astrophys. **16**, 241.

Bailey, M.E. and Clube, S.V.M. 1978, Nature **275**, 278.

Bardeen, J.M. 1970, Nature **226**, 64.

Bartel, N., Herring, T.A., Ratner, M.I., and Shapiro, I.I., and Corey, B.E. 1986, Nature **319**, 733.

Bolton, C.T. 1972, Nature **240**, 124.

Bradt, H.V.D. and McClintock, J.E. 1983, Ann.Rev.Astron.Astrophys. **21**, 13.

Bregman, J., Butler, D., Kemper, E., Koski, A., Kraft, R.P., and Stone, R.P.S. 1973, Ap.J.(Lett.) **185**, L117.

Burbidge, G. 1985, in Active Galactic Nuclei, ed. J.E. Dyson (Manchester: Manchester Univ. Press).

Caporaso, G. and Brecher, K. 1979, Phys.Rev. D, **20**, 1823.

Carter, B. 1979, in General Relativity: An Einstein Centenary Survey, eds. S.W. Hawking and W. Israel (Cambridge: Cambridge Univ. Press).

Chitre, D.M. and Hartle, J.B. 1976, Ap.J. **207**, 592.

Clark, J.P.A., van den Heuvel, E.P.J., and Sutantyo, W. 1979, Astron.Astrophys. **72**, 120.

Conners, P.A., Piran, T., and Stark, R.F. 1980, Ap.J. **235**, 244.

Cowley, A.P. and Crampton, D. 1975, Ap.J.(Lett.) **201**, L65.

Cowley, A.P., Crampton, D., Hutchings, J.B., Remillard, R., and Penfold, J.E. 1983, Ap.J., **272**, 118.

Davis, M.M., Taylor, J.H., Weisberg, J.M., and Backer, D.C. 1985, Nature **315**, 547.

Duncan, M.J. and Wheeler, J.C. 1980, Ap.J.(Lett.) **237**, L27.

Eddington, A.S. 1935, Observatory **58**, 37.

Flaser, F.M. and Morrison, P. 1976, Ap.J. **204**, 352.

Gies, D.R. and Bolton, C.T. 1982, Ap.J. **260**, 240.

Gies, D.R. and Bolton, C.T. 1986, Ap.J. **304**, 371.

Ginzburg, V.I. and Ozernoy, L.M. 1977, Astrophys.Space Sci. **48**, 401 (in Russian).

Gold, T. 1968, Nature **218**, 731.

Goldman, I. 1981, Astron.Astrophys. **97**, 219.

Hawking, S.W. 1974, Nature **248**, 30.

Hewish, A., Bell, S.J., Pilkington, J.D.H., Scott, P.F., and Collins,
    R.A. 1968, Nature **217**, 709.
Hutchings, J.B. 1979, in Mass Loss and Evolution of O-Type Stars,
    eds. P.S. Conti and C.W.H. de Loore (Dordrecht: Reidel), p.3.
Hutchings, J.B., Crampton, D., and Cowley, A.P. 1983, Ap.J.(Lett.),
    275, L43.
Joss, P.C. and Rappaport, S.A. 1984, Ann.Rev.Astron.Astrophys. **22**,
    537.
Kerr, R.P. 1963, Phys.Rev.Lett. **11**, 237.
Kundt, W. 1979, Astron.Astrophys.(Lett.)   **80**, L7.
Laplace, P.S. 1795, Le Système du Monde, Vol. II, Paris (English
    edition: The System of the World, W. Flint, London, 1809).
Long, K.S., Helfand, D.S., and Gabelsky, D.A. 1981, Ap.J. **248**,
    925.
Margon, B., Bowyer, S., and Stone, R.P.S. 1973, Ap.J.(Lett.) **185**,
    L113.
Mason, K.O., Murdin, P.G., Tuohy, I.R., Seitzer, P., and
    Branduardi-Raymont, G. 1982, MNRAS **200**, 793.
Mazeh, T., van Paradijs, J., van den Heuvel, E.P.J., and Savonije,
    G.J. 1986, Astron.Astrophys. **157**, 113.
McClintock, J.E. and Remillard, R.A. 1986, to appear in Ap.J., **308**.
Michell, J. 1784, Philos.Trans. **74**, 35.
Murdin, P., Allen, D.A., Morton, D.C., Wholan, J.A.J., and Thomas.
    R.M. 1980, MNRAS **192**, 709.
Oke, J.B. 1977, Ap.J. **217**, 181.
Oppenheimer, J.R. and Snyder, H. 1939, Phys.Rev. **56**, 455.
Oppenheimer, J.R. and Volkoff, G.M. 1939, Phys.Rev. **55**, 374.
Paczynski, B. 1973, Annals of the N.Y. Academy of Sciences **224**, 233.
Paczynski, B. 1974, Astron.Astrophys. **34**, 161.
Paczynski, B. 1983, Ap.J.(Lett.)   **273**, L81.
Rappaport, S.A. and Joss, P.C. 1983, in Accretion-Driven Stellar
    X-ray Sources, eds. W.H.G. Lewin and E.P.J. van den Heuvel
    (Cambridge: Camb. Univ. Press), p.1.
Rees, M.J. 1980, Contemp.Phys. **21**, 99.
Rhoades, C.E. and Ruffini, R. 1974, Phys.Rev.Lett. **32**, 324.
Romani, R.W. and Taylor, J.H. 1983, Ap.J.(Lett.) **265**, L35.
Ruffini, R. 1973, in Black Holes, eds. C. DeWitt and B.S. DeWitt
    (New York: Gordan and Breach).
Schwarzschild, K. 1916, Sitzungsber Dtsch.Akad.Wiss. Berlin,
    Kl.Math.Phys.Tech., p.189 (in German).
Shafter, A.W., Harms, R.J., Margon, B., and Katz, J.I. 1980, Ap.J.
    240, 612.
Shapiro, S.L. and Teukolsky, S.A. 1983, Black Holes, White Dwarfs,
    and Neutron Stars, (New York: Wiley), pp.264, 337.
Stella, L., White, N.E., Davelaar, J., Parmar, A.N., Blissett, R.J.,
    and van der Klis, M. 1985, Ap.J.(Lett.) **288**, L45.
Tananbaum, H., Gursky, H., Kellogg, E., Giacconi, R., and Jones, C.
    1972, Ap.J.(Lett.) **177**, L5.

Taylor, J.H. and Weisberg, J.M. 1982, Ap.J. **253**, 908.

Thorne, K.S. 1974, Ap.J. **191**, 507.

Treves, A., Chiappetti, L., Tanzi, E.G., et al. 1980, Ap.J. **242**, 1114.

Trimble, V., Rose, W.K., and Weber, J. 1973, MNRAS **162**, 1P.

van den Heuvel, E.P.J. 1983, in Accretion-Driven Stellar X-ray Sources, eds. W.H.G. Lewin and E.P.J. van den Heuvel (Cambridge: Camb. U. Press), p.303.

van der Klis, M., Tjemkes, S., and van Paradijs, J. 1983, Astron.Astrophys. **126**, 265.

Warner, B. 1976, in Structure and Evolution of Close Binary Systems, eds. P. Eggleton, S. Mitton, and J. Whelan (Dordrecht: Reidel) p.85.

Watson, M.G., King, A.R., and Osborne, J. 1985, MNRAS **212**, 917.

Wheeler, J.A. 1966, Ann.Rev.Astron.Astrophys. **4**, 393.

Whelan, J.A.J., Ward, M.J., Allen, D.A., et al. 1977, MNRAS **180**, 657.

White, N.E. 1985, talk presented at the NATO Advanced Workshop on the Evolution of Galactic X-ray Binaries, Tegernsee, W. Germany (EXOSAT Preprint No. 15).

White, N.E. and Marshall, F.E. 1984, Ap.J. **281**, 354.

White, N.E., Kaluzienski, J.L., and Swank, J.H. 1984, in High Energy Transients in Astrophysics, ed. S.E. Wooskey (New York: AIP), p.31.

Young, A., and Wentworth, S.T. 1982, PASP **94**, 815.

VARIABILITY IN ACCRETING BLACK HOLES

A.C. Fabian
Institute of Astonomy
Madingley Road
Cambridge CB3 OHA
United Kingdom

1. Introduction

Variability is a major source of information on the behaviour of accreting black
holes. Ideally, we wish to observe flashes of decreasing period produced as a large
blob spirals into a black hole. Despite early hopes in Cygnus X-1, such an event
has not yet been observed. Instead we are presented with a picture of apparent
chaotic variability. Much more data on this has emerged from EXOSAT (see Barr &
Mushotzky 1986 and references therein, together with other results reported at this
Meeting). Current analyses do not lead to any clearer picture. Some of the
variability looks almost quasi-periodic (especially see Motch et al. 1983 on
GX339-4) whilst other light curves resemble exponential shots (e.g. NGC 4151,
Marshall et al. 1981; Cygnus X-1, Weisskopf & Sutherland 1978). A main theme is the
relatively common occurrence ($\gtrsim 10$ per cent of the time) of rapid variability.

In this review, I attempt to explain some of the arguments that can be used to
obtain physical limits from X-ray light curves. Most of these also apply to
variation at longer wavelengths. It is not generally necessary that the accreting
object be a black hole, although some of the arguments are used to identify black
holes (e.g. the Eddington limit and arguments for high efficiency; see Rees 1984 for
a general review of accreting black holes). The hardness of the spectra of many
candidate accreting black-holes (e.g. Cyg X-1, Nolan et al. 1984; active galaxies,
Rothschild et al. 1983; Perotti et al. 1982) and their compactness suggest that
electron-positron pairs are commonly produced, mostly by photon-photon collisions.
These severely complicate any simple interpretations and represent an area of active
theoretical research. Rapid source variability argues against equilibrium thermal
models for the X-ray emission.

2.   The Origin of Variability

This is perhaps the most important, and the least explored, section of this talk.
All objects considered to be powered by accretion appear to be variable and many
mechanisms have been put forward. It is quite likely that the mass flow onto a
stationary point mass accreting from a perfectly homogeneous medium is unstable.
The accretion luminosity heats and ionizes the accreting gas and so provides
feedback with a time-delay (the infall time). Most studies so far have considered
either perfectly spherical, cylindrical (disks) or linear (accretion lines) systems.
All are unstable (see e.g. Ostriker et al. 1976; Shapiro et al. 1974; Cowie 1977,
respectively). Other geometries are likely to make the situation worse. Perhaps
accretion occurs more through droplets than a steady stream. Some of the mechanisms
proposed to account for aperiodic, periodic and quasi-periodic variations are listed
in Table 1.

Table 1: MECHANISMS

APERIODIC VARIABILITY

- Accretion instabilities (empirical)
- Turbulence (Shakura & Sunyaev 1973); Rothschild et al. 1974)
- Disk instability (Lightman & Eardley 1974; Shakura & Sunyaev
     1976; Pringle 1981; Papaloizou & Pringle 1985)
- Magnetic flares and turbulence (Pringle 1981; Pudritz & Fahlman 1982)
- Heating, radiation pressure feedback (Buff & McCray 1975;
     Ostriker et al. 1976)
- Absorption / scattering / geometry
- Pair heating instability (Moskalik & Sikora 1986)

PERIODIC VARIABILITY (not yet observed from black holes!)

- Orbits in disk (Sunyaev 1973; Bath et al. 1974; Pringle 1981)
- Binary companion
- Precession (see Rees 1984)
- Spin of black hole (Blandford & Znajek 1977)

QUASI-PERIODIC VARIABILITY

- Spin of hole coupling through magnetic fields to disk
- Coronal scattering (Boyle et al. 1986)
- Spiralling blobs

Some advances should be made when power spectra are generally available, although the shapes of the power spectra of Cygnus X-1 (Nolan _et al_. 1981) and GX339-4 (Motch _et al_. 1983) are not yet explained. The identification of black holes through the similarity of their power spectra may be possible then although it is now clear that rapid aperiodic X-ray variability in Galactic X-ray sources is not restricted to black hole candidates. VO332+53 and Cir X-1, for example, are neutron stars; one of them also pulses and the other also shows X-ray bursts (Stella _et al_. 1985; Tennant _et al_. 1986). f^{-1} noise may prove to be as common in accretion as it is in physics generally and it may generate as many different explanations! At the present time, variability in candidate black hole sources is used mainly to restrict mechanisms and hypotheses rather than build up any detailed picture of the accretion process. Such empirical pictures are vague. It is not at all clear, for example, which geometry is relevant to active galaxies. Are disks present ? As there are no clear examples of periodic variability in black hole sources, I now concentrate on aperiodic variability.

3.   The Interpretation of Aperiodic Variability

The characteristic variability timescale, defined as for energy loss timescales, is

$$\Delta t_{\nu} \; = \; \left| \frac{L_{\nu}}{dL_{\nu}/dt} \right| \; = \; \left| \frac{L_{\nu}}{\dot{L}_{\nu}} \right|$$

Generally, it is frequency dependent. Observations of Δt_{ν}, can immediately lead to estimates of some basic physical parameters.

3.1 Masses

It is likely that the size of the emission region, $R \gtrsim c\,t$, unless there is relativistic motion or some special geometry. If most of the luminosity emerges in the observed waveband, then R is probably a few Schwarzschild radii ($R_S = 2GM/c^2$, where M is the mass of the black hole). A 'few' is likely to mean 3 or more, so

$$M \leq \frac{c^3}{6G} \; \Delta t \left(\frac{R}{3R_S} \right)^{-1}$$

If the energy radiated is produced and released locally, then $t \sim t_{dyn} \doteq R^{3/2}/\sqrt{GM}$, the dynamical crossing time. This gives a similar mass limit if R is a few R_S. The application of the Eddington limit, L_E, provides a check.

$$L_{Bol} \lesssim L_E \approx 10^{46} M_8 \text{ erg s}^{-1},$$

where $M = 10_8 M_8 M_\odot$.

These arguments are weak if applied to isolated events. Energy storage followed by sudden release (e.g. magnetic flares) could lead to mass estimates much smaller than actual masses.

3.2 Heating

Repeated rapid variability implies that the heating timescale,

$$t_{heat} \lesssim \Delta t.$$

The time taken for the radiating electrons to establish a Maxwellian velocity distribution (i.e. a thermal distribution) increases with temperature. This means that the primary radiation in rapidly varying <u>hard</u> X-ray sources is not thermal (Cavaliere 1982; Guilbert, Fabian & Stepney 1982). The rapid cooling implied in many variable sources means high energy electrons cannot propagate far inside the emission region. The acceleration (or heating) must then be distributed throughout the emission region. The acceleration mechanism is unclear. Shocks and electromagnetic (collective) processes are often discussed. Somehow, the acceleration mechanism must take electrons to high energy in the presence of a large surrounding radiation density. It is likely that there is strong feedback here so that the maximum electron energy is least in the most compact sources (Cavaliere & Morrison 1980).

3.3 Cooling

Again, we expect that the radiative cooling timescale,

$$t_{cool} \lesssim \Delta t.$$

Spectral variability is expected since t_{cool} usually depends upon the frequency of the emitted radiation, however there could be many independent blobs rapidly heated and cooled which follow some overall input variation. Spectral variations are important if observed (see Cyg X-1; Page 1985), but need not be directly related to the underlying radiation mechanisms.   Payne (1980) and Guilbert, Fabian & Ross (1982) have discussed the spectral variations expected from various Comptonization models where $t_{cool} \propto (\epsilon_{rad})^{-1}$, where ϵ_{rad} is the radiation energy density.   Note that the timescale for the photon distribution to equilibrate with the electrons is typically $\gg t_{cool}$, particularly near $h\nu = kt$. t_{cool} for bremsstrahlung is sufficiently long that it is generally unimportant in rapidly varying sources (Lightman, Giacconi & Tananbaum 1978), except perhaps as a source of photons.

The strong possibility that heating and subsequent cooling is spread throughout the emission region suggests that single temperature, single density models are not necessarily a good guide to the spectrum.   It is much more likely that the region is filled with blobs that are impulsively heated (by shocks or electromagnetic processes) and cool independently. Time-averaged cooling spectra are then relevant (Guilbert, Fabian & Ross 1982).

3.4 Efficiency

Let us define the efficiency of a source, η, by considering the energy, E, released in an outburst of luminosity amplitude ΔL and duration Δt;

$$E = \Delta L \, \Delta t = \eta Mc^2$$

M is the mass involved. η is generally much less than 1, in which case the 'spent' matter is a source of scattering opacity, of optical depth τ.   If the source is physically large, then Δt measures the light crossing time, R/c, but if it is small, photon diffusion limits Δt.   Generally,

$$\Delta t \gtrsim (1 + \tau) \, R/c$$

Then considering a spherical emission region,

$$\tau \gtrsim \frac{3\Delta L \, \Delta t \sigma}{4\pi R^2 \eta m_p c^2}$$

where σ is the scattering cross-section (the Thomson cross-section σ_T at frequencies well below $m_e c^2/h$). For a given mass, Δt is minimised if $\tau \simeq 1$, so that

$$\Delta L \leqslant \frac{4}{3} \pi \eta \frac{m_p c^4}{\sigma_T} \Delta t ,$$

or

$$\Delta L \leqslant 2.10^{41} \eta_{0.1} \Delta t \ \mathrm{erg \ s}^{-1} .$$

the so-called efficiency limit (Fabian 1979). The rate of change of luminosity, $|\dot{L}| = |\Delta L / \Delta t|$, can thus limit the efficiency. In some cases this enables thermonuclear fusion to be ruled out (if $\eta > 0.007$). Special geometry can overcome this limit (see Guilbert, Fabian & Rees 1983) but is unlikely to apply generally to accreting black holes.

4.   The Compactness Parameter

The compactness of a source is parametrized by

$$\ell = \frac{L \sigma_T}{R \ m_e c^3}$$

which is dimensionless.

$$\ell \propto \frac{L}{R} \propto |\dot{L}| .$$

If $R = c \ \Delta t$, then

$$L = \frac{\ell \Delta t \ m_e \ c^4}{\sigma_T}$$

which can be compared with the efficiency limit to give

$$\ell \approx 4\eta \left[\frac{m_p}{m_e} \right] .$$

Consequently if $\ell > 800$, $\eta > 0.1$, about the maximum value expected from accretion, and if $\ell > 4000$ ($3\ R_S/R$), $L > L_E$, the (electron-proton) Eddington limit.

If the source emits hard radiation so that L (> 1 MeV) $\sim L$ then electron-positron ($e^\pm$) pairs are easily produced through photon-photon collision. They begin to dominate the behaviour of the source when $\ell \gtrsim 40$, above which the Thomson depth in the pairs,

$$\tau\ (e^\pm) > 1,$$

(Guilbert, Fabian & Rees 1983). Since the pair production rate $\propto$ photon flux $\propto \ell$, and the pair annihilation rate (at energies $\lesssim 1$ MeV) $\propto$ (pair density)^2 $\propto \tau^2$, then in equilibrium,

$$\tau\ (e^\pm) \approx \sqrt{\ell/40} .$$

4.1 The Influence of Pairs

Electron-positron pairs limit the maximum temperature obtainable by thermal plasmas under any reasonable condition (Bisnovatnyi-Kogan et al 1972; Stepney 1982; Svensson 1986). Very hot thermal plasmas are probably not very important as the primary radiation source in accretion due to the Maxwellianization timescale discussed earlier. It is most likely that the electrons follow some non-thermal distribution and are probably injected at some relativistic energy γ_{max}. Inverse Compton (or synchrotron) cooling of relativistic electrons leads to $\nu^{-1/2}$ energy spectrum for the radiation, which is close (but a little too flat) to that observed from many candidate black holes (e.g. the active galaxies, Mushotzky 1982). Relativistic electrons generally produce γ-rays by Comptonization (or synchro-self-Compton) which can then collide with X-rays and create pairs.   These pairs produce secondary radiation which adds to the primary spectrum.   The observed spectrum is steepened (Bonometto & Rees 1971) to $\nu^{-0.6}$ to $\nu^{-0.8}$ if $\gamma_{max} \sim 10^3$ which is just what is observed (Kazanas 1984; Zdziarski & Lightman 1985; Fabian 1985; Svensson 1986; Fabian et al. 1986).   The pair-production also depletes the γ-rays, so causing a turnover in the emergent spectrum around 1 MeV.   That source spectra do turn down above 1 MeV is a sign that $m_e c^2$ is significant.   Pair-production is the simplest explanation (Fabian 1982).

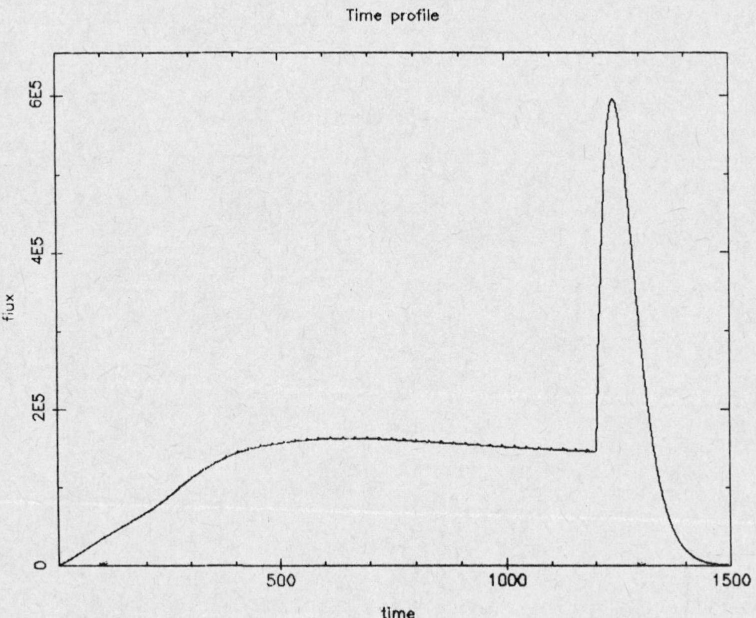

Figure 1.   Computed 2 keV light curve from a source in which electrons of $\gamma = 10^3$ and soft photons at energy $10^{-4}\, m_e c^2$ are continuously injected from time 0 to 1200 (corresponding to 60 light crossing times). $\ell = 4000$. The relativistic electrons Compton cool on the soft photons, thereby generating a hard γ-ray spectrum.   The γ-rays create pairs which also Compton cool before annihilating.   The optical depth so produced in pairs corresponds to $\tau_T \approx 20$.   A large flare occurs when the injected power is switched off.   This is due to photons trapped in the source by electron scattering being suddenly released as no further pairs are produced, while the remaining ones annihilate.

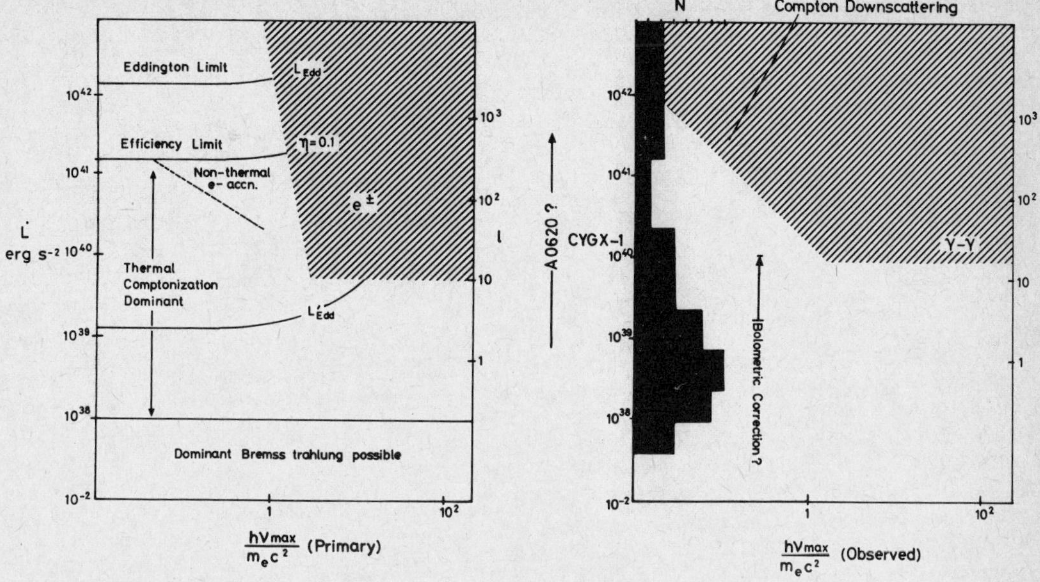

Figure 2. The left panel shows the ranges over which pair production and the efficiency limit dominate as a function of $\dot{L}$ and the maximum frequency of the primary spectrum, ν_{max}. $|\dot{L}|$ is assumed to be equal to Lc/R. This is likely to be a poor approximation when pairs are important. Small amplitude changes in input power are then damped so that an observer underestimates ℓ. Primary γ-rays are destroyed by photon-photon collisions and the remaining hard photons are Compton down-scattered by cooled pairs so that sources are unlikely to be observed above the hatched curve in the right panel. A $\nu^{-1/2}$ primary spectrum is assumed. The varying sources listed by Barr & Mushotzky (1986) and others presented at this Meeting are binned at the left of this panel. If a bolometric correction is made assuming that the spectra extend to $\sim$ 1 MeV, then most sources are observed with $\ell \sim 10$. The importance of ℓ and $|\dot{L}|$ suggests that light curves should be plotted in terms of $|\dot{L}(t)|$. Galactic transient sources such as A0620 undergo large changes in ℓ (see White & Marshall 1984).

238

Sources which are very pair thick ($\ell \gg 40$; $\tau \gg 1$) can store photons. The annihilation time is, however, so rapid that if the input relativistic electron power drops it can lead to a loss of opacity and so a burst of photons. The sense of the variability can be _opposite_ to that of the energy input (Guilbert, Fabian & Rees 1983; Fabian _et al_. 1986). The effect is to damp small changes and amplify large ones. In this way the efficiency (and Eddington) limits may be overcome for a time. It is possible that the rapid variability observed in BL Lac objects is explained by this mechanism.

Note that pairs affect the _whole_ emergent spectrum from the pair-dominated region. Much of the spectral shape and variability informs about the secondary processes of the pairs (where the velocity distribution may be Maxwellian) and not about the primary energy injection.

The photo-production of $e^{\pm}$ pairs may load the acceleration mechanism when $\tau_T > 1$. In other words γ_{max} and the inferred ℓ, ℓ_{obs}, may be limited by the need to accelerate the secondary pairs. This provides a further complication in the interpretation of observed light curves. If R is estimated from $c\Delta t$, $\ell_{obs} \propto \ell^{1/2}$ above $\ell \sim 40$ if bolometric luminosities are used.

The Eddington limit for $e^{\pm}$ pairs,

$$L'_E = L_E \, (m_p/m_e),$$

so pairs exceed L'_E when $\ell \gtrsim 2$. Outflows are thus expected in pair-dominated sources, the rate depending upon, among other factors, the amount of surrounding 'normal' matter.

The work of Barr & Mushotzky (1986) suggests that $\ell_{obs} \sim 10 - 100$ is the preferred operating range for many active galaxies. In accretion terms, $\dot{M} \sim 0.01 \, \dot{M}_E$, the mass accretion rate required to generate L_E. Why sources should limit themselves to that value may imply pair feedback, such as the acceleration loading mentioned above, or perhaps outflows which carry away 'surplus' power in pair winds. The damping effect (due to scattering) of pairs on source variability means that $\ell_{obs} \propto (\Delta L/\Delta t)$ may be significantly less than the $\ell \propto (L/R)$. Compton down-scattering of hard X-rays into the 2-10 keV band tends to increase X-ray estimates of ℓ_{obs}. $\dot{M}$ may also be restricted if accretion takes place from a surrounding medium at the virial temperature of the galaxy ($\sim 10^7$K). Hot gas is common in elliptical galaxies. Its density is limited by cooling such that

$$\dot{M} \sim 0.01 - 0.1 \; \dot{M}_E$$

(Nulsen, Stewart & Fabian 1984; Begelman 1986).

5. Relativistic Effects

Accreting black holes are likely to give observable relativistic effects.

$$\delta = \left(\gamma \left(1 - \beta \cos \theta \right) \right)^{-1},$$

where γ is the usual Lorentz factor of matter moving at velocity $\beta = v/c$ at angle θ to the line of sight. Then the observed luminosity of that matter varies as δ^4 (2 powers from beaming, 1 from blueshift and 1 from time). $\Delta t \propto \delta$, so $(L, \Delta t)$ diagrams are changed by δ^5. Consequently if matter is approaching the observer ($\cos \theta = -1$), then for $0.3 < \beta < 0.5$; $5 < \delta^5 < 15$ (Guilbert, Fabian & Rees 1983). Maximum expected luminosities can be increased by ~ 10 if $\beta \sim 1/2$ which can readily occur from matter swirling into a black hole. Large flares can occur from single blobs in an isotropic velocity distribution. The need for relativistic corrections does not necessarily imply jets. This is compounded by the Compton-Getting effect on the spectral shape, especially important for BL Lac objects.

6. Conclusions

Detailed interpretation of any observed variability in candidate accreting black holes is complicated by uncertainties in the spatial distribution of heating and cooling and by the occurrence of $e^{\pm}$ pairs. It is likely that the emission regions consist of an inhomogeneous mess of $e^{\pm}$ pairs and 'normal' matter. Subregions flash brightly as they temporarily swing along the line of sight, or are heated by shocks or magnetic flares. Accretion is a chaotic process even in the weakest potential wells, and there is no reason to suspect that the black hole case differs. Much of the variability then follows the accretion power and need not directly relate to the cooling processes which may occur on still shorter timescales. Apparent steady emission results when the source is swathed in $e^{\pm}$ pairs which, through scattering, damp out small changes. Outbursts can occur when the power suddenly decreases.

The most dramatic changes observed in a source can still provide strong limits on the underlying processes. Caution is need in applying these limits.

Acknowledgements

Paul Guilbert, Martin Rees and Roland Svensson are thanked for help and discussions.

References

Barr, P. & Mushotzky, R.F., 1986. Nature, 320, 421.

Barr, P., White, N., Sanford, P. & Ives, J., 1977. Mon. Not. R. astr. Soc., 181, 43p.

Bath, G.T., Evans, W.D. & Papaloizou, J., 1974. Mon. Not. R. astr. Soc., 167, 7p.

Begelman, M., 1986. Nature, in press.

Bisnovatnyi-Kogan, G.S., Zeldovich, Ya.B. & Sunyaev, R.A., 1971. Sov. Astr., 15, 17.

Blandford, R.D. & Znajek, R.L., 1977. Mon. Not. R. astr. Soc., 179, 433.

Bonometto, S. & Rees, M.J., 1971. Mon. Not. R. astr. Soc., 152, 21.

Boyle, C.B., Fabian, A.C. & Guilbert, P.W., 1986. Nature, 319, 648.

Buff, J. & McCray, R., 1979. Astrophys. J., 189, 147.

Cavaliere, A., 1982. In Plasma Astrophysics, ESA SP 161, 67.

Cavaliere, A. & Morrison, P., 1980. Astrophys.J., 238, L63.

Cowie, L.L., 1977. Mon. Not. R. astr. Soc., 180, 491.

Fabian, A.C., 1979. Proc. Roy. Soc., 366, 449.

Fabian, A.C., 1982. Highlights of Astronomy, 6, 491.

Fabian, A.C., 1985. Active Gal. Nuclei, ed. Dyson, J., M.U.P., 221.

Fabian, A.C. Blandford, R.D., Guilbert, P.W., Phinney, E.S. & Cuellar, L., 1986. Mon. Not. R. astr. Soc., in press.

Guilbert, P.W., Fabian, A.C. & Rees, M.J., 1983. Mon. Not. R. astr. Soc., 205, 593.

Guilbert, P.W., Fabian, A.C. & Ross, R.R., 1982. Mon. Not. R. astr. Soc., 199, 763.

Guilbert, P.W., Fabian, A.C. & Stepney, S., 1982. Mon. Not. R. astr. Soc., 199, 19P.

Kazanas, D., 1984. Astrophys. J., 287, 112.

Liang, E.P. & Nolan, P.L., 1984. Space Sci. Rev., 38, 353.

Lightman, A.P., Giacconi, R. & Tananbaum, H., 1978. Astrophys. J., 224, 375.

Lightman, A.P. & Eardley, D.M., 1974. Astrophys. J., 187, L1.

Marshall, N., Warwick, R.S. & Pounds, K.A., 1981. Mon. Not. astr. Soc., 194, 987.

Moskalik, P. & Sikora, M., 1986. Nature, 319, 649.

Motch, C., Ricketts, M.J., Page, C.G., Ilovaisky, S.A. & Chevalier, C., 1983.
    Astr. Astrophys., 119, 171.

Mushotzky, R.F., 1982. Astrophys. J., 256, 92.

Nolan, P.L. & Matteson, J.L., 1983. Astrophys. J., 265, 389.

Nolan, P.L., Gruber, D.E., Matteson, J.L., Peterson, L.E., Rothschild, R.E.,
    Doty, J.P., Levine, A.M., Lewin, W.H.G. & Primini, F.A., 1981. Astrophys. J.,
    246, 494.

Ostriker, J.P., McCray, R., Weaver, R. & Yahil, A., 1976. Astrophys. J., 208, L61.

Page, C.G., 1985. Space Sci. Rev. 40, 387.

Papaloizou, J.B. & Pringle, J.E., 1985. Mon. Not. R. astr. Soc., 213, 799.

Payne, D., 1980. Astrophys. J., 237, 951.

Perotti, F., Della Ventura A., Villa, G., Di Cocco, G., Bassani, L., Butler, R.C.,
    Carter, J.N. & Dean, A.J., 1981. Astrophys. J., 247, L63.

Pringle, J.E., 1981. Ann. Rev. Astr. Astrophys., 19, 137.

Pudritz, R.E. & Fahlman, G.G., 1982. Mon. Not. R. astr. Soc., 198, 689.

Rees, M.J., 1984. Ann. Rev. Astr. Astrophys., 22, 471.

Rothschild, R.E., Boldt, E.A., Holt, S.S. & Serlemitsos, 1974. Astrophys., J.,
    189, L13.

Rothschild, R.E., Mushotzky, R.F., Baity, W.A., Gruber, D.E. & Matteson, J.L.,
    1983. Astrophys. J., 269, 423.

Shakura, N.I. & Sunyaev, R.A., 1973. Astr. Astrophys., 24, 337.

Shakura, N.I., Sunyaev, R.A., 1976. Mon. Not. R. astr. Soc., 175, 613.

Stella, L., White, N.E., Davelaar, J., Parmar, A.N., Blissett, R.J. &
    van der Klis, M., 1985. Astrophys. J., 288, L45.

Stepney, S., 1983. Mon. Not. R. astr. Soc., 202, 467.

Sunyaev, R.A., 1973. Sov. Astr., 16, 941.

Svensson, R., 1986. Proc. IAU Coll. 89, in press.

Tennant, A.F., Fabian, A.C. & Shafer, R.A., 1986. Mon. Not. R. astr. Soc.,
    219, 871.

Weisskopf, M.C. & Sutherland, P.G., 1978. Astrophys. J., 221, 228.

White, N.E. & Marshall, F.E., 1984. Astrophys. J., 281, 354.

Zdziarski, A. & Lightman, A.P., 1985. Astrophys. J., 294, L79.

RAPID X-RAY VARIABILITY IN RADIO-QUIET AGN:
A PROBE OF THE INNERMOST REGIONS OF THE ACTIVE NUCLEUS

P. Barr*
EXOSAT Observatory, ESOC
Robert Bosch Str.5, 6100 Darmstadt, FRG

*Affiliated to the Astrophysics Division, Space Science Dept. ESA

1. Introduction

The most rapid variability encountered to date in AGN has been in the X-ray region (0.2-10 keV), with large amplitude (factor two) variations in as short as a few minutes. Thus, the bulk of the X-ray emission must originate in the innermost regions of the active nucleus. In this talk I review briefly the recent results of Barr & Mushotzky (1986) and extend their discussion of the implications of fast X-ray variability for the physical conditions in AGN.

2. The $L_x/\Delta t$ Relationship

Barr & Mushotzky (1986) found, for Seyfert galaxies and quasars, a highly significant correlation between the (logarithm of the) 2-10 keV X-ray luminosity, log L_x, and the (logarithm of the) time for the source intensity to double, log Δt. I consider here only the radio-quiet AGN. Fig. 1 shows the log L_x/log Δt plot for the radio-quiet AGN from the sample of Barr & Mushotzky, plus more recent results for NGC 4593 (Barr et al. 1986), Mkn 766 (Barr & Giommi, this symposium) and NGC 3516 (G. Reichert, private communication). There is a clear trend for more luminous objects to vary more slowly. A linear regression analysis, viz:

$$\log L_x = a \log \Delta t + b \text{ erg s}^{-1}$$

yields a=0.9±0.15, b=39±0.7 and linear correlation coefficient r_c = 0.82. The probability of obtaining this value of r_c by chance from a random sample is 1.8 x 10^{-5}. Within the errors the relationship between L_x and Δt is linear, so $L_x/\Delta t = 10^{39}$ erg s^{-2} to within a factor five.

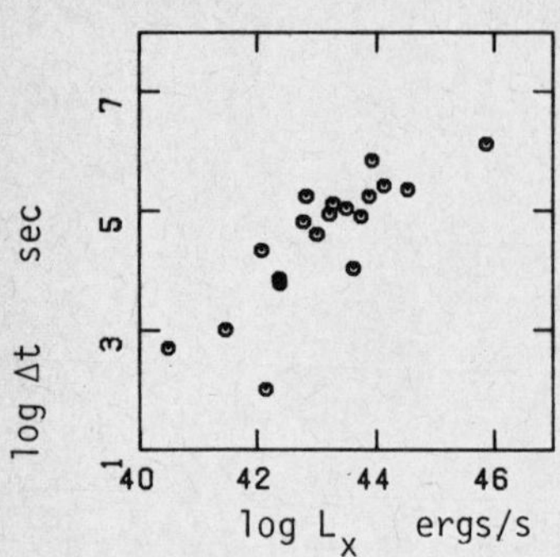

Figure 1.
The two-folding variability
timescale, Δt, versus the mean
2-10 keV rest-frame X-ray
luminosity, L_x

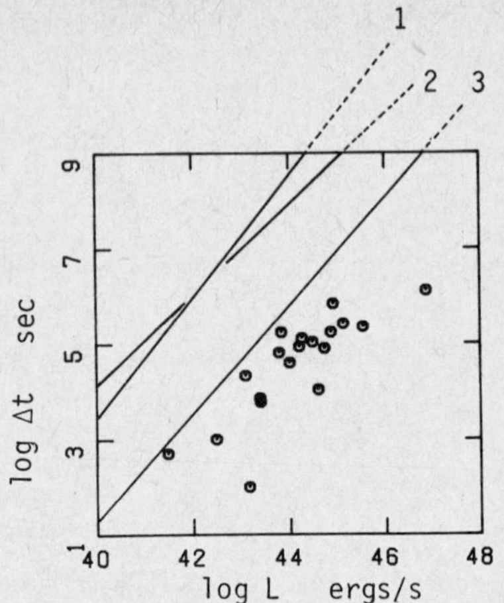

Figure 2.
As fig. 1 but L is now the total hard X-ray
luminosity. The various locii represent the
limits on $L/\Delta t$ for three cases;
1. X-ray triggered thermal instabilities
2. Accretion timescale in α-disk models
3. Subsonic propagation of instabilities in
   α-disk models. (α-disk parameters are for R =
   $10R_s$) The region to the lower right of these
   locii is forbidden. See text for details

3. The Bolometric Correction

The luminosities used here are for the 2.10 keV band. However, this represents
only a small fraction of the total hard X-ray luminosity. The hard X-ray spectra
probably extend out to beyond 100 keV (Rothschild et al. 1983); the handful of soft
γ-ray observations available (Bassini & Dean 1983) indicate that they extend to
$E_{max} \sim 1.3$ MeV before rolling over. Thus, in any treatment involving the energy
output of AGN, a bolometric correction should be applied to the observed luminosity. For energy spectral index of 0.7 and E_{max} of 1 MeV, the bolometric correction, BC, to be applied to L_x is a factor ten. Therefore, for the mean of this
sample, Barr & Mushotzky took the total X-ray luminosity to be L = 10^{40} Δt ergs
s^{-1}.

The observed dispersion around this value is $\sim \pm 0.7$ in log/log space. The
actual BC for individual objects in the sample cannot, of course, be determined
until the relevant soft γ-ray observations are available. Assuming a range of E_{max}
between 100 keV (the lower limit found by HEAO-1 A4) and 3 MeV (the upper limit

found from γ-ray balloon measurements), and an uncertainty of ±0.15 in spectral index (Rothschild et al. 1983), the change in log (BC) would ±0.45; it is surely preferable to have an uncertainty of a factor three in the total X-ray luminosity than to underestimate it by a factor ten.

Thus, taking $L/\Delta t = 10^{40}$ ergs s^{-2} and assuming a source size $R < c \cdot \Delta t$, the 'compactness' of the X-ray source in these AGN is constrained by $L/R > 4 \times 10^{29}$ ergs s^{-1} cm^{-1}.

4.   Discussion

As has been pointed out by many workers (e.g. Svensson 1984 and refs. therein) if the hard X-ray spectrum extends to above 511 keV, electron-positron pair production via two-photon (X-ray plus γ-ray) annihilation may occur. Writing the dimensionless compactness parameter (cf. Fabian these proceedings),

$$\ell = L\,\sigma_T/Rm_e c^3$$

for $\ell > 4\pi$ copious pair production will occur. This limit is strikingly close to the value $\ell \gtrsim 10$ found from the observed $L/\Delta t$ relationship. For higher values of the compactness parameter ($\ell > 40$) the pairs would strongly modify the emergent X-ray spectrum to produce a much steeper continuum than that observed (White, Fabian & Mushotzky 1983). Thus, Barr & Mushotzky postulated that these rapidly variable AGN all lie in the regime $10 < \ell < 40$, and the effects of pairs are very important.

What are the consequences of rapid X-ray variability? We can immediatley rule out bremsstrahlung as the dominant emission mechanism. As pointed out by Lightman, Giacconi & Tannanbaum (1978), energy transfer between protons and electrons via Coulomb scattering is too slow to provide for values of $L/\Delta t > 10^{38}$ ergs s^{-2}. The variability is however not highly constrained either by the classical Eddington limit ($L < L_E$ and $R > R_S$ yield L 0.0011_E) nor by the efficiency of conversion of matter to energy ($\eta > 5 \times 10^{-42} L/\Delta t,\ > 0.5\%$). Taking into account the effects of electron-proton Coulomb coupling on Δt for high (> 22 keV) electron temperatures (Guilbert Fabian & Stepney 1983), $\eta > 0.013\,(kT_e/m_e c^2)3/4, > 2\%$ for 1 MeV electrons.

One striking feature of the observed rapid variability is that it cannot be modulated by processes far out in the accretion flow, and must instead be intrinsically associated with the central few R_S of the AGN. For the standard geometrically thin, optically thick accretion disk models (the α-disks), variations in the accretion rate would imply (Alloin et al. 1985).

$$\log L < 0.8 \log \Delta t - \log(R/R_S) + 38.1 \text{ ergs s}^{-1}$$

In fact, it is very difficult to accommodate the fast X-ray variations within simple α-disk models. For instabilities in the disk to propagate subsonically,

$$\log L < 0.89 \log \quad \Delta t - 1.22 \log (R/R_s) + 40.0 \text{ ergs s}^{-1}$$

In a spherically symmetric, optically thick accretion flow thermal instabilities in the outer regions of the accretion flow can be triggered by X-ray heating from the inner regions (Stellingwurf & Buff 1982) at typical AGN luminosities; however their model predicts

$$\log L < \log \Delta t + 37.0 \text{ ergs s}^{-1}$$

The free-fall time can be related to the (sub-Eddington) luminosity by

$$\log L < \log \Delta t - 1.5 \log (R/R_s) + 42.4 \text{ ergs s}^{-1}$$

so the observed $L/\Delta t$ necessitates $R < 40R_s$. The limits on $L/\Delta t$ set by these phenomena are shown in fig. 2.

However, pair production itself may lead to instabilities. As Svensson (1982) and Sikora & Zbyszewska (1985) have shown, for certain values of ℓ and kT_e, a pair plasma cannot be in equilibrium. Mosalik & Sikora (1986) have shown that for $L \sim 10^{41}$ ($\Delta t/\tau$)(R_s/R) ergs s^{-1} cyclic fluctuations in the electron-positron pair density can lead to repeated high-energy X-ray flares following pair production runaway. Their model predicts exactly the observed $L/\Delta t$ relationship for isolated flares. However the predicted recurrence time between flares is somewhat longer than observed in the recurrent variability of objects such as NGC 4051 and Mkn 766. As Fabian has pointed out, the recurrence interval is determined by the timescale for the protons to transfer their energy to the electrons. Mosalik & Sikora assumed Coulomb scattering. However, as Guilbert, Fabian & Stepney (1983) have noted, rapid hard X-ray variability may necessitate some more efficient energy transfer mechanism.

This raises an interesting question. What, then, of the AGN which have not been observed to vary rapidly? One explanation, suggested by Barr & Mushotzky, is that AGN have a certain 'duty cycle' of variability; since well-known rapidly variable AGN have also been observed to experience periods of quiescence, so 'constant' AGN may at some time exhibit variability. An alternative explanation invokes a selection effect. The sample of Barr & Mushotzky was constructed using the criteria that the observed variability was both time-resolved and the fastest reported for each individual object. (Thus the slow, large amplitude variations seen in, say, 3C390.3 and Cen A were not included). The objects preferentially selected are those like Mkn 766 and NGC 4051 which undergo rapid, repeated variability. These are the sources for which we measure a compactness parameter of 10. As shown by Mosalik & Sikora, pair plasmas with $\ell \sim 10$ may be unstable and experience continuous repeated rapid X-ray variability. However, changing the compactness parameter by a factor ten or so can remove the instability responsible for rapid variations. In fact several Seyferts (e.g. NGC 3783 and NGC 5548) have been extensively observed since the mid-1970's without any evidence for fast variations on the timescale predicted by Barr & Mushotzky.

So far, this treatment has concentrated on radio-quiet AGN; the sample is dominated by Seyfert galaxies. Barr & Mushotzky originally included four radio-loud AGN in their sample. However they did note that exclusion of the radio-loud quasars yielded a considerable improvement in the significance of the correlation. Some radio-loud quasars show evidence (from other wavebands) for bulk relativistic motions; this could substantially increase the observed $L_x/\Delta t$ in the X-ray region. In particular, BL Lac objects do not follow the Seyfert $L_x/\Delta t$ trend.

Acknowledgements

I would like to thank Andy Fabian for pointing out the problems associated with electron-proton Coulomb coupling in applying the Mosalik & Sikora model to the observations.

References

Barr, P. et al. 1986. Proceedings of the 119th IAU Symposium, Bangalore, India. In press.
Barr, P., & Mushotzky, R.F. 1986. Nature 320, 421.
Bassini, L., & Dean, A. 1983. Ast.Ap., 122, 83.
Guilbert, P,W., Fabian, A.C., & Stepney, S. 1982. M.N.R.A.S. 199, 19p.
Lightman, A.P., Giacconi, R., & Tananbaum, H. 1978. Ap.J., 224, 375.
Mosalik, P., & Sikora, M. 1986. Nature 319, 649.
Rothschild, R. et al. 1983. Ap.J., 269, 423.
Sikora, M., & Zbyszweka, M. 1985. M.N.R.A.S. 212, 553.
Stellingwurf, R.F., & Buff, J. 1983. Ap.J., 260, 755.
Svensson, R. 1983. M.N.R.A.S. 209, 175.
Svensson, R. 1984. Proceedings Munich AGN Conf. (MPE Report 184) p.152.
White, N.E., Fabian, A.C., & Mushotzky, R.F. 1984. Ast.Ap., 133, L9.

Iron Lines from Galactic and Extragalactic X-ray Sources

Kazuo Makishima
Institute of Space and Astronautical Science
4-6-1 Komaba, Meguro-ku, Tokyo, Japan 153.

I.   Introduction

Iron is the most abundant heavy element in the universe, believed to be produced in stellar nucleosynthesis. It also has a high fluorescence yield (34%), and its K-line energy (6.4-7 keV) falls right in the middle of the traditional 2-10 keV bandpass of the X-ray astronomy. These facts make iron an ideal "tracer" of the matter distribution and elemental abundance in galactic and extragalactic X-ray sources. However, previous iron-line studies were quite limited due either to a poor energy resolution (proportional counters) or to a low sensitivity (dispersive methods).

This situation has been improved drastically by the gas scintillation proportional counter experiment (GSPC) [KOY84] on board the Tenma satellite [TAN84]. It has a total effective area of 640 cm^2, and a FWHM energy resolution of ~ 9 % at 6 keV which is at least twice as good as that of conventional proportional counters. It has enabled high-sensitivity, wide-band (1.5-35 keV) spectroscopic studies of many galactic and extragalactic X-ray sources [TAN84,MAK84,INO85a,TAN86]. One of the richest results from the Tenma mission is, in fact, the iron line spectroscopy [INO85a, KOY85]. Table 1 schematically summarizes the merit of the GSPC in comparison with the proportional counters.

Table 1   A schematic comparison of proportional counters (PC) and gas scintillation proportional counters (GSPC) in cosmic iron-line spectroscopy.

	Strong line		Weak line	
	PC	GSPC	PC	GSPC
line detection	YES	YES	NO	YES
line center energy	MAYBE	YES	NO	YES
line intensity	NO	YES	NO	MAYBE
intrinsic line width	NO	MAYBE	NO	NO

For   later use, we present in Fig.1 the iron K-edge energy and the iron K-alpha emission line energy as a function of the ionization degree. Also in Fig.2 we present the relative ionization population of iron.

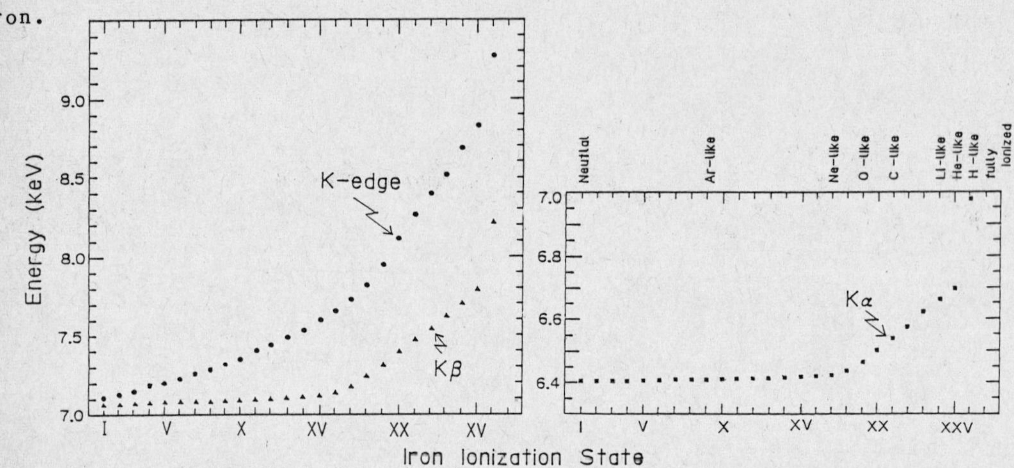

Fig.1   Iron K-edge energy and K-beta line energy (left), and K-alpha line energy (right) as a function of the ionization degree.

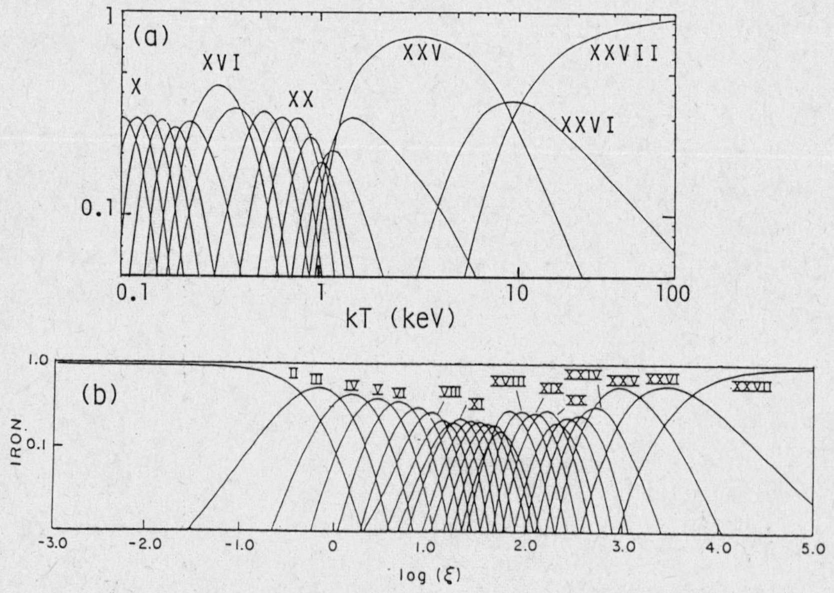

Fig.2   Relative ionization population of iron assuming;
    (a) collisional thermal equilibrium, or
    (b) photoionization-recombination equilibrium in an X-ray irradiated matter with a small optical depth [K&M82].

II. Iron K-edge Absorption

The iron K-edge absorption feature, regarded mostly as local to the source rather than interstellar, was observed by _Tenma_ from the sources listed in Table 2. Examples of such X-ray spectra are shown in Fig.3. In all cases the K-edge energy was consistent with 7.2 – 7.4 keV. Comparison of this with Fig.1 indicates that the absorbing iron is of low ionization degree (Fe III-X) and thus relatively cool (e.g. $kT < 10^5$ K if in collisional equilibrium).

Source		Edge energy (keV)	References
GX301-2	(high absorption)	7.36±0.02	[MLK85]
Vela X-1	(high absorption)	7.24±0.03	[OHA84a,NAG86]
Her X-1	(near turn-off)	∿7.2	[OHA84b]
Cyg X-1	(dips at sup.conj.)	7.18±0.18	[KIT84]
Cen A	(1984 Mar.29–Apr.4)	7.2±0.2[*]	[WAN86]
NGC4151	(1984 January,March)	∿7.2 [*]	[MAT86]

[*] Corrected for cosmological redshift.

Table 2   Energy of the iron K-edge absorption feature measured with _Tenma_.

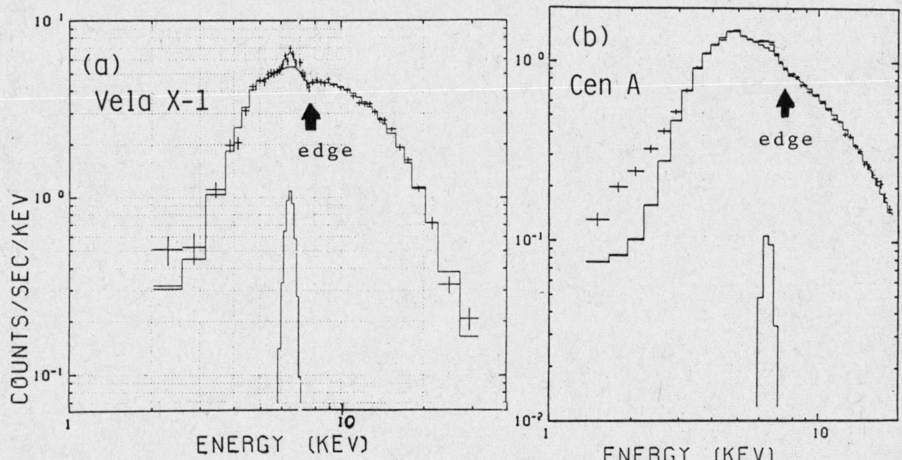

Fig.3   Spectra with iron K-edge absorption feature, observed from Vela X-1 [NAG86] and Cen A [WAN86]. Observed raw spectra (crosses) are fit with a power-law continuum plus a Gaussian line representing the iron K-line emission.

The measured iron edge depth gives the line-of-sight iron column density N_{Fe}. This can be compared directly with the value of N_H determined from the low energy (< 4 keV) absorption of the spectrum, assuming that both N_{Fe} and N_H reflect one and the same material. Here N_H refers to the equivalent hydrogen column density assuming the cosmic abundance [M&M83]. Note that the absorption in the 2-4 keV range is mostly due to O, Ne, Mg, Si, S, Ar, and Ca [M&M83]. Therefore we are effectively comparing the iron abundance relative to these elements. Also note that the inferred ionization degree of iron (Figs.1,2) implies that these medium-light elements also remain in relatively low ionization states, so that the overall photoelectric absorption cross section should not differ significantly from the purely neutral case.

Figure 4 is a scatter plot between the observed values of N_{Fe} and N_H. From this we conclude that the iron abundance is consistent with the "cosmic" value ($N_{Fe}/N_H=3.3\times10^{-4}$ [A&E82]) for all these sources, within an uncertainty of about factor 2.

Fig.4   A comparison between N_{Fe} (determined from the iron K-edge depth) and N_H (from the low-energy absorption). For the chemical composition of the matter contributing to N_H the cosmic abundance is assumed, except for iron which is not included in N_H.

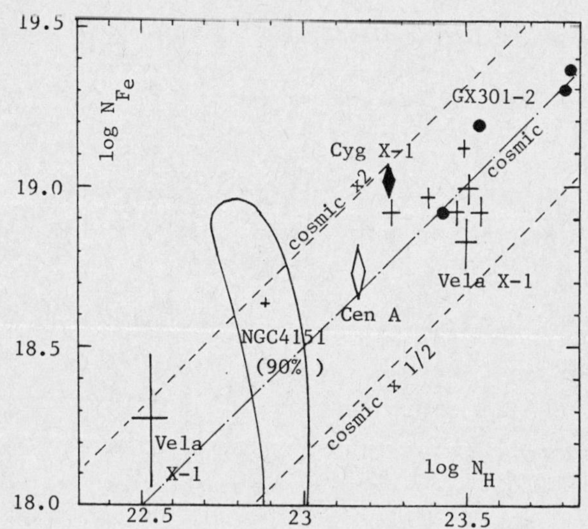

III. Fluorescent Iron K-line Emission from Cool Material

(III-1) Observational overview

We observed iron K-emission lines from many massive galactic binaries and a few AGNs [MAK84, KOY85, INO85a, MAT85, TAN86]. As listed in Table 3, the line-center energy was determined mostly to be 6.4-6.5 keV. From these results and Fig.1 we know that the observed lines are coming from low-ionization (Fe I-XIX), cool ($T<10^5$ K) iron ions. Note that the precise line energy determination was enabled for the first time by the GSPC.

Source	Line energy (keV)	Equivalent width (ev)	References
GX301-2	6.46±0.02	200-2000	[MLK85]
Vela X-1	6.42±0.02	70- 400	[OHA84a,NAG86]
Cen X-3	6.4 - 6.6$	100- 200	[KOY85]
Her X-1	6.41±0.05	100- 200	[OHA84b]
OAO1657-415	6.48±0.05	220	
4U1700-37	6.5±0.1	150- 250	[MUR84]
4U1538-52	6.4±0.2	< 70	[MAK86]
A0535+26	6.6±0.2	∿100	
Cyg X-1	6.53±0.09	50	See §III-6
NGC4151	6.39±0.07*	320	[INO85b,MAT86]
Cen A	6.46±0.08*	90	[INO85b,WAN86]

* Corrected for cosmological redshift.

\$ Line energy varies across pulse phase [KOY85].

Table 3 Summary of observations of iron K-line emission. The
iron line was undetectable in two peculiar pulsars
X0331+53 and 4U1626-67.

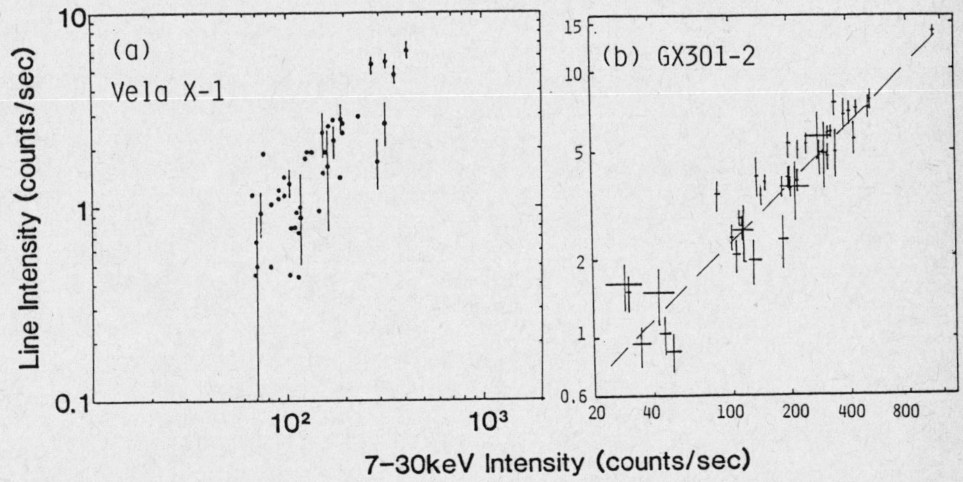

Fig.5 Correlation between the observed iron-line photon flux and the
observed continuum X-ray flux above the neutral iron K-edge
energy (7.1 keV). These results indicate a fluorescence origin
for the iron lines. (a) Vela X-1 [OHA84a]; (b) GX301-2 [MLK85]

The observed line equivalent width varied considerably, from source to source, and even from observation to observation for each individual source. However there exists, as shown in Fig.5, a rough proportionality between the observed line photon flux and the observed continuum flux above 7.1 keV (which can ionize the neutral iron atoms by K-photoionization, followed by fluorescent K-line emission). This indicates that these iron lines are produced through the fluorescent reprocessing of continuum X-rays from the central power source by the cool matter surrounding it. There is no significant difference in this respect between galactic and extragalactic sources.

(III-2) Matter distribution around X-ray sources

We often observed K-line emission and K-edge absorption at the same time (Fig.3), and the line and edge energies always indicated a consistent ionization degree. We thus assume that the same matter is causing the absorption and the fluorescent line emission. Then, the observed X-ray spectrum can be decomposed into three distinct components in reference to Fig.6 [KOY85,INO85a], ignoring other lines;

(i)   direct-beam continuum
$$I(E;N_0) = f(E,\vec{e}_0)\cdot \exp[-\sigma(E)N_0] \tag{1}$$
(ii) Thomson-scattered continuum
$$J(E) = \int_0^\infty dr \int \frac{d\vec{e}}{4\pi}\, \exp[-\sigma(E)(N_1+N_2)]\cdot \sigma_T n(\vec{r})\cdot f(E,\vec{e}) \tag{2}$$

(iii) fluorescent K-line emission
$$L = \delta(E-E_K)\int_0^\infty dr \int \frac{de}{4\pi} \int_{7.1}^\infty dE'\cdot \exp[-\sigma(E')N_1 - \sigma(E_K)N_2]\cdot \alpha\omega\cdot \sigma_{Fe}(E')n(\vec{r})f(E',\vec{e}) \tag{3}$$

Here $f(E,\vec{e})$ is the continuum emitted from the central source with E the photon energy and $\vec{e}$ the directional unit vector, $n(\vec{r})$ is the local matter number density around the source, σ_T is the Thomson cross section, $\sigma(E)$ is the total photoelectric absorption cross section assuming the cosmic abundance [M&M83], σ_{Fe} is that due to neutral iron, $\alpha = 3\times10^{-4}$ is the cosmic iron abundance, $\omega = 0.34$ is the K-fluorescence yield, E_K is the K-line energy, and N_0, N_1 and N_2 are the column densities of the matter as specified in Fig.6.

The second and the third components above were evaluated by the

Monte-Carlo simulations [OHA84a,KOY85,INO85a,MLK85], and a few examples
are presented in Fig.7. The scattered component is negligible as
compared to the direct one only when the Thomson opacity of the cloud is
$\ll 1$. The scattered continuum is somewhat more absorbed than the direct
continuum, reflecting its longer effective path length within the matter
cloud. The scattered continuum, being a superposition of many different
scattering paths with different optical depths, differs in detailed
shape from the direct continuum : notice $\langle \exp(-\tau) \rangle \neq \exp(-\langle \tau \rangle)$ in
general, where τ is the opacity and $\langle \ \rangle$ means the geometrical average.

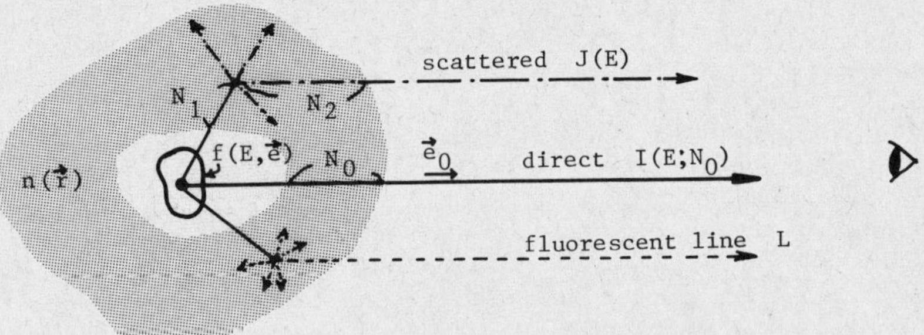

Fig.6 A schematic of the radiation transfer in cold matter around a
compact X-ray source.

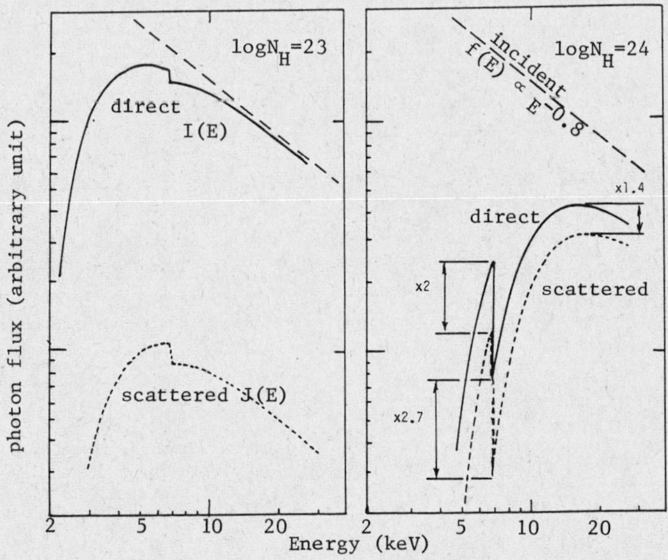

Fig.7 Scattered continuum spectra, calculated using Monte-Carlo method
for an isotropic incident photon spectrum of the form $E^{-0.8}$
[NAK86]. The matter distribution around the central source is
assumed uniform and isotropic, with radial column density N_H. The
direct continuum is also presented for comparison.

(III-3) Line equivalent width -- calculation

To compare with the observation, we calculated by the above Monte-Carlo method the expected line equivalent width (E.W.) as a function of the matter thickness [KOY85,INO85a,MLK85]. For the chemical composition we assume the cosmic abundance. In Fig.8 we show, after [KOY85], the results for several representative cases.

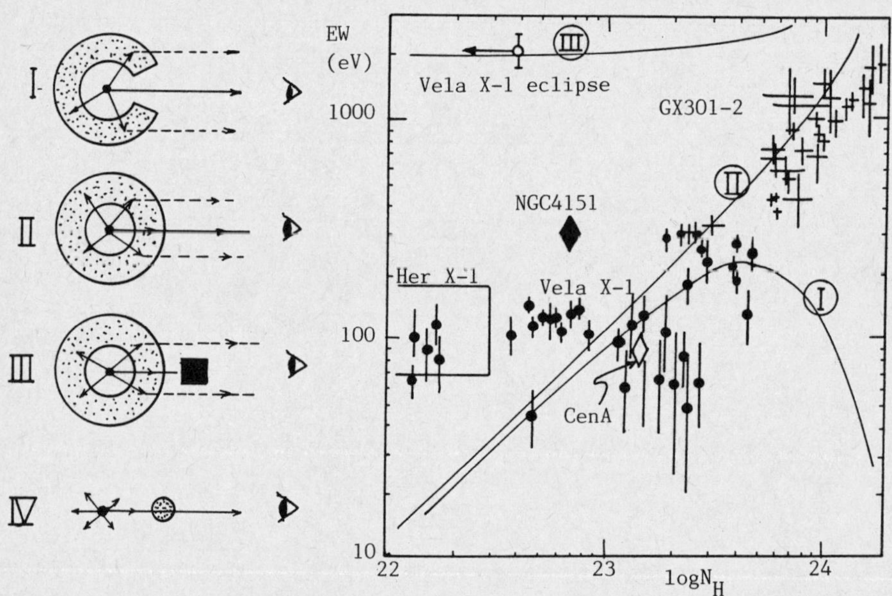

Fig.8   Relation (solid line) between the matter thickness N_H and the iron line equivalent width calculated by Monte-Carlo method for the representative Models I-III illustrated on the left. On the same diagram, the observed E.W. and the observed <u>line-of-sight</u> absorption column are plotted for comparison.

<u>Model I</u> corresponds to the definition of E.W. as
$$E.W. = L/f(E_K,\vec{e}) \quad . \tag{4}$$
This implies a situation when the absorption somehow avoids the line of sight. In this case the line equivalent width reaches the maximum value of $\sim$180 eV at about $N_H=10^{24}$ cm^{-2}, beyond which it decreases due to self absorption. <u>Model II</u> is the simplest one, when an uniformly distributed matter causes the line-of-sight absorption as well as the scattering and fluorescence. This corresponds to the definition as
$$E.W. = L/[\ I(E_k;N_0) + J(E_K)\ ] \quad . \tag{5}$$

The E.W. then increases almost linearly with increasing N_H.

Model III assumes that the direct beam is completely blocked by some thick material (e.g. the companion star). Then the definition of E.W. becomes

$$\text{E.W.} = L/J(E_K) . \qquad (6)$$

In this case E.W. is nearly constant, at ~ 2 keV, for a wide range of N_H. This is easily understood by comparing eqs.(2) and (3), where the radiative transfer relations are very similar between the two processes. Finally in Model IV the matter is assumed to be localized along the line of sight, so that it causes significant absorption but little scattering or fluorescence.

(III-4) Line equivalent width -- observation

The four cases introduced in §III-3 are in fact found in the observation [KOY85]. In Fig.8 we plot the observed E.W. (defined as the ratio between the observed line photon flux and the observed continuum at 6.4 keV) against observed line-of-sight absorption N_H. Shown in Fig.9 are the corresponding spectra. For GX301-2 the observed E.W.$-N_H$ relation is in good agreement with the prediction by Model II [W&S84, MLK85], indicating that this pulsar is imbedded in a dense, almost uniformly distributed cool matter. This matter can be identified with the dense stellar wind from the supergiant primary star Wray 977 [PAR80]. The data for Cen A are also consistent with this case.

Figure 9c shows a series of spectra from Vela X-1 during its eclipse ingress [SAT86]. The ingress is characterized by an increasing absorption and scattering. At the mid eclipse, however, the spectrum is no longer strongly absorbed and the iron line is very prominent (E.W. ~ 2 keV). This exactly corresponds to the Model III in Fig.8; the direct beam is stopped by the supergiant companion, while the stellar wind and the companion's atmosphere produce the scattered continuum and the fluorescent line. See §III-5 for further details.

Figure 9d is a spectrum of Cyg X-1 during "dip" events near the superior conjunction [KIT84]. A deep iron K-edge absorption is seen, but the iron line emission is nearly absent. This is an ideal example of Model IV [KOY85], implying that the dense matter ejected from the primary temporarily came into our line of sight. See also §III-6.

In Fig.8 we see that the E.W. for Her X-1 and Vela X-1 remains 100-150 eV even when the line-of-sight N_H is very small. Figure 9a gives an example of such spectrum. This may be interpreted, in terms of Model I, that there exists some thick "target" material outside our line of sight. The observed E.W. is nearly the same as the maximum value expected from Model I, suggesting that the target covers a significant fraction of the solid angle containing the pulsed X-ray beam.

NGC4151, with it unusually large E.W., is the most puzzling case. We discuss this case later in §III-8.

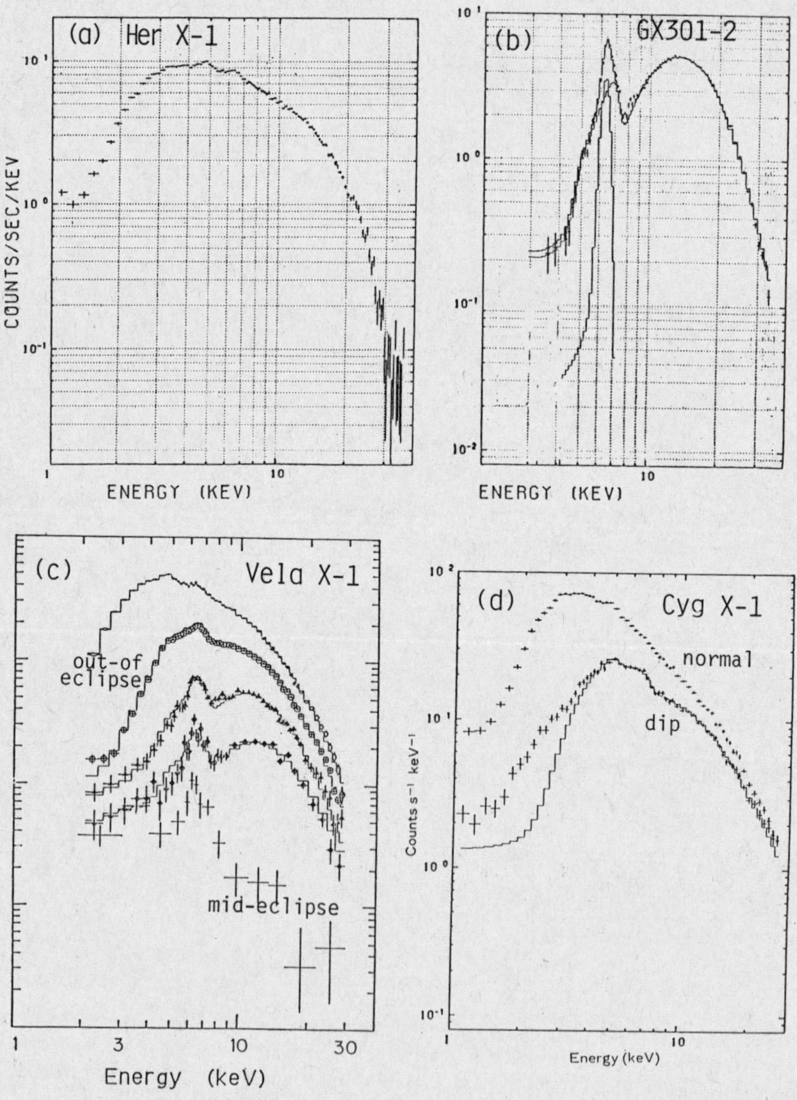

Fig.9 X-ray spectra corresponding to the Models I-IV of Fig.8.
    (a) Her X-1 (Model I) [OHA84b].
    (b) GX301-2 (Model II) [MLK85].
    (c) Vela X-1 during eclipse ingress (Model III) [SAT86]
    (d) Cyg X-1 near the superior conjunction (Model IV) [KIT84].

(III-5) Origin of fluorescent line emissions from X-ray pulsars

Figure 10 shows the orbital-phase variations of the iron line intensity from Vela X-1 [OHA84a,NAG86,SAT86]. As already shown in Fig.9c, we observed a significant (∿15%) residual line emission during the eclipse periods ; a certain fraction of the iron line flux is in fact produced in regions comparable in size to the companion star (~10^{12} cm) [OHA84a]. Such regions may be identified with the stellar wind from the primary and the photosphere of the primary itself.

In Fig.10 we calculated the expected contributions from these two regions, in which the detailed radial structures of the primary's atmosphere and of the stellar wind have been modelled based on the observed changes of the continuum spectrum along the eclipse transitions (Fig.9c) [SAT86]. Although the observation is fairly well simulated, the model still falls short of the observed out-of-eclipse line intensity and fails to reproduce the abrupt jump at the eclipse ingress and egress. These discrepancies are nicely solved if we assume a third component with E.W. ∿80 eV, local to the pulsar thus showing a sharp eclipse transitions. This localized matter should be identified with the fluorescing "target" mentioned in §III-4.

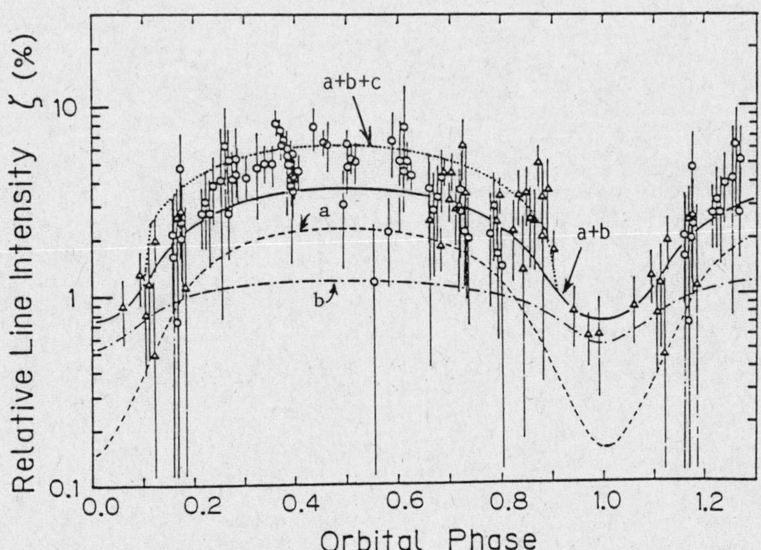

Fig.10   The binary phase dependence of the iron line intensity observed from Vela X-1, in comparison with calculation [SAT86]. The "relative line intensity" is proportional to the E.W. as defined by eq.(4) in the text. Curves a and b represent the calculated contributions from the primary's atmosphere and the stellar wind, respectively. The curve labeled as a+b+c assumes another reprocessing region close to the pulsar.

From these results we conclude that there are three distinct regions in Vela X-1 producing fluorescent iron lines ;
(i)   The atmosphere of the primary supergiant HD77581 (E.W. $\lesssim$ 60 eV);
(ii) The stellar wind from the primary (E.W. $\sim$ 30 eV) ; and
(iii) A matter localized close to the pulsar (E.W. $\sim$ 80 eV).
The case of Her X-1 and Cen X-3 seems similar [KOY85], while in GX301-2 the second component seems dominant. The physical reality of the third component is yet to be investigated, but it is very likely to be matter on the Alfven shell [INO85a].

We found that the iron line intensity from Vela X-1, GX301-2 and Her X-1 were effectively free from pulse modulation [OHA84a,MAK84]. The case of GX301-2 may be explained by the approximate isotropy of the fluorescing matter distribution, which will geometrically average out local pulse modulations in the fluorescent flux. For Vela X-1 and Her X-1, however, we already know that the matter distribution cannot be isotropic, since the component (iii) above must be located outside the line of sight between us and the pulsar. Then the absence of line intensity modulation requires that the local "target" matter corotates with the pulsar [INO85a]. This reinforces the above suggestion that the "target" matter may be identified with the matter on the Alfven shell.

(III-6) Cyg X-1

Cyg X-1 is believed to be a black hole binary but otherwise is very similar to ordinary massive neutron-star binaries. It is therefore very important to study its iron line emission in comparison with those from the X-ray pulsars. The iron K-line intensity from Cyg X-1, observed by EXOSAT [BWP85] as well as Tenma (Table 3), is significantly weaker than those from the binary X-ray pulsars. A self-consistent explanation, then, is that Cyg X-1, containing a black hole instead of a magnetized neutron star, lacks the Alfven surface and hence lacks the reprocessing region (iii) mentioned in §III-5. Thus the results of iron line spectroscopy argue favourably for the black hole scenario for Cyg X-1.

The EXOSAT observations of Cyg X-1 also showed that a single power-law fit to the observed continuum requires a broadened (and possibly red-shifted) line component [BWP85]. Such intrinsic line widths would allow several alternative explanations, including ; (1) blend of many lines from different ionization species; (2) Compton broadening; (2) Doppler broadening; etc. In fact the EXOSAT results were interpreted in terms of Comptonization by the disk corona. However, as shown below, such line broadening should be treated with extreme caution (Table 1).

As shown in Fig.11, the <u>Tenma</u> data for Cyg X-1 closely resemble the <u>EXOSAT</u> data, indicating that the two observations are consistent. Yet, it should be kept in mind that the assumption of a single power-law continuum may not be adequate at the level of data accuracy concerned. Some fraction of the continuum may undergo scattering and may bear a noticeable iron K-edge absorption feature (Fig.7). When this possibility is taken into account, the observed data no longer require the line broadening (Fig.11c). Thus the broad iron line is not an unique interpretation of the Cyg X-1 data, but instead is merely one of several possible alternatives of similar likelihood.

In short, we would easily get false line broadening if the continuum model was only slightly wrong, and such broadening would lead to a serious overestimation of the line equivalent width. This caution should also be applied to the low-level 6.7 keV emission lines from low-mass X-ray binaries [SUZ85,INO85a,WHI86].

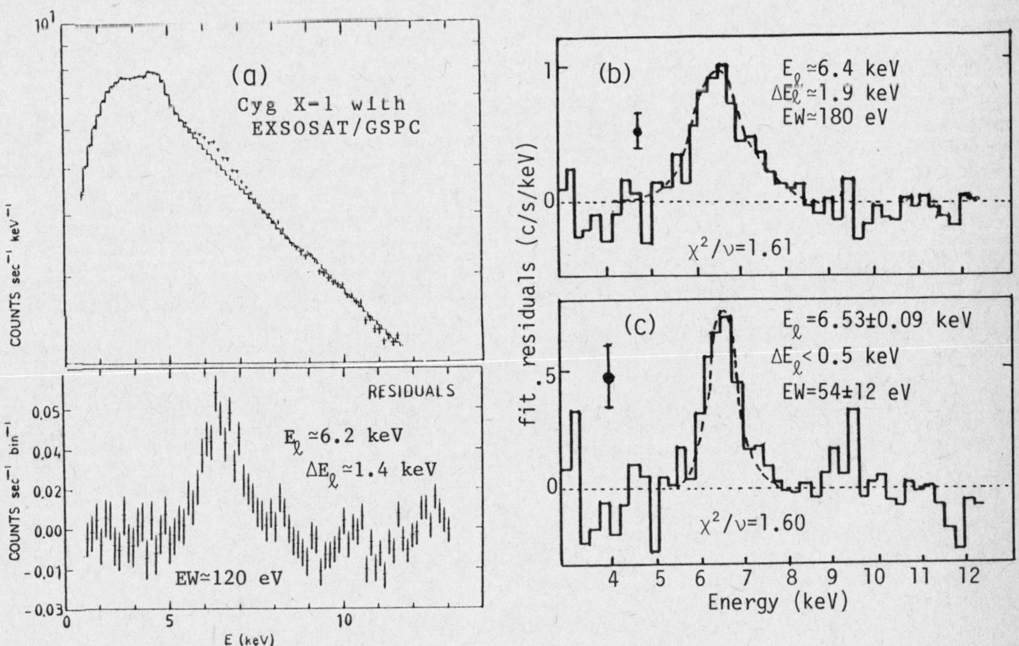

Fig.11 Iron lines from Cyg X-1 observed by <u>Tenma</u> and <u>EXOSAT</u>.
(a) The <u>broadened</u> iron line profile obtained by the <u>EXSOAT</u> GSPC [BWP85], after subtracting the best-fit power-law continuum.
(b) The same as panel (a), but using the <u>Tenma</u> GSPC data. Note that the profile is quite similar to that in panel (a).
(c) The same data as panel (b), but assuming that 20 % of the continuum undergoes absorption by 2.4×10^{23} cm^{-2} and the rest by 1.4×10^{23} cm^{-2}. A narrow line model gives a fit chi-square as good as in panel (b).

(III-7) Anisotropy in the matter distribution : case of GX301-2

Although we have seen that GX301-2 is generally described by the Model II (isotropic matter distribution) of Fig.8, some of its spectra are actually in significant deviation from this situation. Figure 12a is one such example, where the observed line E.W. is in excess of the prediction based on the observed absorption. Furthermore, its spectral shape immediately above the iron K-edge cannot be expressed by a single value of N_H. These imply that the isotropic matter distribution is only an approximation even for GX301-2, and that we must sometimes take into account the possible matter anisotropy. Such anisotropy would first manifest itself as a difference between the line-of-sight column density N_0, contributing to the absorption, and the matter thickness averaged over 4π, denoted $\langle N_H \rangle$, causing scattering and fluorescence.

The above idea may be formulated by referring to eqs.(1)-(3) and Fig.6, and introducing a model photon spectrum of the form [NAK86]

$$I'(E \; ; \; N_0, \langle N_H \rangle \;\;) = I(E;N_0) + \; J(E;\langle N_H \rangle) + L(\langle N_H \rangle) \; . \qquad (7)$$

For the incident continuum $f(E)$ we employed the conventional "power-law plus exponential cutoff" model with 4 free parameters [WSH83]. Including N_0 and $\langle N_H \rangle$, the present model thus involves 6 free parameters. Note that Fig.12a corresponds to the case of $N_0 = \langle N_H \rangle$.

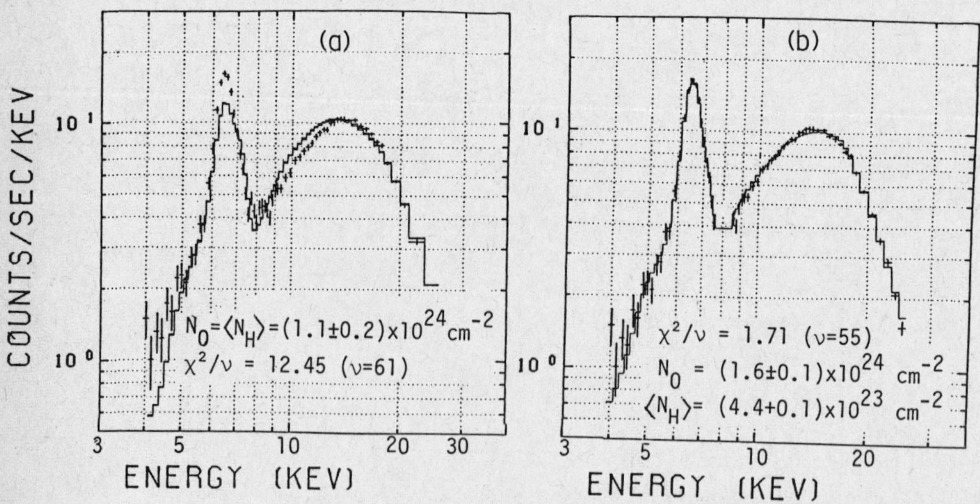

Fig.12   Fitting of the same pectrum from GX301-2 to the model of eq.(7) [NAK86]; (a) under the constraint of $N_0 = \langle N_H \rangle$, where N_0 is the line-of-sight absorption column, and ; (b) When N_0 and $\langle N_H \rangle$ are allowed to vary independently. In either case the iron line intensity is not a free parameter but constrained by $\langle N_H \rangle$, the column density of matter averaged over 4π .

When the spectra of GX301-2 were fit with eq.(7), $N_0/\langle N_H \rangle$ was found mostly in the range between 0.5 and 2 [NAK86]. Such small difference between N_0 and $\langle N_H \rangle$ may be explained by the inhomogeneity in the stellar wind density [NAK86,NAG86]. However we sometimes found much larger ($\gtrsim$ factor 3) difference between them. For example, the spectrum of Fig.12 requires $N_0/\langle N_H \rangle$ = 3.7±0.1. These cases also include an eclipse-like phenomenon of GX301-2 lasting $\lesssim$ 1 hour, when the continuum flux became very weak while the iron line remained very strong [KOY85,MLK85, LEA86,NAK86]. This translates into an abrupt increase in N_0 (from 6×10^{23} to 6×10^{24} cm^{-2}), while $\langle N_H \rangle$ remained roughly constant at $(2-4) \times 10^{23}$. This event may possibly be caused by a dense tail-shock front, which blocked most of the direct continuum thus making the situation similar to the Model III in Fig.8.

(III-8) Active galactic nuclei

As shown in §III-4 and Fig.8, we detected 6.4 keV iron K-lines from two AGNs, Cen A and NGC4151 [INO85b,MAT86,WAN86].   The line E.W. for Cen A is consistent with the prediction based on the observed line-of-sight absorption (Model II of Fig.8), while that for NGC4151 exceeds the prediction so much (by factor 6) that the data cannot be accounted for even by invoking a highly anisotropic matter distribution around [INO85, MAT86].   An ad-hoc assumption of iron overabundance would be inconsistent with the conclusion made in §II.   One possible solution for NGC4151 is to assume that the continuum flux from its nucleus is strongly beamed so that we observe only a small fraction of it ("missing beam" assumption) [INO85a,MAT86].

An alternative explanation was suggested during the conference. Dr. M. Morini, combining the EXOSAT and Tenma observations, showed that the iron line intensity of Cen A may be less variable than its continuum intensity on time scales of months to years. Also Dr. M. Penston pointed out that the case of NGC4151 (Fig.8) may be due to the fact that it was unusually faint in X-rays during the Tenma observations (January and March 1984) [PEN86]: the line intensity may not follow these short-term excursions in the continuum intensity. These suggestions imply that the fluorescent reprocessing region in the AGN has a physical size of 10^{17-18} cm and the line variation is smeared out by the light-travel-time effect. This size nicely coincide with that of the BLR (broad line region), indicating that the flourescence may take place in the BLR.

IV.   Iron K-line Emission from Thin Hot Plasma

(IV-1) Observations

The optically thin, hot plasma is also a very good candidate of the X-ray emission line spectroscopy. This situation has been encountered in a wide variety of celestial objects, including; (1) supernova remnants; (2) clusters of galaxies [OKU86] ; (3) white-dwarf binaries; (4) star-forming regions (ρ-Oph, Orion Nebula [AGR86] etc.); and (5) galactic-ridge excess emission [KOY86, WAR85].   Figure 13 shows several examples.

By measuring accurately the energies of emission lines in these spectra, we can evaluate whether the plasma is in ionization equilibrium or not. If so then the iron line should show up at about 6.7 keV for a wide range of the plasma temperature concerned, due to the predominant stability of the helium-like iron ions (Fig.2a).   To examine this we plot in Fig.14a the observed iron line energies against the measured plasma temperature (assuming a thermal bremsstrahlung continuum). It indicates that these sources are more or less in the ionization equilibrium, except for the two SNRs Cas A and Tycho.

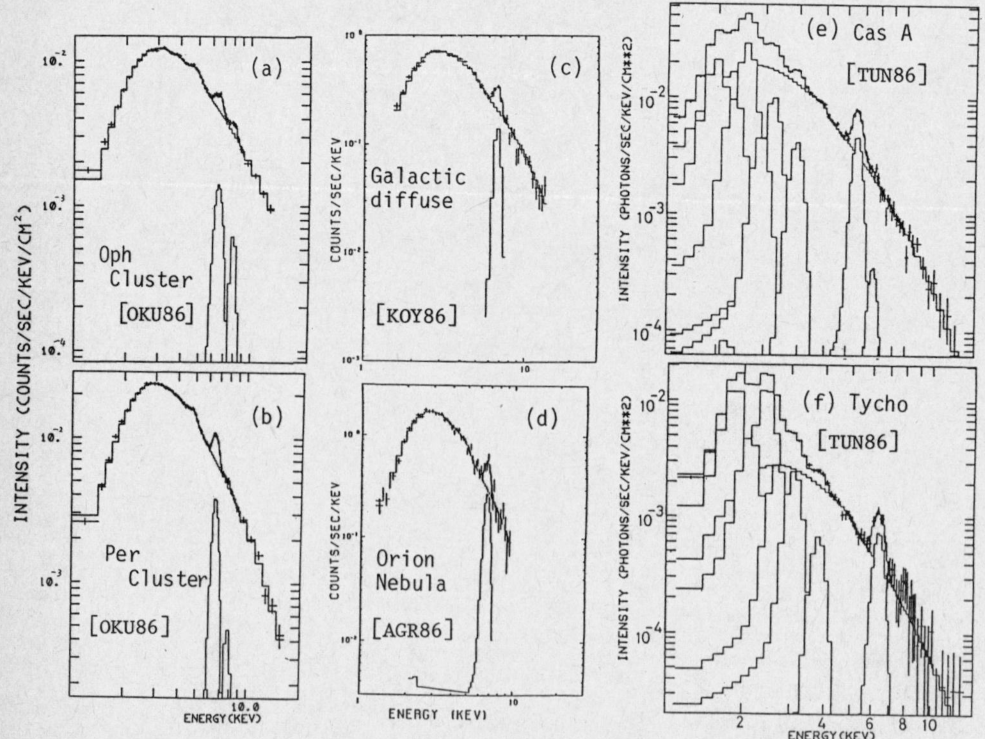

Fig.13 Examples of emission from optically thin, hot plasma.

(IV-2) Case of ionization equilibrium

When the plasma is in the ionization equilibrium, we can estimate
the abundances of elements such as Si, S, Ar, Ca and Fe by comparing the
observed line E.W. with the calculations [R&S77]. As shown in Fig.14b
the iron in these sources is roughly consistent with, or somewhat less
than, the cosmic abundance.

The spectrum of the Coma cluster of galaxies was examined closely
for any non-isothermal indication, but the result was rather negative in
the sense that the data are consistent with both the isothermal and
adiabatic models [HUG86]. This contradicts the previous claim for the
non-isothermality in the Coma cluster [H&M86].

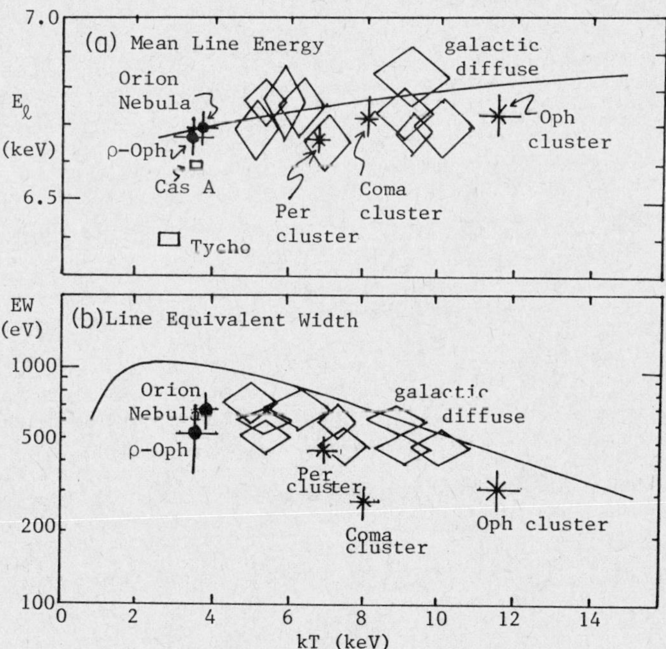

Fig.14   The observed iron K-alpha line energy (panel a) and line E.W.
(panel b) plotted against the observed plasma temperature. In
both panels the solid curve shows the calculation assuming the
ionization equilibrium and the cosmic abundance of iron.

(IV-3) Case of ionization non-equilibrium

In Figs.13e and 13f we present the spectra of two SNRs, Cas A and
Tycho, observed from Tenma [TUN86]. In addition to the iron K-alpha
line, many other emission lines are identified; see Table 4. The Tycho's
SNR exhibits systematically lower emission line energies than Cas A,

despite little difference in the plasma temperature. This indicates that the plasma in Tycho is less ionized as compared to that in Cas A. In fact, these observations have been successfully interpreted in terms of the ionization non-equilibrium model (Fig.15) [MAS84].

These two SNRs exhibit much stronger silicon and sulfur lines as compared to the other sources in Fig.13. This is partly due to the ionization non-equilibrium condition, and partly to the enhanced abundance of heavy elements in these sources as a consequence of nucleosynthesis in the progenitors. Detailed discussion and comparison with the Einstein SSS data are found in [TUN86].

		Cas A	Tycho	helium-like	hydrogen-like
Si	K-α	1.94±0.02	1.84±0.02	1.86	2.01
	K-β	2.23±0.08	2.08±0.04	2.18	2.25
S	K-α	2.53±0.03	2.38±0.02	2.46	2.62
	K-β	2.88±0.15	2.65±0.06	2.88	3.11
Ar	K-α	3.18±0.03	3.06±0.02	3.14	3.32
Ca	K-α	3.89±0.02	3.76±0.04	3.91	4.11
Fe	K-α	6.59±0.02	6.40±0.03	6.71	6.93
	K-β	7.67±0.15	---	7.90	8.21

Table 4   Best-fit emission line energies for Cas A and Tycho [TUN86]. K-line energies from helium-like and hydrogen-like ions are also listed.

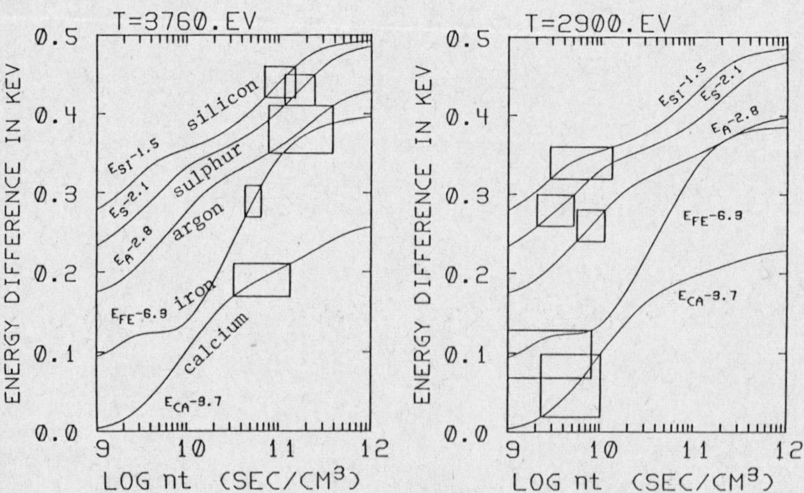

Fig.15 Evolution of the mean energies of various Kα lines, as a function of nt, where n is the plasma density and t is the time after the supernova explosion. Solid curves show the calculation [MAS84] and the boxes represent the observation summarized in Table 4. The line energies are displayed with a constant offset subtracted [TUN86].

V. Summary

Results of the iron line spectroscopy of the galactic and extragalactic X-ray sources, using the Tenma GSPC, are reviewed.

The iron K-edge absorption features, detected in several massive galactic binaries as well as a few AGNs, indicate that the iron abundance in these sources is consistent with the "cosmic" value.

These objects also emit 6.4 keV iron K-lines, which arises due to the fluorescent reprocessing of the continuum X-rays from the central power source by a surrounding cool material. In X-ray pulsars the re-processing material is identified with the photosphere and stellar wind of the primary star. In some cases, additional matter local to the pulsar (e.g. near the Alfven shell) is required. When matter distributed around the source is dense, scattered continuum must be taken into account to explain the observed X-ray spectra. The fluorescent lines from AGNs may be produced in the broad line region (BLR). These results show that the iron line spectroscopy provides a powerful tool to study the matter distribution around accreting compact objects.

Iron emission lines from optically-thin hot plasma was also widely observed in clusters of galaxies, supernova remnants, star-forming regions etc., as well as in the galactic-ridge excess emission. These data are used to examine the iron abundance and the ionization equilibrium.

Finally there are several topics which could not be included in the present talk. They are; the weak 6.7 keV emission lines from the low-mass neutron-star binaries [SUZ84,KOY85,WHI86]; the enigmatic iron emission lines from Cyg X-3 and Cir X-1 [INO85]; and the case of SS433.

ACKNOWLEDGEMENTS

The present talk has been based on the total activity of the Tenma team rather than the author's personal contribution. He is thus deeply indebted to all the members of the Tenma team. He is above all thankful to Drs. Hajime Inoue and Katsuji Koyama, and to Profs. Masaru Matsuoka, Fumiyohsi Makino, Fumiaki Nagase and Koujun Yamashita, for their valuable discussions and for giving him liberty to use their work (published and unpublished) in the present talk. Also he thanks Dr. E. Fenimore of LANL for the careful reading of the manuscript and valuable comments.

[REFERENCES]

(PASJ = Publ. Astr. Soc. Japan)
(ISAS RN = Research Note of Inst. Space. Astronautical Science)
[A&E82] Anders,E. and Ebihara,M. 1982, Geochim. Cosmchim. Acta **46**, 2363.
[AGR86] Agrawal,P.C, Koyama,K., Matsuoka,M. and Tanaka,Y. 1986, submitted to PASJ (ISAS RN #300).
[BWP85] Barr,P., White,N.E. and Page,C.G. 1985, MNRAS **216**, 65p.
[H&R86] Henriksen,M.J. and Moshotzky,R.F. 1986, Ap.J. **302**, 287.
[HUG86] Hugues,J.P. et al. 1986, manuscript in preparation.
[INO85a] Inoue,H. 1985, Space Science Reviews **40**, 317.
[INO85b] Inoue,H. 1985, Proc. Japan-US Seminar on Galactic and Extra-Galactic Compact X-ray Sources, ed. Y.Tanaka and W.H.G.Lewin (January 1985, Tokyo), Published from ISAS, p.283.
[KIT84] Kitamoto,S., Miyamoto,S., Tanaka,Y., Ohashi,T., Kondo,Y., Tawara,Y. and Nakagawa,M. 1984, PASJ **36**, 731.
[K&M82] Kallman,T.R. and McCray,R. 1982, Ap.J. Suppl. **50**, 263.
[KOY84] Koyama,K. et al. 1984, PASJ **36**, 659.
[KOY85] Koyama,K. 1985, Proc. Japan-US Seminar on Galactic and Extra-Galactic Compact X-ray Sources, ed. Y.Tanaka and W.H.G.Lewin (January 1985, Tokyo), Published from ISAS, p.153.
[KOY86] Koyama,K., Makishima,K., Tanaka,Y. and Tsunemi,H. 1986, PASJ **38**, 121.
[LEA86] Leahy,D.A. et al. 1986, manuscript in preparation.
[MAK84] Makishima,K. 1984, Proc. Symposium "X-ray Astronomy '84", ed. M.Oda and R.Giacconi (June 1984, Bologna), published from ISAS, p.165.
[MAK86] Makishima,K. et al. 1986, Ap.J. submitted (ISAS RN #321).
[MAS84] Masai,K. 1984, Astrophys. Space Sci. **98**, 367.
[MAT85] Matsuoka,M. 1985, Proc. ESA Workshop on Cosmic X-ray Spectroscopy Missions (June 1985, Lyngby, Denmark), p.149.
[MAT86] Matsuoka,M., Ikegami,T. and Koyama,K. 1986, PASJ, in press (ISAS RN #309).
[MLK85] Makino,F., Leahy,D.A. and Kawai,N. 1985, Space Sci. Rev. **40**, 421.
[M&M83] Morrison,R. and McCammon,D. 1983, Ap.J. **270**, 119.
[MUR84] Murakami,T. et al. 1984, PASJ **36**, 691.
[NAG86] Nagase,F. et al. 1986, submitted to PASJ (ISAS RN #306).
[NAK86] Nakajo,M. 1986, Master Thesis, University of Kyoto.
[OHA84a] Ohashi,T. et al. 1984, PASJ **36**, 699.
[OHA84b] Ohashi,T. et al. 1984, PASJ **36**, 719.
[OKU86] Okumura,Y., Tsunemi,H., Yamashita,K. et al.1986,in preparation.
[PAR80] Parkes,G.E., Mason,K.O., Murdin,P.G. and Culhane,J.L. 1980, MNRAS **191**, 547.
[PEN86] Penston,M.V. 1986, this volume.
[R&S77] Raymond,J.C. and Smith,B.W. 1977, Ap.J. Suppl. **35**, 419.
[SAT86] Sato,N. et al. 1986, submitted to PASJ (ISAS RN #316).
[SUZ84] Suzuki,K. et al. 1984, PASJ **36**, 761.
[TAN84] Tanaka,Y. et al. 1984, PASJ **36**, 641.
[TAN86] Tanaka,Y. 1986, an invited talk presented at IAU Colloquium No.89, "Radiation Hydrodynamics in Stars and Compact Objects" (June 1985, Copenhagen : ISAS RN #303).
[TUN86] Tsunemi,H. et al. 1986, Ap.J. in press.
[WAN86] Wang,B., Inoue,H., Matsuoka,M., Tanaka,Y., Hirano,T. and Nagase,F. 1986, PASJ in press (ISAS RN #317).
[WAR85] Warwick,R.S., Turner, M.J.L., Watson,M.G. and Willingale,R. Nature **317**, 1985.
[WHI86] White,N.E., Peacock,A., Hasinger,G., Mason,K.O., Manzo,G., Taylor,B.G. and Branduardi-Raymont,G. 1986, MNRAS **218**, 129.
[W&S84] White,N.E. and Swank,J.H. 1984, Ap.J. **287**, 856.
[WSH83] White,N.E., Swank,J.H and Holt,S.S. 1983, Ap.J. **270**, 711.

$$\text{Discrete Spectra of Accreting Compact Sources}$$

T. R. Kallman
Laboratory for High Energy Astrophysics
NASA/Goddard Space Flight Center
Greenbelt, MD, 20771, USA

(I) Introduction

The study of line and edge features has a unique importance in aiding understanding of accreting sources. This is because these features are likely to be formed in gas related to the accretion flow. If so, their strengths depend on both the conditions in this gas and on the broad band properties of the continuum from the accreting object. Furthermore, discrete spectral features in wavelength bands such as the ultraviolet are often probes of the continuum in wavelength bands which are not accessible to direct observation, such as the extreme ultraviolet.

This review will discuss the formation of discrete spectral features resulting from illumination of accreting gas by an X-ray source and their importance in the study of accreting objects. Emphasis will be given to the formation of UV and optical emission lines and their use as diagnostics, since this is the area which has received the most theoretical attention so far. Mention will also be made of absorption spectra in some illustrative situations. The formation of lines intimately associated with the site of the release of accretion energy, such as cyclotron lines, will not be discussed. Iron fluorescence lines will also not be discussed, since they are treated elsewhere in this volume. Following a brief discussion of the physics of spectral formation, some examples of what can be learned from the discrete spectra of each of several classes of accreting compact objects will be discussed: low mass X-ray binaries (LMXRB's), massive X-ray binaries (MXRB's) and active galactic nuclei (AGN). Cataclysmic variables (CV's) share some properties with LMXRB's and MXRB's; they are discussed elsewhere in this volume.

(II) The Reprocessing Problem

The physics of heating of gas by a source of ionizing radiation, and the attendant absorption and emission by the gas, has been discussed in detail by a variety of authors (Tarter, Tucker, and Salpeter, 1969; Tarter and Salpeter, 1969; Kallman and McCray 1982; Kallman, 1984); hence it will be reviewed only briefly here. The main assumptions which will apply in all situations of interest in this review are as follows: (1) A distribution of gas (the reprocessor) with given properties (pressure or density and elemental composition) is illuminated by a radiation source which produces a smooth continuum with given flux and spectral shape (the source spectrum). (2) The physical processes affecting the emitted spectrum are in a steady state. (3) The elemental abundances are cosmic (Withbroe 1971) or nearly so. These assumptions allow a number of simple scaling rules to be deduced.

If the reprocessing gas is optically thin to the incident continuum photons at all energies of interest then the shape of the spectrum is independent of position in the gas. Under these conditions the temperature and degree of ionization in the gas depend only on the continuum shape and on the ratio of the continuum flux to the gas density or pressure, the ionization parameter. We adopt the following definitions for the ionization parameter: $\zeta = 4\pi F/n$ (Tarter, Tucker, and Salpeter 1969) if the gas density is prescribed, and $\Xi = F/(cP)$ (Krolik, McKee and Tarter 1981) if the gas pressure is prescribed. Here and in what follows we define F as the radiation energy flux between 1 and 1000 Ry, n as the gas number density, and P as the pressure.

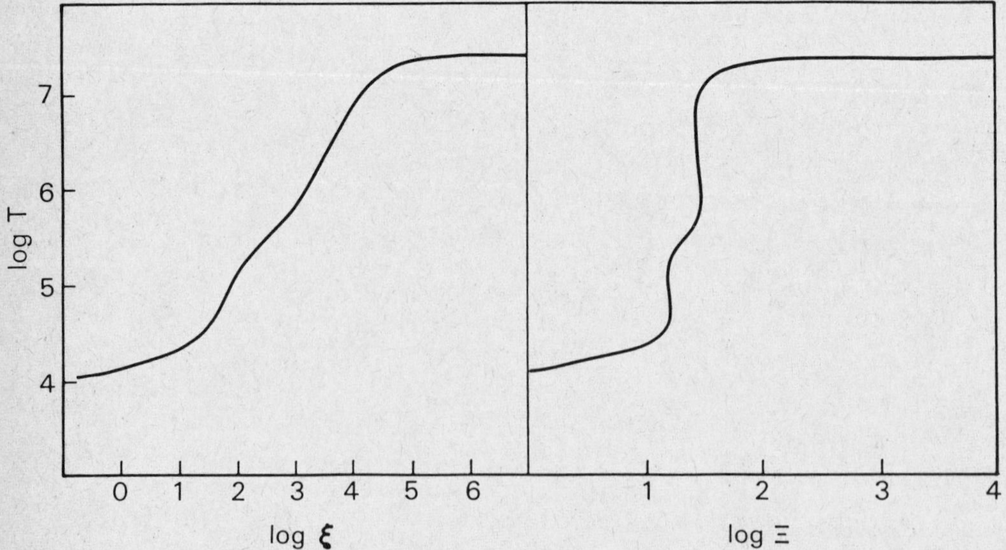

Figure 1. Gas temperature vs. ionization parameter for optically thin gas illuminated by a 10 KeV thermal bremsstrahlung spectrum. Panel (a) shows the constant density case, and panel (b) shows the constant pressure case.

Figure 1(a) shows the dependence of gas temperature on ξ for constant density gas illuminated by a 10 KeV thermal bremsstrahlung spectrum. The behavior of the gas in this case is representative of what is expected for a variety of plausible choices of incident spectrum. At large ionization parameter, $\xi > 10^4$ erg cm s^{-1}, the gas is fully ionized and the thermal properties are dominated by Compton scattering. In this region the temperature, denoted T_{IC}, depends only on the shape of the incident spectrum. In the non-relativistic limit, $T_{IC} = <\varepsilon>/4k$, where $<\varepsilon>$ is the mean photon energy; exact relativistic values of T_{IC} have been calculated by Guilbert (1986). At small ionization parameter values $\xi < 1$ erg cm s^{-1}, the gas is nearly neutral and the thermal properties are dominated by atomic physics: photoionization heating and cooling due to recombination and collisional excitation. The temperature in this region is $T \sim 10^4$ K, corresponding to the excitation thresholds of the resonance transitions of hydrogen, helium, and the abundant trace elements. At intermediate ionization parameters, 1 erg cm s^{-1} $< \xi < 10^4$ erg cm s^{-1}, the thermal balance is dominated by the atomic processes in multiply ionized trace elements.

The dependence of the gas temperature on Ξ, the constant pressure ionization parameter, can be derived from Figure 1(a) by a simple transformation. As shown in Figure 1(b), the constant pressure gas can have more than one temperature at a given value of the ionization parameter; hot gas at a temperature T_{IC} can coexist at the same Ξ with warm 10^4K gas. This is a manifestation of a thermal instability analogous to that which allows for the coexistence of the warm and cool phases of the interstellar medium (Field, Goldsmith and Habing, 1969). The extent of the unstable region depends primarily on the value T_{IC}, and hence on the shape of the illuminating spectrum. For $T_{IC} < 10^6$K, Compton heating is never as strong as photoelectric heating and the thermally unstable region is evanescent (Krolik, McKee, and Tarter, 1981; Guilbert, Fabian, and McCray, 1983; Kallman, 1984; Kallman and Mushotzky, 1985). Such low values of T_{IC} occur, for example, in the case of thermal bremsstrahlung spectra with characteristic energies less that 1 KeV, or in the case of single power law spectra with (energy) indeces greater than ≈ 1.4.

The local emissivity of optically thin gas depends only on the local temperature and the degree of ionization and so also depends only on ionization parameter and spectral shape. The reprocessed spectrum depends on both the distribution of emissivities and on the total number of ions in the reprocessor (the emission measure). In situations where the optically thin assumption is appropriate fitting of observed discrete spectra to models can in principle be used to constrain the range of emission measures and ionization parameters in gas near compact sources. This is similar to the procedure which has been used to constrain the temperature distributions in supernova remnants using model spectra of collisionally ionized gases (e.g. Shull, 1981).

In the interior of a reprocessor which is optically thick to the incident continuum the ionizing spectrum depends on position. Although the radiative transfer process in principle couples the temperature and ionization state at every point in the reprocessor with that at every other, when the optical depths are moderate the local spectrum at a given point depends to lowest order only on the optical depth along the ray connecting that point with the continuum source. The optically thin criterion can then be derived using arguments familiar from the study of Stromgren spheres (e.g. Spitzer, 1978), for various choices for cloud geometry. If the reprocessor is a sphere surrounding the source the column density of a given ion can be shown to scale according to the quantity $(Ln)^{\alpha}$ if the density is constant, or $(LP)^{\alpha}$ if the pressure is constant, where $\alpha \approx 1/2 - 1/3$. For a plane-parallel slab reprocessor the column densities scale according to Ξ^{β}, where $\beta \approx 1$. Ion column densities for a given situation can be derived using these scaling rules and photoionization models. If the column densities and cloud sizes derived using this formula exceed the cloud size or column density inferred by other means, then the optically thin assumption is valid. For example, for $L=10^{38}$ erg s^{-1} and $n=10^3$ cm^{-3}, the H II column density derived by this method for a spherical cloud is 10^{21} cm^{-2}, which is of the same order of magnitude as the total column density of such a cloud at an ionization parameter $\xi \approx 10^{-1}$ erg cm s^{-1}. Thus, this cloud represents the boundary between optically thin and optically thick parameter values for spherical clouds.

A further consequence of radiative transfer which affects the emitted spectrum is the fate of resonance line photons emitted by the reprocessor. Although the cross section for resonance line scattering greatly exceeds the continuum absorption cross section for a strong line of a given ion, resonance line transfer is not important in determining the total emitted line flux unless the line is "effectively thick", that is, unless the line photons are likley to be destroyed by collisions or absorption while they diffuse out of the reprocessor. An approximate criterion for the importance of photon destruction by collisions may be derived by assuming that the escape probability at each resonance scattering is proportional to $1/\tau$ (τ is the line center optical depth; see Hummer and Kunasz, 1980, for a discussion of the validity of this assumption), and using an approximate form for the collisional rate coefficient given by Van Regemorter (1962):

$$\frac{\tau_{therm}}{\tau_{cont}} = \frac{1.4}{T_4^{1/2}} \frac{P(W/kT)}{\Delta\epsilon_{eV}^3} \tag{1}$$

In this formula τ_{therm} is the value of the line center optical depth at which thermalization is important, τ_{cont} is the continuum optical depth at the corresponding ionization threshold, $P(W/kT)$ is a number in the range 0.2 - 1.0, tabulated by Van Regemorter (1962), T_4 is the gas temperature in units of 10^4K, and $\Delta\epsilon_{eV}$ is the line energy in units of eV. It is apparent that for moderate continuum depths, $\tau_{cont} < \tau_{crit}$, $\tau_{crit} = \Delta\epsilon_{ev}^3$, thermalization of lines will not be important. For continuum depths greater than this value the local line emissivities will approach Planck values. The emitted line spectrum will be determined by the temperature gradient in the reprocessor according to the Eddington-Barbier relation (e.g. Mihalas, 1978). If so, the reprocessing efficiency depends on the stellar (or accretion disk local) gravity and effective temperature in addition to the incident continuum flux and spectral shape (e.g. Milgrom and Katz, 1976). In the absence of resonance scattering thermalization will be important if the gas density exceeds the critical value,

$$n_{crit} = 2.5_{14} \; T_4^{1/2} \; \frac{\Delta\epsilon_{eV}^3}{P(W/kT)} \tag{2}$$

We can define the reprocessing efficiency for a given line as the ratio of the reprocessed line flux to the absorbed continuum flux. This quantity may be as large as 0.5 in the case of a strong line which is effectively thin. For effectively thick lines the reprocessing efficiency is proportional to the smaller of n/n_{crit} or τ/τ_{therm} for values of these quantities in the range 0.1 to 1.

A convenient way of summarizing the dependence of reprocessing physics on the properties of the reprocessor is to plot the various constraints and scaling relations in the flux-density plane. As shown in Figure 2 surfaces of constant ionization parameter are diagonal lines in this representation. Also shown is the density at which thermalization will dominate for UV lines ($\Delta\epsilon_{eV}=10$) in the absence of line scattering (the vertical line labeled $\tau=0$), and in the presence of scattering with line center optical depths $\tau=10^2$ and 10^4. Estimated loci of the reprocessors of each of the major classes of compact sources are also shown in Figure 2 . These will be discussed in turn in the following sections.

(III) Low Mass X-ray Binaries

Reprocessing in low mass X-ray binaries remains relatively poorly understood, in spite of recent advances in understanding the gas flows in these systems. With typical X-ray luminosities in the range 10^{36}-10^{38} erg s^{-1}, and orbital separations of 10^{11}cm or less (Bradt and McClintock, 1985), the reprocessors in LMXRB's are subject to an intense incident continuum flux. The spectra (see elsewhere in this volume) are typically hard enough to cause a thermal instability to occur at

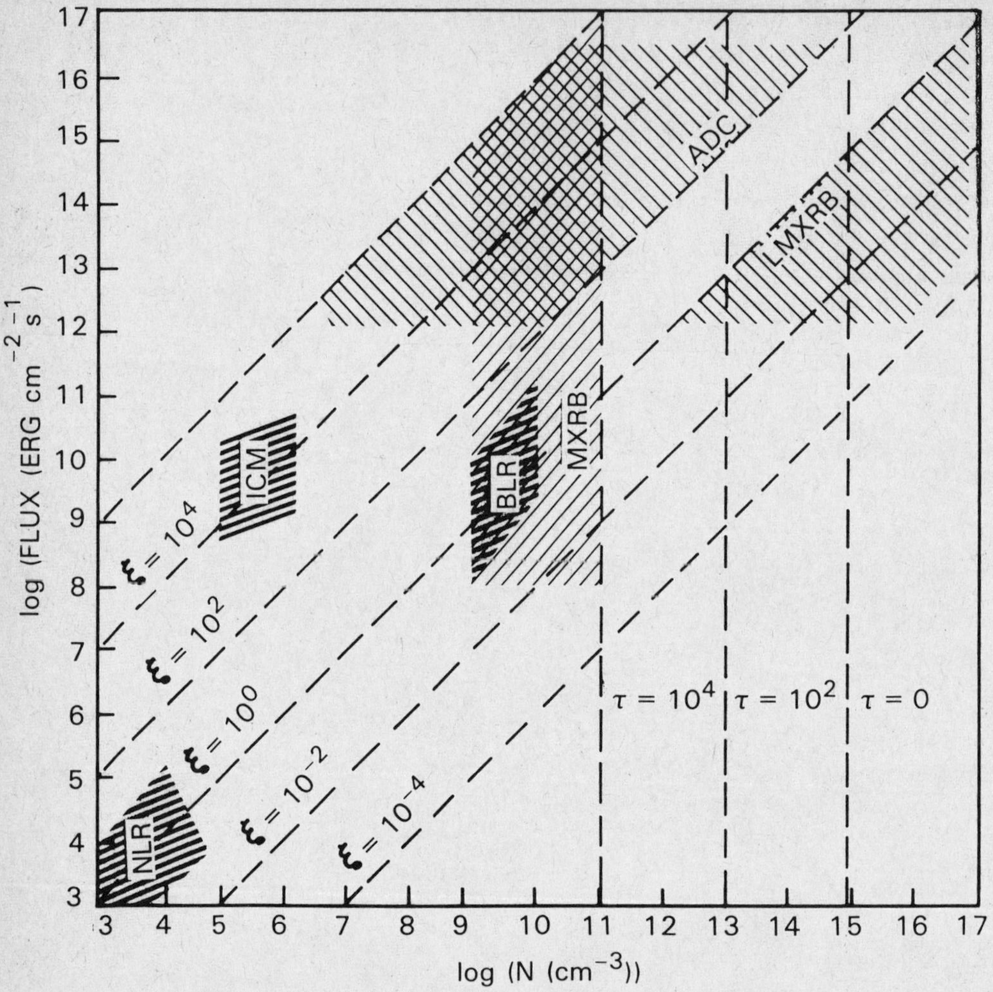

Figure 2. Loci of reprocessors in the flux - density plane. Diagonal dashed lines are surfaces of constant ionization parameter, ξ. Vertical dashed lines define onset of thermalization for resonance lines with line center optical depth τ. Shaded regions denote approximate parameter values for low mass X-the accretion disk or binary companion in low mass X-ray binaries (LMXRB), an accretion disk corona (ADC), the stellar wind in massive X-ray binaries (MXRB), and the broad line region (BLR), intercloud medium (ICM), and narrow line region (NLR) of active galaxies.

constant pressure.   Possible reprocessing sites include the companion star, accretion disk, or accretion disk corona.

Since both companion star and the surface of an accretion disk may be assumed to be in hydrostatic equilibrium, X-ray illumination will have similar effects on these two sites. As a consequence of the monotonic decrease of gas pressure with height the ionization parameter will increase past the critical value above which warm 10^4K gas cannot exist in   thermal equilibrium. The transition to coronal temperatures occcurs at a pressure, denoted P_{min} (London, McCray, and Auer, 1978), which is determined by the incident flux. Thus, the conditions in the warm 10^4K reprocessing gas depend primarily on the orbital separation and X-ray source luminosity; the effects of thermal instability serve to regulate the reprocessor at $\Xi \leq 10$.   As shown in Figure 2, the incident flux is 10^{13}-10^{17} erg cm^{-2} s^{-1}, corresponding to values of P_{min} in the range 10^1-10^5dyne cm^{-2} and densities 10^{13}-10^{17} cm^{-3} at 10^4K. Densities at the upper end of this range are sufficient to cause most strong resonance lines to be thermalized in the absence of multiple scatterings; only moderate optical depths are required to thermalize lines at the lower density limit.

Conditions in the coronal gas above an X-ray illuminated star or accretion disk (White and Holt, 1982) can be estimated under the assumptions of thermal equilibrium and neglect of radiative transfer effects: The density, Compton optical depth, and emission measure are then given by

$$n \leq 2.0_{13} \ L_{38} \ R_{10}^{-2} \ \Xi_{10}^{-1} \ T_8^{-1} \quad cm^{-3} \tag{3}$$

$$\tau_{Comp} \sim 0.1 \ L_{38} \ R_{10}^{-1} \ \Xi_{10}^{-1} \ T_8^{-1} \tag{4}$$

$$n^2 V \sim 8.0_{56} \ L_{38}^2 \ R_{10}^{-1} \ \Xi_{10}^{-2} \ T_8^{-2} \quad cm^{-3} \tag{5}$$

where L_{38} is the source luminosity in units of 10^{38} erg s^{-1}, R_{10} is the distance from the source in units of 10^{10} cm, and T_8 is the Compton temperature in units of 10^8K.   An estimate for the emissivity of such coronal gas in a line such as O VIII Lα is $\sim 10^{-14}$ cm^3 s^{-1} corresponding to an equivalent width less than 0.01 eV for the emssion measure given in (5).   Such estimates can be compared with recent high resolution X-ray spectra of LMXRB using objective grating spectrographs on the <u>Einstein Observatory</u> (HEAO-2) and EXOSAT satellites, which show strong line emission in lines such as N VII Lα with equivalent widths greater than 1 eV (Kahn, Seward, and Chlebowski, 1984; Vrtilek, et al., 1986a,b). Furthermore, these lines were found to be variable on the 10^4 sec timescale of the observation, and to have anomalous line ratios such as O VIII Lα/N VII Lα. Measurements of the residual X-ray fluxes during eclipse also suggest scattering optical depths greater than those predicted by equation (4) (McClintock, et al., 1982). One possible explanation for this discrepancy is that the true source luminosity in these sources is much greater than what is observed directly. On the other hand, these estimates neglect many potentially important effects, such as departures from thermal and hydrostatic equilibrium, Compton scattering of the incident X-rays, and the details of the disk geometry . These issues are treated in more detail by Begelman, McKee and Shields (1983), Begelman and McKee (1983), London, McCray and Auer (1981), and London and Flannery (1982), Fabian, Guilbert, and Ross (1982), but further modelling is needed to predict the X-ray emission line spectra and to attempt more detailed comparisons with observations.

(IV) Massive X-ray Binaries

This class contains the best studied accreting sources, a large fraction of which are X-ray pulsars. X-ray luminosities in the range 10^{35}-10^{38} erg s^{-1} and binary separations 10^{12}-10^{13} cm result in X-ray fluxes significantly smaller than in the LMXRB's. Furthermore, the luminosity of the early-type primary in these systems typically equals or exceeds that of the X-ray source, so that Compton cooling by the primary will suppress any thermal instability throughout most of the volume of the binary system in most sources. The reprocessing site in MXRB's which distinguishes them from other classes of accreting sources is the strong stellar wind emanating from the early-type primary.

Although considerable uncertainty remains about the detailed conditions in an early-type stellar wind (see e.g. Owocki and Rybicki, 1984), for the purposes of the discussion here wind material will be assumed to be monotonically accelerated by radiation pressure to a terminal velocity which is determined by the stellar gravity. The wind velocity depends on R, the distance from the center of the primary according to :

$$v(R) = v_T \, (1-R_\star/R)^\alpha \tag{6}$$

where $R_\star$ is the radius of the star, and v_T is the terminal velocity. The exponent α in this relation is in the range 1/2 - 1 (Castor and Lamers, 1978). Conditions in the stellar wind are typically as follows:

$$n(R) = 2.1_{10} \; \dot{M}_6 \; R_{12}^{-2} \; v_8^{-1} \; cm^{-3} \tag{7}$$

$$\xi(R) = 4.7_3 \; L_{38} \; r_{12}^{-2} \; R_{12}^{2} \; v_8 \; \dot{M}_6^{-1} \; erg \; cm \; s^{-1} \tag{8}$$

where R_{12} and r_{12} are the distances from the point in question to the center of the primary and to the X-ray source, respectively, in units of 10^{12} cm, L_{38} is the X-ray source luminosity in units of 10^{38} erg s^{-1}, $\dot{M}_6$ is the wind mass loss rate in units of 10^{-6} $M_\odot$ yr^{-1} and v_8 is the wind terminal velocity in units of 1000 km s^{-1}. A consequence of the dependence of the ionization parameter on the distances from both members of the binary system is that surfaces of constant ionization parameter are approximate spheres containing the X-ray source for greater ξ greater than a critical value, $\xi_X = (4.7_3) L_{38} v_8 M_6^{-1}$ erg cm s^{-1}, and open sufaces containing the primary for $\xi < \xi_X$ (Hatchett and McCray, 1977). The volumes of the closed surfaces are approximately

$$V \sim 1.2_{37} \; D_{12}^{3} \; \frac{q^{3/2}}{(q-1)^3} \; cm^3 \tag{9}$$

and the emission measure within each surface is

$$n^2 V \sim 2.4_{47} \; \dot{M}_6 \; D_{12} \; v_8^{-1} \; \frac{q^{1/2}}{(q-1)} \; cm^{-3} \tag{10}$$

where $q = \xi/\xi_X$. A large fraction of the wind volume ($q \approx 1$) must therefore emit in order to produce strong X-ray line emission. As shown in Figure 2 the winds in MXRB's with moderate luminosities are in a region of the flux-density plane where the reprocessing physics is generally free from the complicating effects of thermalization. In addition, the large velocity gradients in the wind preclude the buildup of large resonance line optical depths. The fact that the UV/optical radiation is generally dominated by the star in these systems means that reprocessed emission is not directly observable. However, the effects of X-ray ionization can be observed indirectly.

The scattered resonance lines formed in the winds from early-type stars are the strongest discrete features in their UV spectra. The presence of a compact X-ray source in the wind can affect these lines by destroying the scattering ions.

The P-Cygni line profile will be suppressed at wavelengths corresponding to the wind velocity near the X-ray source, the suppression being greatest at orbital phase 0.5, when the X-ray source is along the line of sight to the primary, and absent at phases when the X-ray source is behind the primary. This effect, first predicted by Hatchett and McCray (1977), has been observed from all the massive binaries which are accessible to current UV instruments (Hammerschlag-Hensberge 1980). Detailed observations of 4U 0900-40 (Vela X-1) have been compared with models by Dupree, et al. (1980), and by McCray et al. (1984). Modelling of UV line profiles is uncertain owing to the lack of simultaneous X-ray and UV observations -- the X-ray luminosity, and hence the extent of the X-ray ionized zone, is known to vary by factors of order 2 - 10 on a timescale of hours to days.   This introduces a corresponding uncertainty into the degree of modulation in the profile expected as a function of orbital phase. However, the observed wavelength at which the effects of X-ray ionization (evidenced as a bump on the line profile) will be greatest depends only on the wind velocity near the X-ray source. The orbital separation of systems such as 4U0900-40 are accurately known, so that the position of the X-ray ionized bump on the profile is a measure of the wind law v(R) (c.f. equation (6)).   The observed bump in the case of 4U0900-40 lies at significantly higher velocities, $V=800$ Km s^{-1}, than can be accounted for by the wind law of equation (6) with either $\alpha=0.5$ or $\alpha=1$.   In this functional form, a velocity exponent $\alpha=0.25$ would be required to explain this observation, corresponding to a more rapidly accelerated wind than has been previously suggested.   One possible explanation for this discrepancy concerns the effect of the X-ray source on wind dynamics. As shown by Friend and Castor (1982), the gravitational attraction of the X-ray source can enhance the stellar wind mass flux in a direction along the line of centers.

A further consequence of X-ray ionization of the stellar wind in MXRB 's is the time variability associated with X-ray pulsations. The recombination time in the stellar wind is ~10s, for gas near $T=10^4$K and $n=10^{10}$ /cm^3, shorter than the pulse periods in several cases.   If so, the UV lines may pulse for the same reason that orbital phase variations in these lines occur. Furthermore, the one-to-one correspondence between wind velocity (and therefore position) and observed wavelength in the line causes the line profile shape to depend on orientation and opening angle of the pulsar X-ray beam. Figure 3 shows a simulation of C IV $\lambda1549$ line pulsations for the 4U0900-40 system. The dashed curve corresponds to the profile at orbital phase 0, in the absence of X-ray ionization.   The solid curve corresponds to orbital phase 0.5, and the beam is assumed to be a filled cone (pencil beam) with opening angle 45°. The different panels correspond to different values of $\vec{p}\cdot\hat{n}$, the orientation of the beam relative to the observer's line of sight. For most orientations of the neutron star beamed X-rays cause two "bumps" to occur on the profile, corresponding to ionization by the two different X-ray

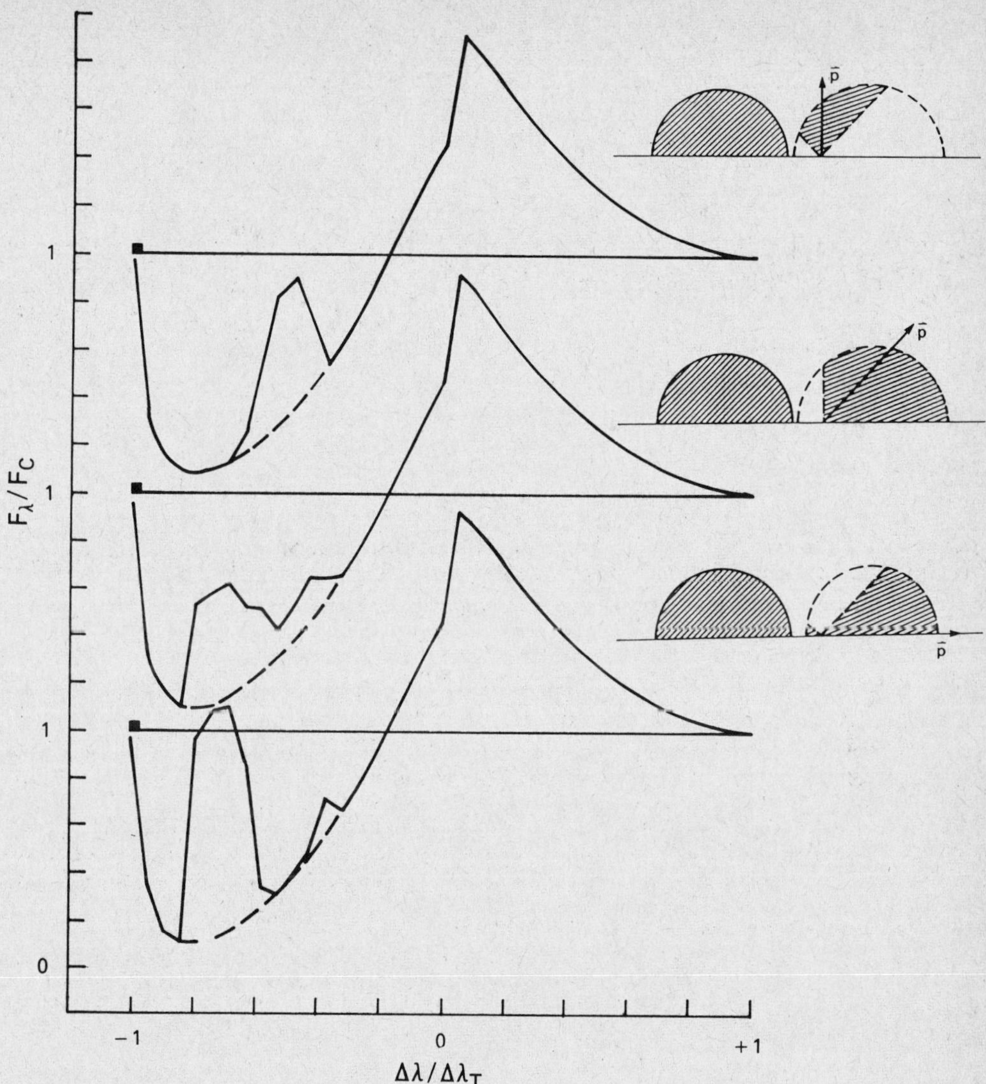

Figure 3. Model spectra in the vicinity of the C IV λ1549 resonance line for the X-ray pulsar 4U0900-40. Sketches of the binary system display the relative orientation of the magnetic axis and the line of centers. The orbital phase is assumed to be 0.5. Dashed curves denote spectrum in the absence of x-ray ionization. Flux is given in units of the continuum, wavelength in units of the terminal wavelength.

beams. When the beams are both perpendicular to the line of sight they ionize material at the same velocity, and the bumps merge into one. Observations of pulsed UV lines will be accessible to the next generation of UV instruments. Such observations will allow the relative orientation of the magnetic axis, rotational axis and observer's line of sight to be constrained, since the maximum angle between the line of sight and the magnetic axis is i+b, where i is the inclination of the rotational axis relative to the line of sight, and b is the angle betwen the magnetic and rotational axes. The excursions of the bumps in wavelength can be used to constrain i+b. This information about the orientation of the X-ray pulsar cannot be obtained from X-ray observations alone; it is a direct consequence of the proximity of the reprocessor to the X-ray source in MXRB's.

Although reprocessed X-ray emission is important in only a narrow volume around the compact object in MXRB's, X-ray absorption is a probe of a large volume of the stellar wind. If the emitted spectrum of the compact object can be smoothly extrapolated from high (>3 KeV) to low (<1 KeV) X-ray energies, then fitting observed spectra to this extrapolation + photoelectric absorption can be used to infer the column density of material along the line of sight to the X-ray source. If the spectrum fits adequately to absorption by neutral material (e.g. Brown and Gould, 1971), then the column density as a function of orbital phase can in principle be combined with the assumption of a spherically symmetric wind in order to infer the wind density as a function of distance from the center of the primary. This is analogous to the eclipse mapping method of studying CV accretion disks (Horne, elsewhere in this volume). However, low energy X-ray spectra from 4U0900-40 typically shows an excess of soft flux at the lowest energies relative to that predicted by pure neutral absorption (e.g. Nagase, et al., 1986, Kallman and White, 1983). If this excess is interpreted as being due to partial ionization of the wind by the X-ray source (e.g. Krolik and Kallman, 1984), then fits to observed spectra will provide a measure of both the column density and of the mean ionization parameter (or range of ionization parameters) along the line of sight to the X-ray source. Since the ionization parameter depends primarily on the position of the partially ionized material along the line of sight, such a procedure can constrain the distribution of gas in the binary in two dimensions, and can be used to test for departures from spherical symmetry in the wind.

(V) Active Galactic Nuclei

The study of reprocessed emission has long been an important clue to the nature of active galactic nuclei. Although these objects are generally fainter than the galactic sources discussed so far, their greater numbers permit statistical arguments about their properties which are not possible for galactic objects. The AGN continuum source produces an approximately power law spectrum extending from the infrared to the gamma rays, with a wide range of luminosities. Reprocessing sites include the broad line clouds, hot gas in the broad line region, and the gas responsible for the narrow emission lines. In this section we will focus on the reprocessing by the broad line gas.

The hard power law continuum spectra which are characteristic of AGN can produce a broad zone of thermal instability (Krolik, McKee, and Tarter, 1982). Therefore, broad line clouds, which typically have temperatures $T \approx 10^4$K and densities $n \approx 10^{10}$ cm^{-3}, will be able to form at distances from the source such that $1 < \Xi < 10$, or $R_C = (L_{46}/P_{-2}\Xi)^{1/2}$ pc, where L_{46} is the continuum luminosity in units of 10^{46} erg s^{-1} and P_{-2} is the pressure in units of 10^{-2} dyne cm^{-2}. Since clouds are likely to have line center optical depths as large as 10^8 for Lα, for example, at densities much greater than this value the reprocessing efficiency will be greatly reduced owing to thermalization. Thus, although denser clouds may exist closer to the source than R_C, they may provide only a small contribution to the total UV/optical emission line spectrum. Compton scattering of line photons by hot intercloud gas may also provide a lower limit on R_C (Kallman and Krolik, 1986). A lower bound on the allowed broad line cloud density, and hence an upper bound on R_C, may be provided by the effects of internally generated line radiation pressure (Ferland and Elitzur, 1985). As shown in Figure 2, the AGN broad line clouds have fluxes and densities such that thermalization is of marginal importance for resonance lines of moderate optical depth. The optical depth of Lα is large enough to necessitate an accurate treatment of line transfer.

Other properties of broad line clouds have been discussed in detail by a variety of authors (e.g. Davidson and Netzer, 1978; Mushotzky and Ferland, 1984; Kwan and Krolik, 1981; Kwan, 1984), and are illustrated in the results given in Figure 4(a). In this figure are plotted the values of some of the important observable line ratios as a function of ionization parameter and pressure for model broad line couds resembling those of Kwan and Krolik (1981). The ionization parameter and pressure are expressed in units of $\log\Xi = 0.$, $P = 0.027$ dyne cm^{-2}, the column density is chosen to be large, $N = 10^{24}$ cm^{-2}, the carbon abundance is reduced by a factor 0.6 relative to solar, and the incident continuum spectrum is assumed to be a hard broken power law with the optical-to-X-ray power law index $\alpha_{OX} \approx 1.2$. The hatched regions delineate the values of these line ratios observed in an ensemble of bright quasars; the contour levels are evenly spaced in the logarithm of the relevant line ratio. These results may be understood in terms of the principles discussed in section II. Resonance lines formed in the ionized part of the cloud depend on ionization paramter with a sensitivity which depends inversely on the abundance of the parent ion. Thus, the O VI λ1034 line is much more sensitive to ionization parameter than are C IV λ1549, N V λ1240, or Lα, and this is reflected in

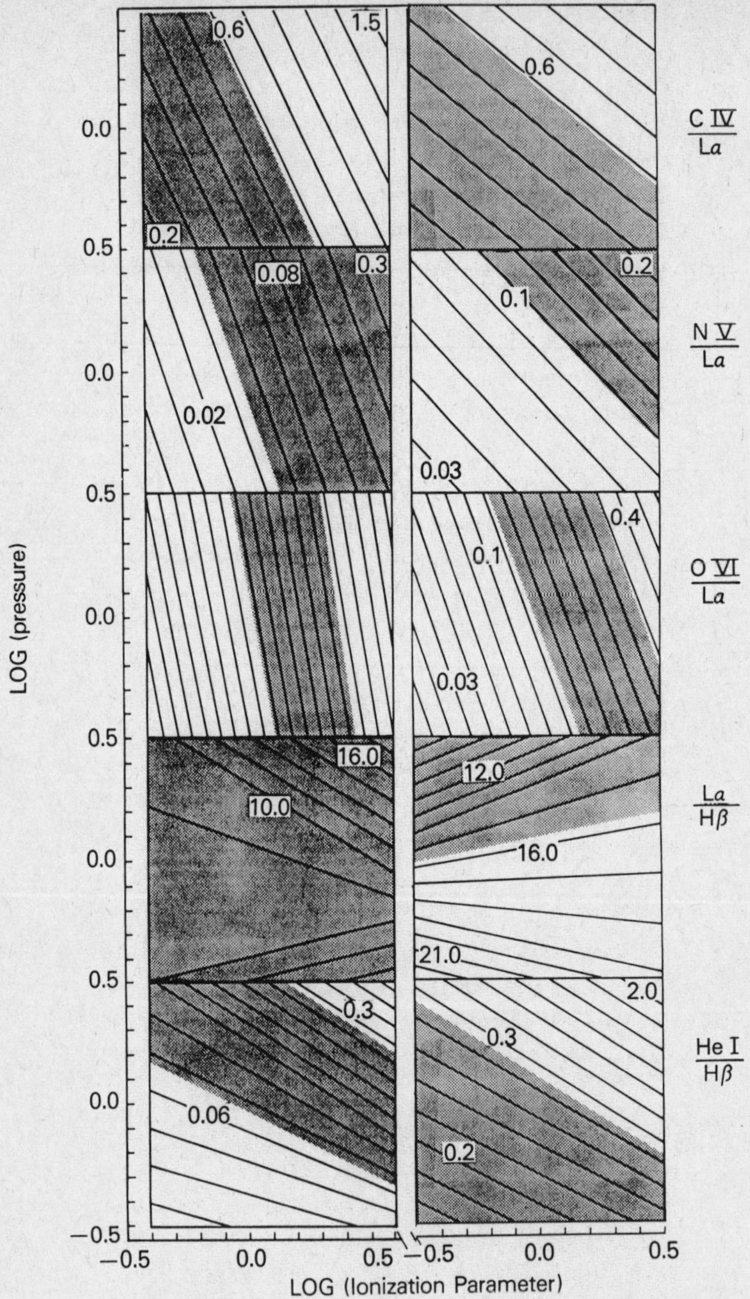

Figure 4. Contours of constant line ratios as functions of gas pressure and ionization parameter. Hatched regions denote the values observed for an ensemble of quasars and compiled by Kwan and Krolik (1981). Panel (a) shows the results for models which assume a hard broken power law continuum with $\alpha_{OX} \approx 1.2$. Panel (b) shows the results for models which include a ~30,000 K thermal component in addition to the power law continuum, the sum having $\alpha_{OX} \approx 1.4$.

the O VI/Lα contours compared with the C IV/Lα contours. Subordinate lines such as He I λ5876 are most sensitive to the continuum flux, PΞ (Kwan, 1984), leading to roughly diagonal constant ratio contours for the ratio He I λ5876/Hβ. The Lα/Hβ ratio depends weakly on pressure. It is apparent from Figure 4 that the observationally allowed line ratio regions overlap at values of P and Ξ corresponding to the fiducial values chosen above.

Recent studies of the continuum spectra of active galaxies (e.g. Zamorani, et al. 1982, Malkan, elsewhere in this volume) have revealed that the spectra are likely to be considerably softer than that considered by Kwan and Krolik (1982). The effects of such a spectrum, consisting of a thermal component superimposed on a hard power law, the sum having $\alpha_{OX} \approx 1.4$, are displayed in Figure 4(b). Comparison with 4(a) reveals the following: (1) The steeper decrease in the EUV leads to smaller values of C IV/Lα,   N V/Lα, and O VI/Lα at a given P and Ξ. (2) The relatively weaker hard (>1KeV) X-rays leads to larger Lα/Hβ at all values of P and Ξ. Only at high pressures can the values of this ratio in the observed range be obtained. (3) Such large pressures are not consistent with the observed range of He I λ5876/Hβ ratios. Thus, reprocessing models are not consistent with both the soft spectra observed from many active galaxies and the line ratios ranges in the ensemble of sources given in Kwan and Krolik (1981). This suggests that either objects with soft spectra ($\alpha_{OX} \approx 1.4$) will have systematically different line ratios from those with harder spectra ($\alpha_{OX} \approx 1.2$), or that spectral variability prevents us from knowing the ionizing spectrum at the time when the lines are formed, or that geometrical effects shield the clouds preferentially from EUV continuum photons (e.g. Fabian, et al. 1986). These results are discussed in more detail by Krolik and Kallman (1986).

Although the broad UV and optical emission lines are generally the most readily observable features in AGN spectra, reprocessing in the hot intercloud medium (ICM) associated with the broad line clouds is likely to provide important information about the gas flow. Although the ICM is likely to be highly ionized, its column density may exceed 10^{24} cm^{-2}. Future X-ray spectroscopy experiments, such as those planned for the AXAF satellite, will be able to detect features such as absorption by the Fe XXVI Lα line (Kallman and Mushotzky, 1985) if the ICM properties are similar to those predicted by thermal instability theory.

REFERENCES

Begelman, M., McKee, C. E., and Shields, G. B., 1983, Ap. J. 271, 70.

Begelman, M. and McKee, C. E., 1983, Ap. J. 271, 89.

Bradt, H., and McClintock, J. E., 1983, Ann. Rev. Ast. and Ap., 21, 13.

Brown, R. L., and Gould, R. J., 1970, Phys. Rev. D., 1, 2252.

Davidson, K. and Netzer, H., 1978, Rev. Mod. Phys., 51, 715.

Dupree, A. K., et al., 1980, Ap. J., 238, 969.

Fabian, A., Guilbert, P., and Ross, R.R., 1982, MNRAS, 199, 1045.

Fabian, A., et al., 1986, MNRAS, 8, 457.

Ferland, G. J. and Elitzur, M., 1985, Ap. J. Lett., 285, L11.

Field, G. G., Goldsmith, D. W., and Habing, H. J., 1969, Ap. J. (Letters), 155, L149.

Friend, D. B. and Castor, J. I., 1982, Ap. J. 261, 293.

Gathier, R., Lamers, H. J., and Snow, J. P., 1981, Ap. J., 247, 12.

Guilbert, P. W., Fabian, A. C., and McCray, R., 1983, Ap. J., 266, 486.

Guilbert, P. W., 1986, MNRAS, 218, 171.

Hammerschlag-Hensberge, G., 1980, ESA SP-157.

Hatchett, S. P. and McCray, R. A., 1977, Ap. J., 211, 552.

Hummer, D. B., and Kunasz, P. B., 1980, Ap. J., 236, 609.

Kahn, S. M., Seward, F. D., and Chlebowski, T., 1984, _Ap. J._, _283_, 286.

Kallman, T. R., and McCray, R. A., 1982, _Ap. J. Suppl._, _50_, 263.

Kallman, T. R., and White, N. E., 1982, _Ap. J. (Letters)_, _261_.

Kallman, T. R., 1984, _Ap. J._, _280_, 269.

Kallman, T. R., and Mushotsky, R. A., 1985, _Ap. J._, _292_, 49.

Kallman, T. R., and Krolik, J. H., 1986, _Ap. J._, in press.

Krolik, J. H., McKee, C. F., and Tarter, C. B., 1981, _Ap. J._, _249_, 422.

Krolik, J., and Kallman, T. R., 1984, _Ap. J._, _286_, 366.

Krolik, J., and Kallman, T.R., 1986, _Ap. J._, submitted.

Kwan, J., and Krolik, J., 1982, _Ap. J._, _250_, 478.

Kwan, J., 1984, _Ap. J._, _283_, 70.

London, P. B., McCray, R. A., and Auer, L., 1981, _Ap. J._, _243_, 970.

London, R. A. and Flannery, B., 1982, _Ap. J._, _258_, 260.

McClintock, J., London, R., Bond, H.E., and Grauer, A., 1982, _Ap. J._, _258_, 245.

McCray, R. A., Kallman, T. R., Castor, J. I., and Olson, G. L., 1984, _Ap. J._, _282_, 295.

Mihalas, D., 1978, _Stellar Atmospheres_ (San Francisco: Freeman).

Milgrom, M. and Katz, J. J., 1976, _Ap. J._, _250_, 545.

Mushotsky, R. and Ferland, G., 1984, _Ap. J._, _250_, 478.

Nagase, F., et al., 1986, _PASJ_ in press.

Owocki, S. P. and Rybicki, G. B., 1984, _Ap. J._, _284_, 337.

Shull, J. M., 1981 in X-ray Astronomy in the 1980s, ed. S. Holt, (NASA TM83848).

Spitzer, 1978, Physical Processes in the Interstellar Medium (New York: Wiley).

Tarter, C. D., Tucker, W., and Salpeter, E. E., 1967, Ap. J., 156, 943.

Tarter, C. D., and Salpeter, E. E., 1969, Ap. J., 156, 953.

Van Regemorter, H., 1962, Ap. J., 136, 906.

Vrtilek, S. D., Helfand, D. J., Halpern, J. R., Kahn, S. M., and Seward, F. D.,
    preprint.

Vrtilek, S. D., Kahn, S. M., Grindlay, J. E., Helfand, D. D., and Seward, F. D.,
    preprint.

Withbroe, G. L., 1971, in The Menzel Symposium on Solar Physics, Spectra, and
    Gaseous Nebulae, ed. K. B. Gebbie (NBS SP353).

White, N. E. and Holt, S. S., 1982, Ap. J., 257, 318.

Zamorani, G., et al., 1982, Ap. J., 253, 504.

<u>RADIO EMISSION FROM X-RAY BINARIES AND THE PROTO-TYPE JETS OF SS433</u>

R.M. Hjellming

National Radio Astronomy Observatory, Socorro, N. M. 87801

K.J. Johnston

E.O. Hulburt Center for Space Research

Naval Research Laboratory, Washington, D.C.   20375

ABSTRACT

We summarize the X-ray binaries that are known to exhibit radio emission and concentrate on SS433 as the proto-type of X-ray binaries with radio and optical emission from jet flows perpindicular to accretion disks. In the context of accretion onto compact objects, SS433 provides the most detailed information about high velocity (v = 78,000 km/sec), gas dynamical flows moving outward along accretion disk axes. A working hypothesis whereby the radio and optical jets are produced in an expanding sheath environment around flows perpindicular to an accretion disk is discussed, and the minor deviations from a simple twin-jet kinematical model are summarized. The longest time scale deviations from the simple kinematical model are shown to be consistent with a "beat model" in which the average 162.5^d period of precession of the accretion disk is the harmonic mean of two periods of 175^d and 152^d, and the long time scale deviations from the kinematic model exhibit a beat difference period of ~ 2300^d.

I. INTRODUCTION AND SUMMARY KNOWN OF RADIO-EMITTING X-RAY BINARIES

Approximately 13 X-ray binaries have been found to be associated with observable radio emission, whereas approximately twice that number have not yet been detected as radio sources. Table 1 provides a brief summary of the known radio emitting X-ray binaries. While we have chosen to concentrate on SS433 in this paper, it is useful to provide a summary of related objects, even though only SS433, Cyg X-3 and Sco X-1 have been found to exhibit extended radio emission.

II. THE KINEMATIC MODEL FOR THE JETS OF SS433

The foundation of considering SS433 as the proto-type for radio-emitting x-ray binaries is the success of a simple twin-jet doppler shift model in explaining the radial velocities changes of optical

TABLE 1

SUMMARY OF X-RAY BINARIES WITH ASSOCIATED RADIO SOURCE(S)

SS433	Variable radio and optical jets with v = 0.26c	Ryle et al. (1979) Seaquist et al. (1979) Hjellming and Johnston (1981b)
Cyg X-3	Strong synchrotron flaring events, variable quiescent levels, occasion jets in with velocities ~0.2-0.4c	Braes and Miley (1972) Gregory et al. (1972) Geldzahler et al. (1983) Spencer et al. (1986)
Sco X-1	Variable radio on XR source surrounded by radio double	Ables (1969), Hjellming and Wade (1971a), Fomalont et al. (1983)
Cyg X-1	Mostly stable radio source, rare flaring, radio changes with XR state, associated with OB + B.H.(?) binary	Hjellming and Wade (1971b) Braes and Miley (1971) Hjellming et al.(1975)
Cyg X-2	Weak, stable radio source	Hjellming & Blankenship (1973) Hjellming et al. (1987)
GX5-1	Weak, variable radio source	Braes et al. (1972) Geldzahler (1983) Grindlay and Seaquist (1986)
GX13+1	Weak radio on IR object	Grindlay and Seaquist (1986)
GX17+2	Weak, variable radio source associated with G star	Hjellming and Wade (1971b) Hjellming (1978) Geldzahler (1983) Grindlay and Seaquist (1986)
A0620-003	Transient radio & XR source	Owen et al. (1976) Geldzahler (1983)
Cen X-4	Transient radio & XR source	Hjellming (1979)
LSI+61°303	Periodic radio source	Gregory and Taylor (1978) Taylor and Gregory (1982)
Cir X-1	Periodic and flaring radio	Whelan et al. (1977) Preston et al. (1983)
NGC 6624	Weak radio source(s) near globular cluster center	Geldzahler (1983) Grindlay and Seaquist (1986)

emission lines (Margon 1984), the proper motion of extended radio emission (Hjellming and Johnston 1981a, 1981b, 1982), and doppler-shifted x-ray lines (Watson *et al.* 1983, Band and Grindlay 1986).

The equations describing twin-jet doppler shifts in three dimensions can be briefly summarized as follows. Let the x-y-z coordinate system (cf. Figure 1a) be the "natural" coordinate system of a rotating ejection vector $\mathbf{v}$, where z is the axis about which $\mathbf{v}$ rotates at an angle ψ with a time dependent phase Φ, and the y-axis is located in the tangent place of the sky. The origin of this coordinate system is coincident with the location of the center of ejection. Since ψ, which corresponds to one of the angular coordinates of the vector $\mathbf{v}$, and the angle Φ, which is a function of both time t and a reference time t_{ref} (when the vector lies in the x-z plane), describe the spherical polar coordinates of $\mathbf{v}$, we can write

$$\mathbf{v} = (v_x, v_y, v_z) = s_{jet} v[\sin \psi \cos \Phi(t), \sin \psi \sin \Phi(t), \cos \psi] \quad (1)$$

with s_{jet} = -1 and +1 identifying the "blue" (approaching) and "red" (receding) jets, respectively. For the purposes of this paper we use

$$\Phi(t) = s_{rot}[\Omega(t-t_{ref})] \quad (2)$$

where the angular rotation rate of the vector $\mathbf{v}$ is $\Omega = 2\pi/P$, P is the rotation period, and s_{rot} identifies the sense of rotation, with s_{rot} = +1 meaning counter-clockwise (right-handed) rotation about the z-axis, and the opposite sign meaning clockwise rotation. The x'-y'-z' coordinate system is defined by the x-axis being pointed in the direction of the observer, perpindicular to the plane of the sky. It is obtained by rotating the x-y-z system an angle θ_y about the y-axis (cf. Figure 1b). The velocity vector in this (primed) coordinate system is given by

$$\mathbf{v} = \begin{pmatrix} v'_x \\ v'_y \\ v'_z \end{pmatrix} = s_{jet} v \begin{pmatrix} \cos \theta_y \sin \psi \cos \Phi(t) - \sin \theta_y \cos \psi \\ \sin \psi \sin \Phi(t) \\ \sin \theta_y \sin \psi \cos \Phi(t) + \cos \theta_y \cos \psi \end{pmatrix} \quad (3)$$

One can then obtain the velocity vector with components in the radial (line of sight), declination, and right ascension directions by rotating the primed coordinate system by an angle θ_x counter-clockwise about the x-axis (cf. Figure 1b) so that

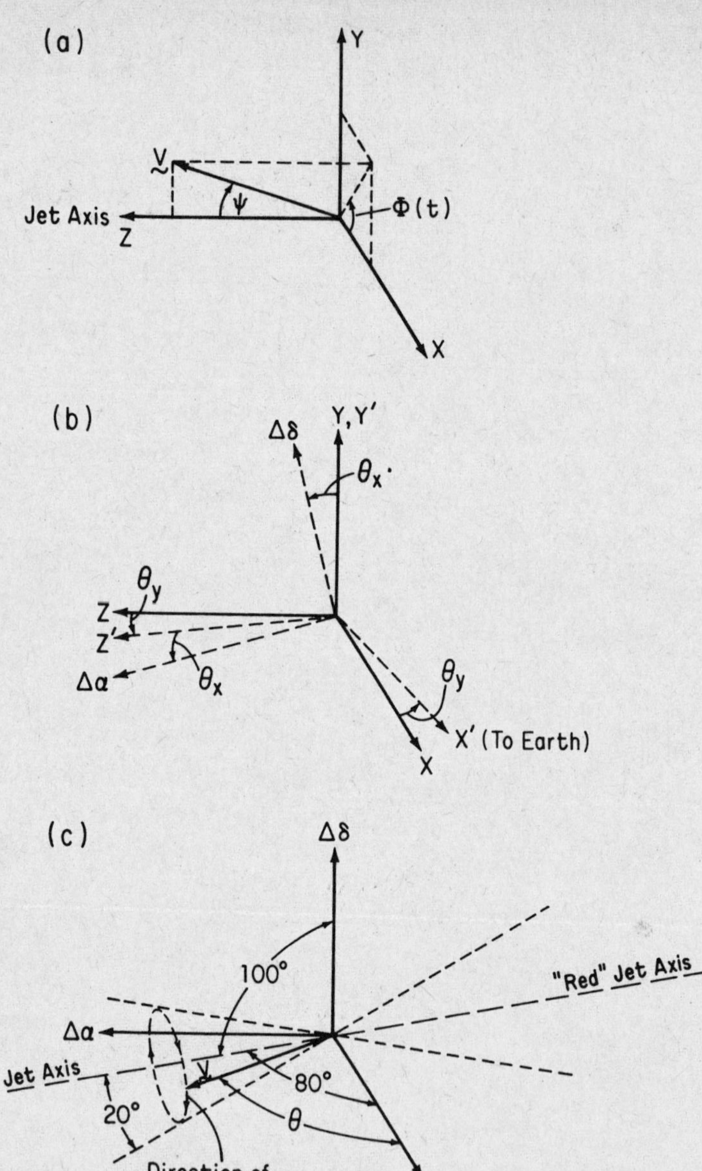

Figure 1 - The geometry of twin-jet ejection vectors in which:
(a) the ejection vector **v** rotates with angular displacement $\Phi(t)$ about
the z-axis at an angle ψ in an x-y-z coordinate system; (b) a rotation
of θ_y about the y-axis produces the x'-y'-z' coordinate system in
which the x'-axis points towards the observer and a subsequent
rotation of θ_x about the x'-axis produces the a coordinate system in
which the y'-axis becomes the declination-offset axis and the z'-axis
becomes the right ascension-offset axis; and (c) the final coordinate
system is shown with the approximate parameters of the rotating
ejection vectors of the SS433 system.

291

$$\mathbf{v} = \begin{pmatrix} v_r \\ v_\delta \\ v_\alpha \end{pmatrix} = \begin{pmatrix} v'_x \\ v'_y \cos \theta_x + v'_z \sin \theta_x \\ -v'_y \sin \theta_x + v'_z \cos \theta_x \end{pmatrix} \qquad (4)$$

and one can then describe the motion of a jet feature ejected at a time t_{eject}, when observed at a time t_{obs}, by a distance vector $\mathbf{D}$ where

$$\mathbf{D} = (D_r, D_\delta, D_\alpha) = (v_r, v_\delta, v_\alpha/\cos \delta)(t_{obs} - t_{eject})/(1 + v_r/c) \, . \quad (5)$$

The apparent proper motion vector (μ) for a particular piece of ejected material is

$$\mu = \mathbf{D}/d \qquad (6)$$

where d is the distance from the observer to the center of jet ejection.

Figure 2 is a plot of $z = \gamma(1 + v_r/c)$ (where $\gamma = 1/(1 - \mathbf{v} \cdot \mathbf{v}/c^2)$) for the moving optical emission lines of SS433, based upon the 1978-1983 data summary of Margon (1984), with crosses identifying the receding or red-shifted jet and triangles identifying the approaching or blue-shifted jet. The solid curves in Figure 2 are based upon the twin-jet model parameters listed in the following Table 2 with the values of ψ, inclination, v/c, P, and t_{ref} corresponding to parameters given by Margon (1984), and with the other parameters determined from radio observations of the proper motion in the jet as determined by Hjellming and Johnston (1981b).

TABLE 2

Summary of Twin-Jet Parameters of SS433

ψ	19.80°	Angle jet makes with jet axis
θ_y	-11.18°	Inclination of jet axis to plane of sky (=i-π/2, i=inclination to line of sight)
θ_x	-10°	Corresponds to 100° positon angle (= -χ, χ of Hjellming and Johnston 1981b)
s_{jet}(East)	-1	For $\mu_\alpha > 0$, "blue" (approaching) jet
s_{jet}(West)	+1	For $\mu_\alpha < 0$, "red" (receding) jet
s_{rot}	-1	Clockwise (LH) rotation about z-axis
v/d	(3.0" ± 0.02")/yr	Proper motion scale factor
v/c	0.2601 ± 0.02	Jet velocity
d	(5.5 ± 0.3) kpc	Distance to SS433
t_{ref}	JD 2443501.48	Reference time for jet rotation
P	162.532 days	Period of rotation about jet axis

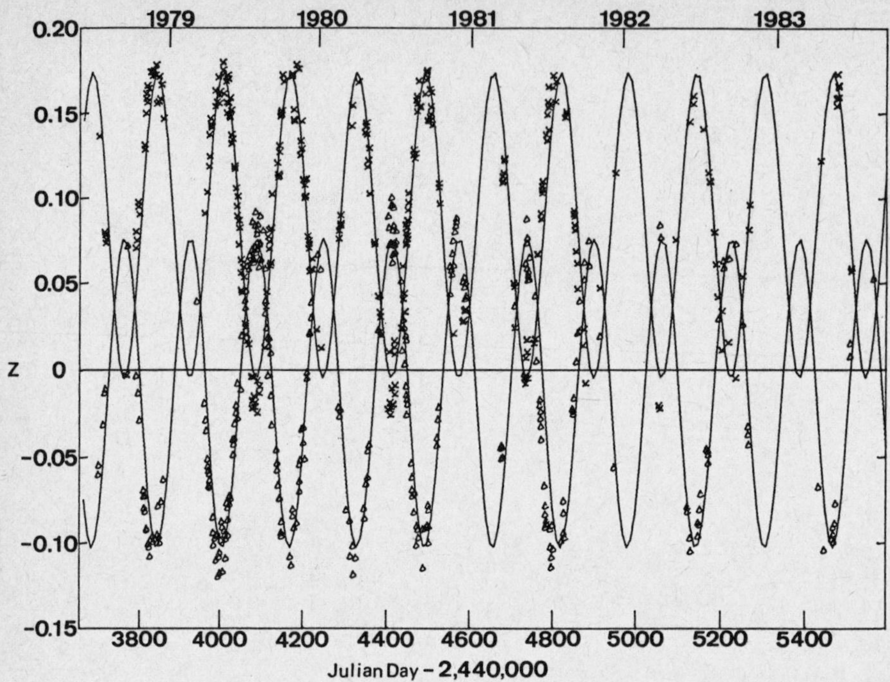

Figure 2 - Optical data (Margon 1984) for the jets of SS433 for the
period 1978-1983 are plotted as a function of Julian Day - 2,440,000
with the receding or red-shifted jet data identified with crosses and
the approaching or blue-shifted jet data identified with triangles.
The solid curves correspond to a constant period model with P =
162.532^d and the other parameters listed in Table 2.

A close examination of the major differences between data and
model reveals that there are systematic deviations in the sense that
shifts in time, velocity, or both are need locally to bring data and
model into systematic agreement. We will discuss these deviations,
which are clearly only second order effects when compared with the
major agreement between data the the simple kinematic model, in
sections III and IV.

The first complete determination of the jet parameters that
cannot be determined from optical data, by Hjellming and Johnston
(1981b), was based upon four epochs (December 1979 to June 1980) of
4.9 GHz VLA radio images with 0.3" resolution. Since then Hjellming

and Johnston (1987) have systematically observed SS433 at 4.9 and/or
15 GHz with resolutions of 0.3" and 0.1", respectively, for 23 more
epochs from December 7, 1980 through June 8, 1986. These data provide
a wealth of information about both the physical evolution of SS433 radio
jets and the proper motions of radio-emitting material that we will
discuss in more detail in the next section.

III. DEVIATIONS OF SS433 FROM THE SIMPLE KINEMATIC MODEL

Since 1980 there has been ample evidence for significant
deviations of SS433 optical and radio data from the simple kinematic
model. Using optical data taken before 1982 Collins and Newsome
(1980, 1982), Margon (1981), and Newsome and Collins (1981) noted that
the fit of the kinematic model could be improved by assuming that the
period, intially taken to be 164.0^d, was decreasing monotonically at a
rate of ~0.01 days per day. Anderson et al. (1983) later showed that
with even more data the deviations were more complex: a constant P =
162.7^d gave a mean residual index of 0.00806; P = 164.73^d with a
monotonic decrease of -0.0024 days/day gave a mean residual index of
0.00757; P = 172.56^d with a monotonic decrease of -0.024 days/day and
a finite second derivative term gave a mean residual index of 0.00664;
and P = 163.62^d with a sinusoidal term of period 1625^d gave the lowest
mean residual index of all, 0.0066.

Although the proper motion patterns of radio emission ejected
from the central system in SS433, which have the appearance of
relativistically distorted corkscrews, are not as accurate an
indicator of the SS433 "clock" as optical radial velocities, an
analogous story can be derived from the 1979-1986 radio data of
Hjellming and Johnston (1987). Although the 15 epochs of SS433
"corkscrews" observed at both 4.9 and 15 GHz through 1982 could be
reasonably fit by the constant period parameters of Hjellming and
Johnston (1981b), using P = 164^d, there was a small but consistent
drift between the model and the data for the 15 GHz images of 1982.
By the fall of 1983 these deviations were even more drastic, and the
best fit for the 1979-1983 radio data seemed to require P = 164^d and a
monotonic period decrease of -0.002 days/day, which was consistent
with the equivalent model of Anderson et al. (1983). Extensive
observations in the next A-array "season" of the VLA, in the winter of
1984-1985, did not confirm such a large monotonic decrease, but with a
monotonic decrease of -0.001 days/day the fits were marginally
"acceptable". This trend continued when the early 1986 data were
available since the best solution required an even smaller monotonic
drift of the period. With 27 epochs of data from 1979 to 1986 the

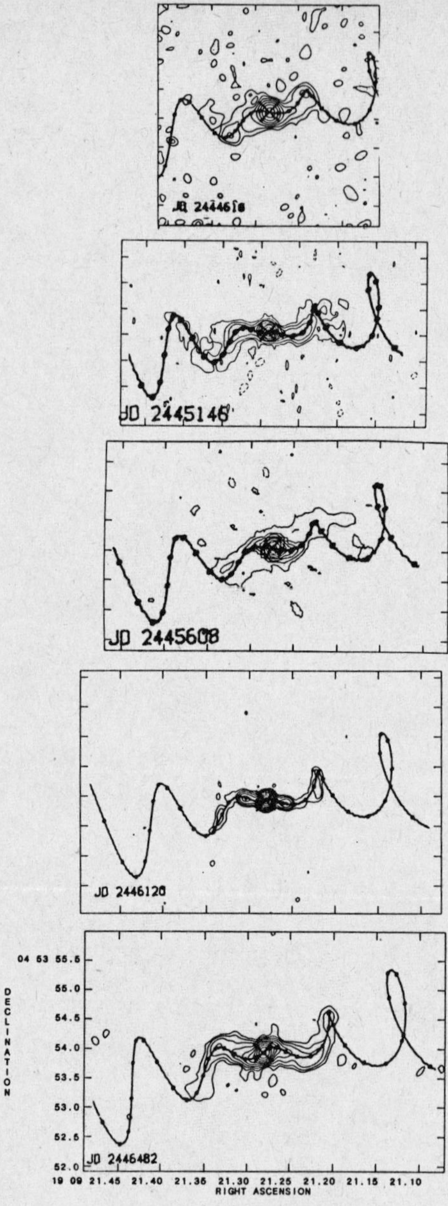

Figure 3 - A sample of five of the 20 epochs of 15 GHz image of SS433 obtained by Hjellming and Johnston (1987), corresponding to Julian Days of 2444616, 2445146, 2445608, 2446120, and 2446482, which covers the time period between January 1981 and February 1986. The predicted proper motion corkscrews are for a model with constant $P = 162.5^d$.

"best fit" to the proper motion corkscrews of SS433 involves a constant period of the order of 162^d, and for this reason we have adopted the Margon (1984) period of 162.532^d. Figure 3 shows a sample of five of the 20 epochs of 15 GHz VLA images of SS433 covering a five year period between January 1981 and February 1986.

The global agreement of 1979-1986 radio images of SS433 with a constant period model with $P = 162.532^d$ days is achieved at the price of accepting transient and minor deviations between the model corkscrews and the radio data. These deviations are analogous to the complex fluctuation found in the optical data.

IV. A WORKING HYPOTHESIS ABOUT THE OPTICAL AND RADIO JETS OF SS433

Since 1983 we have been pursuing a simple working hypothesis about the relationship between the radio and optical jets of SS433. This working hypothesis has led us to re-examine the optical data.

Figure 4 is a schematic diagram illustrating a working hypothesis about the origin of the 0.26c jets as outflows perpindicular to a (thick) accretion disk, which is precessing with a period of $\sim162^d$, in a binary system with a period of 13.1^d. The relative location of optical- and radio-emitting material is shown in terms of laterally expanding jets.

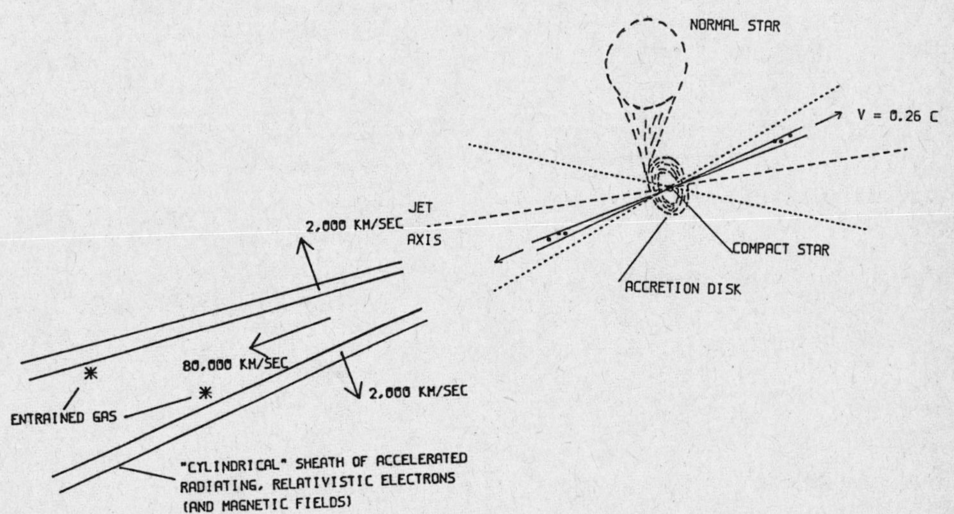

Figure 4 - A schematic diagram illustrating, on the right, the relationship between the SS433 jets and a binary system with a thick accretion disk, and, on the left, the laterally expanding jet hypothesis in which the optically emitting gas is entrained material while the radio emission is in a sheath environment analagous to the cylindrical equivalent of an expanding supernova shell.

Since 1983 we have been asking ourselves whether it was possible that the optically emitting gas was entrained material in jets expanding laterally with speeds of the order of 2000 km/sec, corresponding to the shortest time-scale scatter in the optical data. The lateral expansion velocity would then be identified with the kinetic temperature of the jet flows, when they become free of the accretion disk environment, indicating a temperature of the order of 3×10^8K. The lateral expansion of the jets causes the obvious environment for the location of radio-emitting material to be in a sheath outside the expanding jet with particle acceleration in a expanding cylindrical environment of shocks, instabilities, etc. Proper physical modeling of the SS433 radio jets would then involve computations in a cylindrical coordinate system analagous to the spherical explosion geometry of supernova (Chevalier 1981, 1982a, 1982b).

The main initial reason for adopting the hypothesis of an expanding cyindrical sheath was the need to explain why the SS433 radio emission always has essentially the same optically thin, non-thermal spectral index of the order of -0.7, and never shows significant signs of the self-absorption effects that are difficult to avoid where radio emission arises from compact stellar environments. The entrained gas hypothesis for the origin of the optical emission lines would imply that the optically emitting elements have a "random" lateral velocity component with values up to the actual lateral velocity of expansion. It was obvious to search for this effect in the deviations between the optical red shift data (z) and the kinematic model (with redshifts denoted by z_{eph}) which assumes no lateral velocity component for the optical emission.

A plot of $(z - z_{eph})$ vs Julian Day is shown in Figure 5, which is nothing more than the difference between the data in Figure 1 and the theoretical ephemeris plotted by the solid curves in Figure 1, except for Figure 5 we have ploted $(z - z_{eph})$ for the approaching (blue) jet and $-(z - z_{eph})$ for the receding (red) jet. As can be seen from Figure 5 the result is reasonably good superposition of the deviations from the twin-jet ephemeris, for both red and blue jets, and clear indication that the deviations have a periodicity that matches the 162.5^d period and an amplitude that varies systematically on time scales much longer than this period. It is obvious that, to first order, Figure 5 has the approximate nature of a "beat curve". Because short time scale fluctuation with time scales of ~6^d and ~13^d are known or expected in these data, we re-plotted the data of Figure 5 in the form of a running mean average with an averaging time of 20

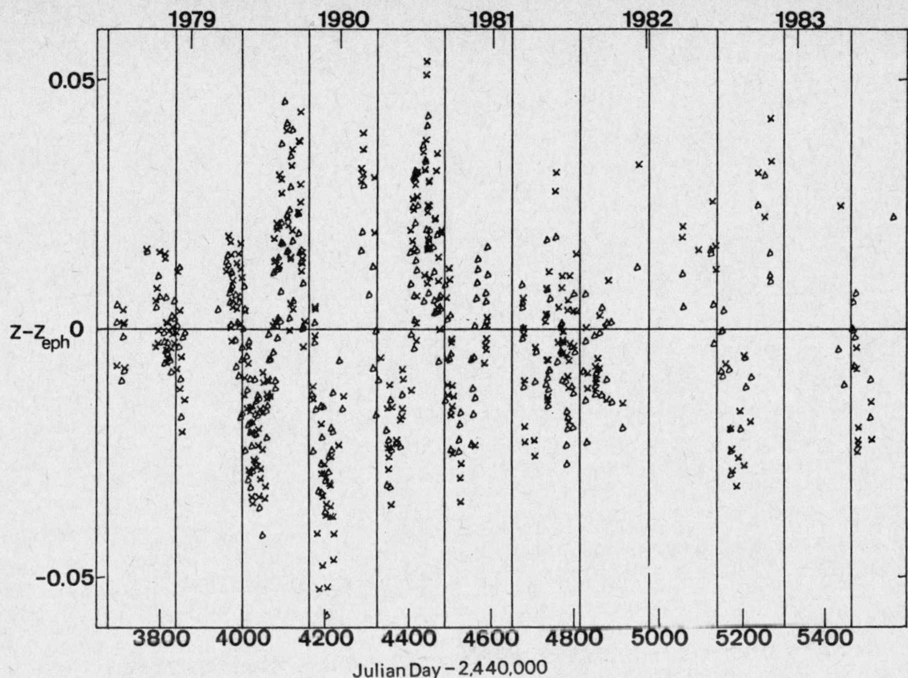

Figure 5 - A plot of $(z - z_{eph})$ for SS433 vs modified Julian day, where z is the redshift data for the moving lines (Margon 1984), with the triangles indicated blue jet data and the crosses indicating red jet data, and z_{eph} is the best fit twin-jet doppler shift model used in Figure 2 and Table 2.

days, with results shown in Figure 6, together with simple beat curves based upon the "beat" model

$$z - z_{eph} = \pm 0.03\{ \cos[\pi(1/151.73^d + 1/175^d)(t - t_{ref})] \times \qquad (7)$$
$$\sin[\pi[1/151.73^d - 1/175^d)(t - t_{ref}) - 2\pi(0.1)]\} .$$

The "precession" period of 162.532^d is now the harmonic mean of two periods, $P_1 = 175^d$ and $P_2 = 151.73^d$, and these have a beat period of $\sim2282^d$. It is obvious from Figure 4b that, while the beat formula (Equation (7)) fits the residuals surprisingly well, significant deviations remain. The applicability of a "beat model" is obviously related to the fact that the "best" fit of Anderson et al. (1983) was a constant period solution with a sinusoidal component of period 1625^d.

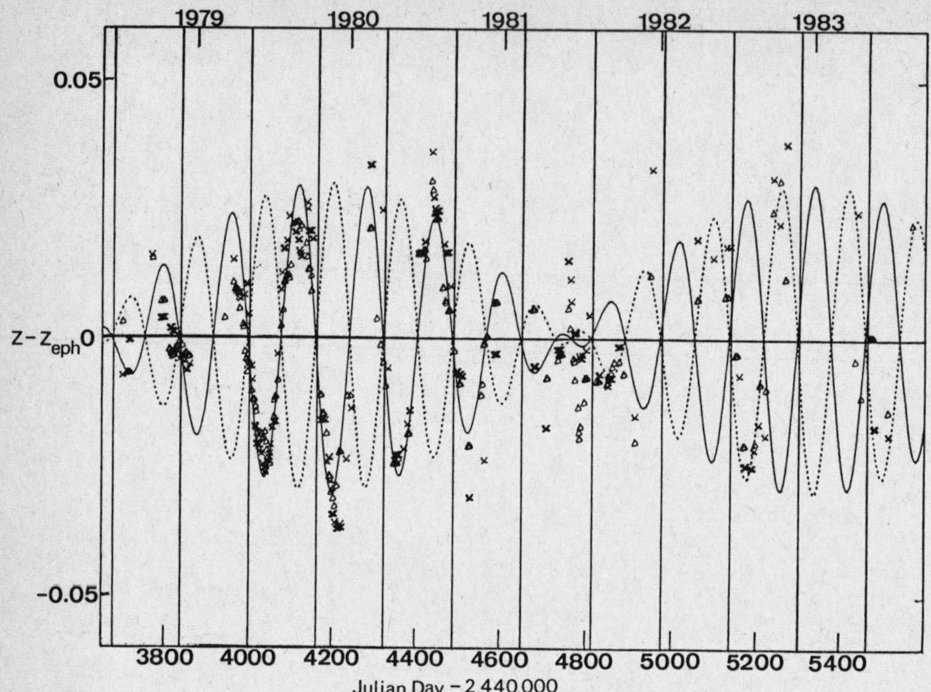

Figure 6 - A plot of $(z - z_{eph})$ vs modified Julian Day from data of Margon (1984), as discussed in Figure 5, excepted the plotted $(z - z_{eph})$ is a running mean with an averaging time of 20 days, and the dashed or solid lines show the "beat" model based upon Equation (7).

The implication of applicability of Equation (7) is that the x-coordinate's dependence of **v** upon time is changed from

$$\cos s_{jet}\Omega(t-t_{ref})$$

to

$$\cos[s_{jet}(\Omega_1+\Omega_2)(t-t_{ref})] \{1 \pm 0.03 \sin[(\Omega_2-\Omega_1)(t-t_{ref}) - 2\pi(0.1)]\} \quad (8)$$

so the quantity in angular brackets represents the first two terms in a fourier series in $(\Omega_2-\Omega_1)(t - t_{ref})$. The rotation of the twin-jet ejection vector in the x-y plane does not have a constant frequency, but is the result of two "waves" of different frequencies. This should be considered as a possibly major clue to the poorly understood mechanism responsible for the precession of the thick accretion disk in the SS433 system (Margon 1984).

V. CONCLUSIONS

The occurence of radio emission in many x-ray binaries is well established, but in most cases we know very little about the underlying mechanisms for producing the relativistic electrons and magmetic fields. Only SS433 and Cyg X-3 (Geldzahler _et al._ 1983, Spencer _et al._ 1986) have been observed to show moving radio jets with velocities of ~0.26c. We have concentrated on SS433 in this paper because it is most likely that its jets arise from 0.26c flows perpindicular to a thick accretion disk, therefore it is most certain that it is directly related to the subject of this workshop. If the "moving" optical emission lines of SS433 are due to entrained gas in laterally expanding jets, and the radio emission is due to a lower energy, cylindrical equivalent of a supernova explosion, our unprecedented knowledge of the detailed geometry of the jets provides us with working models for the physics of the observed material.

An examination of available optical data on the moving lines of SS433 shows that the long time scale differences between the simple kinematic model and the observations look remarkably, but not exactly, like a beat between two periods. The observed period of 162.5^d, which has been ascribed to precession of the accretion disk, is the harmonic mean of two periods of 175^d and 152^d, and the beating between "waves" with these two periods causes the $\sim2300^d$ beat period in the long time scale deviations from the simple kinematic model. These facts may provide us with major clues about the poorly understood mechanisms underlying the "precession" of the SS433 accretion disk.

REFERENCES

Ables, J.G. 1969, Proc. Astron. Soc. Australia, 1, 237.

Anderson, S.F., Margon, B., and Grandi, S.A. 1983, Ap.J., 273, 697.

Band, D.L. and Grindlay, J.E. 1984, Ap.J., 285, 702.

Braes, L.L.E. and Miley, G.K. 1971, Nature, 232, 246.

__________. 1972, Nature, 237, 507.

Braes, L.L.E., Miley, G.K., and Schoenbaker, A.A. 1972, Nature, 236, 392.

Chevalier, R.A. 1981, Ap.J., 251, 259.

__________. 1982a, Ap.J., 258, 790.

__________. 1982b, Ap.J., 259, 302.

Collins, G.W. and Newsom, G.H. 1980, I.A.U. Circular 3547.

__________. 1982, Ap. Space Sci., 81, 199.

Fomalont, E.B., Geldzahler, B.J., Hjellming, R.M., and Wade, C.M.

1983, Ap.J., 275, 802.

Geldzahler, B.J. 1983, Ap.J. (Letters), 264, L49.

Geldzahler, B.J. 1983. Johnston, K.J., Spencer, J.H., Lepczynski,
   W.J., Josties, F.J., Angerhofer, P.E., Florkowski, D.R., McCarthy,
   D.D., Matsakis, D.N., and Hjellming, R.M. 1983,
   Ap.J.(Letters), 273, L65.

Gregory, P.C., Kronberg, P.P., Seaquist, E.R., Hughes, V.A.,
   Woodsworth, A.A., Viner, M.R., Retalleck, D., Hjellming, R.M., and
   Balick, B. 1972, Nature Phy. Sci., 239, 114.

Gregory, P.C., and Taylor, A.R. 1978, Nature, 272, 704.

Grindlay, J.E. and Seaquist, E.R. 1986, Ap.J. (Nov. 1, 1986 issue).

Hjellming, R.M. 1979, IAU Circular No. 3369.

Hjellming, R.M. 1978, Ap.J., 221, 225.

Hjellming, R.M. et al. 1987, in preparation.

Hjellming, R.M. and Blankenship, L.C. 1973, Nature Phy. Sci., 243, 81.

Hjellming, R.M., Gibson, D.M., and Owen, F.N. 1975, Nature, 256, 111.

Hjellming, R.M. and Johnston, K.J. 1981a, Nature, 290, 100 (Paper I).

__________. 1981b, Ap.J.(Letters), 246, L141 (Paper II).

__________. 1982, "Extragalactic Radio Sources", eds. D.S. Heeschen
   and C.M. Wade, 197 (Paper III).

__________. 1987, in preparation.

Hjellming, R.M. and Wade, C.M. 1971a, Ap.J.(Letters), 164, L1.

__________. 1971b, Ap.J.(Letters), 168, L21.

Johnston, K.J., Spencer, J.H., Simon, R., Waltman, E.B., Pooley, G.,
   Spencer, R.E., Swinney, R.W., Angerhofer, P.E., Florkowski, D.R.,
   McCarthy, D.D., Matsakis, D.N., and Hjellming, R.M. 1986
   Ap.J., (October 15 Issue)

Margon, B. 1981, in Proceedings of the Tenth Texas Symposium on
   Relativistic Astrophysics (Ann. NY Acad. Sci.., 375, 403).

__________. 1984, Ann. Rev. Astron. Ap., 22, 507.

Margon, B., Anderson, S.F., Aller, L.H., Downes, R.A., and Keyes, C.D.
   1984, Ap.J., 281, 313.

Margon, B., Grandi, S.A., and Downes, R.A. 1980, Ap.J., 241, 306.

Margon, B., Anderson, S., Grandi, S.A., and Downes, R.A.
   1981, I.A.U. Circular 3626.

Newsome, G.H. and Collins, G.W., II. 1981, A.J., 86, 1250.

Owen, F.N., Balonek, T.J., Dickey, J., Terzian, Y., and Gottesmann, S.
   1976, Ap.J.(Letters), 293, L15.

Preston, R.A., Morabito, D.D., Wehrle, A.E., Jauncey, D.1., Baty,
   M.J., Haynes, R.F., and Wright, A.E. 1983, Ap.J.(Letters), 268, L23.

Ryle, M., Caswell, J.L., Hine, G., and Shakeshaft, J. 1979,
   Nature, 276, 271.

Seaquist, E.R., Garrison, R.E., Gregory, P.C., Taylor, A.R.,
    and Crane, P.C. 1979, <u>A.J.</u>, 84, **1037.**
    <u>Ap.J.</u>, 260, **220.**
Spencer, R.E., Swinney, R.W., Johnston, K.J., and Hjellming, R.M.
    1986, <u>Ap.J.</u>, (October 15 issue).
Taylor, A.R. and Gregory, P.C. 1982, <u>Ap.J.</u>, 255, **210.**
Watson, M.G., Willingale, R., Grindlay, J.E., and Seward, F.D. 1983,
    <u>Ap.J.</u>, 273, **688.**
Whelan, J.A.J. et al. 1977, <u>M.N.R.A.S.</u>, 181, **259.**

X-RAY OBSERVATIONS OF THE JETS IN SS433

G.C. Stewart and M.G. Watson

X-ray Astronomy Group, Physics Department,
University of Leicester,
Leicester LE1 7RH, U.K.

ABSTRACT

EXOSAT observations of SS433 have revealed the presence of an emission line whose energy varies with time. The variation is consistent with that predicted for a line with a rest energy of 6.7 keV with shifts in energy analogous to those of the moving optical emission lines, i.e. the material emitting the X-rays is in jets moving at 0.26c and precessing about a fixed axis. Only one X-ray line is seen at any one time, however. Using this fact, and observations of an X-ray eclipse, constraints are derived on the dimensions of the components of the SS433 system. In conjunction with previously derived optical results this leads to a mass estimate for the compact object in the system of $> 10 M_\odot$. The energetics of the material responsible for the X-ray emission from the jet also require a mass of $> 10 M_\odot$ if the system is to be sub-Eddington.

1. INTRODUCTION

Since SS433 was first catalogued as an X-ray, radio and optical emission line source, and the discovery of the anomalous 'moving' emission lines in the optical made the object unique, a canonical picture of the system has emerged (e.g. Margon 1984) which has been successful in explaining the principal observational features. In this model SS433 is a binary system with a binary period of 13.08 days. The components of the system are a compact object of unspecified nature and a massive, early-type stellar companion. Mass transfer between the star and the compact object is occurring and there is an accretion disk around the compact object. Two jets of material are being ejected from the compact object at a velocity of 0.26c. These jets are precessing with a period of ~ 162 days and with a cone angle, α, of 19°. The angle between the binary plane of the system to the line of sight is 11°.

Previous X-ray observations have shown that SS433 is a relatively undistinguished X-ray source. For a distance of 5 kpc the 2-10 keV luminosity of SS433 inferred from UHURU, ARIEL-V and HEAO-1 observations is $\approx 5 \times 10^{35}$ erg s^{-1}. Either a power law or a high temperature ($kT \geq 10$ keV) bremsstrahlung spectrum provided adequate fits to the observed X-ray continuum spectrum of those observations, and there was also evidence of an emission line feature at 6.7 keV with an equivalent width of ≈ 600 eV (Marshall *et al.*, 1979, Ricketts *et al.*, 1981).

The most extensive series of X-ray observations prior to the EXOSAT observations discussed here were made with the EINSTEIN Observatory. Imaging observations of SS433 made with the IPC showed X-ray emitting lobes and a central point source; this established unequivocally the relationship of SS433 with W50, the surrounding radio supernova remnant (Seward *et al.*, 1980, Watson *et al.*, 1983). Grindlay *et al.* (1984) have analysed a series of 34 MPC observations ranging in duration from 100 s and 3200 s with a total exposure time of $\approx$ 50000 s. Flux variations on all timescales greater than 500 s were observed and the total range in flux being a factor of 5. Flux increases were reported that were coincident in time with two radio flares (Seaquist *et al.*, 1982). The MPC spectra showed no evidence for an emission line, with a 2σ upper limit marginally below the equivalent width reported by Marshall *et al.* The continuum was best fit with a power law the slope of which varied in time from a photon index of 0.3 to -1.4. No emission lines were detected in the 0.5-4.0 keV band using the SSS although comparison of the SSS and MPC spectra suggested the existence of a soft excess.

2. EXOSAT OBSERVATIONS

The EXOSAT observations discussed here comprise a series of 15 pointings made between 1983 and 1985. Individual exposure times range from $\approx$ 5000 s to $\approx$ 30000 s and the total exposure time is $\approx$ 400000 s. Details of the observations are given in Table 1. Included amongst these observations are a set of 7 made over one cycle, contemporaneously with photometric observations in the V band at 4 different observatories worldwide. The optical observations are discussed in Kemp *et al.*, 1985, while the X-ray results are discussed in more detail by in Watson *et al.*, 1986, and Stewart *et al.*, 1986.

Table 1: Summary of the EXOSAT observations. The energy of the emission line, E_{Line}, is given only once for each group of observations. Within in any group of observations E_{Line} was constant to within the observational errors.

Observation	Date	U.T. (mid-exp)	Duration (s)	Orbital Phase θ_{13}	Precession Phase ϕ_{162}	E_{Line} keV
A1	83::287	06:15	45000	.732	.67	7.68
A2	83::288	08:30	40000	.820	.68	
A3	83::289	08:00	43000	.895	.68	
Z1	84::108	23:15	45000	.004	.82	
B1	84::265	17:15	12000	.981	.78	7.32
B2	84::268	13:00	10000	.197	.80	
B3	84::269	12:15	22000	.271	.81	
B4	84::271	04:00	16000	.398	.82	
B5	84::273	14:00	16000	.582	.83	
B6	84::275	23:00	4000	.764	.85	
B7	84::278	22:45	10000	.992	.87	
C1	85::138	02:00	43000	.195	.25	6.49
C2	85::148	01:30	40000	.960	.31	
D1	85::202	01:00	42000	.470	.64	7.81
E1	85::278	12:45	48000	.003	.11	6.66

The X-ray flux levels recorded were typically 50% higher than those observed by the Einstein MPC and ranged between $1 - 3 \times 10^{-10}$ erg s^{-1}cm^{-2}. Despite the longer overall observation time no X-ray flare events similar to those seen by the EINSTEIN satellite were detected. This may in part be due to the different sampling technique used.

2.1 The X-ray Spectrum

During the period that the EXOSAT measurements were made the shape and behaviour of the X-ray spectrum differs from that reported by Grindlay *et al.*, 1984. In both the GSPC and the ME data there is clear evidence for the existence of an emission line at all times. There is also little gross variation of the continuum spectrum between different observations.

Neither a simple power law nor an optically thin thermal bremsstrahlung spectrum provide a statistically adequate representation of the spectrum, even with the inclusion of an emission line. A good representation of the continuum spectrum at all times is achieved, however, with a two component model where one component is optically thin bremsstrahlung and the other is a blackbody. The temperatures found for each component are then typically 2 - 3 keV. An equivalent width of approximately 750 eV is found for the emission feature, roughly what would be expected from a thermal plasma in equilibrium with a temperature of 2 keV and with a solar abundance of iron.

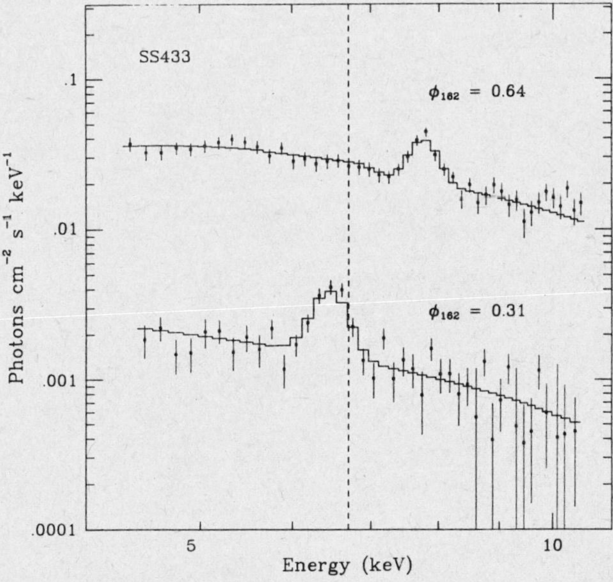

Figure 1. The photon spectrum of SS433 on day 202, ϕ=0.64 , and day 148 of 1985 as inferred from the GSPC data showing the shift in energy of the emission line. The vertical dashed line is at an energy of 6.7 keV. (An offset in flux of a factor of 10 has been applied to the day 202 data.)

The most striking feature of the X-ray spectrum is the behaviour of the energy centroid of the emission line. As is illustrated in Figure 1 this varies from observation to observation and the range of measured line centre energies over all the observations is ~ 6.5 and ~ 7.8 keV. For all the observations the line energies measured with both the GSPC and ME are in excellent agreement, ruling out an instrumental effect.

In Figure 2 the measured line energy is shown plotted as a function of the 162 day phase of SS433 for all the EXOSAT observations, and also for one observation made by the TENMA satellite (Matsuoka *et al.*, 1986). Also shown in the Figure are the predicted energies which would have been observed for emission lines with rest energies of 6.6, 6.7 and 6.8 keV if they had been Doppler shifted according to the canonical model for the shifts of the moving optical emission lines. The good agreement between the observed energy and that expected for a 6.7 keV line, when coupled with the agreement between the observed line equivalent width and that for a thermal plasma with solar abundances is strong evidence that the X-ray emission in SS433 originates in the jets.

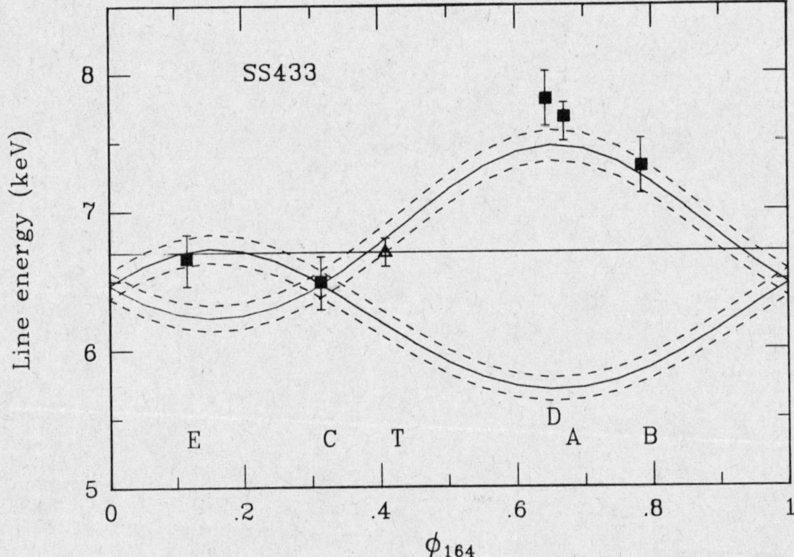

Figure 2. Energy of the X-ray emission line as a function of 162-day phase. Individual lines are labelled according to the groupings in Table 1. The point labelled 'T' is a measurement by the Tenma satellite (Matsuoka *etal.*, 1985). The solid curve shows the predicted line energy in the kinematic model for a line with a rest energy of 6.7 keV. The dashed curves are predictions for rest line energies of 6.6 and 6.8 keV.

A major difference between the optical moving lines and the X-ray result, however, is that in the X-ray data only one line is visible at any time. This is despite the fact that over most of the 162 day period the lines would easily have been resolved. The 'missing' line cannot be explained by Doppler boosting of the radiation from the jets, at least for most of the observations, as the gain in strength of the observed line relative to the unobserved line is too small to cause the line to be undetectable.

The most natural explanation for the non-observation of one line is that the line of sight to that jet is such that the jet is obscured by material which also responds to the 162 day clock in the system. The most natural location for this material is in the accretion disc. For the relevant jet to be obscured at a particular phase, the relative sizes of the jet and the accretion disc are then constrained. This constraint will be discussed in section 3.

2.2 13-Day X-ray Light Curve

Figure 3 shows the results obtained from the seven observations made over one binary cycle together with the results from the optical monitoring. There is a reduction in the X-ray flux of approximately 40% in each of the energy ranges coincident with the deeper optical eclipse (at MJD 49567.8). In addition there is evidence for a gradual decrease of $\approx 15\%$ over the observation period.

Both the ratios of fluxes in the different energy bands and the equivalent width of the iron line suggest that within the observational errors the X-ray spectrum of SS433 remains constant on timescales longer than a few thousand seconds. This implies that the flux reduction is the eclipse of the single X-ray emitting region which can be identified as the jets because of the behaviour of the line emission.

3. SYSTEM DIMENSIONS AND COMPONENT MASSES

Since the location of the X-ray emission is known, the parameters of the X-ray eclipse can be used to constrain the dimensions of the the emitting region. In particular the depth and duration of the X-ray eclipse and the absence of emission from the jet which is directed away from the observer provide information on the relative sizes of the jets, accretion disk, stellar radius and the orbital separation.

Full details of the constraint equations as derived for the assumed geometry of SS433 (Figure 4) are given in Stewart *et al.*, 1986. The results are summarised in Figure 5 which shows the $l_j - R_\star$ plane where l_j and $R_\star$ are respectively the length of the X-ray emitting jet and the radius of the optical star, both in units of the orbital separation.

Briefly, as the eclipse lasts for a minimum of 1.2 days, or 0.09 of the orbital period, the radius of the eclipsing star must satisfy

$$R_\star > 0.34$$

A minimum length for the X-ray jet can be derived from the fact that the maximum duration of totality during the eclipse is 0.7 days, or 0.05 of the period. The minimum length is the function of $R_\star$ shown in the figure.

For values of $R_\star < 0.38$, the eclipse depth of at least 40% implies that

$$l_j < 2.5R_\star - 0.48$$

independent of the geometry of the accretion disc. For $0.38 < R_\star < 0.58$

$$l_j < 0.30 + 0.56R_\star$$

while for $R_* > 0.58$,

$$l_j < 1.42(1 - R_*)$$

The latter two criteria are maxima for the jet length under the assumptions that the accretion disk has a radius which would contact the optical star and a height which is less than its radius. These are conservative assumptions in that at any value of the stellar radius they maximise the possible jet length.

It is clear that only a limited region of the $l_j - R_*$ plane would satisfy the observed eclipse parameters. The radius of the optical star is constrained such that $0.34 < R_* < 0.65$ while for most values of R_* the jet length lies between 0.1 and 0.6. (For a very restricted range of R_*, l_j could be smaller).

In an analysis of the optical light curve Leibowitz, 1984, has also derived limits to the radius of the optical star and found $0.14 < R_* < 0.40$ with a best fit value of $R_* = 0.27$. While the best fit value is excluded by the X-ray solution, the overlap of the allowed regions gives $0.34 < R_* < 0.40$, implying $l_j < 0.5$.

3.1 Absolute Dimensions

To convert the derived dimensions to absolute values, a value for the orbital separation a must be found. The value for R_* can be converted to a value of $q = M_x/M_*$ under the assumption that the optical star is filling its Roche lobe and using the appropriate Roche formulae (Paczynski, 1971). This scaling is also shown in Figure 5. Using the K value of 195 km s^{-1} for the $HeII$ lines determined by Crampton and Hutchings, 1981, $a \approx 2.5 \times 10^{12}(1 + q)$ cm. This implies a lies in the range of $6 - 9 \times 10^{12}$ cm and that the length of the X-ray emitting jet is between 5×10^{11} cm and 3×10^{12} cm for all allowed values of q.

Using the K velocity and the optical/ X-ray constraints on the allowed value of q, the derived mass ranges for the compact object and optical star are then $25M_\odot < M_x < 100M_\odot$ and $30M_\odot < M_* < 70M_\odot$. For the compact object to have a mass of $1.4M_\odot$ (i.e. a neutron star) would require a value of $q \approx 0.1$, which is marginally allowed by the X-ray data but excluded by the optical result or the true K velocity to be ≈ 60 km s^{-1}.

4. JET ENERGETICS

A length for the X-ray emitting portion of the jets of $\approx 10^{12}$cm was derived in the previous section with an upper limit of 3×10^{12}cm while the optical and radio emission from the jets comes from radii of $\approx 10^{14} - 10^{15}$cm. This implies that the material in the jets is cooling to temperatures lower than those observable in the X-ray region in the flow time required to reach a radius of 10^{12}cm which is ≈ 100s at 0.26c.

For a plasma at a temperature of $\approx 10^{7.5}$ K, the requirement that the radiative cooling timescale be ~ 100 s then implies a density of $n_e \approx 10^{13}$cm^{-3}. For a total observed luminosity of $\approx 10^{36}$erg s^{-1} the emitting volume, V, is then

$$V \approx 8 \times 10^{32} f^{-1} n_{13}^{-2} D_5^{2} \text{ cm}^3$$

where D_5 is the distance in units of 5 kpc, f is the volume filling factor of the X-ray emitting gas and n_{13} is the electron density in units of 10^{13}cm^{-3}.

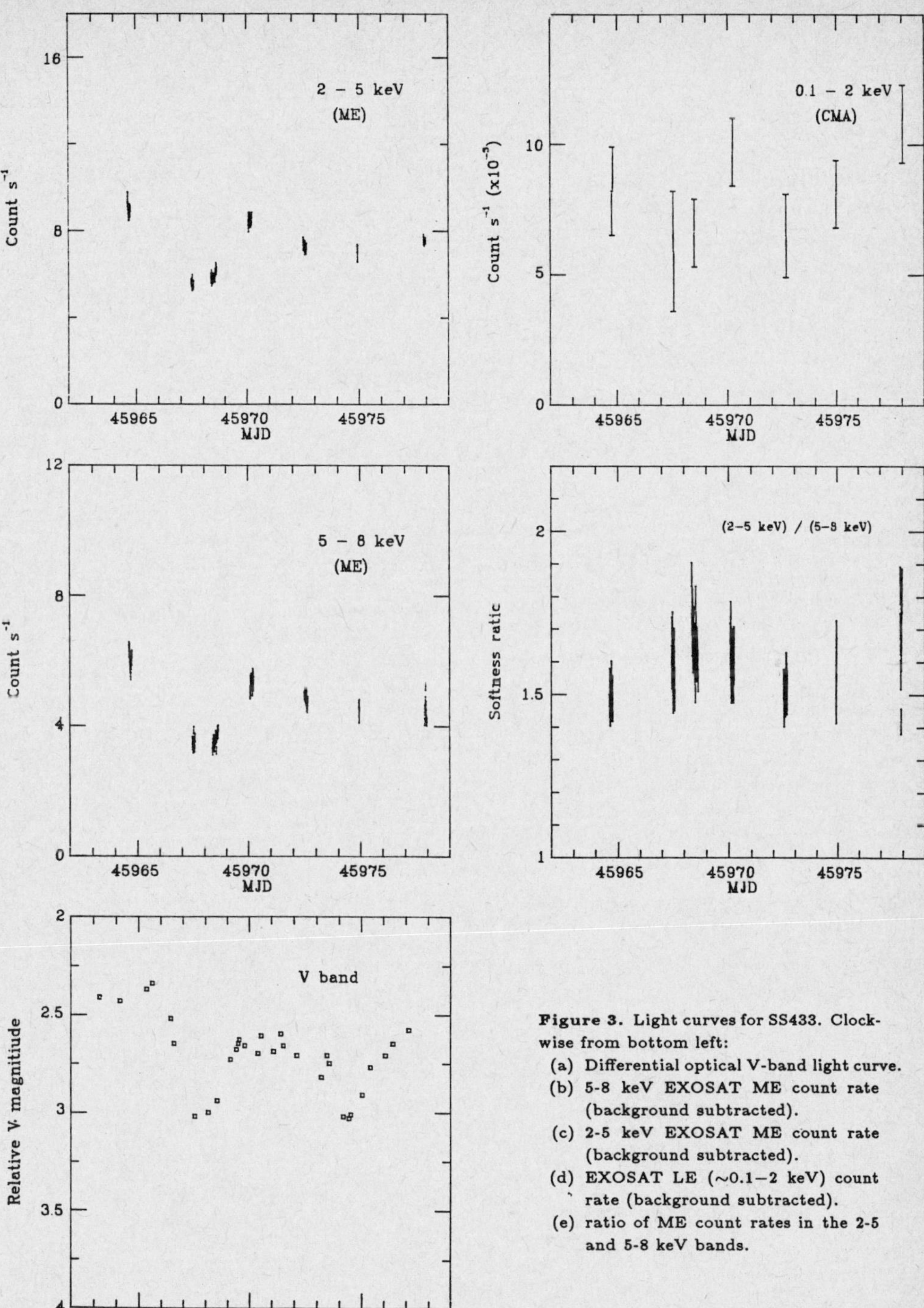

Figure 3. Light curves for SS433. Clockwise from bottom left:
  (a) Differential optical V-band light curve.
  (b) 5-8 keV EXOSAT ME count rate (background subtracted).
  (c) 2-5 keV EXOSAT ME count rate (background subtracted).
  (d) EXOSAT LE ($\sim$0.1$-$2 keV) count rate (background subtracted).
  (e) ratio of ME count rates in the 2-5 and 5-8 keV bands.

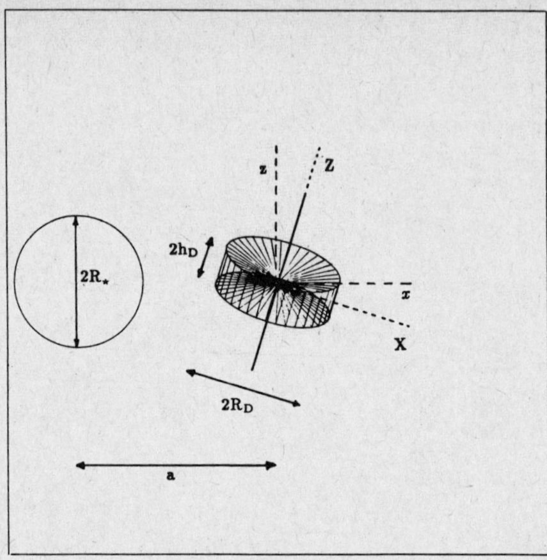

Figure 4. Schematic illustration of the model of the SS433 system used in this analysis showing the co-ordinate systems used.

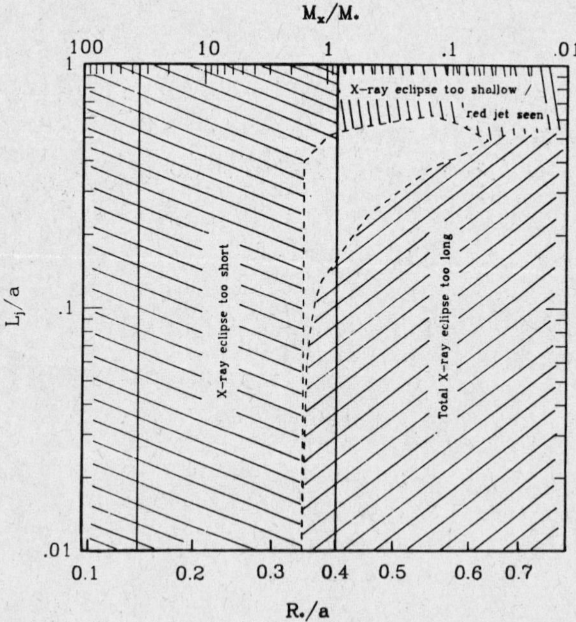

Figure 5. Constraints on the length of the X-ray emitting regions of the jets of SS433 and the radius of the optical star as fractions of the binary separation. Dashed lines and shaded regions define the areas of the $R_* - l_j$ plane excluded by the X-ray observations as derived in the text. Solid lines show the allowed range of parameter space derived by Leibowitz from analysis of the optical light curve.

The implied cylindrical radius of the jets r_x is then

$$r_x \approx 10^{10} f^{-0.5} n_{13}{}^{-1} D_5 \ \text{cm}$$

which is consistent with the opening angle expected for free expansion. The mass flow rate, $\dot{M}$, and kinetic luminosity, L_j, for the jet are :

$$\dot{M} \approx 2 \times 10^{-6} f^{-1} n_{13}{}^{-1} D_5{}^2 \ \text{M}_\odot \text{yr}^{-1}$$

$$L_j \approx 3 \times 10^{39} f^{-1} n_{13}{}^{-1} D_5{}^2 \ \text{erg s}^{-1}$$

The kinetic energy in the jets inferred from these observations is thus much larger than the Eddington luminosity for a $1 M_\odot$ compact object, but is compatible with the values derived less directly from observations of the 'moving' optical lines (e.g. Milgrom, 1981, Begelmann, 1980) or of the X-ray lobes (Watson *et al.*, 1983). As the distance to SS433 is well determined the only way to reduce the required luminosity would be if the density in the jets was $\gg 10^{13} \text{cm}^{-3}$ and some unknown mechanism were keeping the gas at X-ray temperatures for ~ 100 s. This cannot be the central luminosity as the Thompson optical depth along the jet is already > 1 for $n_e \approx 10^{13} \text{cm}^{-3}$.

5. CONCLUSIONS

The EXOSAT observations have shown that the bulk of the X-ray emission from SS433 originates in the relativistic jets and is thermal in nature. Clear evidence for an X-ray eclipse is found and the jets are shown to be $\approx 10^{12} \text{cm}$ long. Variations in the relative strength of the emission from the two jets with 162 day phase are consistent with the geometrical obscuration of the X-ray emission regions by an accretion disk in the system.

The kinetic luminosity of the material in the jets is larger than the Eddington luminosity for a neutron star. The mass of compact object required for this luminosity to be sub-Eddington is consistent, howver, with the constraints derived for the mass ratio of the system, with the value of the K velocity found by Crampton and Hutchings and also with the indirect evidence that the optical star has a spectral type of OB.

References

Begelmann, M.C., Sarazin,C.L., Hatchett, S.P., McKee, C.F. & Arons, J., 1980. *Astrophys. J.*, **238**, 722.

Crampton, D., & Hutchings, J.B., 1981. *Astrophys. J.*, **251**, 604.

Grindlay, J.E., Band, D.L., Seward, F., Leahey, D., Weisskopf, M.C., & Marshall, F.E., 1984. *Astrophys. J.*, **277**, 286.

Kemp, J.C., Henson, G.D., Kraus, D.J.,Carroll, L.C., Beardsley, I.S., Takagishi, K., Jugaku, J., Matsuoka, M., Leibowitz, E.M., Mazeh, T. & Mendelson, H., 1985. *Astrophys. J.*, **305**, 805.

Leibowitz, E.M., 1984. *Mon. Not. R. astr. Soc.*, **210**, 279.

Margon, B., 1984. *Ann. Rev. Astron. Astrophys.*, **22**, 507.

Marshall, F.E., Swank, J.H., Boldt, E.A., Holt, S.S. & Serlemitsos, P.J., 1979. *Astrophys. J. (Lett.)*, **230**, L145.

Matsuoka, M. et al., 1986. *Preprint*

Milgrom, M., 1981. *Vistas Astron.*, **25**, 141.

Paczynski, B., 1971. *Ann. Rev. Astr. Astrophys.*, **9**, 183.

Ricketts, M.J., Hall, R., Page, C.G., Pounds, K.A. & Sims, M.R., 1981. *Vistas Astron.*, **25**, 71.

Seaquist, E.R., Gilmore, W., Johnston, K.J. & Grindlay, J.E., 1982. *Astrophys. J.*, **260**, 220.

Seward, F.D., Grindlay, J.E., Seaquist, E. & Gilmore, W. 1980, *Nature*, **287**, 806.

Stewart, G.C.,Watson, M.G.,Matsuoka, M., Brinkmann, W., Jugaku, J., Takagishi, K.,Omodaka, T., Kemp, J.C., Kenson, G.D., Kraus, D.J., Mazeh, T. & Leibowitz, E.M., 1986. *Mon. Not. R. astr. Soc.*, in press.

Watson, M.G., Willingale, R., Grindlay, J.E. & Seward, F., 1983. *Astrophys. J.*, **273**, 688.

Watson, M.G., Stewart, G.C., Brinkmann, W. & King, A.R., 1986. *Mon. Not. R. astr. Soc.*, **222**, 261.

Cygnus X-3

Lawrence A. Molnar
Harvard-Smithsonian Center for Astrophysics
60 Garden Street
Cambridge, MA 02138

ABSTRACT

The low-mass X-ray binary Cyg X-3 is unusual for its high radio, infrared, and X-ray luminosities. This paper reviews the observational situation at all wavelengths, and reports the first results of a series of coordinated observations at all wavelengths.

Our observations of low-level radio emission from Cyg X-3 show that this emission is characterized by successive flares whose spectral evolution is similar to the giant radio flares observed in Cyg X-3. As with the giant flares, we interpret these flares to be expanding sources of electron synchrotron radiation. We present VLBI data at 1.3 cm wavelength that confirm this interpretation. We find that the timing of the flare onsets is periodic with a period several percent longer than the binary orbital period. We suggest the radio period is the result of periodic modulation of the binary separation. We present two models for this period that are consistent with the existing data, while making differing predictions about what more extensive data should show. Modulation at the radio period may be caused by precession of the line of apsides in an eccentric orbit. Alternatively, three-body simulations show that the modulation of the binary separation in a triple system may mimic the radio period much of the time, but would also show well-defined phase jumps twice per orbit of the third star.

We find that the effect of X-ray scattering by the interstellar medium on the observed X-ray light curve is to decrease the depth of modulation from its intrinsic value for photon energies less than 6 keV. These data are not consistent with either cocoon or stellar wind models for the light curve. They are consistent with the accretion disk corona model of White and Holt, which we have elaborated to take into account the density stratification of the disk.

Our results show that Cyg X-3 is an important source for the study of binary evolution, accretion disk structure, and the formation and evolution of radio jets.

I. Introduction.

A. X-ray.

Cyg X-3 was first discovered as an X-ray source in a rocket flight in 1966 (Giacconi *et al.* 1967), and is among the brightest of Galactic X-ray sources. The average X-ray intensity varies over months by as much as a factor of 7 (cf. Priedhorsky and Terrell 1986). The spectrum cuts off rapidly below ~ 2 keV, indicating an absorbing column density of 10^{22}–10^{23} cm^{-2}. Brinkman *et al.* (1972) found the X-ray intensity to be periodic at 4.8 h. The exact shape of the light curve varies from cycle to cycle (van der Klis and Bonnet-Bidaud 1982), but the average light curve (averaging 5 or more cycles) has remained unchanged over 12 years of observations. Figure 1 shows the normalized light curve of 30 hours of *Einstein* data we obtained in June, 1984 and of 7 days of *COS-B* data obtained by van der Klis and Bonnet-Bidaud (1981) in May, 1980. It can be described as quasisinusoidal, rising more slowly than it falls. The ephemeris is well-described with a quadratic fit. (The most recent value for the period derivative is $+7.8\times10^{-10}$, K. Mason private communication.) While the shape of the light curve is not typical of an eclipsing binary, the stability of the period suggests the X-ray period is a binary orbital period. Assuming further that the system consists of a compact object and a Roche-lobe filling companion, the period constrains the mass of the companion (cf. Faulkner, Flannery, and Warner 1972). For a hydrogen companion, the ZAMS mass-radius relationship constrains the mass to be 0.5 $M_\odot$. As is likely in this system, a high rate of mass loss from the companion may drive it out of thermal equilibrium, in which case 0.5 $M_\odot$ is a good upper limit on its mass. For a helium companion, the equivalent assumption yields a companion mass of 4 $M_\odot$ (van den Heuvel and De Loore 1973).

Several classes of models have been suggested to account for the dependence of the X-ray light curve on phase and on energy. In the stellar wind model, the X-rays are scattered by a wind centered on the companion of the X-ray source and modulated by the varying optical depth to scattering along the line of sight (Davidsen and Ostriker 1974; Hertz, Joss, and Rappaport 1978; Becker *et al.* 1978; Ghosh *et al.* 1981; Willingale, King, and Pounds 1985). In the cocoon model, the X-rays are scattered by a large shell surrounding a binary and modulated by the changing aspect of the shadow of the companion of the X-ray source on the surrounding shell (Milgrom 1976; Milgrom and Pines 1978; Hertz *et al.* 1978; Ghosh *et al.* 1981; and Bonnet-Bidaud and van der Klis 1981). In the accretion disk corona (ADC) model, the X-rays are scattered by an ADC and modulated by structure on the edge of the accretion disk (White and Holt 1982). The applicability of this model to Cyg X-3 rests largely on the similarity of the X-ray light curve and spectrum to those of the X-ray binaries 4U2129+470 and 4U1822–371, for which the ADC model has also been suggested. The model is more fully developed for these sources as they also display an eclipse by the secondary and optical spectral lines.

Serlemitsos *et al.* (1975) discovered strong iron line emission from Cyg X-3 at ~ 6.7 keV. Current studies dispute the line characteristics. In particular, Kitamoto (1985) finds the flux density in the line to be independent of orbital phase, while van der Klis (1985) finds it to show a dependence similar to that of the continuum emission.

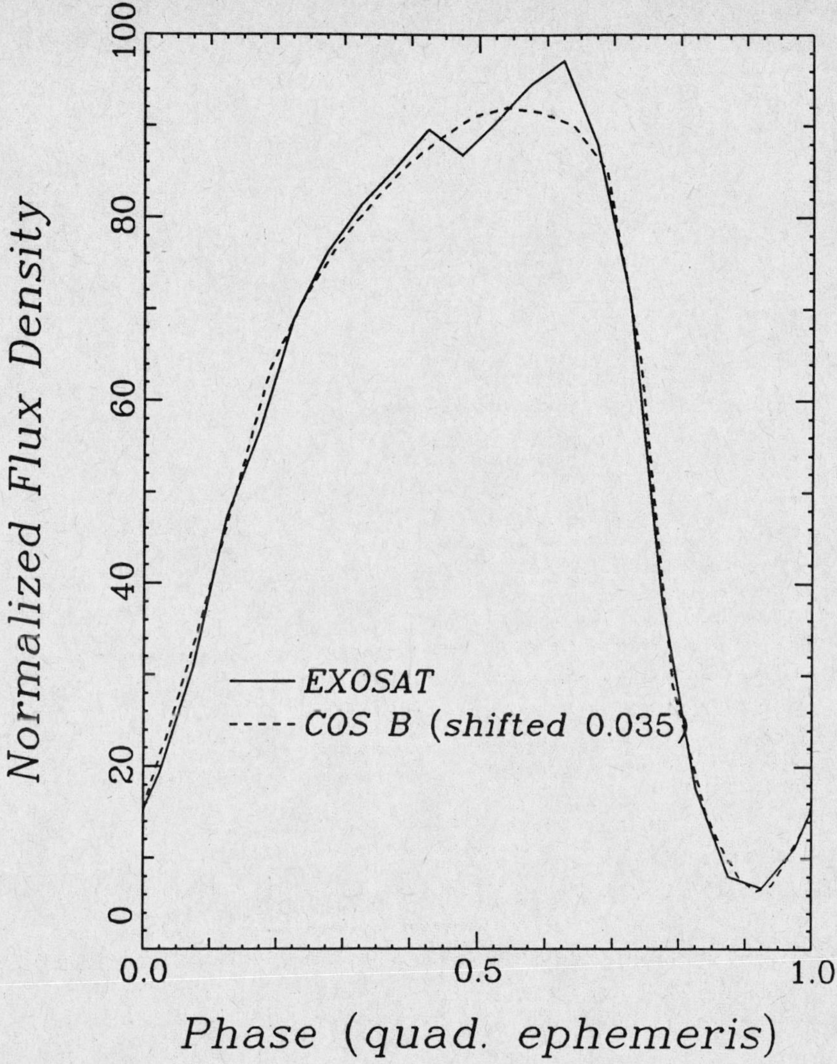

Fig. 1—The solid curve is the 1–15 keV *EXOSAT* data folded on the quadratic ephemeris of van der Klis and Bonnet-Bidaud (1981). The dashed curve is the *COS-B* data of van der Klis and Bonnet-Bidaud on the same ephemeris shifted back 0.035 cycles. The scales and zero offsets are normalized for comparison.

Several other periods have been suggested. Molteni *et al.* (1980) suggested a 34.1 d period in the X-ray intensity of Cyg X-3, but further observations failed to confirm this (Priedhorsky and Terrell 1986). Bonnet-Bidaud and van der Klis (1981) suggested an 18.7 d modulation of the timing of X-ray minima. No additional observations have been made with the sensitivity and time coverage needed to confirm or disprove this period. Finally, Priedhorsky and Terrell (1986) find nonperiodic fluctuations with a timescale of $\sim$ 100 d.

A hard X-ray tail has been reported by several groups, with energies up to 250 keV (cf. Meegan *et al.* 1979). These data are reasonably fit with power laws, although power law indices range from 2.2 to 3.6, probably indicating time variability (Meegan *et al.* 1979; Reppin *et al.* 1979).

B. Radio.

Braes and Miley (1972) discovered radio emission from Cyg X-3 in 1971 during a radio survey of X-ray sources. They found the 20 cm wavelength flux density to vary from 20 mJy to 200 mJy on time scales of months, already making Cyg X-3 one of the brightest and most variable radio stars. In 1972, Gregory (1972) discovered a giant radio flare from Cyg X-3, with flux densities reaching 20 Jy. In the extensive observations following this discovery, it was found that the spectrum of the giant flare evolved with time as would be expected for a relativistically expanding synchrotron source (cf. van der Laan 1966). A VLBI nondetection at the time implied that the source had indeed become extended (Hinteregger *et al.* 1972). Absorption measurements at 21 cm wavelength (Dickey 1983) imply a lower limit on the distance to Cyg X-3 of 8 kpc for an assumed Galactic center distance of 7 kpc (Molnar and Mauche 1986). Subsequent monitoring (Johnston *et al.* 1986) has shown that giant flares typically occur in groups once or twice a year. Spatial mapping of giant flares has shown Cyg X-3 to become elongated in the north-south direction in the weeks following flare onset (Geldzahler *et al.* 1983; Spencer *et al.* 1986). Estimates of apparent transverse expansion velocities in these flares are conflicting. With limited angular resolution and the confusion of multiple flares, deriving velocities from these data is an uncertain business.

D. Florkowski (private communication) found upper limits to the 6 cm wavelength flux densities of the other ADC systems (4U1822–371 and 4U2129+470) of 0.1 mJy with the VLA on 20 February, 1986. This is a factor of 100 lower than the lowest flux densities measured for Cyg X-3. As these other systems are thought to be closer than Cyg X-3, it appears their radio properties must differ fundamentally from those of Cyg X-3.

C. Optical and Infrared.

Westphal (1972) placed a limit of 23.9 mag on the visual brightness of Cyg X-3. Becklin *et al.* (1972) discovered an infrared counterpart with a flux density of 18 mJy at 2.2 μm and 9 mJy at 1.6 μm. Subsequent observations have shown that the infrared source generally has the same 4.8 h light curve as the X-ray source (Becklin *et al.* 1973; Mason *et al.* 1976; Mason *et al.*

1986). The infrared source also shows occasional flares lasting for only a few minutes and often coming in groups. No correlated X-ray flares occur, however (Mason *et al.* 1976; Mason *et al.* 1986).

Since the discovery of the infrared emission it has been thought to be thermal. Figure 1 shows the normalized light curve of 30 hours of data we obtained with *EXOSAT* in June, 1984. Infrared (JHK) photometry we performed with the Multiple Mirror Telescope in June, 1984 is inconsistent with a nonthermal power law for any amount of interstellar reddening. These observations are consistent with reddened thermal bremsstrahlung spectra with optical extinction between 20 and 27 mag (assuming extinction $A_\lambda \propto \lambda^{-1.8}$, Willner 1984). This range is roughly consistent with interstellar absorption at this distance through Cygnus and with the high X-ray absorption column.

D. Gamma Rays and Secondary Muons.

Many detections and equally sensitive nondetections of gamma rays from Cyg X-3 with photon energies from 10^7–10^{16} eV have been reported (cf. Weekes 1986 for a review). These reports claim a variety of source characteristics, including aperiodic signals and periodic signals with periods of 12.6 ms, 4.8 h, and 34 d and duty cycles that vary from observation to observation. The differences between the various results at each energy are commonly attributed to source variability with time. As no two reports have ever claimed the same positive result for the same time at any energy, the validity of any of the gamma ray detections remains to be established.

Two groups have recently reported detections of muons (in proton decay experiments) from the direction and with the orbital period of Cyg X-3 (Soudan group: Marshak *et al.* 1985a and 1985b; NUSEX group: Battistoni *et al.* 1985). These authors note that if the detections are real, the progenitor particles can be either photons, but only if some previously unknown interactions produce muons with high probability, or a new type of stable neutral particle emitted by Cyg X-3. While numerous possibilities for the new particle have been pursued, none is yet theoretically satisfactory (Berezinsky, Ellis, and Ioffe 1986). Also, subsequent analyses of data obtained with three other decay experiments yield nondetections with upper limits below the Soudan and NUSEX values (Kamioka: Oyama *et al.* 1986; Fréjus: Berger *et al.* 1986; Harvard-Wisconsin Worstell 1986). The resolution of the conflicting data probably lies in the statistical methods used by the Soudan and NUSEX groups. Molnar (1986) shows that the statistics quoted by these groups reflect neither the number of independent searches made nor the nonrandom character of the background.

E. Coordinated Observing Program.

Since 1983, the author, M. Reid, and J. Grindlay have, along with collaborators at various observatories, carried out a series of coordinated observations of Cyg X-3 (Molnar 1985). We have five epochs of observations (September, 1983; December, 1983; June, 1984; February, 1985; and October, 1985) each of which include some subset of radio, millimeter, infrared, X-ray, and gamma ray wavelength measurements. The purpose is to utilize a new generation of telescopes to better define the observational characteristics of Cyg X-3, and especially to identify characteristics that are correlated among the wavelength regimes. The hope is to use this information to understand the various emission mechanisms and to develop a single self-consistent model of the whole source. The remainder of this paper reports on the first new results of the program. Section II describes new radio results on the low-level radio emission. Section III describes new X-ray results on the light curve.

II. Observations of Low-level Radio Emission.

The radio observations make up the most extensive portion of the coordinated program data set, totaling 127 h of coverage over the five epochs. Already after the first two epochs observation (23 h of data) it was apparent that the radio emission of Cyg X-3 could be characterized at all times as entirely a series of flares each of which has the spectral evolution expected for a relativistically expanding synchrotron source (Molnar et $al.$ 1984). For each flare the flux density peaks sooner and at a greater value at higher frequencies (e.g., data in Figure 2.) The spectral index (α, where $S_\nu \propto \nu^\alpha$) goes from a positive value to a negative value. The overlapping of successive flares will diminish the range of variation of the spectral index, but the data that are least affected by this overlap consistently indicate that the optically thin spectral index goes to a minimum value of –0.5. Figure 3 shows examples of the two extremes of the spectrum. Typically, flare amplitudes vary by less than a factor of $\sim$ 2 within each epoch, although the average amplitude varies from epoch to epoch ranging from 10s of milliJanskys to a Jansky in our data. Hence the "giant flares" discussed in the introduction seem to be just the high amplitude end of a continuum of flare sizes.

The preceding interpretation of the low-level radio emission of Cyg X-3 predicts that high resolution observations will show significant changes in the source size over the lifetime of a single flare. Therefore, we made VLBI measurements during our fourth and fifth epochs to confirm our interpretation. We present here the results of the fourth epoch, the details of which will be published elsewhere. The fifth epoch has not yet been analyzed. For this experiment we chose to observe at 1.3 cm wavelength to minimize the scattering due to the interstellar medium and to minimize the overlapping of successive flares (thereby permitting a simple model fit to the data). We took data on 5 and 8 February, 1985, with a five element array consisting of Effelsberg, Haystack, Green Bank, the phased VLA, and Owens Valley. We selected the VLBI data for one flare on each of those days for which the overlapping of previous flares could be ignored to first order. We determined the starting times of the flares from the light curve. We fit all of the VLBI data to a single three-parameter model: an elliptical Gaussian source with a fixed intrinsic axial ratio and a constant rate of expansion since flare beginning, convolved with a fixed scattering size. The smallest value of χ_ν^2, 0.51, was obtained with an expansion velocity (of the major axis FWHM) of 0.41 mas/h, an intrinsic axial ratio of 2.2, and a FWHM scattering size of 0.65 mas. In Figure 4 we plot this point and 1- and 3-σ contours of the statistical uncertainty on the axial

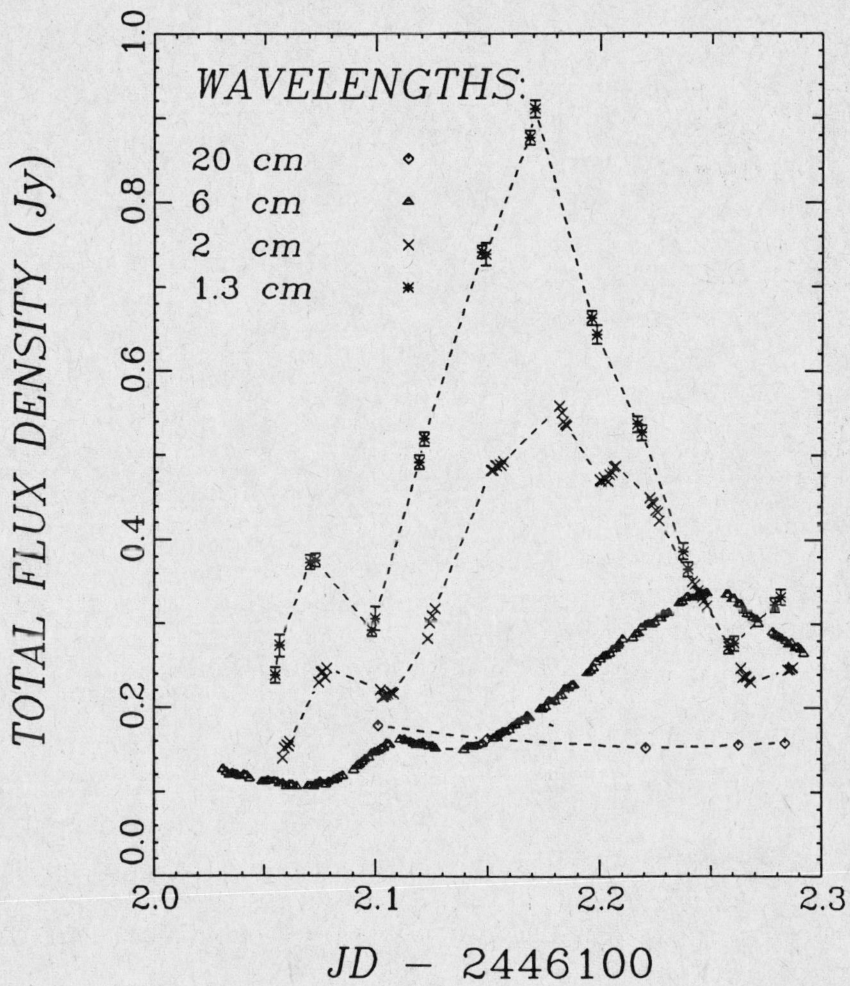

Fig. 2—Total flux density versus time on February 5, 1985 measured at the VLA.

Cyg X–3 Spectra

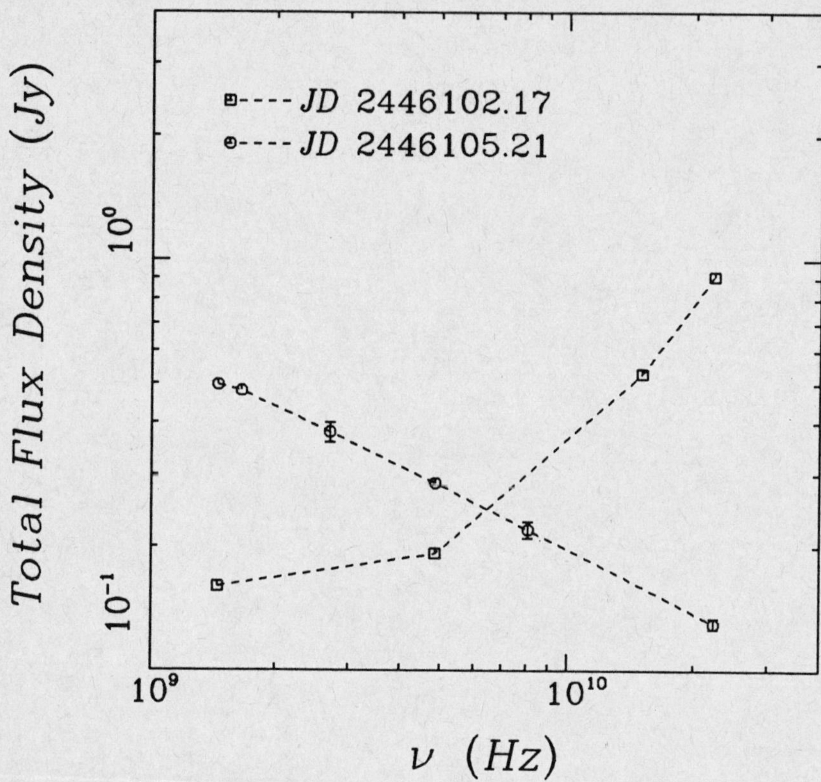

Fig. 3—Total flux density versus frequency at JD 2446102.17 and at JD 2446105.21. The 8.1 and 2.7 GHz values are from the 3-element interferometer at Green Bank, West Virginia (K. Johnston, private communication); all other values are from the VLA.

ratio-expansion velocity plane. By 1-σ (3-σ) contour, we mean the locus of grid points with χ^2_ν (i.e., the χ^2_ν obtained by optimizing the scattering size) such that the fractional probability is 0.32 (.0027) of obtaining a higher value. Dashed lines indicate the optimal value of the scattering size. We can conclude that our simple model does provide a good fit, and that no simple symmetric model (axial ratio equal to 1) or simple static model (expansion velocity equal to 0) provides a good fit. In other words, the source is elongated and expands with time. If we assume that the expansion rate corresponds to the transverse velocity of a double-sided jet at 8 kpc, that velocity is 0.23c. The more extensive observations of our fifth epoch will be able to confirm this result. As the scattering size is still large even at 1.3 cm wavelength, only the shortest VLBI baselines show detections. Hence no closure phases could be obtained with our data. Significant advances in detailing source structure will require a greater density of short spacings such as will be afforded by the first two VLBA stations (Pietown and Los Alamos) when combined with Owens Valley, the VLA, and Goldstone.

After the first epoch of observations, Molnar *et al.* (1983) suggested that the timing of the flares may be periodic. Based on the first two epochs of observations, Molnar *et al.* (1984) suggested that the best value for the period is slightly longer than the orbital period. Using the longer time baseline of the third epoch, Molnar *et al.* (1985) found a best value of 4.95 h, 3.3% longer than the orbital period, although one of the eight cycles observed in that epoch deviated significantly from the period. Figure 5 shows 2 cm wavelength VLA data of 14 and 16 June, 1984, folded on both the suggested 4.95 h period and the 4.79 h X-ray period. While the shape of each individual light curve seems somewhat more complicated than a single rise and fall, it repeats in phase for each cycle up to a DC offset for the 4.95 h folding. By contrast is does not phase up for the 4.79 h folding. The flare timing in the fourth and fifth epochs were again consistent with the 4.95 h period.

Assuming the validity of the radio period and its distinctness from the X-ray period, to what physical mechanism might the period be attributed? Two other radio sources have well-established periodic radio flaring: Cir X-1 (Nicolson *et al.* 1980) and LS I+61°303 (Taylor and Gregory 1982). The radio periods matches the X-ray periods in these more widely separated binary systems. It has been suggested that these systems are highly eccentric and that the radio flares are triggered at periastron passage. We adapt this suggestion to Cyg X-3 by hypothesizing that the radio period is the result of periodic modulation of the binary separation. We present two models for this period that are consistent with the existing data, while making differing predictions about what more extensive data should show.

Modulation at the radio period may be caused by precession of the line of apsides in an eccentric orbit. The precession period required to yield the observed radio period when beat with the orbital period is 6 d. Papaloizou and Pringle (1979) give an approximate expression for the precession due to the primary acting on the quadropole moment of a hydrogen secondary that agrees to within a factor of $\sim$ 2, which is within the limits of the uncertainties of the derivation. A more exact expression requires a detailed knowledge of the structure of the secondary star. Elsner *et al.* (1980) and Ghosh *et al.* (1981) show that the precession period for a helium secondary in

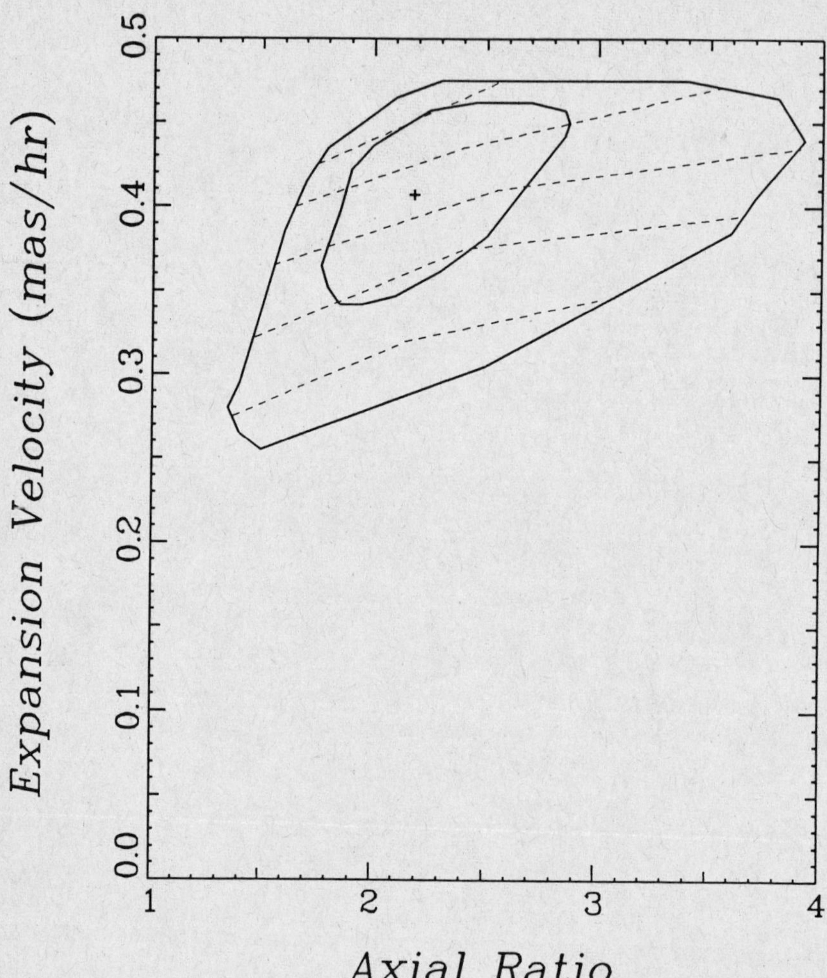

Fig. 4—Expansion velocity versus axial ratio of the expanding elliptical Gaussian model. A cross marks the best fit values. Solid curves denote 1- and 3-σ error contours. Dashed lines from top to bottom denote the loci of points for which 0.5, 0.6, 0.7, 0.8, and 0.9 mas, respectively, are the optimal FWHM scattering sizes.

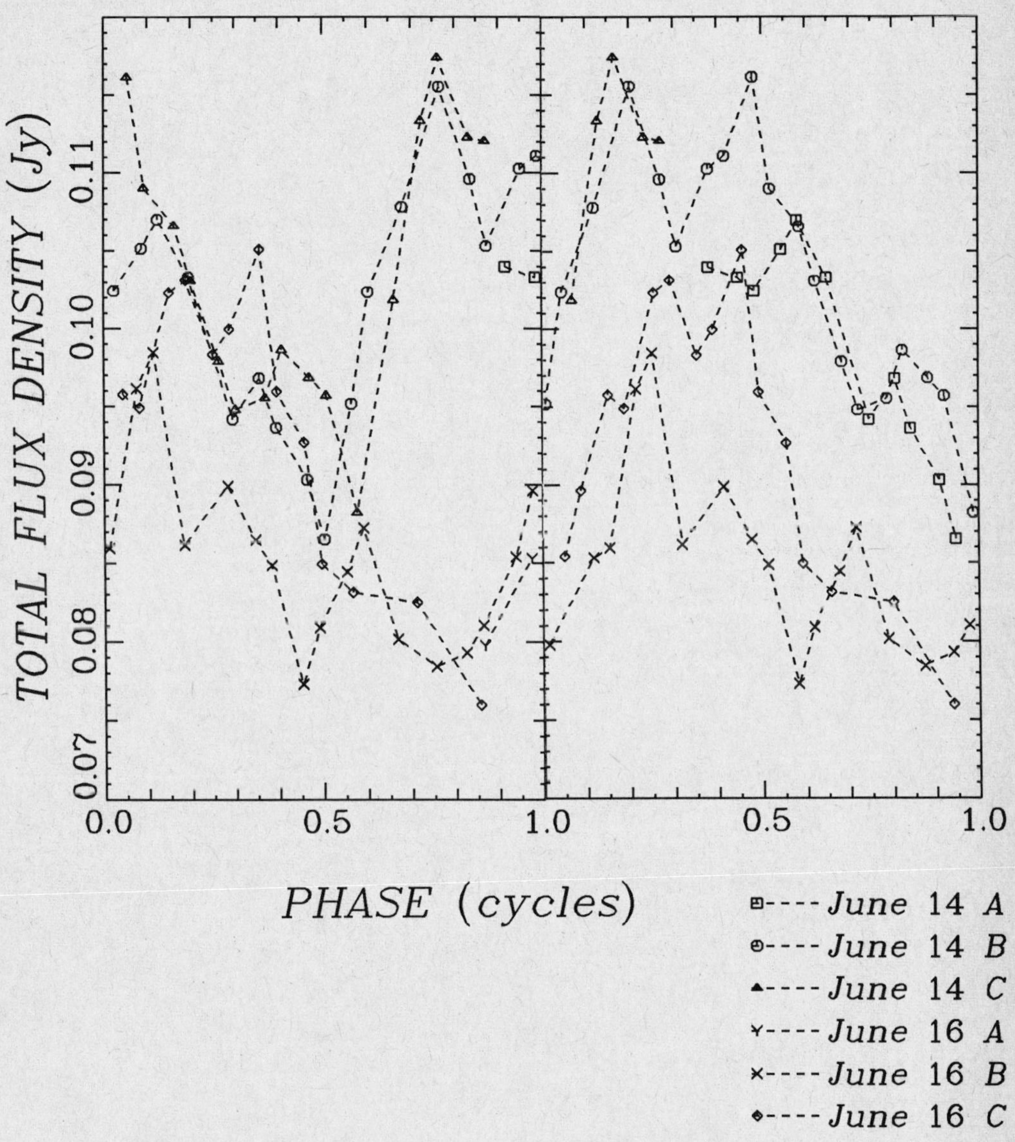

Fig. 5—Flux density at 2 cm wavelength plotted as a function of phase with a 4.95 h period and with a 4.79 hr (with arbitrary zero points) for the data of June 14–16. Successive cycles are distinguished by differing symbols given in the key (dates are followed by a letter denoting wrap—A first, B second, C third).

Cyg X-3 would be many years, and hence is not a possible explanation for the radio period.

We have undertaken a numerical study of the effects of a third body on the orbital separation of a close binary. In brief, we have found that for a third body in a prograde orbit (orbital inclination to the plane of the binary orbit less than $\sim 50°$) with a small eccentricity (less than $\sim .01$) the modulation of the binary separation may mimic the radio period much of the time, but would also show well-defined phase jumps twice per orbit of the third star. The period of the third body (in this case ~ 4 d) is determined by the deviation of the apparent radio period from the orbital period. The relative timing of the two phase jumps is determined by the mass ratio of the inner binary. The inner binary separation as a function of time over one orbit of the third body is shown in figure 6. The time between successive minimum separations as a function of time for the same model is shown in figure 7.

Although three body models of other close binary systems have been suggested, none has yet made a prediction that was sufficiently distinctive to demonstrate the validity of the model. Furthermore, the evolutionary path of close binaries is only poorly understood, and no plausible evolutionary scenario has been suggested that would result in a close triple. The importance of the three body model suggested here is that it relies on an aspect of three body systems that has been previously overlooked and that it makes a distinctive prediction about flare timing for observations longer than about 4 days.

III. The X-ray Light Curve.

As we noted in the introduction, the X-ray light curve of Cyg X-3 is not characteristic of an eclipse of the compact object by the secondary, although we assume the period stability indicates that the X-ray period is the binary orbital period. We noted the classes of models that have been suggested to account for the shape of the light curve. In this section we consider recent data on the dependence of the light curve on photon energy and the implications for the three classes of models.

Early studies with non-imaging detectors of the energy dependence of the light curve found it to be largely energy-independent out to 10 keV (Becker *et al.* 1978, Blissett, Mason, and Culhane 1981). *EXOSAT* data showed, however, that the depth of modulation decreases steadily at energies above 10 keV, until it is undetectable at 50 keV (Willingale, King, and Pounds 1985). Mauche and Gorenstein (1986) reported that $\sim 30\%$ of the X-rays in the *Einstein* bandpass ($\sim$0.5–3.5 keV) are in an extended halo due to the scattering of X-rays by interstellar grains. Molnar and Mauche (1986) noted that the delay in arrival time of the scattered photons with respect to the unscattered ones will diminish the apparent depth of modulation of the light curve in an energy-dependent fashion and derived an expression to correct for the effect. The corrected depth of modulation decreases as the log of the energy from nearly 100% modulation at ~ 1 keV to near zero at 50 keV, except for a dip at the iron line. (They also noted that the scattering will significantly affect spectral fits made separately at different parts of the light curve.)

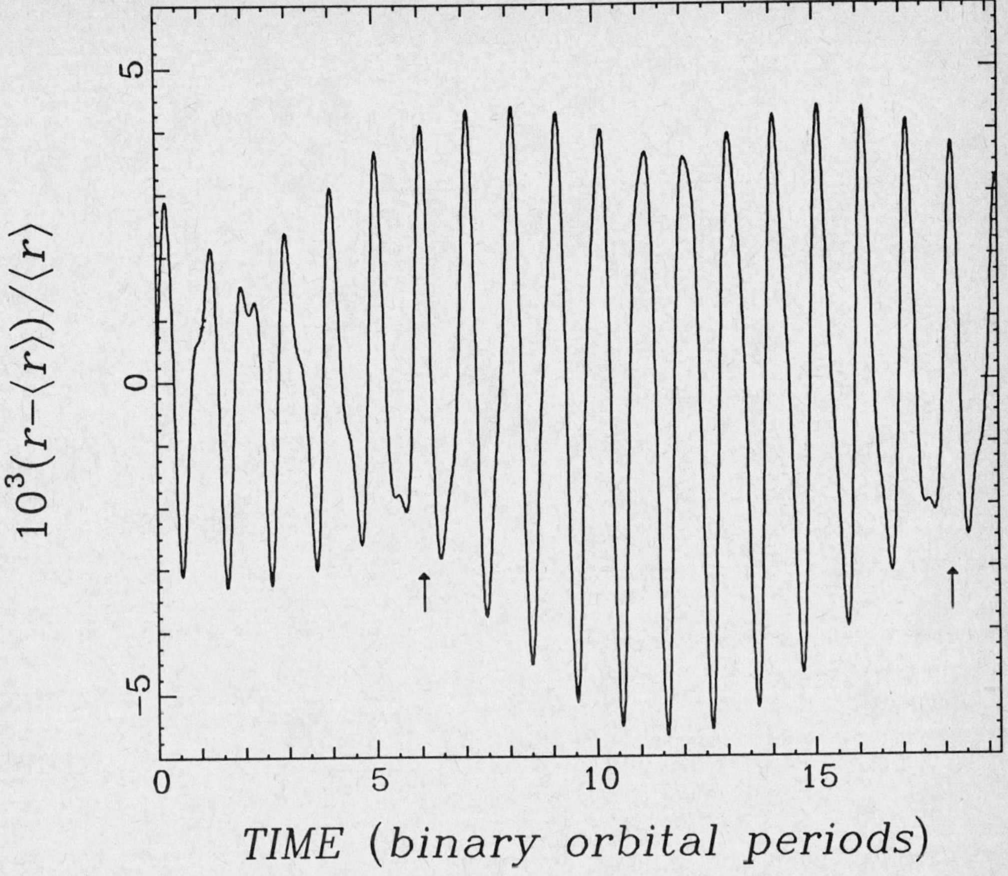

Fig. 6—Deviation of binary separation from the mean separation in units of thousandths of a mean separation versus time in units of the binary orbital period over one orbital period of the third star for parameters appropriate to Cyg X-3. Arrows indicate the two short cycles.

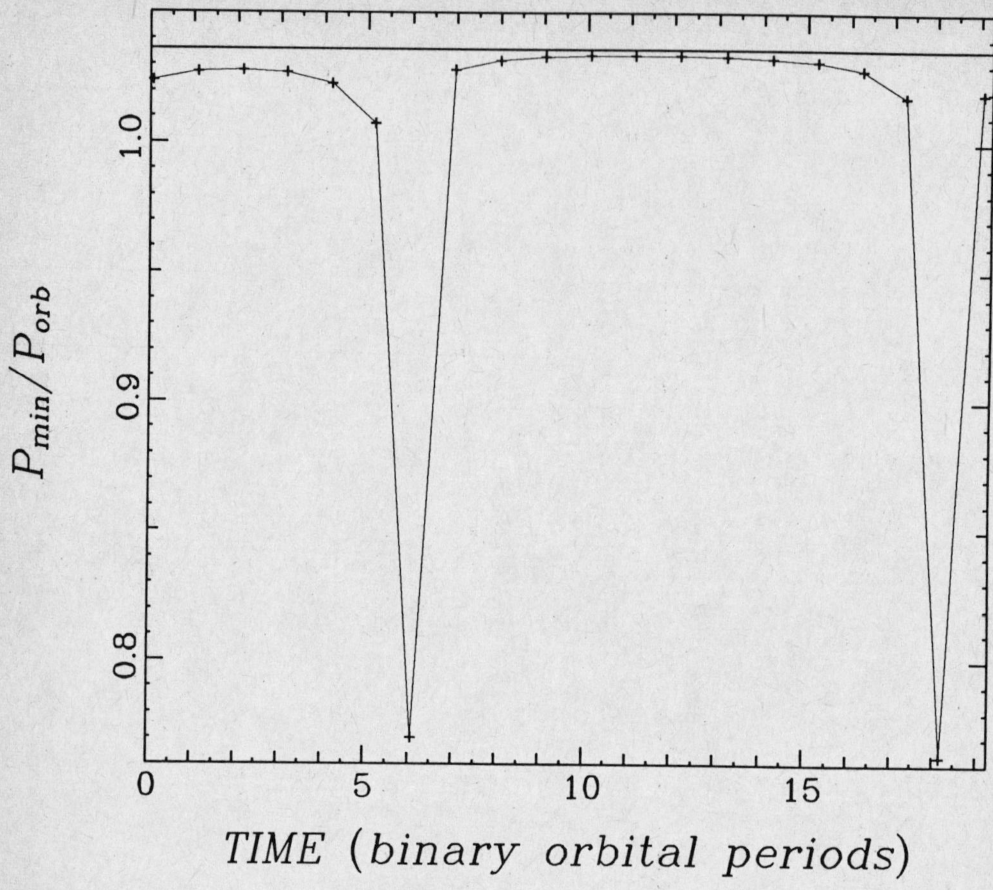

Fig. 7—P_{min}/P_{orb} values for the model orbit in Figure 6. Each datum represents the difference in time of two minima; times are taken to be midway between the times of the minima.

The stellar wind and cocoon models depend on Compton scattering to modulate the flux density with orbital phase. As this process is energy-independent for the energy range of interest, these models fail to reproduce the energy-dependent depth of modulation that is observed. In the ADC model, on the other hand, the flux density is modulated by varying amounts of absorption, an energy-dependent process. The details of this dependence were not considered in the original model, so we briefly develop them here. The absorption cross section depends on the −2.5 power of the photon energy (Cruddace *et al.* 1974). If the absorption could be characterized by a single column density which is modulated over the orbit, the depth of modulation would vary by 90% over only a factor of ∼ 5 in energy, a dependence to strong to match the data. However, the density of the absorbing material should be stratified according to hydrostatic equilibrium. This stratification weakens the dependence, permitting a good fit to the data (Figure 8). The specific parameters required depend on the unknown inclination of the system to the line of sight. Hence we conclude that the ADC model for the light curve is the only model yet suggested that may be able to reproduce the data. However, the model needs more development to confirm that the necessary parameters are self-consistent with the size corona that would be expected.

IV. Conclusions.

The unusually high radio, infrared, and X-ray luminosities of Cyg X-3 make it a good source for in depth observational study. Early results of our observational study show that Cyg X-3 is an important source for theoretical study of several types of astrophysical phenomena. Our study of the depth of modulation of the X-ray light curve with energy has provided a new kind of observational support for the accretion disk corona model. This extra dimension of information along with the infrared and X-ray spectral information makes Cyg X-3 a system for which a more complete and self-consistent ADC model of this source can be severely tested. Our study of the low-level radio emission shows that the radio emission can be characterized at all times as a succession of flares, each of which are expanding sources of synchrotron radiation. The timing of these flares suggests that they are triggered by small variations in the binary separation. Hence in Cyg X-3 we understand (far more than usual) the environment and the circumstances under which a radio jet forms, making it uniquely suited for detailed modelling of radio jet formation. Finally, if further observations bear out the three body model of Cyg X-3, we will need to develop evolutionary scenarios that result in close triples.

The VLA and Green Bank are facilities of the National Radio Astronomy Observatory, which is operated by Associated Universities, Inc., under contract with the National Science foundation.

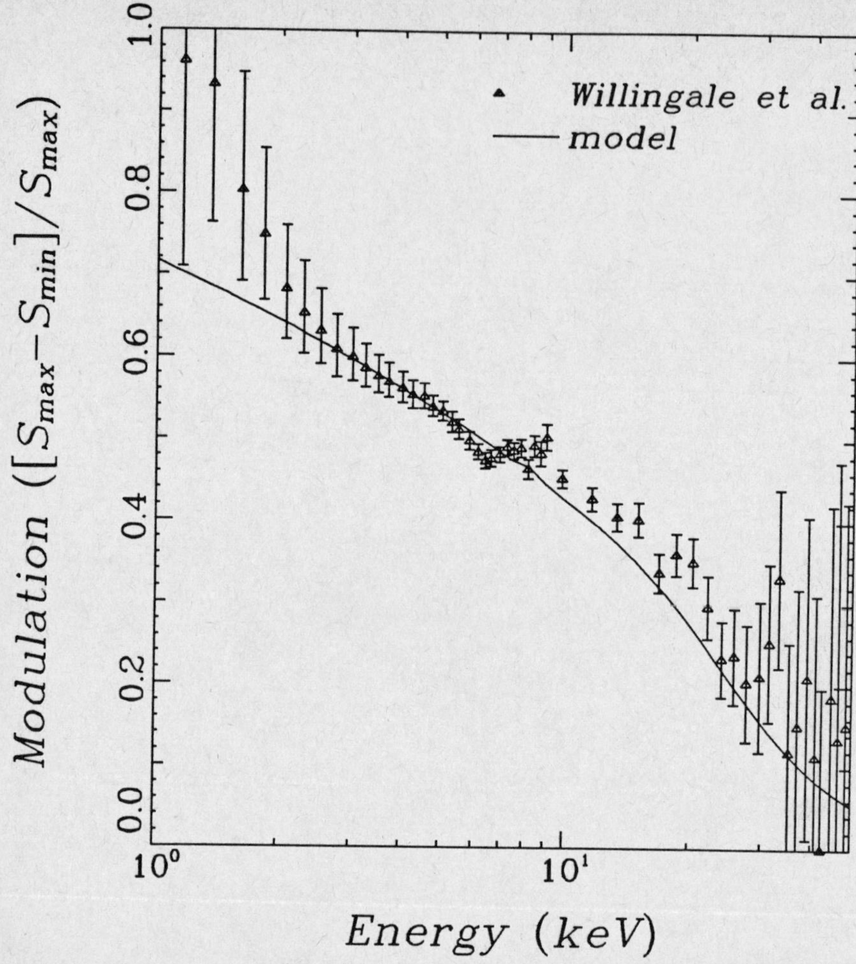

Fig. 8—Depth of modulation at X-ray minimum as a function of photon energy. The triangles are the data of Willingale, King, and Pounds (1985) corrected for the presence of X-ray scattering. The model curve is for an accretion disk corona as described by White and Holt (1982), but with a hydrostatic equilibrium density stratification for the accretion disk bulge.

REFERENCES

Battistoni, G. *et al.* 1985, *Phys. Lett.*, **155B**, 465.

Becker, R. H., Robinson-Saba, J. L., Boldt, E. A., Holt, S. S., Pravdo, S. H., Serlemitsos, P. J., and Swank, J. H. 1978, *Ap. J. (Lett.)*, **224**, L113.

Becklin, E. E., Kristian, J., Neugebauer, G., Wynn-Williams, C. G. 1972, *Nature Phys. Sci.*, **239**, 130.

Becklin, E. E., Neugebauer, G., Hawkins, F. J., Mason, K. O., Sanford, P. W., Matthews, K., and Wynn-Williams, C. G. 1973, *Nature*, **245**, 302.

Berezinsky, V. S., Ellis, J., and Ioffe, B. L. 1986, *Phys. Lett.*, **172B**, 423.

Berger, Ch. *et al.* 1986, submitted to *Phys. Lett. B*.

Blissett, R. J., Mason, K. O., and Culhane, J. L. 1981, *M. N. R. A. S.*, **194**, 77.

Bonnet-Bidaud, J. M., and van der Klis, M. 1981, *A. A.*, **101**, 299.

Braes, L. L. E., and Miley, G. K. 1972, *Nature*, **237**, 506.

Brinkman, A., Parsignault, D., Giacconi, R., Gursky, H., Kellogg, E., Schreier, E., and Tananbaum, H. 1972, *IAU Circ.*, No. 2446.

Cruddace, R., Paresce, F., Bowyer, S., and Lampton, M. 1974, *Ap. J.*, **187**, 497.

Davidsen, A., and Ostriker, J. P. 1974, *Ap. J.*, **189**, 331.

Dickcy, J. M. 1983, *Ap. J. (Lett.)*, **273**, L71.

Elsner, R. F., Ghosh, P., Darbro, W., Weisskopf, M. C., Sutherland, P. G., and Grindlay, J. E. 1980, *Ap. J.*, **239**, 335.

Faulkner, J., Flannery, B.P., and Warner, B. 1972, *Ap. J. (Lett.)*, **175**, L79.

Geldzahler, B. J., Johnston, K. J., Spencer, J. H., Klepczynski, W. J., Josties, F. J., Angerhofer, P. E., Florkowski, D. R., McCarthy, D. D., Matsakis, D. N., and Hjellming, R. M. 1983, *Ap. J. (Lett.)*, **273**, L65.

Ghosh, P., Elsner, R. F., Weisskopf, M. C., and Sutherland, P. G. 1981, *Ap. J.*, **251**, 230.

Giacconi, R., Gorenstein, P., Gursky, H., and Waters, J. R. 1967, *Ap. J. (Lett.)*, **148**, L119.

Gregory, P. C. 1972, *Nature*, **239**, 439.

Hertz, P., Joss, P. C., and Rappaport, S. 1978, *Ap. J.*, **224**, 614.

Hinteregger, H. F. *et al.* 1972, *Nature Phys. Sci.*, **240**, 159.

Johnston, K. J., *et al.* 1986, *Ap. J.*, in press.

Kitamoto, S. 1985, in *Japan–U.S. Seminar on Galactic and Extragalactic Compact X-ray Sources*, ed. Y. Tanaka and W. H. G. Lewin (Tokyo: Institute of Space and Astronautical Science), p. 201.

Marshak, M. L. *et al.* 1985a, *Phys. Rev. Lett.*, **54**, 2079.

Marshak, M. L. *et al.* 1985b, *Phys. Rev. Lett.*, **55**, 1965.

Mason, K. O., *et al.* 1976, *Ap. J.*, **207**, 78.

Mason, K. O., Cordova, F. A., and White, N. E. 1986, *Ap. J.*, in press.

Mauche, C. W., and Gorenstein, P. 1986, *Ap. J.*, **302**, 371.

Meegan, C. A., Fishman, G. J., and Haymes, R. C. 1979, *Ap. J. (Lett.)*, **234**, L123.

Milgrom, M. 1976, *A. A.*, **51**, 215.

Milgrom, M., and Pines, D. 1978, *Ap. J.*, **220**, 272.

Molnar, L. A. 1985, Ph.D. thesis, Harvard University.

Molnar, L. A. 1986, submitted to *Physical Review Letters*.

Molnar, L. A., and Mauche, C. W. 1986, *Ap. J.*, in press.

Molnar, L. A., Reid, M. J., and Grindlay, J. E. 1983, *IAU Circ.*, No. 3885.

Molnar, L. A., Reid, M. J., and Grindlay, J. E. 1984, *Nature*, **310**, 662.

Molnar, L. A., Reid, M. J., and Grindlay, J. E. 1985, *Radio Stars*, (R. M. Hjellming and D. M. Gibson, eds.), Astrophysics and Space Science Series, Reidel, Dordrecht, p. 329.

Molteni, D., Rapisarda, M., Robba, N. R., and Scarsi, L. 1980, *A. A.*, **87**, 88.

Nicolson, G. D., Feast, M. W., and Glass, I. S. 1980, *M. N. R. A. S.*, **191**, 293.

Oyama, Y. *et al.* 1986, *Phys. Rev. Lett.*, **56**, 991.

Papaloizou, J., and Pringle, J. E. 1979 *M. N. R. A. S.*, **189**, 293.

Priedhorsky, W., and Terrell, J. 1986, *Ap. J.*, **301**, 886.

Reppin, C., Pietsch, W., Trümper, J., Voges, W., Kendziorra, E., and Staubert, R. 1979, *Ap. J.*, **234,**, 329.

Serlemitsos, P. J., Boldt, E. A., Holt, S. S., Rothschild, R. E., and Saba, J. L. R. 1975, *Ap. J. (Lett.)*, **201**, L9.

Spencer, R. E., Swinney, R. W., Johnston, K. J., and Hjellming, R. M. 1986, *Ap. J.*, in press.

Taylor, A. R., and Gregory, P. C. 1982, *Ap. J.*, **255**, 210.

van den Heuvel, E. P. J., and De Loore, C. 1973, *A. A.*, **25**, 387.

van der Klis, M. 1985, in *Japan–U.S. Seminar on Galactic and Extragalactic Compact X-ray Sources*, ed. Y. Tanaka and W. H. G. Lewin (Tokyo: Institute of Space and Astronautical Science), p. 195.

van der Klis, M., and Bonnet-Bidaud, J. M. 1981, *A. A.*, **95**, L5.

van der Klis, M., and Bonnet-Bidaud, J. M. 1982, *A. A. (Suppl.)*, **50**, 129.

van der Laan, H. 1966, *Nature*, **211**, 1131.

Weekes, T. C. 1986, The Proceedings of the VIth Astrophysics Meeting (on Accretion Processes in Astrophysics) of the XXIth Rencontre de Moriond, held at Les Arcs, Savoie, France.

Westphal, J. A., Kristian, J., Huchra, J. P., Schectman, S. A., Brucato, R. J. 1972, *Nature Phys. Sci.*, **239**, 134.

White, N. E., and Holt, S. S. 1982, *Ap. J.*, **257**, 318.

Willingale, R., King, A. R., and Pounds, K. A. 1985, *M. N. R. A. S.*, **215**, 295.

Willner S. P. 1984, *Galactic and Extragalactic Infrared Spectroscopy*, M. F. Kessler and J. P. Phillips (eds.), D. Reidel, p. 37.

Worstell, W. A. 1986, Ph.D. thesis, Harvard University.

MORE SURPRISES FROM NGC 4151

M.V. Penston

Royal Greenwich Observatory

Hailsham, East Sussex

BN27 1RP/UK

1. Pre-1981 ultraviolet behaviour or NGC 4151

Because the Type I Seyfert galaxy, NGC 4151, seems to have behaved differently before and after 1981, it is useful first to recall the nature of the variations prior to 1981. The ultraviolet variations from the launch of IUE in 1978 until that date are recorded and discussed in a series of papers from the "European Extragalactic Collaboration".

Firstly Perola _et al_. (1982) described the continuum variations. They found that the optical and long wavelength ultraviolet (2500Å) continua are excellently correlated and that the long and short wavelength (1455Å) ultraviolet continua were also generally correlated except at what Perola _et al_. termed "anomalous epochs". Perola _et al_. attempted a decomposition of the continuum into a long wavelength power law and a short wavelength excess but this seems to be mistaken since the "power law" does not continue to rise into the optical region. It seems more likely that the long wavelength region contains a substantial pseudo-continuum component from blended FeII lines and Balmer continuum (Wills, Netzer and Wills 1985) and that the underlying non-thermal component dominates the flux in the short wavelength region. Nonetheless the character of the variability still supports the existence of anomalous epochs: see Fig.1 where the raw fluxes at 2500 and 1455Å are plotted against each other. While I am not sure the collaboration is unanimous on this, in my view this figure shows the anomalous epochs do not fit the correlation defined by the remaining points and thus they are not simply an artefact of the decomposition in Perola _et al_. Two components with different patterns of variation contribute to the flux at 1455Å. As we shall see the anomalous epochs also show other differences from the remaining data.

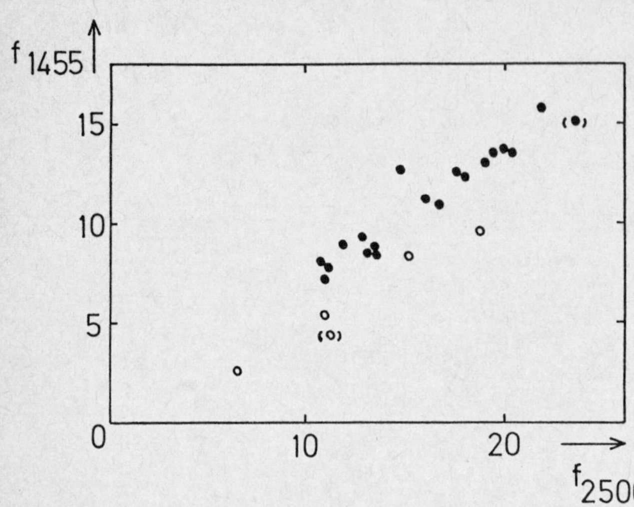

Fig.1. The raw fluxes of NGC 4151 at 1455 and 2500Å prior to 1981 plotted against each other. The open circles are the "anomalous" epochs defined in Perola _et al_. (1982). The fact that they do not fit the correlation defined by the other points implies the "anomalous" epochs are not an artefact of the continuum decomposition used in that paper.

The behaviour of the absorption lines in NGC 4151 prior to 1981 were discussed by Bromage _et al_. (1985). The equivalent width of NV absorption is positively correlated with the continuum while SiII is anticorrelated. The anticorrelation is markedly improved if the anomalous epochs are ignored. These correlations and anticorrelations are understood if the absorbing material is primarily in ionization states corresponding to C^{++} and Si^{++} and this is borne out by the strength of the excited line of CIII at 1176Å which arises from a meta-stable level. From high-dispersion spectra, the line-width is $\sim$ 1000 km s^{-1} and the column density may reach $\sim 10^{21}$ a.m.u. cm^{-2}, although naturally this figure is uncertain since the dominant ionization stages are not observed by their resonance lines (CIII 977Å is outside the IUE range and awaits the LYMAN satellite and SiIII 1206Å is blended with $L\alpha$). This value for the column, however, is an order of magnitude less than the X-ray column.

Observations of the emission lines in NGC 4151 led to one of the most exciting results on active galaxies in recent years (Ulrich _et al_ 1984a). The Broad Line Region (BLR) emission lines vary in a way which lags behind continuum changes. If this lag is attributed to travel time effects, it tells the distance, _r_, of the line emitting clouds from the continuum source (and not merely an upper limit as the time-scale of variation alone does - Terrell 1967). When this is combined with the line width, $\underline{v}$ in velocity units, this provides the combination $\underline{v^2 r/G}$, an estimate of the central mass if the line-emitting clouds move on Keplerian orbits. In NGC 4151 and other Seyferts, unlike in quasars (Baldwin and Netzer 1978), the emission lines do not all have the same width (except at "anomalous" epochs when they are all relatively narrow e.g. ± 4000 km s^{-1} FWHM) : In fact the broader lines all agreed on a "weight" for any central black-hole of 10^9 M$_\theta$ (when an initial numerical error was corrected! - see Ulrich et al. (1984b)).

2. Post - 1981 ultraviolet behaviour of NGC 4151

Since 1981 on the contrary, NGC 4151 has always appeared faint when viewed by IUE (although, in fact, there have been occasions when it has brightened up in this period). This has both prevented any checks or amplifications being made on the conclusions reached above but did permit the discovery of narrow, rapidly-variable "satellite" lines to CIV 1550Å (Ulrich et al. 1985). These lines lie at 1518 and 1594 Å in the rest frame of NGC 4151 and have no satisfactory identifications in line lists from laboratory or cosmic sources, including the Sun and RR Tel. Their outstanding peculiarity is that they are both narrow and rapidly-variable, being of order 1500 km s^{-1} wide and doubling in ~ 4 days.

Ulrich et al. identify them as moving CIV but it is highly embarassing that they are not seen astride other strong lines in the ultraviolet or the optical Hα and Hβ lines (Perez, private communication). The fact that they vary together, within lags of order 1 day, limits their separation perpendicular to the plane of the sky. The radial velocity difference, however, would produce separations of order 1 light day since they were first observed (Davidson and Hartig 1978). Thus they appear to represent a stationary flow rather than ballistically moving blobs.

It is tempting to make an analogy with the "jet" of SS433 and if this analogy is carried through one is led to an equal and opposite velocity of ejection of 28 500 km s^{-1} on an axis inclined at 75° to the line of sight.

3.   EXOSAT Observations

3.1 New LE component

EXOSAT observations of NGC 4151 were reported in the journals first by Pounds _et al_. (1986) and later with a slightly more detailed discussion by Perola _et al_ (1986). Warwick's paper earlier in this volume also presents some of the data.

One outstanding EXOSAT discovery is that of a new almost constant low-energy (LE) component, seen in NGC 4151 and not in other systems (perhaps because of the high N_H column). If this is interpreted as thermal, fits to the EXOSAT filter data give an emission measure of 2.10^{64} cm^{-3} and a temperature of 1.8 10^6 K. These figures are in fact remarkably similar to those deduced from the optical [FeX] 6374 Å luminosity (Penston _et al_. 1984) and suggest an identification of the LE and 6374 Å emitting regions.

One intriguing point from the identification is related to the probable variabilty of [FeX] 6374Å in NGC 4151 over periods of order 3 years discussed by Penston _et al_. If the LE source is limited to this size (1pc) and is spherical, then the column of emitting gas computed from the emission measure is 4.10^{22} a.m.u. cm^{-3}, very close to the medium-energy X-ray absorbing column.

The idea that the X-ray absorption arises in the FeX emission region is also consistent with two other observational results from TENMA described in the paper by Makishima in this volume. Firstly the energy of the fluorescent Fe line in the X-rays is consistent with an ionization state of iron between Fe^{3+} and Fe^{9+}. Secondly the excess strength of this line over that expected from the Fe K-edge absorption (Inoue 1985, Matsuoka _et al_. 1986) can well be understood if the region emitting the X-ray Fe line lies a light year from the nucleus. The medium energy X-ray light curve (Warwick - this volume) shows TENMA observed at a deep minimum. Six months before, the source was some three times brighter and given the lag in response expected from travel time arguments, this can account for the excess strength in the fluorescent line. Earlier data by Holt _et al_. (1980) were also taken at an X-ray minimum (Perola _et al_. 1982), albeit not such a deep one, but their relatively larger errors mean this data, which also shows the Fe line too strong, may be explained in the same way.

On the other hand problems remain with this enticing picture. Plainly the LE source cannot be optically thin if it is so small. In addition there is the correlation between optical FeVII and FeX line intensities among different Seyfert galaxies which suggests photo - rather than collisional ionization for Fe^{9+} (Penston et al. 1984). Finally the LE source may be seen to be extended on much larger scales by Einstein HRI observations. (Elvis et al. 1983).

3.2 X-ray/ultraviolet correlation

One of the important questions in quasar and active galaxy research concerns the relationship between the continua in different wavebands. Perola et al. (1986) reported an excellent correlation between the medium-energy X-rays observed by EXOSAT and simultaneous ultraviolet continuum fluxes from IUE. This correlation applies only to the 1983/4 data and earlier simultaneous observations (Perola et al. 1982) did not fit the same regression line.

Since the optical flux is well correlated with the ultraviolet, it is possible to include more epochs in a related diagramme plotting X-ray against optical flux using the optical light curve of Gill et al. (1984). This is displayed in Fig.2 where optical fluxes computed from the Gill et al. light curves are plotted against various X-ray continuum measurements in the literature. There is no single cor-relation but there may be two different types of behaviour present, divided either by date (before and after 1981) or by optical brightness.

It is tempting to see an analogy with the behaviour of low mass X-ray binaries (LMXRBs) described earlier in this volume by van der Klis and Priedhorsky. This analogy would divide Fig.2 into a "quiescent" part in which the optical/ultraviolet is faint, there is an ultraviolet/X-ray correlation, the BLR emission lines are relatively narrow ($\pm$ 4000 km s^{-1} FWHM) and SiII absorption is absent (i.e. the "anomalous" epochs)and an "active" part when NGC 4151 show its optical/ultraviolet source to be bright, no ultraviolet/X-ray correlation, broader emission lines and SiII absorption present.

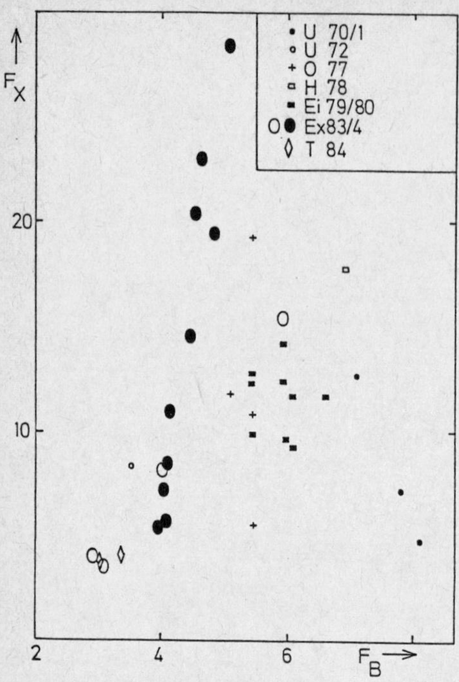

Fig.2.   X-ray flux between 2-10 kev in units of 10^{-11} ergs cm^{-2} s^{-1} plotted against blue flux in units of 10^{-14} ergs cm^{-2} s^{-1} Å^{-1}. This figure seems to have two parts - an "active" part with $\underline{F}_B \gtrsim 5.10^{-14}$ and a "quiescent" part at smaller optical fluxes.

If we assume that the optical/ultraviolet continuum comes from a disc and the X-ray emission from inside $\underline{r}=6\underline{m}$, one is led to the picture shown in Fig.3. This cannot be described as a "model", much less a "theory", but, as a "cartoon", it is more in the style of van de Klis' explanation for LMXRBs than that of Priedhorsky. X-ray emission from inside $\underline{r}=6\underline{m}$ still requires a detailed mechanism but the bulk of the potential (rest-mass?) energy is still to be extracted after material leaves the inner edge of the accretion disc (Lynden-Bell 1969) in accord with the fact that, in NGC 4151, the X-ray luminosity exceeds that in the optical/ultraviolet.

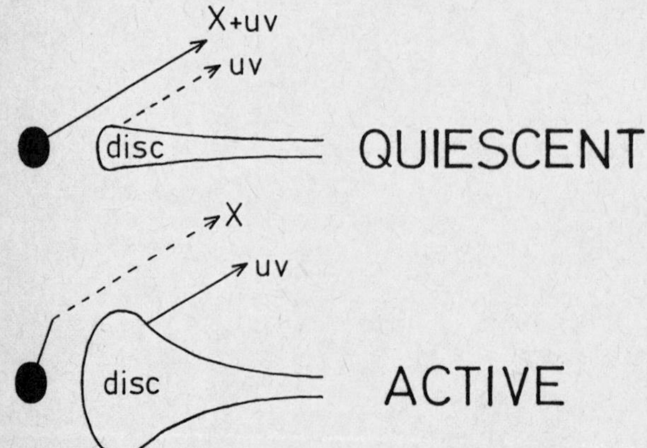

Fig.3. Cartoons of the NGC 4151 nucleus in "active" and "quiescent" states, illustrating a possible occultation model rather similar to van der Klis' picture of LMXRBs.

4. The 18 year optical light curve

Gill _et al_. (1984) published a compilation of optical magnitude of NGC 4151 from 1978 and this has been extended to the present by Snijders (private communication). If one believes disc models apply to both active galaxies and stellar binaries, it is interesting to apply scaling laws to the time scale, _t_, and wavelength, λ, to see to what this corresponds in terms of the binaries which are the main subject of this volume. Using the scalings:

$$t \propto R/c \propto M$$
$$\text{and} \quad \lambda \propto 1/T \propto M^{1/4},$$

one learns that 18 years data at 4000Å for a 10^8 $M_\odot$ nucleus corresponds to 6 seconds at 0.2 kev for one solar mass! Unsurprisingly therefore we know less about active galaxy nuclei than about active binaries and will find it tough to catch up!

Nonetheless Snijders and I are in the process of analysing the light curve. In particular, there is plentiful structure in the light-curve which we are trying to understand by examining its auto-correlation function (ACF).

There are gross changes of level in the light-curve at intervals typically of order 1000 days. During such intervals the object is never observed bright (or faint if the level is bright). There do not seem, however, to be any particularly preferred levels. Transitions between levels are sharp and there are occasions on which transitions seem to start but "fail" and the light-curve returns to its original value. A crude period-finding program applied to the light curve finds "periods" of 1068 and 1566 days (!) but these are simply intervals between particular sharp transitions and it seem unlikely any true periodic behaviour is present.

For smaller lags, the ACF shows "flickering" with rather well gaussianly distributed changes day-to-day with a root mean square rate of change of 0.10 mag day^{-1}. There is little power in such flickering for lags greater that about 50 days.

The ACF goes significantly negative for lags near 500 days. For an infinite data string this would imply that not all the light from NGC 4151 can originate in pulses randomly distributed in time and all of the same shape. However it is not so clear that this conclusion can be drawn from the present finite data set (Tennant private communication) and this point is still under investigation.

Acknowlegement

This review would have been impossible without the involvement of my friends in the "European Extragalactic Collaboration", particularly Toon Snijders, who have contributed key observations, data reduction, analysis and discussions leading to the ideas in this paper.

References

Baldwin, J. and Netzer, H., 1978. Ap.J., 226, 1.

Bromage, G.E., et al., 1985. M.N.R.A.S., 215, 1.

Davidsen, A.F. and Hartig, G.F., 1978. Paper read at COSPAR/IAU
             Symposium on X-ray Astronomy, Innsbruck, June 1978.

Elvis, M., Briel, U.G. and Henry, J.P., 1983. Ap.J., 268, 105.

Gill, T.R., et al., 1984. M.N.R.A.S., 211, 31.

Holt, S.S., et al., 1980. Ap.J.Letts., 241, L13.

Inoue, H., 1985. Proc Japan/US Seminar on X-ray Astronomy, Jan.1985.

Lynden-Bell, D., 1969. Nature, 223, 690.

Matsuoka, M., 1986. P.A.S.J., 38, in press.

Penston, M.V., et al., 1984. M.N.R.A.S., 208, 347.

Perola, G.C., et al., 1982. M.N.R.A.S., 200, 293.

Perola, G.C., et al., 1986. Ap.J., in press.

Pounds, K., et al., 1986. M.N.R.A.S., 218, 685.

Terrell, J., 1967. Ap.J., 147, 827.

Ulrich, M-H., et al., 1984a. M.N.R.A.S., 206, 221.

Ulrich, M-H., et al., 1984b. M.N.R.A.S., 209, 479.

Ulrich, M-H., et al., 1985. Nature, 313, 745.

Wills, B.J., Netzer, H. and Wills, D., 1985. Ap.J., 288, 94.

THE PROMISE OF HIGH RESOLUTION UV SPECTROSCOPY FOR
UNDERSTANDING THE WINDS OF CATACLYSMIC VARIABLE STARS

France A. Córdova
Los Alamos National Laboratory
M.S. D436
Los Alamos, NM 87545

I. Introduction

One of the best places to study accretion onto compact objects is
in cataclysmic variable stars (CVs). These are close binary systems
in which a low mass, approximately main-sequence, "companion" star
transfers matter onto a degenerate dwarf star. The binary orbital
period of a CV is typically a few hours. Since the mass of each of
the component stars is about 1 $M_\odot$, this implies a binary separation on
the order of a solar radius. The gas from the Roche lobe-filling
companion star falls onto the compact star after circulating in a
Keplerian accretion disk (provided that the magnetic field of the
compact star is not so large as to channel the inflowing material
radially onto its surface). Optical and ultraviolet radiation are
emitted from the disk as the matter is viscously transported inward;
the conversion of the gravitational potential energy of the matter
into kinetic energy near the degenerate dwarf results in X-ray
emission.

There are a couple of hundred of CVs that are easily observable
with present-day optical and IR ground-based instruments, and more
than 50 are detectable with X-ray and UV satellites. Most of these
binary stars are within a few hundred parsecs.

In this paper I focus on one feature of CVs, their high velocity
winds. As described elsewhere in this volume, winds and jets are
ubiquitous where accretion is important, both in galactic and
extragalactic systems. It was with the International Ultraviolet
Explorer (IUE) satellite that CV winds were discovered, and the IUE
has continued to contribute data for modeling the wind.

Until recently, only low resolution (~6 - 9 Å) ultraviolet spectroscopy was made of CVs, a situation forced by their short timescale variability and relative faintness. A recent review of IUE low resolution observations of CVs and other accreting, compact binaries is given by Cördova and Howarth (1986), who also discuss the observations and interpretations of the wind line profiles. Drew (1986b) presents model UV resonance line profiles for a constant ionization wind flowing from the center of a disk and compares these to low-resolution IUE data on CVs to derive information about the physical conditions in the wind, its geometry, and the mass loss rate.

The subject of the present paper are very recent <u>high</u> resolution IUE observations (i.e., $\lambda/\Delta\lambda \sim 10^4$) which potentially offer much new information on CV winds. The work reported here represents the first results of a study undertaken by the author and Keith O. Mason. The data were taken by us or are from the National Space Science Data Center. I present a subset of the available high-resolution data, that is, the portions of the spectrum around the highly ionized lines of N V $\lambda1240$, Si IV $\lambda1400$, C IV $\lambda1549$, and He II $\lambda1640$. I consider three aspects of the data: the shapes of the broad-line profiles and what they imply about the nature of the high-velocity winds; the origin of a number of both lowly and highly ionized narrow absorption features in the spectra; and the promise of time-resolved, high-resolution data as evidenced by one attempt at acquiring such data. The discussion is general and preliminary; a more detailed analysis of all the available high-resolution data will appear in a later paper.

II. <u>Data Acquisition and Reduction</u>

The data were obtained with the high-resolution echelle mode of the IUE. A detailed description of the satellite is given in Boggess et al. (1978). The data were reduced at the IUE data reduction facility at the Goddard Space Flight Center, using the echelle ripple and background corrections provided by the IUE Observatory. The wavelength calibration, which includes corrections for thermal and temporal variations on the SWP camera, is accurate to about 3 km s^{-1}. Correction for the motion of the IUE and the Earth have also been applied to the reduced wavelengths.

III. The Observations

The CVs selected for high-resolution IUE observations are among the brightest of their class: the dwarf novae SS Cygni and VW Hydri during optical outbursts; and the novalike stars CD-42° 14462 (also called V3885 Sgr), RW Sex, CPD-48° 1577, and TT Ari (the latter during an optical high state). A spectrum of a dwarf nova during quiescence is also of interest for comparison, and therefore one such spectrum of SS Cygni, the brightest of the quiescent dwarf novae, is also presented. These objects are viewed at relatively low inclination angles with respect to the rotation axis of the disk (i <60°).

Table 1 contains the following information for each of the several objects discussed here: the date of the observation, exposure time, binary orbital period, and apparent visual magnitude during the observation. Note that in most cases the exposure duration is at least as long as the orbital period.

Figure 1 shows, for six observations of five different CVs, portions of the spectra within a few tens of Angstroms around the high ionization lines of N V at λ1240, Si IV at λ1400, C IV at λ1549, and He II at λ1640. Figure 2 represents the one example we have of relatively rapid, sequential, high-dispersion spectra of a CV: three spectra of CPD-48° 1577 taken with a temporal resolution of ~1 hr. For this case only the portions of the spectra around C IV are plotted.

Table 1: HIGH DISPERSION OBSERVATIONS

Object	Date of Observation		$t_{exposure}$	P_{orb}	m_V
SS Cygni(o)	1985	Nov 9	2.5 hr	6.6 hr	9.6
VW Hydri(o)	1982	Jan 5	3.1 hr	1.8 hr	10.8
CD-42° 14462	1983	Aug 12	5.0 hr	(4.9 hr)	10.3
RW Sex	1985	Nov 12	4.0 hr	(5.9 hr)	10.9
TT Ari	1981	Mar 3	5.7 hr	3.3 hr	11.5
SS Cyg(q)	1984	Feb 14	10.7 hr	6.6 hr	12.0
CPD-48° 1577	1985	Mar 14	1,1,0.6 hr	(4.5 hr)	9.8

Notes to Table 1: "(o)" and "(q)" refer to outburst and quiescent states, respectively, of dwarf novae. Parentheses around the orbital period indicate an uncertain value.

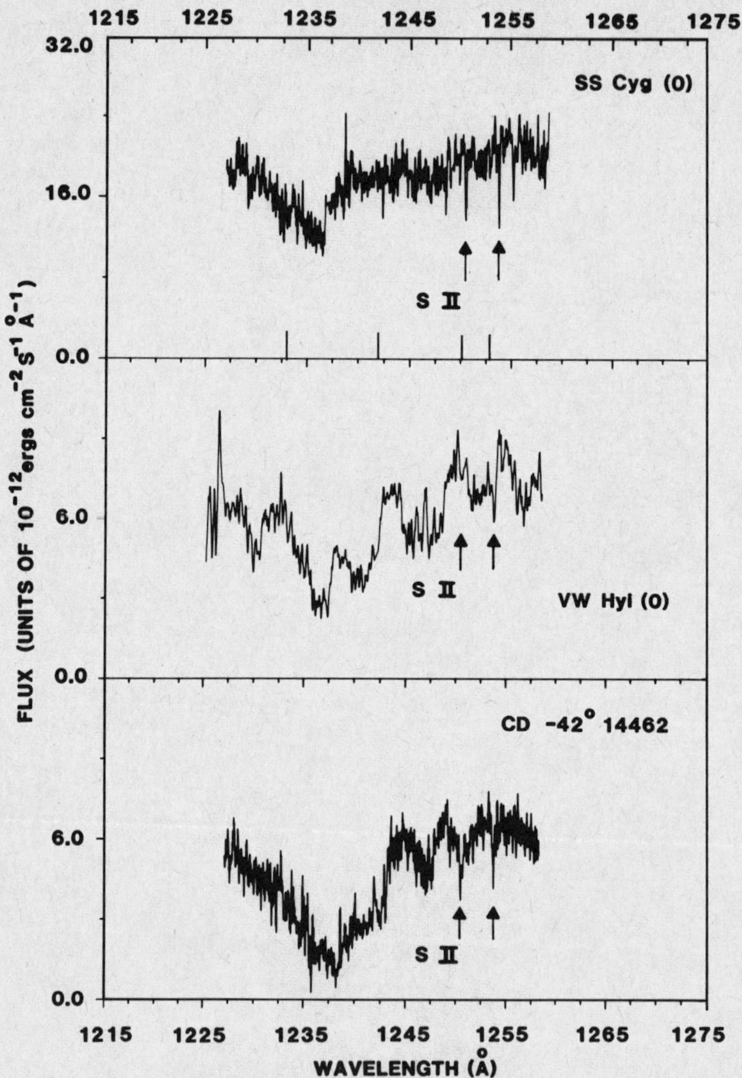

Figure 1. Twenty-four plots showing portions of high-resolution UV
spectral data ($\lambda/\delta\lambda \sim 10^4$) on five cataclysmic variable stars: the
dwarf novae SS Cygni and VW Hydri during optical outbursts, the
luminous novalike stars CD-42° 14462 and RW Sex, and the novalike star
TT Arietis during an optical high state. Also included is a spectrum
of SS Cygni during an optical low ("quiescent") state. All data were
taken in the echelle mode using the IUE's SWP camera. The wavelength
regions shown are in the vicinity of the prominent resonance doublets
N V, Si IV, and C IV, and the vicinity of He II. An "X" shows data
corrupted by a camera reseau; the approximate locations of reseaux in
all the spectra are noted by the small vertical lines on the bottom of
each of the plots of SS Cygni outburst data.   Possible narrow
absorption lines suggesting an interstellar or circumstellar origin
are indicated by arrows.

N V PROFILES

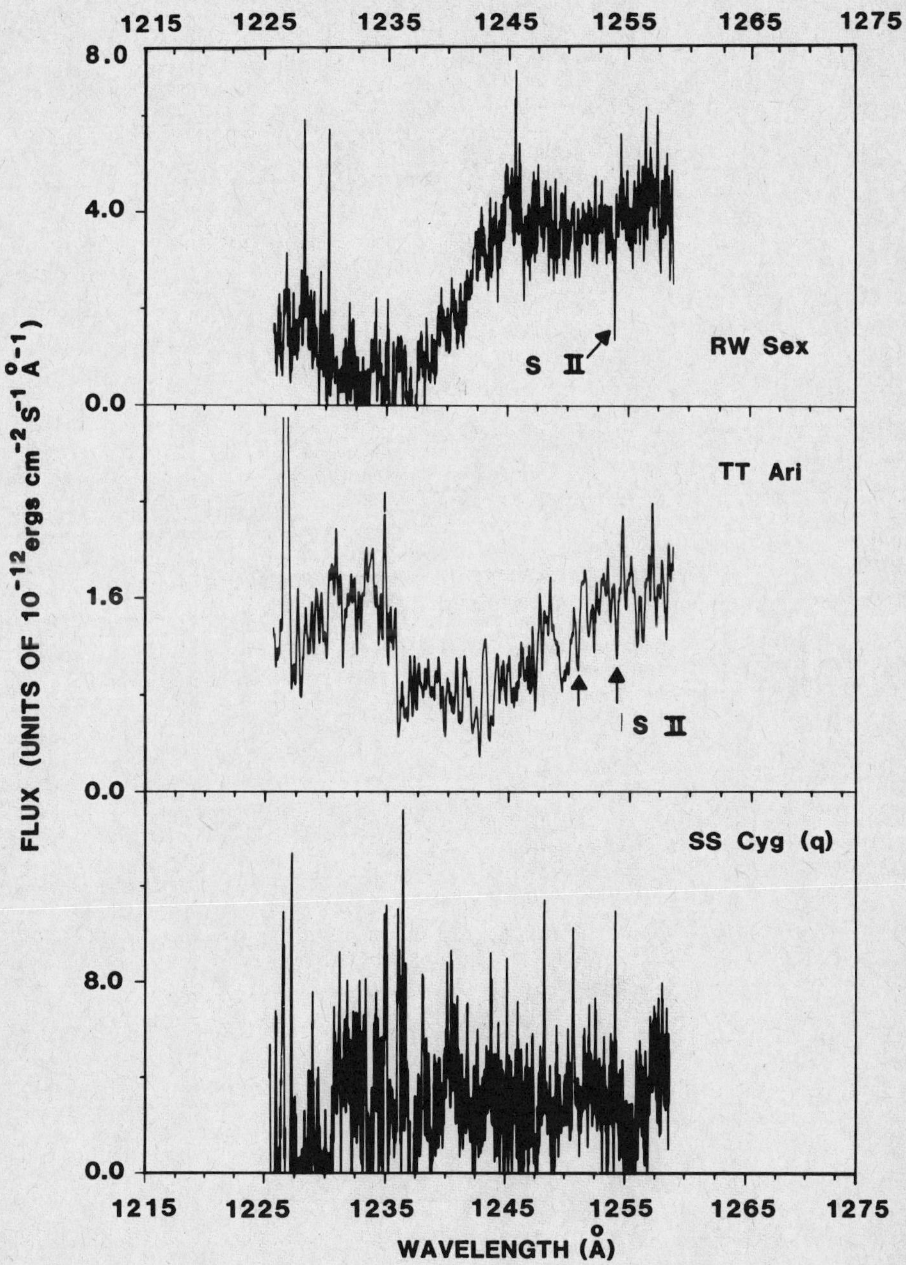

Figure 1 continued

Si IV PROFILES

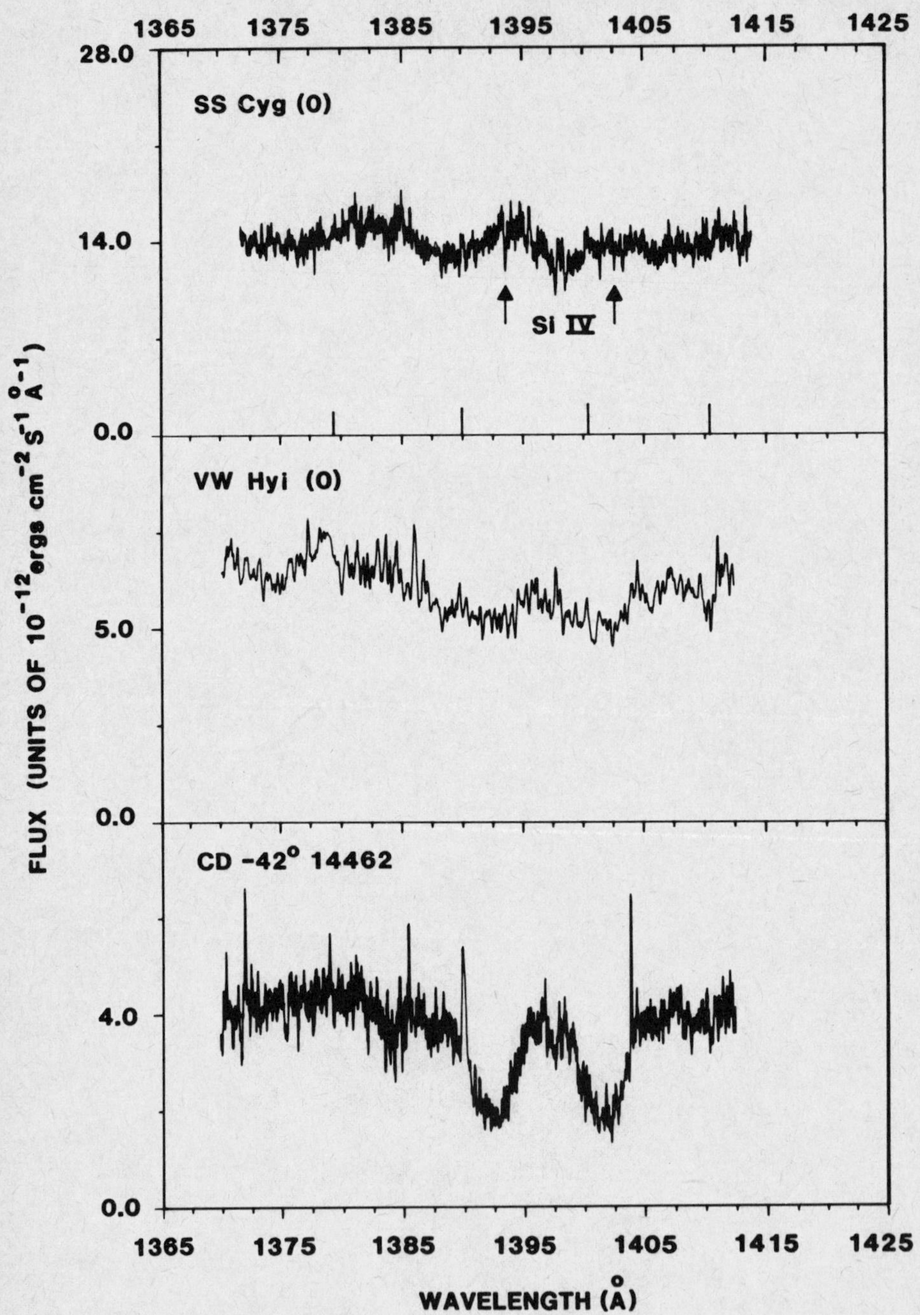

Figure 1 continued

Si IV PROFILES

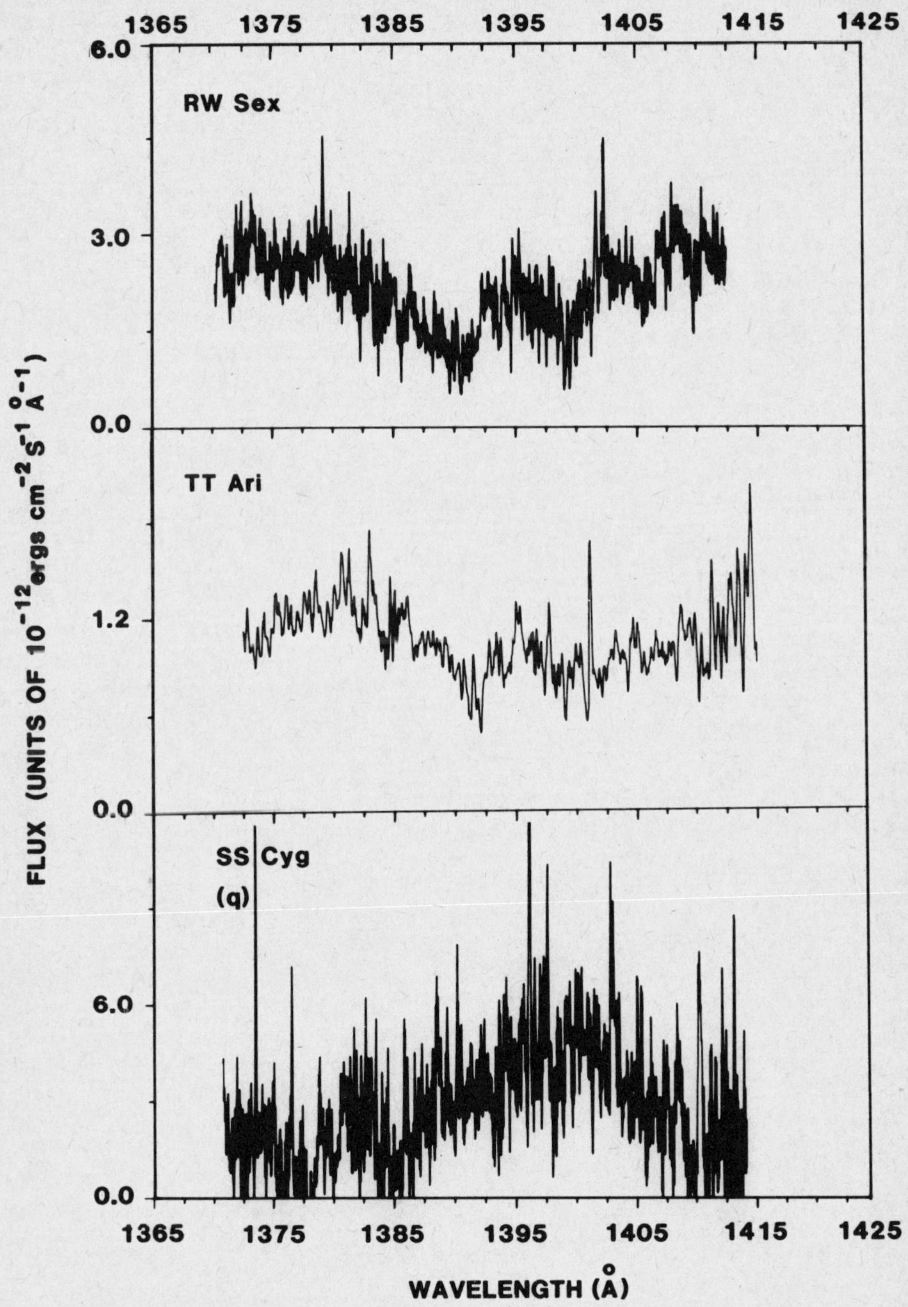

Figure 1 continued

C $\underline{\text{IV}}$ PROFILES

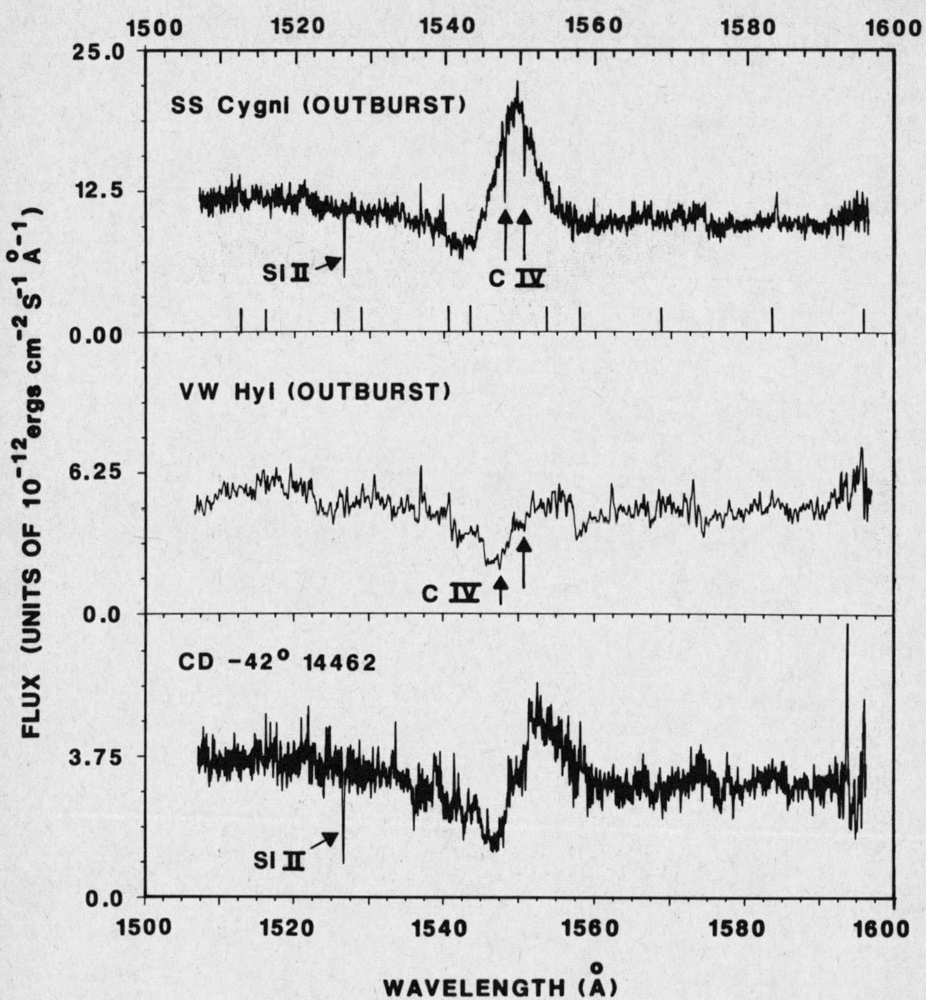

Figure 1 continued

C $\overline{IV}$ PROFILES

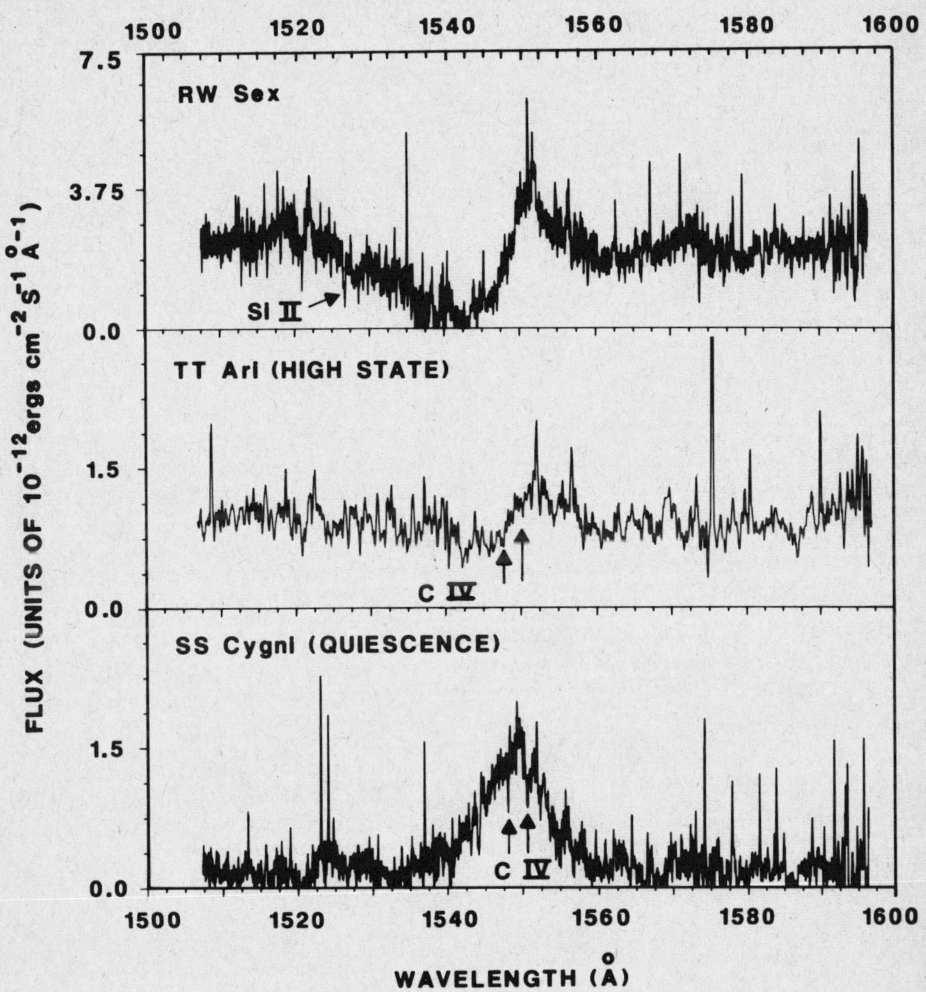

Figure 1 continued

He II PROFILES

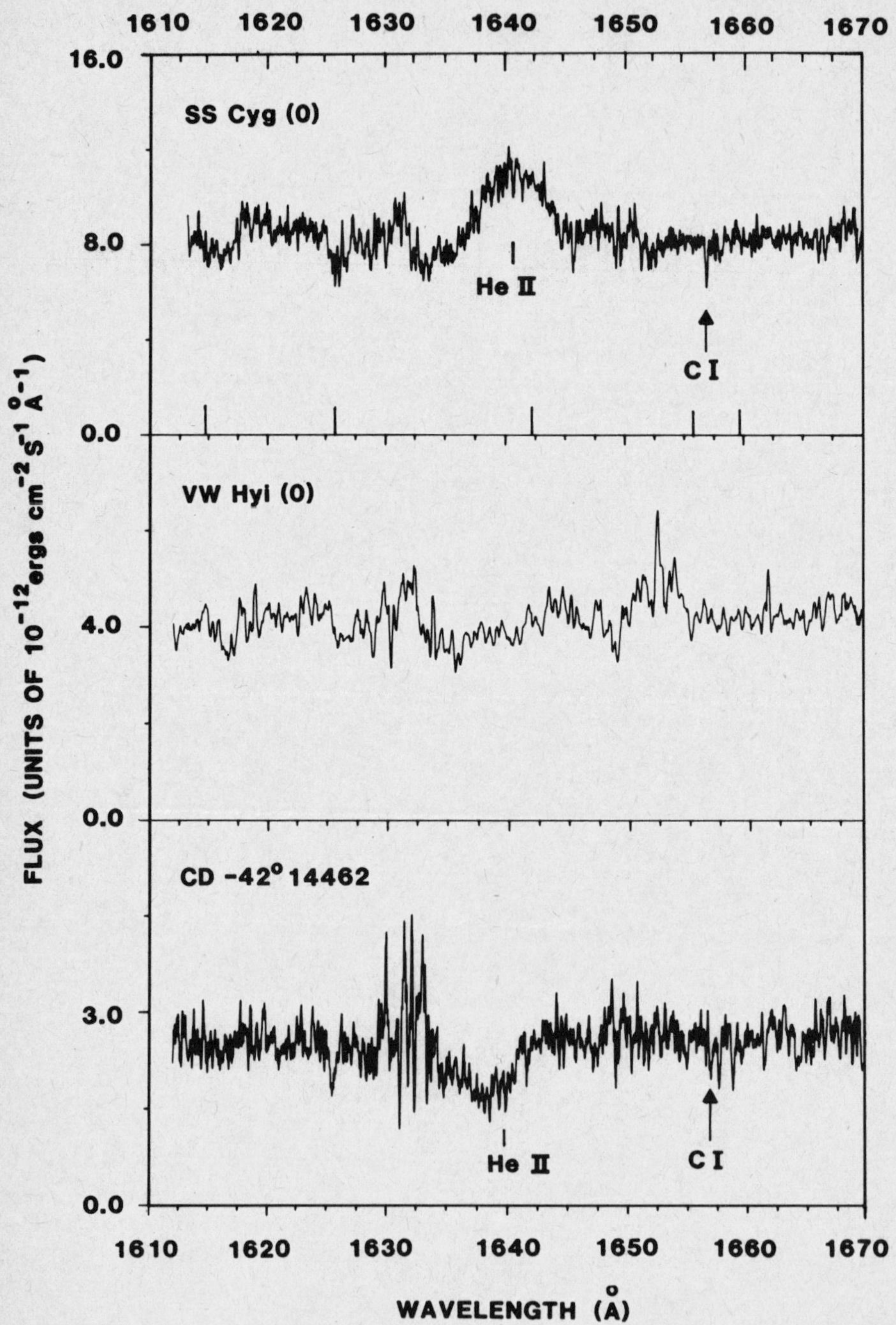

Figure 1 continued

He II PROFILES

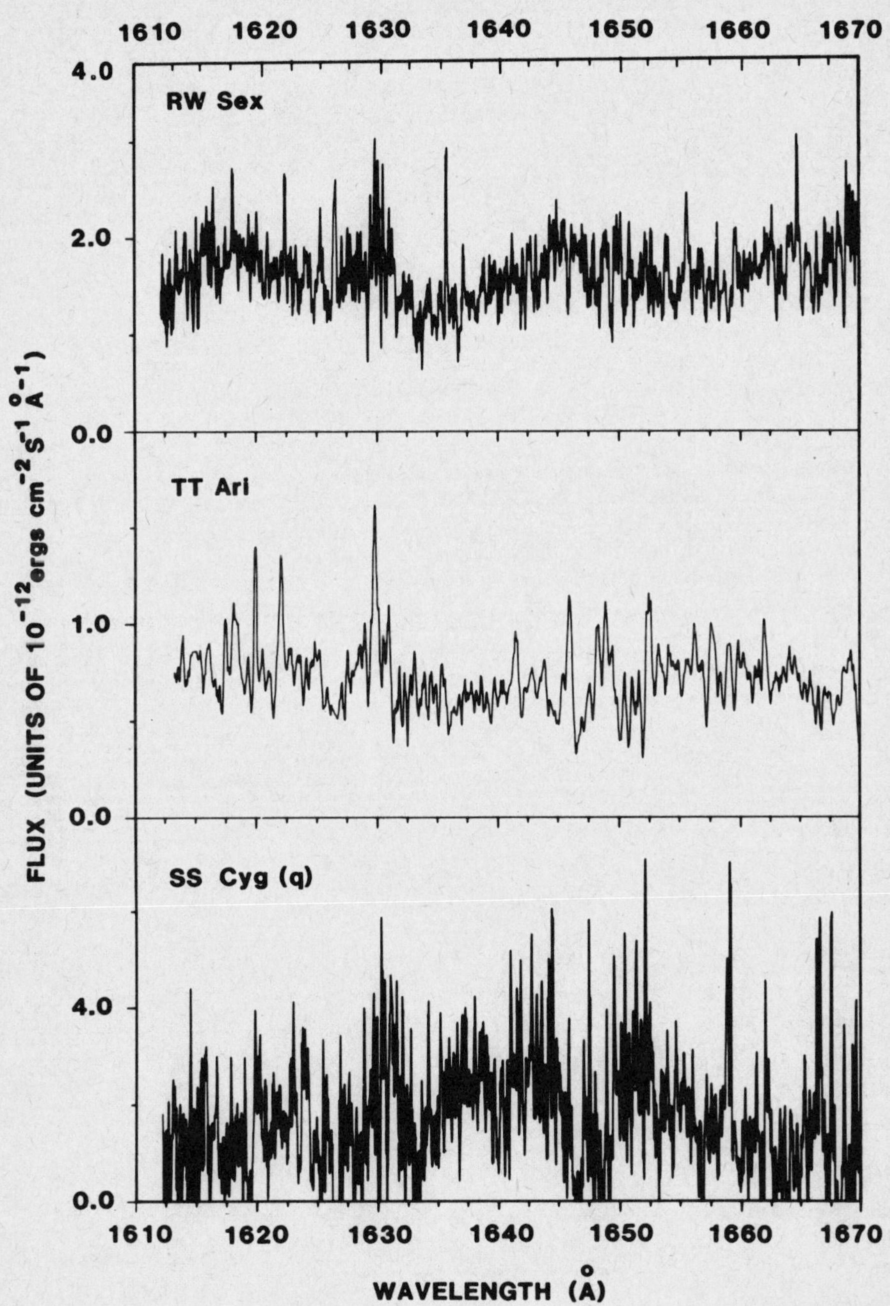

Figure 1 continued

IV. Discussion

a) The broad-line profiles

Figure 1 shows that there is no qualitative difference between the spectrum of the dwarf nova SS Cygni in outburst and the novalike objects. (The spectrum of the dwarf nova VW Hydri during outburst, however, is markedly different.)

In Figure 1 the resonance lines, N V, Si IV, and C IV, are present as broad, violet-shifted absorptions in the spectra of all the objects that are accreting at a high rate (i.e., all the spectra in the figure except that of SS Cygni in quiescence). The C IV profiles of all those objects consists of, in addition to the shortward-shifted absorption, an emission component near the rest wavelength. For SS Cygni during low mass accretion, the C IV profile consists only of broad emission centered at about the rest wavelength.

He II is present with both emission and shortward-shifted absorption in the spectrum of SS Cyg during outburst, and as shortward-shifted absorption only in the spectra of CD-42° 14462 and RW Sex.

The full-width at zero-intensity of SS Cyg's quiescent C IV line is ~21 Å. In the outburst spectrum, the red edge wavelength of the emission component is about the same as the red edge wavelength of the emission line during quiescence, i.e., ~1560 Å.

In all of the C IV profiles the deepest part of the absorption is at relatively low velocities: a few hundred km s^{-1} for CD-42° 14462 and CPD-48° 1577 (Figure 2) up to 1000 km s^{-1} for RW Sex. Without model fitting, it is difficult to estimate the blue edge wavelength of the absorption components, and hence a lower limit to the terminal velocity in the wind. In the case of RW Sex the meeting of the continuum and the absorption edge is more distinct and appears to be at ~1520 Å (implying an edge velocity of ~5000 km s^{-1}). In the other spectra the C IV absorption profiles seem to extend to similarly high velocities, but with much smaller fluxes at high negative velocities. Drew (1986b) has shown that to retain the absorption minimum near the line center, and to explain the observations of objects viewed at high inclination, the wind needs to be _gradually_ accelerated, such that the terminal velocity is not approached until the wind is at a distance of about 20 times the radius of the white dwarf.

Most of the shortward-shifted absorption lines are shallow. RW Sex, however, shows very deep absorptions in N V and C IV; in fact, there is very little residual flux left in these lines. Si IV is deepest in the spectrum of CD-42° 14462. Drew (1986a) has shown that the shallowness of the violet-shifted absorption lines may be a consequence of the accretion disk's shape and size, which affects the continuum radiation that a central wind sees; in a disk whose radius is comparable to the length scale of the wind's acceleration the absorption lines may be optically thick even while appearing shallow. This affects the derivation of the mass loss rate: a higher rate is allowed than under the assumption that the acceleration is due to a small, spherical continuum source.

The ratio of emission to absorption equivalent width in C IV differs greatly between systems; low-resolution IUE spectra of these and other CVs (see also Figure 2) show that it differs, too, within the same system. The fact that C IV has emission, but Si IV and N V do not, suggests that the C^{3+} abundance could be orders of magnitude greater than the abundances of Si^{3+} or N^{4+}, which could imply that the mass loss rate is $> 10^{-10}$ $M_\odot$ yr^{-1} (Drew 1986b).

A comparison of the CV spectral lines with those of O-type spectra (cf. Wallborn, Nichols-Bohlin, and Panek 1985) shows little similarity. The CV spectra have some features in common with luminosity class I early O star spectra; namely, the presence of He II and C IV with P Cygni-like profiles, and the near absence of Si IV and N V emission. In contrast to hot-star spectra, the violet-shifted absorption components in CVs are not saturated, and minimum residual flux occurs at a lower velocity.

b) CVs and the Interstellar Medium

Because of their proximity to Earth and their intense UV radiation, the CVs provide an opportunity to study the local interstellar medium (ISM) in the UV. Such studies have been done with IUE on a few nearby white dwarfs (Bruhweiler and Kondo 1982; Dupree and Raymond 1983), but CVs offer a larger sample of objects with which to explore more lines of sight, and hence a potentially greater volume of space.

Figure 1 shows several examples of narrow absorption lines which suggest a possible interstellar origin. Both high and low ionization species are evident. The strongest such features are the resonant doublet of C IV ($\lambda 1548.2$, $\lambda 1250.8$ are the laboratory rest wavelengths), which appears in the broad violet-shifted C IV absorptions, and Si II ($\lambda 1526.7$). Also evident in the spectra of Figure 1 are sharp S II lines ($\lambda 1250.6$ and $\lambda 1253.8$). The Si IV doublet ($\lambda 1393.8$, $\lambda 1402.8$) may be present in the outburst spectrum of

SS Cygni, and narrow lines of He II (λ1640.4) and C I (λ1656.9) may be present in the spectra of many objects. Not shown here, but prominent in a high-dispersion long wavelength IUE spectrum of CPD-48° 1577, are deep, narrow absorptions of Mg II (λ2796.6, λ2802.7).

The clearest example of the narrow absorption features is provided by both C IV spectra of SS Cygni. The C IV doublet appears at λ1548.04, λ1550.63, a shift of 0.155 Å to the blue of the laboratory rest wavelengths (or -30 km s^{-1}), and the Si IV doublet appears at λ1393.63, λ1402.67, a shift of -0.13 Å (-28 km s^{-1}). However, the Si II line appears at 1526.65, a shift of only -0.06 Å (-12 km s^{-1}), and other low ionization lines such as S II and C I have similarly small blueward shifts (-0.07 Å and -0.05 Å respectively) with respect to the rest wavelengths of these lines.

The Si II and S II lines of CD-42° 14462 and RW Sex have no velocity shifts, while the S II line of TT Ari (if real: the spectrum is extremely noisy) is shifted 0.155 Å to the red (+30 km s^{-1}).

The difference in velocity between the lowly and highly ionized lines implies physically distinct regions for these species. Although it is attractive to assume that the lowly ionized lines are interstellar, their velocity of ~-12 km s^{-1} in SS Cygni is similar to the gamma velocity of this system (-15 km s^{-1}: Hessman <u>et al.</u> 1984), and thus a circumstellar origin for these lines should be considered. The highly ionized lines (C IV and Si IV) are shifted by about -15 km s^{-1} relative to the lowly ionized species. Their origin is almost certainly circumstellar rather than interstellar.

In the case of white dwarfs sharp resonance lines of N V, C IV, and Si IV, which are shifted substantially relative to the lowly ionized species, have been attributed to the stellar photosphere (Bruhweiler and Kondo 1982; Dupree and Raymond 1982; Sion and Guinan 1983). For the white dwarfs, however, these high ionization lines are shifted <u>longwards</u> 15 - 20 km s^{-1} with respect to the low ionization lines, while in the case of SS Cygni, at least, the C IV doublet is shifted <u>shortwards</u> by the same amount. The highly ionized species in SS Cygni could result from heating, and subsequent expansion, of circumstellar gas by the radiating star (cf. Raymond 1984).

A future paper will contain an analysis of all the narrow absorption lines in the complete IUE spectra of these CVs (i.e., from 1150 Å to 3000 Å), including derivation of the equivalent widths and column densities of the lines and what these imply about the lines of sights to these systems, and the local environment of the objects.

c) <u>Time-resolved spectroscopy</u>

In trying to model the wind line profiles of Figure 1, it must be remembered that the exposure lengths are of the order of the binary orbital periods of these stars. Shorter timescale variability in the profiles could produce distortions in such long exposures.

One CV, CPD-48° 1577, is bright enough that we attempted to make much shorter observations with high spectral resolution. The result is shown in Figure 2 for the C IV line alone. The first two observations each lasted 1 hr, while the third observation lasted 36 minutes. The observations are separated by about 30 minutes.

The three profiles differ substantially, but close inspection shows that it is the <u>low-velocity</u> part of the profile, i.e., longwards of 1545 Å, that varies the most. The absorption profile appears to broaden with time, but this is due to enhanced absorption at increasingly low velocities. The additional absorption at low velocities has the effect of reducing the equivalent width of the broad emission component.

The constant, central wind model examined by Drew (1986b) does not allow for significant orbital-phase related line profile variations in a low inclination system like CPD-48° 1577. The observed change in the C IV line profile is unlikely to be due to a variation in the effective temperature of the inner disk because there is no corresponding continuum change. If the velocity field of the wind is constant, the wind may be strongly asymmetric and geometrical effects (as the binary components rotate) will be important.

One possibility is that the effect results from an additional contribution to the absorption profile which is narrower than the wind line contribution; in this scenario this added absorption component is centered at or near the rest wavelength of the doublet, and its appearance is orbital-phase dependent.

A different possibility is that episodic mass ejection, on the timescale of hours, might create such line profile variability because of changes in the density of the emitting region.

There is a hint of variable, fine structure (at the level of a few Angstroms) in the broad absorption component; this structure is also promising for further investigation.

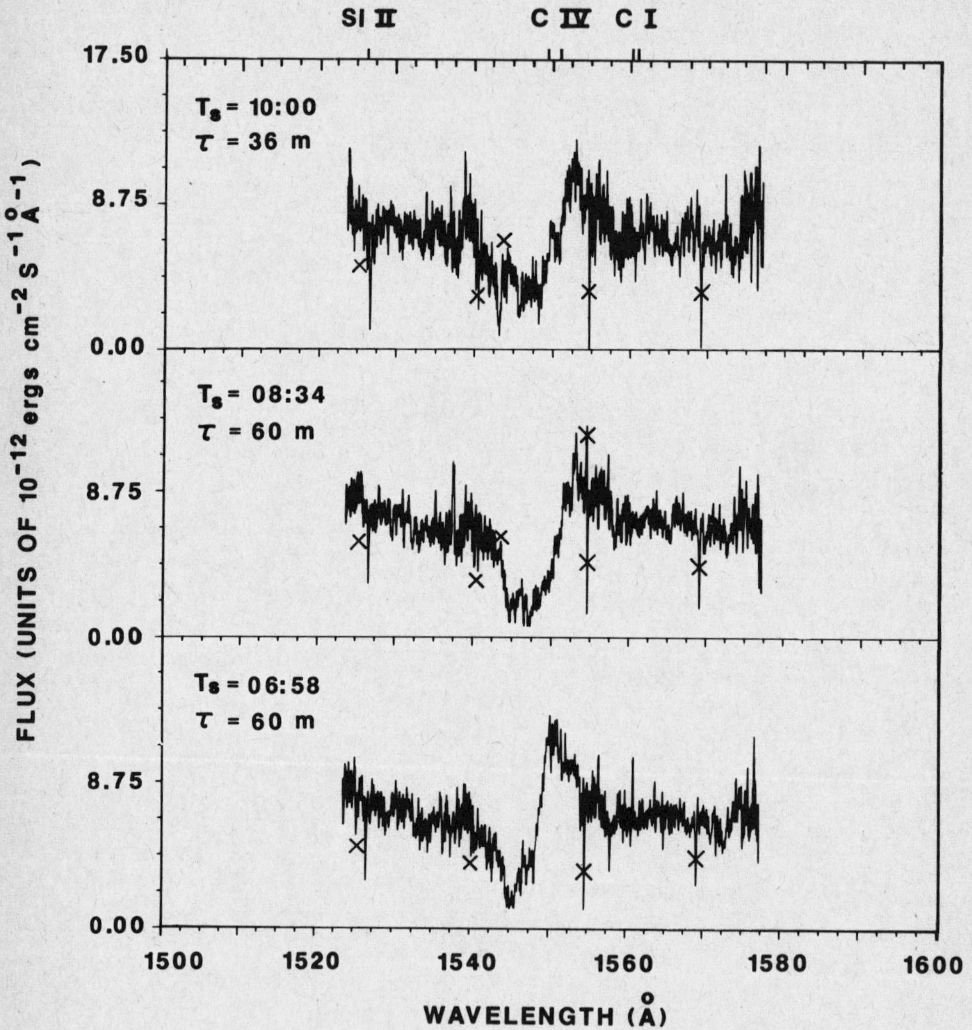

Figure 2.   Time-resolved   data, taken in the high spectral resolution mode of the IUE, on the   C IV line of the bright novalike star CPD-48° 1577.   "T$_s$" gives   the UT start time of each observation on 1985 March 14, and "τ" gives the exposure duration. As in Figure 1 an "X" marks contamination by a reseau. Narrow absorptions at Si II, C IV, and C I may be present; their wavelengths are indicated above the top plot.

V. Concluding Remarks

The purpose of this contribution has been to demonstrate the potential usefulness of high-resolution UV spectroscopy of cataclysmic variables for understanding the nature of winds in disk-accreting, close binary systems, and the nature of the local interstellar medium and circumstellar environment of these stars. Drew (1986b) has shown that for low inclination systems the UV resonances line profiles are sensitive to the mass loss rate, the velocity law, and the wind geometry (i.e., size of the line-forming region and the cone-angle of the wind). Hence, detailed fitting of these resolved line profiles could begin to constrain these quantities.

Interpretation of the data, however, is complicated by the possibility of short timescale variability; the relative faintness of these stars requires that the exposure durations are long compared to the binary orbital period. Nevertheless, for the very brightest of the known wind-emitting CVs, CPD-48°1577, there is hope that further progress in understanding its wind can come from IUE observations made rapidly (i.e., ~40 minute exposures) and sequentially for at least two binary orbital periods. Such an observation is scheduled for the IUE during the Fall of 1986.

This work was funded in part by the US Department of Energy.

References

Boggess, A. _et al._ 1978, _Nature_, _275_, 387.

Bruhweiler, F. C., and Kondo, Y. 1982, _Ap. J._, _259_, 232.

Cordova, F. A., and Howarth, I. D. 1986, _Scientific Accomplishments of the IUE_, ed. Y. Kondo, _et al._ (D. Reidel Publishing Co.: Dordrecht, Holland), in press.

Drew, J. 1986a, _M.N.R.A.S._, _218_, 41P.

Drew, J. 1986b, _M.N.R.A.S._, in press.

Dupree, A. K., and Raymond, J. C. 1982, _Ap. J. Letters_, _263_, L63.

Dupree, A. K., and Raymond, J. C. 1983, _Ap. J. Letters_, _275_, L71.

Hessman, F. V., Robinson, E. L., Nather, R. E., and Zhang, E.-H. 1984, _Ap. J._, _286_, 747.

Raymond, J. 1984, _IAU Colloq. 81, The Local Interstellar Medium_, ed. Y. Kondo, F. C. Bruhweiler, and B. D. Savage (NASA Conference Publ. 2345), p. 311.

Sion, E. M., and Guinan, E. F. 1983, _Ap. J._, _265_, L87.

Wallborn, N. R., Nichols-Bohlin, J., Panek, R. J. 1985, _IUE Atlas of O-Type Spectra from 1200 to 1900 Å_, NASA Ref. Publ. 1155.

BL LACERTAE OBJECTS:
ACCRETION, JETS, AND WINDS

C. Megan Urry

Center for Space Research, Massachusetts Institute of Technology

Cambridge, Massachusetts 02139

ABSTRACT

Observed properties of BL Lacertae objects, in particular X-ray spectral shape, overall broadband spectral shape, and rapid variability in intensity, are reviewed briefly. A number of promising scenarios for the energy-producing mechanisms are then discussed in light of the observations. These include accretion disks in the vicinity of a massive black hole, and synchrotron and inverse Compton emission from a relativistic jet. Finally, the possible presence of winds, suggested by broad X-ray absorption features in the spectra of a handful of BL Lac objects, is described.

I. INTRODUCTION

BL Lacertae objects may be well-studied but they are not well-understood. One common assertion is that they are "naked quasars", with a central engine unobscured by the partially ionized gas observed in quasars. A different suggestion is that BL Lac objects are dominated by emission from a relativistic jet, in which case the observed emission reveals little or nothing about the central engine. At this time, the validity of neither picture has been demonstrated.

In some respects BL Lac objects do resemble an extreme version of quasar, often being highly polarized, rapidly variable, and highly luminous. However they lack the strong emission lines characteristic of quasars. Fewer than a third of the ~120 known BL Lac objects exhibit any discrete spectral features at all, and most of those are absorption features from a surrounding galaxy (for nearby objects) or from intervening absorption systems (in the case of a few high redshift objects). A few have weak, narrow emission lines, typically [O II], [O III], [N II], Hα, and Hβ (Miller *et al.* 1978). Naturally the dearth of redshifts makes it difficult to generalize about the properties of these objects, which may explain why they are sometimes ignored in discussions of active galactic nuclei (AGN). Another reason is that they differ from quasars and other AGN in several significant ways.

This paper summarizes our current understanding of BL Lacertae objects, with emphasis on those topics of particular interest to this workshop: accretion, jets, and winds. First, the spectral and temporal properties of BL Lac objects are briefly reviewed. Then theories of the underlying physics are discussed in light of the observations. This is necessarily a selective discussion that stresses the more promising models and the observations that test them.

II. OBSERVATIONS

(a) Radio-Loud or Radio-Quiet?

Most BL Lac objects were discovered in large radio surveys; two recent, slightly differing catalogs are given by Urry (1984) and Ledden and O'Dell (1985). About a dozen have recently been found as optical counterparts to X-ray sources. The defining characteristics of BL Lac objects include (*e.g.* Stein *et al.* 1976): absence of emission lines; rapid variability in intensity; and strong, variable polarization of radio and optical emission. No known BL Lac object is radio-quiet, according to the criterion F_ν (5 GHz) $< F_\nu$ (5000 Å). This is not suprising for sources found in radio surveys. Furthermore, those found in X-ray surveys are biased toward radio loudness by the process of optical identification: first a candidate blue stellar object is found to have a featureless optical spectrum, then one looks for radio emission, and then polarization. It is not clear that polarization studies of a radio-quiet candidate would be undertaken, or that it would be monitored for variability, particularly if alternative candidate objects exist. Thus, selection effects may well be the reason for the lack of radio-quiet BL Lac objects. Perhaps they will be found in a survey unbiased toward radio emission, such as one based on polarization properties.

(b) X-Ray Spectra

The X-ray spectra of BL Lac objects are typically atypical. Unlike Seyfert galaxies, they exhibit a wide range of power law spectral indices, $0 \leq \alpha_x \leq 5$ (with α defined by $F_\nu \propto \nu^{-\alpha}$), and the mean spectral index tends to be steeper on the average than for other kinds of AGN, $<\alpha_x>=1.1$-1.6, depending on sample (Worrall *et al.* 1981, Urry 1984, Singh and Garmire 1985, Madejski 1985). The column density of cold gas along the line-of-sight, as inferred from the lowest energy X-ray measurements (near $E \sim 0.1$-0.2 keV), is always consistent with the line-of-sight column density of interstellar material in the Galaxy, implying there is no intrinsic cold gas in the BL Lac objects themselves (Agrawal and Riegler 1979, Riegler *et al.* 1979, Singh and Garmire 1985, Madejski 1985, Warwick *et al.* 1986). Observations of the five X-ray-brightest BL Lac objects indicate that they have a two-component spectral shape, with a steep, low energy spectrum turning up around 10 keV to a flat, high-energy tail. Figure 1a shows the composite spectrum of one of the more variable BL Lac objects, PKS 2155-304, derived from observations made over the past ten years. The hard tail is apparent, but is also clearly variable. The X-ray spectra of the four other X-ray bright BL Lac objects, PKS 0548-322, H 1218+304, Mrk 421, and Mrk 501, are similar: soft at low energies with a variable hard tail (see Fig. 1). Experiments that are at least as sensitive above 10 keV as the *HEAO1 A2* proportional counters are needed to detect these features. If BL Lac objects commonly have hard tails that extend to more than ~50 keV (the highest energy at which the tail is detected in PKS 2155-304), then the bulk of the radiative energy from these sources may be at very high, rarely observed energies. It remains to be seen whether these five bright, well-studied sources are typical of all BL Lac objects.

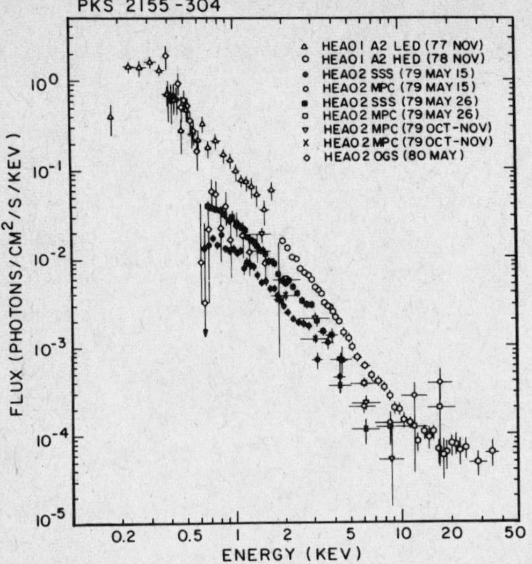

Figure 1a. X-Ray Spectra of PKS 2155-304
*HEAO*1 *A2 LED*, Agrawal and Riegler 1979; *HEAO*1 *A2 HED*, Urry and Mushotzky 1982; *HEAO*2 *SSS* and *MPC*, Urry *et al.* 1986; *HEAO*2 *MPC* (79 Oct-Nov), Agrawal *et al.* 1983; *HEAO*2 *OGS*, Canizares and Kruper 1984.

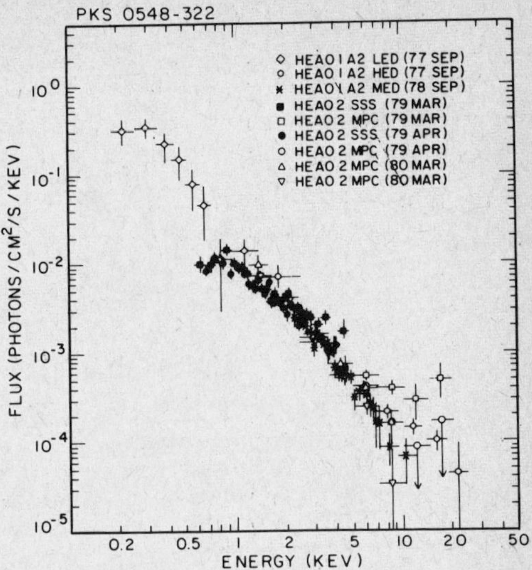

Figure 1b. X-Ray Spectra of PKS 0548-322
*HEAO*1 *A2 LED*, *HED*, Riegler *et al.* 1979; *HEAO*1 *A2 MED*, Worrall *et al.* 1981; *HEAO*2 *SSS* and *MPC*, Urry *et al.* 1986; *HEAO*2 *MPC* (80 Mar), Agrawal *et al.* 1983.

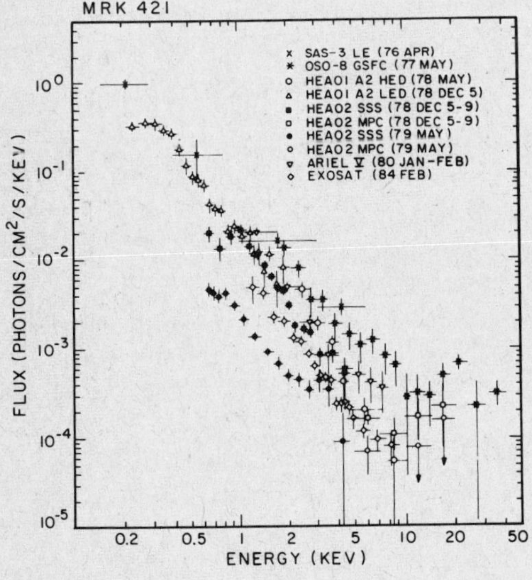

Figure 1c. X-Ray Spectra of Mrk 421
*SAS*3 *LE*, Hearn *et al.* 1979; *OSO*8 *GSFC*, Mushotzky *et al.* 1978; *HEAO*1 *A2 HED*, Mushotzky *et al.* 1979; *HEAO*1 A2 *LED*, Singh and Garmire 1985; *HEAO*2 *SSS* and *MPC*, Urry *et al.* 1986; *ARIEL V*, Hall *et al.* 1981; *EXOSAT*, Warwick *et al.* 1986.

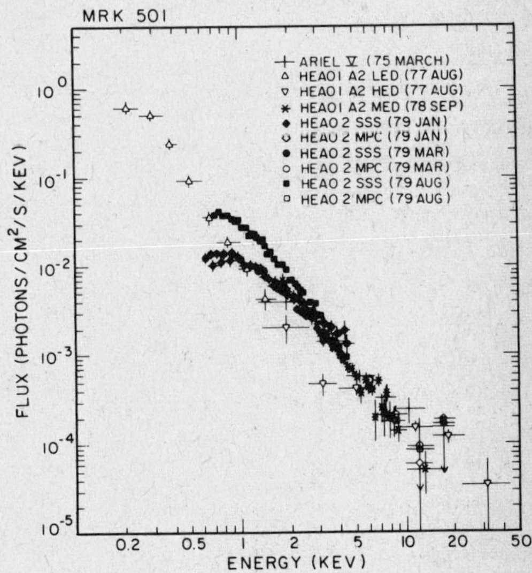

Figure 1d. X-Ray Spectra of Mrk 501
ARIEL V, Snijders *et al.* 1979; *HEAO*1 *A2 LED*, Singh and Garmire 1985; *HEAO*1 A2 *MED*, Worrall *et al.* 1981; *HEAO*1 *A2 HED*, Mushotzky *et al.* 1978; *HEAO*2 *SSS* and *MPC*, Urry *et al.* 1986.

(c) Broad Band Spectra

Simultaneous multifrequency observations have been carried out for nearly 30 BL Lac objects and Optically Violently Variable quasars (OVVs; Bregman 1986; Bregman *et al.* 1982, 1984; Feigelson *et al.* 1986; Glassgold *et al.* 1983; Kondo *et al.* 1981; Landau *et al.* 1986; Maraschi *et al.* 1985; Mufson *et al.* 1984; Worrall *et al.* 1982, 1984a,b,c, 1986). A typical broad-band spectrum is shown in Figure 2. For comparison, the broad-band spectrum of a well-known synchrotron source, the Crab Nebula, is plotted on the same figure. The gross characteristics of the broad-band spectra of BL Lacs and OVVs (collectively called blazars) across the ten decades from radio frequencies to X-ray energies are very similar, and are not too different from the Crab Nebula spectrum. In a given energy band, the spectrum is roughly a power law, the index of which gradually steepens with increasing energy band. The radio-through-infrared spectrum is very flat and in some cases inverted. If it arises from synchrotron emission, then the self-absorption turnover is never unequivocally seen. However, the extreme flatness of the radio spectrum argues for partially optically-thick regions, either in separate clumpy components or in some inhomogeneous jet-like structure. Near the optical band, the spectrum steepens more sharply and continues smoothly into the ultraviolet with $\alpha \gtrsim 1$. The X-ray spectrum is often a smooth extrapolation of the ultraviolet spectrum, although in some cases the X-ray emission lies above or below the extrapolation (Bregman *et al.* 1986).

No "blue bumps" (blue and ultraviolet emission in excess of the usual

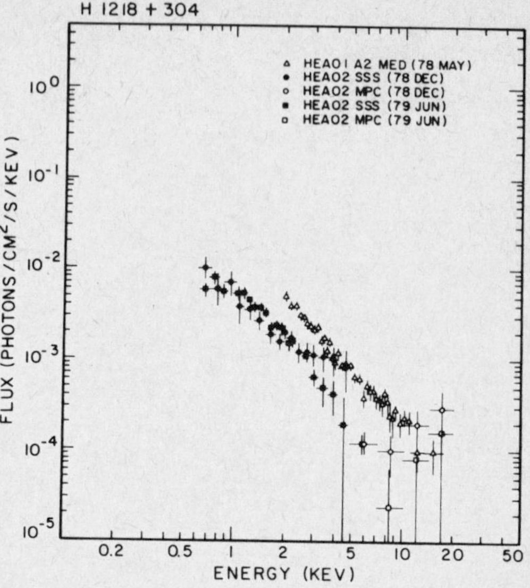

Figure 1e. X-Ray Spectra of H 1218+304
*HEAO*1 A2 *MED*, Worrall *et al.* 1981; *HEAO*2 *SSS* and *MPC*, Urry *et al.* 1986.

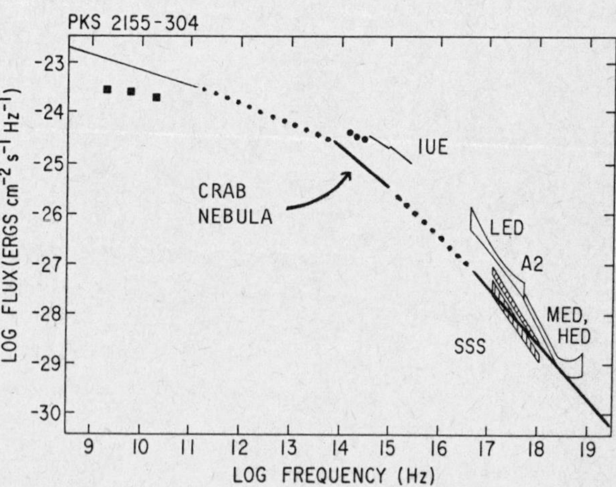

Figure 2. Broad-Band Spectrum of PKS 2155-304
Nonsimultaneous radio, infrared, optical, ultraviolet, and X-ray spectra, (Urry and Mushotzky 1982 and references therein, with added X-ray data from Urry *et al.* 1986). For comparison, the Crab Nebula synchrotron spectrum, reduced a factor of 1000 in intensity but otherwise unchanged, is also plotted (Baldwin 1971; dotted lines connect observed parts of Crab spectrum).

power law) are seen in BL Lac spectra, but a small blue bump (less than 1% of the total luminosity) has been seen in the spectrum of the OVV 3C 446 (Bregman 1986). This is significant because ordinarily BL Lac and OVV continuum spectra are indistinguishable. The similarity in continuum spectra is puzzling in light of the very different optical spectral line characteristics. OVVs have prominent emission lines, while BL Lac objects do not. This implies that either the observed continua of OVVs and BL Lacs do not represent the primary photoionizing photons, as in a relativistic jet picture (see § IIIb), *or* the physical conditions near the region of continuum production are very different, with OVVs having plenty of cold gas to be photoionized and BL Lacs having little.

(d) Variability

The shortest timescale variability must indicate a lower limit to the size of the emitting region via $\Delta R \lesssim \delta \, c \, t_{var}$, where the possibility of relativistic motion in a source is accommodated via the kinematic Doppler factor $\delta = (1 - \beta^2)^{1/2}/(1 - \beta \cos\theta)$, which is a function of the bulk velocity of the emitting region, βc, and the angle θ between the bulk motion and the line-of-sight. Note that ΔR need not be the linear dimension of a massive black hole; it could be the thickness of a thin planar shock at some distance out along a jet (*e.g.* Marscher and Gear 1985) or the thickness of the optically-thin boundary layer of an optically thick emission region (Guilbert *et al.* 1983b). If the variable emission does arise at the inner radius of an accretion disk, $R_{in} \sim 3R_S$, where $R_S = 2GM/c^2$, then the variability timescale defines an upper limit to the black hole mass, $M \lesssim c^3 \delta \, t_{var}/6G$ $\sim 1.2 \times 10^8 \, \delta \, t_{var} \, M_\odot$ with t_{var} in hours, and an associated Eddington limit, $L_{Edd} \equiv 4\pi GM m_p c/\sigma_T$ $\sim 1.5 \times 10^{46} \, \delta \, t_{var}$ erg/sec. This rough estimate is ~ 20 times higher than the more rigorous limit considered by Cavallo and Rees (1978) and Fabian (1979), who argued that a luminosity change ΔL on a timescale t_{var} should satisfy the limit $\Delta L \lesssim 7 \times 10^{44} \, \eta_{0.1} \, t_{var}$, or $\Delta L/\Delta t \lesssim 2 \times 10^{41} \, \eta_{0.1}$ ergs/sec^2 (for an efficiency $\eta_{0.1} = 0.1$ of converting matter to energy), based on the ability of primary radiation to escape the production region without appreciable scattering.

For BL Lac objects, as for other AGN, variability timescales tend to be shorter at higher energies. In the radio, the most variable objects have timescales as short as weeks, but most sources change intensity by only $\sim 20\%$ over several years, so that $t_{var} \sim 1$-10 years (Aller *et al.* 1981, 1985; Altschuler 1982; Altschuler and Wardle 1976; Andrew *et al.* 1978; Dent and Hobbs 1973; Dent and Kapitzky 1976; Dent *et al.* 1974; Medd *et al.* 1972; Waltman *et al.* 1986). In the optical, timescales of days are not uncommon, and several objects have had half-magnitude changes in hours (McGimsey *et al.* 1975, Scott *et al.* 1976, Pollock *et al.* 1979, Pica *et al.* 1980, Webb and McHardy 1986, Webb *et al.* 1986). An extreme case was a flare in OJ 287, which varied by a factor of ~ 2 in the J band ($\lambda \sim 1.25$ μm) in 50 seconds (Wolstencroft *et al.* 1982). In X-rays, most BL Lacs that have been monitored vary by factors greater than two on timescales shorter than a year, and many also vary on shorter timescales of a few hours to a day (Marshall *et al.* 1981, Worrall *et al.* 1981, Snyder and Wood 1984, Madejski 1985, Morini *et al.* 1985, Pollock *et al.* 1985, Singh and Garmire 1985, Brodie *et al.* 1986, Giommi *et al.* 1986, Staubert *et al.* 1986b, Urry *et al.* 1986). In the most dramatic event, the ~ 1 keV X-ray flux of the BL Lac object H 0323+022 decreased by a factor greater than 3 in 30 seconds (Doxsey *et al.* 1983, Feigelson *et al.* 1986). As with the optical event in OJ 287, this kind of ultra-rapid variability is difficult to explain without invoking relativistic bulk motion. Both H 0323+022 and OJ 287 strongly violate the $\Delta L/\Delta t$ criterion of Fabian (1979), requiring efficiencies $\eta \gg 1$; other events observed from the BL Lac object PKS 2155-304 (*e.g.* Morini *et al.* 1985) and the OVV 3C 446 (McHardy 1985) have $\eta \sim 1$. Efficiencies much greater than 1 can result from sudden release of stored energy, and may occur in pair-dominated atmospheres (Guilbert *et al.* 1983b), but $\eta \gtrsim 100$ is not easily explained.

One of the most interesting aspects of BL Lac variability in recent years is the question of periodicity. Several periods have been found in the optical light curve of OJ 287 (Visvanathan and Elliot 1973, Frohlich *et al.* 1974, Carrasco *et al.* 1985) and in its radio emission (Valtaoja *et al.* 1985), all on the order of 20 minutes. Much longer periods, $P \gtrsim 5$ years, were seen in the light curve of 3C 446 (Barbieri *et al.* 1985). Very recently there has been a suggestion of X-ray periodicity in Mrk 421 ($P \sim 4$ hours, Brodie *et al.* 1986), but this is not well established. None of the periods in any source appears to be maintained over long times. Even if this periodic behavior is only transitory, it is still more easily explained in the context of accretion onto a massive black hole, with the periodicity resulting from orbital motion or from cyclic regulation of the accretion rate, than in the context of the relativistic jet scenario.

These observations add up to a picture of violently variable, highly luminous objects. BL Lacs and OVVs have similar continuum spectra, relatively flat and highly polarized from radio to optical frequencies. It is interesting that the leading models for blazars, each of which has some successful features, are so entirely different from one another. Either we are looking directly at the fundamental energy-producing machine, unmodified by any large amount of surrounding gas, or we are fooled into incorrectly inferring high luminosities and short timescales by the special relativistic consequences of the random alignment of a jet with our line-of-sight. The following section discusses some of the virtues and some of the drawbacks of these competing pictures.

III. THEORIES AND MODELS

(a) Accretion

In contrast to quasars, BL Lac objects show no direct evidence of accretion disks: there are no blue bumps in the broad-band spectra, and periodic behavior, if real, involves only a small fraction of the total intensity. However, accretion in BL Lac objects may be relevant because of the analogy to accreting galactic sources. White *et al.* (1984) first noted a striking similarity between the 2-50 keV X-ray spectra of AGN, which are thought to be powered by massive black holes, and galactic binary black hole candidates. In particular, the X-ray spectrum of the BL Lac object PKS 2155-304, probably radiating near its Eddington limit, closely resembles the soft high-state X-ray spectrum of LMC X-3, (see Figure 1 of White *et al.*), a black hole candidate that has an X-ray luminosity on the order of 10% of its Eddington limit. At the same time, a typical Seyfert galaxy has a hard X-ray spectrum, with energy index $\alpha_x \sim 0.7$, much like the low-state X-ray spectrum of Cyg X-1. Also, like Cyg X-1 in the low state, Seyfert galaxies do seem to be emitting less than 1% of their Eddington luminosities (Wandel and Mushotzky 1986). The transition from low state to high state, which has been observed only for the galactic sources, seems to occur near $L_x \sim 0.1 L_{Edd}$. White *et al.* suggested that in both the galactic and extragalactic cases, X-ray emission results from accretion onto a black hole, and that the high-state/low-state transition seen in galactic sources corresponds in the extragalactic case to the difference between Seyfert galaxies and BL Lac objects. Presumably the timescale for the identity change in AGN is much longer than for galactic sources, since it is not observed. As the accretion rate in these sources is cranked up, the luminosity increases until, because of the compactness of the emitting region, runaway pair production occurs, assuming the spectrum extends to ~ 1 MeV (Guilbert *et al.* 1983b). This has the effect of redistributing hard X-ray photons to lower energies, thereby steepening the X-ray spectrum. However, theoretical calculations do not suggest an abrupt transition at $L_x \sim 0.1 L_{Edd}$ (Fabian *et al.* 1986; Fabian, this volume).

Despite the attractive simplicity of this picture, the scaling from a 10 $M_\odot$ galactic black hole to a $10^8 M_\odot$ black hole at the center of an active galaxy may not preserve the spectral homology between the respective systems. If we introduce the dimensionless accretion rate $\dot{m} = \dot{M}/\dot{M}_{Edd}$, where $\dot{M}_{Edd} = L_{Edd}/c^2$, then for many accretion disk models, the temperature of the disk at a dimensionless radius $r = R/3R_S$ scales as $T \propto \dot{m}^{1/4} m^{-1/4} r^{-3/4}$ (Shakura and Sunyaev 1973). The value of the proportionality constant depends on the details of the model (it is on the order of $\sim 10^8$ K for a black hole mass m in units of $M_\odot$), and the powers of m and $\dot{m}$ vary slightly, but the basic scaling law is $T \propto m^{-p}$, where $p \sim O(1/4)$ under many different assumptions. Then if $\dot{m}$ is fixed (*i.e.* the source luminosity is a fixed fraction of the Eddington luminosity, preserving the "low-state" or "high-state" character), the temperature of an AGN disk will be a factor of ~ 100 lower than the galactic binary disk. If the disk radiation is actually observed, as was suggested by Makishima *et al.* (1986) to explain the high-state spectrum of the galactic black hole candidate GX 339-4, then the spectra of galactic and extragalactic objects should look very different. Actually, White *et al.* (1984) proposed that the shape of the high-state spectrum results from pair processes. If the primary [pair-producing] radiation comes directly from the accretion disk, the 1 MeV photons in the galactic binary are only ~ 10 keV photons in the AGN, and no pairs will be created. If instead we scale up in mass while preserving the temperature of the system, then the dimensionless accretion rate is proportional to the mass, and AGN are wildly super-Eddington. This destroys the high-state/low-state transition picture. Of course the energy in the hard primary spectrum may not be closely related to the temperature of the accretion disk. But in that case, pair processes rather than accretion disks are of greatest importance in these objects.

It is also important to note that the X-ray and optical characteristics of BL Lac objects and galactic binary sources do differ in certain key ways. In particular, BL Lacs are highly polarized, while the binaries are not. The X-ray and optical spectra of BL Lacs are lineless; those of the binaries are not. Indeed, there is no evidence at all of thermal (*i.e.* accretion disk) emission in the spectra of BL Lac objects. In all observed bands, from radio through ultraviolet, and at medium X-ray energies and above, there is no evidence for thermal bumps or cutoffs. It is possible that, in analogy to the soft X-ray spectra of some quasars (*e.g.* Elvis *et al.* 1985), the soft low energy X-ray spectra of BL Lac objects could be a thermal component, but in contrast to quasars, the BL Lac thermal component cannot extend longward to a blue bump. The peak of the thermal spectrum must be hidden in the unobserved far-ultraviolet, meaning $T \geq 2 \times 10^5$ K.

Since quasars do show evidence of thermal emission, possibly from an accretion disk, it is worth noting the possible relevance of the temperature scaling described above. If one scales the $T \sim 1$ keV soft spectral component of an $m=10$ high-state galactic binary to an $m=10^9$ AGN, again fixing $\dot{m}$, the equivalent emission appears at $T \sim 10$ eV, or 1240 Å, just where the big blue bump is in quasars. Thus the homology between galactic binary sources and AGN may rightly pertain to quasars.

Lin and Shields (1986) explored another galactic-extragalactic analogy, applying the physics of limit cycles in accretion disks of dwarf novae to supermassive disks in AGN. They found that timescales for the analogous cycles in AGN are quite long, 10^4-10^7 years, and that the zone of instability, a region with $T \sim 2000$-5000 K, is typically far from the hot, highly luminous, inner edge of the accretion disk (except for systems with extremely massive black holes or very low accretion rates). The quasi-periodic dwarf-nova-type instability cycle can still regulate the accretion rate, affecting the luminosity produced at the inner part of the accretion disk. Lin and Shields estimated that the high- and low-state accretion rates differ by less than an order of magnitude, so the difference in luminosity between a high-state and a low-state object is not dramatic. The identification of classes of AGN with high- or low-state accretion, á la White *et al.* (1984), seems unlikely. Lin and Shields (1986) proposed instead that quasiperiodic events in *individual* objects, such as knots in jets, can be explained by the cyclic process, and warned that emission from high luminosity sources with $L \gtrsim 10^{45}$ ergs/s, was undoubtedly dominated by instabilities in the hotter, inner region of the accretion disk, not the relatively cool, dwarf-nova-cycle region.

Do BL Lac objects have galactic counterparts at all? Until more is understood about BL Lac objects, the existence and identity of galactic analogies cannot be established. The importance to BL Lacs of accretion processes in general, and of pair atmospheres in particular, needs to be examined further. One crucial observation involves going to higher energies with greater sensitivity, in order to measure the extent of a hard tail in BL Lac X-ray spectra. The hard tail in these "high-state" objects is supposedly a fraction of the primary hard spectrum that was down-scattered by pairs rather than absorbed in γ-γ collisions. The luminosity near $E \sim m_e c^2$ might indicate the role of pair production in the source. Moreover, the hard tail may also have a significant effect on the ionization balance in BL Lac objects. Guilbert *et al.* (1983a) have suggested that the featureless optical spectra of BL Lac objects result from a high ionization parameter, $\Xi \equiv L/4\pi R^2 cnkT \geq 30$, which in turn results from photoionization by a steep X-ray spectrum. The addition of substantial luminosity in a hard tail could alter this ionization balance. Unfortunately, the presence of high-energy photons may not be detectable at all if they are efficiently transferred to lower energies by pairs. Thus the importance of pair production in AGN may never be directly observed in X-rays, although there is the possibility of observing an annihilation feature in γ-rays.

(b) Jets

In contrast to accretion disks, jets have been discussed extensively in connection with BL Lac objects. Relativistic jets aligned with the line-of-sight would produce radiation with many of the characteristics of BL Lacs: rapid variability, high polarization, high luminosity, and large implied efficiencies (as deduced from $\Delta L / \Delta t$ arguments). In addition, relativistic electrons in the presence of a magnetic field could naturally produce a continuum spectrum like those in Figures 1 and 2 via synchrotron and inverse Compton processes. This section concentrates on the spectral evidence for jets.

Many groups have spent many hours of telescope time obtaining simultaneous multifrequency spectra of BL Lac objects and OVVs (see § IIc). Almost universally these spectra have been interpreted in terms of a synchrotron self-Compton (SSC) model, wherein a single relativistic electron population radiates synchrotron photons and also Compton-scatters them to higher energies. But there the similarity of approach ends. There are almost as many plausible ways to fit a spectrum with an SSC model as there are people who do the fitting. As an example, a number of SSC calculations for PKS 2155-304 are summarized in Table 1. The SSC theory is outlined briefly below and the different approaches of Table 1 are explained.

TABLE 1. SSC CALCULATIONS FOR PKS 2155-304

Calculation	ν_A (GHz)	t_{var} (hours)	δ	B (Gauss)	t_s (hours)	Energy Density
Urry and Mushotzky 1982 (simplest model)	1	60	20-45	10^{-5}	10^8	$u_e \gg u_B$
Urry and Mushotzky 1982 (Königl model)	-	-	10	10^{-4}	10^{11}	$u_e \gg u_B$
Madejski 1985	3×10^4	6	1	6.5×10^3	10^{-4}	$u_e \ll u_B$
	3×10^4	240	1	1.7×10^{10}	10^{-14}	$u_e \ll u_B$
	15	6600	1	6.5×10^2	10^5	$u_e \ll u_B$
Ghisellini *et al.* 1985	-	-	1	10^1-10^2	10^{-8}	$u_e \approx u_B$
Present work	10^4	6	3	4×10^2	10^{-2}	$u_e \ll u_B$

The basic model comprises a stationary volume filled with a uniform density of relativistic particles radiating isotropically (Jones *et al.* 1974a,b; Gould 1979). If the particle distribution is a power law in energy, $n(E)=E^{-p}$, then the emitted flux will be a power law in frequency (to first order), $F_\nu \propto \nu^{-(p-1)/2}$, with steepening above a so-called break frequency, ν_b, due to energy-dependent losses.

According to this model, where the X-ray spectrum is a smooth extrapolation of the ultraviolet spectrum, it is due to synchrotron radiation, with $\alpha_x > (p-1)/2$ because the X-rays are at frequencies above ν_b. Where the observed X-rays lie below an extrapolation of the ultraviolet, then further steepening must have occurred, perhaps due to a cutoff in electrons above some energy. Where the X-rays lie above the extrapolation, they are due to the self-Compton tail.

In fitting SSC models to the data, one generally tries to determine magnetic fields, electron densities, and electron energies such that the observed synchrotron spectrum is produced *and* the inverse Compton emission is not over-produced. That is, the level of self-Compton (SC) emission is predicted directly from the parameters of the synchrotron model, and compared to the observed hard tail or an upper limit to a hard tail. The predicted SC flux depends most strongly on the observed angular size of the synchrotron source, Θ; its self-absorption turnover frequency (below which photons do not get out of the emitting volume), ν_A; the fiducial flux at this frequency, F_A, such that the optically-thin spectrum is given by $F(\nu)=F_A(\nu_A/\nu)^\alpha$; and the Doppler factor δ defined earlier:

$$F_{SC}^{pred} \propto \Theta^{-4\alpha-6} \; \nu_A^{-3\alpha-5} \; F_A^{2\alpha+4} \; \delta^{-2\alpha-4} \tag{1}$$

(Marscher *et al.* 1979). If the angular size is estimated from the variability timescale via $\Theta \sim \delta c t_{var}/D$, where D is the distance to the source, then

$$F_{SC}^{pred} \propto t_{var}^{-4\alpha-6} \; \nu_A^{-3\alpha-5} \; F_A^{2\alpha+4} \; \delta^{-6\alpha-10} \; . \tag{2}$$

To the extent that the predicted SC flux is greater than the observed flux, one can infer a value of δ greater than 1, thus solving the so-called "Compton Catastrophe" (Rees 1966). As Fabian (1983) has pointed out, this ignores the possibility that substantial hard radiation is actually produced but never observed because pair production and related processes efficiently transfer the hard photons to lower energies.

The simplest assumption in fitting an SSC model is that the radio-through-infrared spectrum is due to optically-thin synchrotron radiation. The self-absorption frequencies are then very low, below the lowest radio frequencies observed, so these models invariably predict copious SC X-ray emission that is not observed. This was one interpretation ofy the broad-band spectrum of PKS 2155-304 by Urry and Mushotzky (1982), who inferred relativistic motion with $\delta \sim 25$-40. Values of the parameters ν_A, Θ, δ, magnetic field strength B, and the synchrotron lifetime t_s, as well as the relative energy densities of electrons and magnetic field u_e and u_B, are listed in Table 1. For the simple models, δ, B, t_s, u_e, and u_B are derived quantities. In the more complicated jet models discussed below, they are not well constrained by the available data.

The very flat radio-through-infrared spectrum observed is more likely due to opacity effects than to an intrinsically flat electron distribution with a low self-absorption frequency. Therefore the next step in sophistication is to assume the presence of several distinct radio components, say one dominating every one to two decades in energy, each with some canonical optically-thin synchrotron slope ($\alpha \sim 0.5$) and a different turnover frequency. The superposition of the different turnover frequencies then produces a flat spectral index. This picture is supported by multifrequency VLBI maps of spatially resolved flat-spectrum radio sources (although not BL Lac objects) that reveal a number of separate clumps, each with different synchrotron characteristics. Each component is then tested independently for SC flux level, and thus for the necessity of bulk relativistic motion. This approach was used for the final entry in Table 1, which has a derived Doppler factor $\delta \sim 3$ and a magnetic field of several hundred Gauss. Worrall and Bruhweiler (1982) have used a nicely self-consistent version of the component approach, adding the constraint that the synchrotron lifetime of radiating particles be comparable to the light crossing time in each component. They found that relativistic motion (*i.e.* $\delta > 1$) was required for many of the BL Lac objects considered; PKS 2155-304 was not in their sample.

An often-ignored subtlety of performing SSC calculations on arbitrary synchrotron components is that SC photons are not necessarily *X-rays*. Synchrotron photons will be upscattered by their parent electron population to energies in the range E_{min} to E_{max}, where

$$E_{min} \sim 10^{-4} \, \Theta^{-4} \, v_A^{-3} \, F_A^2 \left[\frac{1+z}{\delta} \right]^3 \text{ keV} \tag{3}$$

and

$$E_{max} \sim 0.4 \left[\frac{v_b}{v_A} \right]^2 E_{min} \tag{4}$$

(Urry 1984, with the exponent of the factor $(1+z)/\delta$ corrected.) These component models frequently produce optical photons rather than X-rays, and the appropriate limit on observed SC flux is then an optical flux.

It is of interest to investigate which synchrotron component in a multiple-component model produces the most SC emission. The components with the highest turnover frequencies are the most compact since the angular size depends most strongly on the turnover frequency:

$$\Theta \propto v_A^{-5/4} \, B^{1/4} \, F_A^{1/2} \, \delta^{-1/4} \, . \tag{5}$$

Combining equations (1) and (5) it is easy to show that, in the absence of strong variation in magnetic field, the predicted SC emission increases with frequency, or equivalently, decreases with increasing angular size:

$$F_{SC}^{pred} \propto v_A^{(4\alpha+5)/2} \, B^{-(2\alpha+3)/2} \, F_A \, \delta^{-(2\alpha+5)/2} \, , \tag{6}$$

or

$$F_{SC}^{pred} \propto \Theta^{-(8\alpha+10)/5} \, B^{-(3\alpha+5)/5} \, F_A^{(4\alpha+10)/5} \, \delta^{-(7\alpha+15)/5} \, . \tag{7}$$

Consider two synchrotron components with self-absorption frequencies v_1 and v_2. From Eqn. (6), the ratio of SC emission from component 1 to component 2 is

$$\frac{F_{SC}^{pred,1}}{F_{SC}^{pred,2}} = \left[\frac{v_{A,1}}{v_{A,2}} \right]^{2\alpha+\frac{5}{2}} \left[\frac{B_2}{B_1} \right]^{\alpha+\frac{3}{2}} \left[\frac{F_1}{F_2} \right] \, . \tag{8}$$

Since the observed radio-through-infrared spectrum is relatively flat, $F_1 \sim F_2$. In the *ad hoc* case $B \sim$ constant, the most compact component clearly produces the most SC emission. In order for a low-frequency component to produce as much SC flux as a high-frequency component, then, its magnetic field must be much smaller. That is, for $F_{SC}^{pred,1} \gtrsim F_{SC}^{pred,2}$,

$$\frac{B_1}{B_2} \lesssim \left[\frac{v_1}{v_2}\right]^{\frac{4\alpha+5}{2\alpha+3}} \sim \left[\frac{v_1}{v_2}\right]^2 , \qquad (9)$$

or $B_1/B_2 \lesssim 4\times10^{-4}$ for realistic assumptions. Thus if the magnetic field is several orders of magnitude stronger in the more compact synchrotron components, as might be expected (see jet models below), the lowest-frequency, largest component dominates the SC emission. Whether this SC emission appears as X-rays is another matter, since very high electron energies ($\gamma_e \sim 10^5$) are required.

One can generalize the finite number of components to a continuous inhomogeneous SSC model; in other words, a simple jet model. Such a model necessarily has more free parameters than the simplest SSC model, but it is still more simple than the physical reality. In general, jet models specify magnetic field and electron density as decreasing power law functions of a radial dimension, in geometries that are cones or parabolas, or a combination of the two. Such models have been discussed by Blandford and Königl (1979), Marscher (1980), Königl (1981), Reynolds (1982), Ghisellini et al. (1985), and Hutter and Mufson (1986). The geometries and underlying physical assumptions, while all perfectly plausible, differ slightly from one model to the next, leading to the very different derived source parameters (magnetic field strength, radiative lifetimes of particles, departure from equipartition, etc.) discussed in the literature. For example, Table 1 illustrates that the early application by Urry and Mushotzky (1982) of a Königl jet model to the broad-band spectrum of PKS 2155-304 gave reasonable agreement for a Doppler factor $\delta \sim 10$, although that was by no means a unique solution. With a different jet model, Ghisellini et al. (1985) fit the same spectrum with $\delta=1$.

Inhomogeneous SSC models do fit the broad-band spectra of BL Lac objects well, but even the simplest jet models have too many free parameters to be well-constrained. The jet models of Königl (1981) do tend to require, or at least easily accomodate, relativistic motion (e.g. Worrall et al. 1986). The models of Ghisellini et al. (1985) do not. The conclusion must be that relativistic beaming is certainly possible in BL Lac objects, and in many ways very attractive, but it is not strictly required by the broad-band spectra.

The similarity of OVV and BL Lac continuum spectra might be an important clue. The observed continuum in OVVs is sufficient to produce the observed emission lines, assuming it is unbeamed (Bregman 1986). Then since the continua are so like those of BL Lacs, the implication is that BL Lac objects cannot be beamed either. This argues against the relativistic jet hypothesis for BL Lac objects. However, it is still possible that the observed blazar continuum is highly beamed, is therefore intrinsically weak, intercepts little of the available emission line gas, and is therefore *not* the photoionizing continuum. Another, probably isotropic continuum component would be needed to account for the emission lines in OVVs. Either this second component would be missing in BL Lacs, or they would have less cold gas. If observations of a large number of OVVs show a close "match" between the observed continuum and the emission line strength, then the jet hypothesis is in trouble, since it is unlikely that the intensity of the beamed radiation (which depends on aspect as well as velocity) would always track the intensity of the photoionizing component.

(c) Winds

There are a number of ways in which the presence of winds in BL Lac objects might be detected. For example, the observation of bremsstrahlung continuum emission from the wind (Beltrametti and Drew 1982) may be feasible with the moderate energy resolution and broad bandwidth of future X-ray detectors. Here we concentrate on the detection of winds via Doppler-broadened absorption features in X-ray spectra that can be detected with energy resolution $E/\Delta E \gtrsim 10$. There is tentative evidence for a wind in one BL Lac object, PKS 2155-304, based on a broad absorption feature near 0.6 keV, or 21 Å (Canizares and Kruper 1984). This observation was made with the Objective Grating Spectrometer (OGS) of the *Einstein Observatory*, which had $E/\Delta E \sim 50$. The observed feature has a width of ~100 eV, which is too narrow to be resolved by current nondispersive X-ray techniques, although it is broad in terms of velocity if Doppler-broadened. Despite the difficulties of identifying a single spectral feature, Canizares and Kruper were able to narrow the possibilities to either an edge or an absorption line of O VIII, largely because of the lack of additional features in the X-ray or ultraviolet. A subsequent measurement of the redshift of PKS 2155-304 (Bowyer *et al.* 1984) suggested that a broadened absorption line was the most likely explanation. The ~100 eV broadening, to the blue side of the $E_{obs} \sim (1+z)^{-1} (654$ eV$) \approx 600$ eV, could be due either to a distribution of absorbers in redshift along the line-of-sight (appropriate for a hot intergalactic medium extending out to $z \sim 0.12$) or to a velocity broadening of the gas in the BL Lac itself (in which case $\beta \sim 0.1$). This last explanation of course fits in well with the relativistic jet picture, as the absorption could be due to a cloud that drifted into the $\beta \sim 1$ beam but had not yet accelerated to the jet flow velocity. In contrast to a popular picture of BL Lac objects as having no surrounding gas, this explanation requires large column densities ($N \sim 10^{23}$-10^{24} atoms/cm^2) of material, almost completely ionized.

Krolik *et al.* (1985) investigated in greater detail the idea of absorption in PKS 2155-304 using a photoionization code with a realistic continuum spectrum, and incorporating an outflowing galactic wind. They were able to produce a feature of the observed strength, but not without postulating a very high mass flux in the wind ($\gtrsim 1000$ M$_\odot$/yr, as compared to ~ 10-50 M$_\odot$/yr in Broad Absorption Line quasars). This mass flux can be reduced if the wind is clumped or anisotropic. Therefore, from this point of view as well, some sort of jet-like structure might be expected.

Unfortunately, PKS 2155-304 is the only extragalactic source ever observed in the X-ray band with such high spectral resolution. No other source was bright enough to give an equivalent spectrum with the OGS. No grating observations of extragalactic sources were done with the EXOSAT experiment. Probably not until AXAF will it be possible to resolve this kind of absorption feature again. However, observations of five BL Lacertae objects with the Solid State Spectrometer (SSS) of the *Einstein Observatory* (Urry *et al.* 1986) do shed further light on the picture. While the O VIII absorption feature was not resolved by the SSS, both because it was at the lower edge of the SSS bandpass and because the energy resolution of the SSS is roughly 10 times worse than that of the OGS, there is nonetheless clear evidence of similar absorption. Five very high quality SSS spectra of PKS 2155-304 and Mrk 501 require absorption in excess of that expected from cold gas in the interstellar medium of our Galaxy. Since measurements at lower energies (*e.g.* Madejski 1985; Agrawal and Riegler 1979; Singh and Garmire 1985; Warwick *et al.* 1985; Morini *et al.* 1985; Staubert *et al.* 1986a) rule out such a large column density of cold gas, the absorption seen in the SSS data must be due to hot material, in direct analogy to the OGS absorption feature. The three other BL Lac objects observed with the SSS, PKS 0548-322, Mrk 421, and H 1218+304, also show evidence of the same kind of excess absorption, although it is not required for an acceptable spectral fit. Therefore it appears that absorption features like that seen by Canizares and Kruper (1984) may not be unusual in BL Lac objects, and may indicate the presence of massive winds. The question remains, does the feature arise in the BL Lac object itself or in intergalactic matter distributed along the line-of-sight? If the absorption is due to a hot intergalactic medium, the same feature should appear in the X-ray spectra of quasars, and it should probably broaden with the redshift of the quasar. Unfortunately, the prospects for testing this possibility are years away.

IV. SUMMARY AND FUTURE PROSPECTS

Several important results have become obvious in the past few years:

(1) The analogy between extragalactic black hole candidates (AGN), and galactic black hole candidates (like Cyg X-1, LMC X-3, and GX 339-4), however intriguing, probably fails to explain BL Lac objects. It may be relevant to quasars, or any AGN with a blue bump. Pair production is likely to be important in at least some AGN, but the origin of the primary high-energy radiation spectrum is unknown and potentially undiscoverable.

(2) Accretion scenarios, where $\dot{m}$ is much larger for BL Lac objects than for other AGN, and relativistic jet scenarios, where $t_{var}^{observed} = \delta\, t_{var}^{rest\,frame}$, both predict that for an object with a given luminosity, BL Lacs vary more rapidly than other AGN. Or equivalently, for a given black hole mass, $t_{var}^{BL\,Lac} < t_{var}^{AGN}$. Both these relations may be true (Bassani *et al.* 1983, Barr and Mushotzky 1986, Wandel and Mushotzky 1986).

(3) Relativistic jets in BL Lac objects are certainly possible and are useful in explaining observed BL Lac properties, but they are not strictly required by broad-band spectral observations.

(4) Special relativistic effects may be required if efficiencies inferred from $\Delta L/\Delta t$ arguments are greater than ~100; the two most extreme BL Lac events imply efficiencies $\eta \sim 70$ (H 0323+022) and $\eta \sim 200$ (OJ 287).

(5) There is evidence of absorption by hot gas in the X-ray spectra of a few BL Lac objects. The absorbing gas is very highly ionized, represents a large column density of material, and is either distributed across a large volume of intergalactic space or is moving at near-relativistic velocities within the BL Lac object. Detailed models need to be pursued further, as do additional observations, but the implications for the structure and energetics of BL Lac objects are profound.

There are clearly observations that will address some of the outstanding questions described above. X-ray polarization measurements would indicate whether the X-rays are synchrotron emission ($P{\sim}O(10\%)$), or result from scattering off a hot corona around an accretion disk ($P{\sim}O(1\%)$), or are due to some other process (unpolarized). VLBI measurements have the potential to detect superluminal motion in BL Lac objects, which would be more direct evidence of relativistic motion than SSC modeling. Multifrequency VLBI would also be very useful for detecting separate synchrotron-emitting clumps and for modeling them more accurately. High energy observations are potentially very important, as the strength and extent of the putative high-energy tail has all-important consequences for pair production scenarios and ionization balance calculations. At the very least, the bulk of the energy in BL Lac objects may be emitted at as yet unobserved hard X-ray and γ-ray energies. Further monitoring of BL Lac objects at all frequencies in order to explore variability timescales is useful. One might expect to detect periodicities due to orbital motions around the central black hole; periods of ≥ 1 hour are expected for a $10^8\,M_\odot$ black hole. Finally, the observation of absorption features in the X-ray spectra of BL Lac objects not only could indicate the presence of large amounts of gas, but could strongly constrain the geometry and ionization structure of the source. These features are likely to be variable, so repeated observations will be necessary. High resolution X-ray spectra of quasars will help determine the origin of the O VIII absorption.

I am grateful to Al Levine and Greg Berthiume for help in generating Figure 1, and to Martin Elvis, Al Levine, Tom Markert, Roland Vanderspek, Peter Vedder, Belinda Wilkes, and Andrzej Zdziarski for helpful discussions. This work was supported by NASA grant NAG 8-494.

REFERENCES

Agrawal, P. C., and Riegler, G. R. 1979, *Ap. J. (Letters)*, **231**, L25.

Agrawal, P. C., Singh, K. P., and Riegler, G. R. 1983, in *Proceedings of 18th International Cosmic Ray Conference*, **1**, 3.

Aller, H. D., Aller, M. F., and Hodge, P. E. 1981, *A. J.*, **86**, 325.

Aller, H. D., Aller, M. F., Latimer, G. E., and Hodge, P. E. 1985, *Ap. J. Suppl.*, **59**, 513.

Altschuler, D. R. 1982, *A. J.*, **87**, 387.

Altschuler, D. R., and Wardle, J. F. C. 1975, *Nature*, **255**, 306.

Andrew, B. H., MacLeod, J. M., Harvey, G. A., and Medd, W. J. 1978, *A. J.*, **83**, 863.

Baldwin, J. E. 1971, in *The Crab Nebula*, eds. R. D. Davies and F. G. Smith, (Dordrecht: D. Reidel), p. 22.

Barbieri, C., Cristiani, S., Omizzolo, S., and Romano, G. 1985, *Astr. Ap.*, **142**, 316.

Barr, P., and Mushotzky, R. F. 1986, *Nature*, submitted.

Bassani, L., Dean, A. J., and Sembay, S. 1983, *Astr. Ap.*, **125**, 52.

Beltrametti, M., and Drew, J. 1982, *Astr. Ap.*, **106**, 153.

Blandford, R. D., and Königl, A. 1979, *Ap. J.*, **232**, 34.

Bowyer, S., Brodie, J., Clarke, J. T., and Henry, J. P. 1984, *Ap. J. (Letters)*, **278**, L103.

Bregman, J. N. 1986, in *Quasars, Proceedings of IAU Symposium No. 119*, eds. G. Swarup and V. K. Kapahi, (Dordrecht: D. Reidel), in press.

Bregman, J. N., Glassgold, A. E., Huggins, P. J., Aller, H. D., Aller, M. F., Hodge, P. E., Rieke, G. H., Lebofsky, M. J., Pollock, J. T., Pica, A. J., Leacock, R. J., Smith, A. G., Webb, J., Balonek, T. J., Dent, W. A., O'Dea, C. P., Ku, W. H.-M., Schwartz, D. A., Miller, J. S., Rudy, R. J., and LeVan, P. D. 1984, *Ap. J.*, **276**, 454.

Bregman, J. N., Glassgold, A. E., Huggins, P. J., Pollock, J. T., Pica, A. J., Smith, A. G., Webb, J. R., Ku, W. H.-M., Rudy, R. J., LeVan, P. D., Williams, P. M., Brand, P. W. J. L., Neugebauer, G., Balonek, T. J., Dent, W. A., Aller, H. D., Aller, M. F., and Hodge, P. E. 1982, *Ap. J.*, **253**, 19.

Bregman, J. N., Maraschi, L., and Urry, C. M. 1986, to be published in *The Scientific Accomplishments of IUE*, ed. Y. Kondo, (Dordrecht: D. Reidel).

Brodie, J., Bowyer, S., and Tennant, A. 1986, preprint.

Canizares, C. R., and Kruper, J. 1984, *Ap. J. (Letters)*, **278**, L99.

Carrasco, L., Dultzin-Hacyan, D., and Cruz-Gonzalez, I. 1985, *Nature*, **314**, 146.

Cavallo, G., and Rees, M. J. 1978, *M.N.R.A.S.*, **183**, 359.

Dent, W. A., and Hobbs, R. W. 1973, *A. J.*, **78**, 163.

Dent, W. A., and Kaptizky, J. E. 1976, *A. J.*, **81**, 1053.

Dent, W. A., Kapitzky, J. E., and Kojoian, G. 1974, *A. J.*, **79**, 1232.

Doxsey, R., Bradt, H., McClintock, J., Petro, L., Remillard, R., Ricker, G., Schwartz, D., and Wood, K. 1983, *Ap. J. (Letters)*, **264**, L43.

Elvis, M., Wilkes, B. J., and Tananbaum, H. 1985, *Ap. J.*, **292**, 357.

Fabian, A. C. 1979, *Proc. Roy. Soc.*, **366**, 449.

Fabian, A. C. 1983, *Proceedings of 18th International Cosmic Ray Conference*, **12**, 109.

Fabian, A. C., Blandford, R. D., Guilbert, P. W., Phinney, E. S., and Cuellar, L. 1986, preprint.

Feigelson, E. D., Bradt, H., McClintock, J., Remillard, R., Urry, C. M., Tapia, S., Geldzahler, B., Johnston, K., Romanishin, W., Wehinger, P. A., Wyckoff, S., Madejski, G., Schwartz, D. A., Thorstensen, J., and Schaefer, B. E. 1986, *Ap. J.*, **302**, 337.

Frohlich, A., Goldsmith, S., and Weistrop, D. 1974, *M.N.R.A.S.*, **168**, 417.

Ghisellini, G., Maraschi, L., and Treves, A. 1985, *Astr. Ap.*, **146**, 204.

Giommi, P., Barr, P., Gioia, I. M., Maccacaro, T., Schild, R., Garilli, B., and Maccagni, D. 1986, *Ap. J.*, **303**, 596.

Glassgold, A. E., Bregman, J. N., Huggins, P. J., Kinney, A. L., Pica, A. J., Pollock, J. T., Leacock, R. J., Smith, A. G., Webb, J. R., Wisniewski, W. Z., Jeske, N., Spinrad, H., Henry,

R. B. C., Miller, J. S., Impey, C., Neugebauer, G., Aller, M. F., Aller, H. D., Hodge, P. E., Balonek, T. J., Dent, W. A., and O'Dea, C. P. 1983, *Ap. J.*, **274**, 101.

Gould, R. J. 1979, *Astr. Ap.*, **76**, 306.

Guilbert, P. W., Fabian, A. C., and McCray, R. 1983a, *Ap. J.*, **266**, 466.

Guilbert, P. W., Fabian, A. C., and Rees, M. J. 1983b, *M.N.R.A.S.*, **205**, 593.

Hall, R., Ricketts, M. J., Page, C. G., and Pounds, K. A. 1981, *Space Sci. Rev.*, **30**, 47.

Hearn, D. R., Marshall, F. J., and Jernigan, J. G. 1979, *Ap. J. (Letters)*, **227**, L63.

Hutter, D. J., and Mufson, S. L. 1986, *Ap. J.*, **301**, 50.

Jones, T. W., O'Dell, S. L., and Stein, W. A. 1974a, *Ap. J.*, **188**, 353.

Jones, T. W., O'Dell, S. L., and Stein, W. A. 1974b, *Ap. J.*, **192**, 261.

Kondo, Y., Worrall, D. M., Mushotzky, R. F., Hackney, R. L., Hackney, K. R. H., Oke, J. B., Yee, H., Feldman, P. A., and Brown, R. L. 1981, *Ap. J.*, **243**, 690.

Königl, A. 1981, *Ap. J.*, **243**, 700.

Krolik, J. H., Kallman, T. R., Fabian, A. C., and Rees, M. J. 1985, *Ap. J.*, **295**, 104.

Landau, R., Golisch, B., Jones, T. J., Jones, T. W., Pedelty, J., Rudnick, L., Sitko, M. L., Kenney, J., Roellig, T., Salonen, E., Urpo, S., Schmidt, G., Matthews, K., Elias, J. H., Neugebauer, G., Impey, C., Clegg, P., Rowan-Robinson, M., and Harris, S. 1986, preprint.

Ledden, J. E., and O'Dell, S. L. 1985, *Ap. J.*, **298**, 630.

Lin, D. N. C., and Shields, G. A. 1986, *Ap. J.*, **305**, 28.

Madejski, G. M. 1985, Ph.D. Thesis, Harvard University.

Makishima, K., Maejima, Y., Mitsuda, K., Bradt, H. V., Remillard, R. A., Tuohy, I. R., Hoshi, R., and Nakagawa, R. 1986, *Ap. J.*, in press.

Maraschi, L., Schwartz, D., Tanzi, E. G., and Treves, A. 1985, *Ap. J.*, **294**, 615.

Marscher, A. P. 1980, *Ap. J.*, **235**, 386.

Marscher, A. P., and Gear, W. K. 1985, *Ap. J.*, **298**, 114.

Marscher, A. P., Marshall, F. E., Mushotzky, R. F., Dent, W. A., Balonek, T. J., and Hartman, M. F. 1979, *Ap. J.*, **233**, 498.

Marshall, N., Warwick, R. S., and Pounds, K. A. 1981, *M.N.R.A.S.*, **194**, 987.

McGimsey, B. Q., Smith, A. G., Scott, R. L., Leacock, R. J., Edwards, P .L., Hackney, R. L., and Hackney, K. R. 1975, *A. J.*, **80**, 895.

McHardy, I. 1985, *Space Sci. Rev.*, **40**, 559.

Medd, W. J., Andrew, B. H., Harvey, G. A., and Locke, J. L. 1972, *Mem. R.A.S.*, **77**, 109.

Miller, J. S., French, H. B., and Hawley, S. A. 1978 in *Pittsburgh Conference on BL Lac Objects*, ed. A. M. Wolfe, (Pittsburgh: University of Pittsburgh), p.176.

Morini, M., Maccagni, D., Maraschi, L., Molteni, D., Tanzi, E. G., and Treves, A. 1985, *Space Sci. Rev.*, **40**, 601.

Mufson, S. L., Hutter, D. J., Hackney, K. R., Hackney, R. L., Urry, C. M., Mushotzky, R. F., Kondo, Y., Wisniewski, W. Z., Aller, H. D., Aller, M. F., and Hodge, P. E. 1984, *Ap. J.*, **285**, 571.

Mushotzky, R. F., Boldt, E. A., Holt, S. S., Pravdo, S H., Serlemitsos, P. J., Swank, J. H., and Rothschild, R. E. 1978, *Ap. J. (Letters)*, **226**, L65.

Mushotzky, R. F., Boldt, E. A., Holt, S. S., and Serlemitsos, P. J. 1979, *Ap. J. (Letters)*, **232**, L17.

Pica, A. J., Pollock, J. T., Smith, A. G., Leacock, R. J., Edwards, P. L., and Scott, R. L. 1980, *A. J.*, **85**, 1442.

Pollock, A. M. T., Brand, P. W. J. L., Bregman, J. N., and Robson, E. I. 1985, *Space Sci. Rev.*, **40**, 607.

Pollock, J. T., Pica, A. J., Smith, A. G., Leacock, R. J., Edwards, P. L., and Scott, R. L. 1979, *A. J.*, **84**, 1658.

Rees, M. J. 1966, *Nature*, **211**, 468.

Reynolds, S. P. 1982, *Ap. J.*, **256**, 38.

Riegler, G. R., Agrawal, P. C., and Mushotzky, R. F. 1979, *Ap. J. (Letters)*, **233**, L47.

Scott, R. L., Leacock, R. J., McGimsey, B. Q., Smith, A. G., Edwards, P. L., Hackney, K. R., and Hackney, R. L. 1976, *A. J.*, **81**, 7.

Shakura, N. I., and Sunyaev, R. A. 1973, *Astr. Ap.*, **24**, 337.

Singh, K. P., and Garmire, G. P. 1985, *Ap. J.*, **297**, 199.

Snijders, M. A. J., Boksenberg, A., Barr, P., Sanford, P. W., Ives, J. C., and Penston, M. V. 1979, *M.N.R.A.S.*, **189**, 873.

Snyder, W. A., and Wood, K. 1984, in *X-Ray and UV Emission from AGN,* eds. W. Brinkmann and J. Trümper, (Munich: Max Planck Institute), p. 114.

Staubert, R., Bazzano, A., Ubertini, P., Brunner, H., Collmar, W., and Kendziorra, E. 1986a, *Astr. Ap.*, in press.

Stein, W. A., O'Dell, S. L., and Strittmatter, P. A. 1976, *Ann. Rev. Astr. Ap.*, **14**, 173.

Staubert, R., Brunner, H., and Worrall, D. M. 1986b, *Ap. J.*, submitted.

Urry, C. M. 1984, Ph.D. Thesis, The Johns Hopkins University.

Urry, C. M., and Mushotzky, R. F. 1982, *Ap. J.*, **253**, 38.

Urry, C. M., Mushotzky, R. F., and Holt, S. S. 1986, *Ap. J.*, **305**, 369.

Valtaoja, E., Lehto, H., Teerikorpi, P., Korhonen, T., Valtonen, M., Terasranta, H., Salonen, E., Urpo, S., Tiuri, M., Piirola, V., and Saslaw, W. C. 1985, *Nature*, **314**, 148.

Visvanathan, N., and Elliot, J. L. 1973, *Ap. J.*, **179**, 721.

Waltman, E. B., Geldzahler, B. J., Johnston, K. J., Spencer, J. H., Angerhofer, P. E., Florkowski, D. R., Josties, F. J., McCarthy, D. D., and Matsakis, D. N. 1986, *A. J.*, **91**, 231.

Wandel, A., and Mushotzky, R. F. 1986, *Ap. J.*, submitted.

Warwick, R. S., George, I. M., McHardy, I., and Pounds, K. A. 1986, *M.N.R.A.S.*, **219**, 39.

Warwick, R. S., McHardy, I., and Pounds, K. A. 1985, *Space Sci. Rev.*, **40**, 597.

Webb, J., and McHardy, I. 1986, in *Proceedings of the Tucson Conference on Continuum Emission from Active Galactic Nuclei*, ed. M. Sitko, (held in Tucson, Arizona, January 1986).

Webb, J., *et al.* 1986, in preparation.

White, N. E., Fabian, A. C., and Mushotzky, R. F. 1984, *Astr. Ap.*, **133**, L9.

Wolstencroft, R. D., Gilmore, G., and Williams, P. M. 1982, *M.N.R.A.S.*, **201**, 479.

Worrall, D. M., Boldt, E. A., Holt, S. S., Mushotzky, R. F., and Serlemitsos, P. J. 1981, *Ap. J.*, **243**, 53.

Worrall, D. M., and Bruhweiler, F. C. 1982, in *Advances in Ultraviolet Astronomy: Four Years of IUE Research*, eds. Y. Kondo, J. M. Mead, and R. D. Chapman, (Greenbelt, Maryland: NASA Conference Publication 2238), p. 181.

Worrall, D. M., Puschell, J. J., Bruhweiler, F. C., Miller, H. R., Rudy, R. J., Ku, W. H.-M., Aller, M. F., Aller, H. D., Hodge, P. E., Matthews, K., Neugebauer, G., Soifer, B. T., Webb, J. R., Pica, A. J., Pollock, J. T., Smith, A. G., and Leacock, R. J. 1984a, *Ap. J.*, **278**, 521.

Worrall, D. M., Puschell, J. J., Bruhweiler, F. C., Sitko, M. L., Stein, W. A., Aller, M. F., Aller, H. D., Hodge, P. E., Rudy, R. J., Miller, H. R., Wisniewski, W. Z., Cordova, F. A., and Mason, K. O. 1984b, *Ap. J.*, **284**, 512.

Worrall, D. M., Puschell, J. J., Rodriguez-Espinosa, J. M., Bruhweiler, F. C., Miller, H. R., Aller, M. F., and Aller, H. D. 1984c, *Ap. J.*, **286**, 711.

Worrall, D. M., Puschell, J. J., Jones, B. Bruhweiler, F. C., Aller, M. F., Aller, H. D., Hodge, P. E., Sitko, M. L., Stein, W. A., Zhang, Y. X., and Ku, W. H.-M. 1982, *Ap. J.*, **261**, 403.

Worrall, D. M., Rodriguez-Espinosa, J. M., Wisniewski, W. Z., Miller, H. R., Bruhweiler, F. C., Aller, M. F., and Aller, H. D. 1986, *Ap. J.*, **303**, 589.

X-ray Spectral Formation in Low Mass X-ray Binaries

N.E. White

EXOSAT Observatory/ESOC
Robert Bosch Str 5
6100 Darmstadt
W. Germany

ABSTRACT

The spectral properties of the non-pulsing low mass X-ray binaries, LMXRB, are reviewed using recent EXOSAT observations. The spectra of the persistent emission from a sample of the lower luminosity, 10^{37} erg/s systems (usually X-ray burst sources), are well modelled by a simple power law attenuated at high energies by an exponential cutoff. Their spectra do not show a strong dependence on inclination. The higher luminosity 10^{38} erg/s systems require an additional blackbody component, whose luminosity ranges from 10 to 40% the total. Various models for the inner region of an accretion disk and its interaction with a neutron star are considered. Blackbody disk models are not appropriate in this case because of the effects of electron scattering. A modified blackbody disk model does not have the correct spectral shape to reproduce the spectra from the lower luminosity sources. It does give a reasonable fit to the spectra of the higher luminosity systems, however the inferred mass transfer rates exceed the Eddington limit by a factor of 3 or more, and are implausibly large. A model where the disk spectrum results from the Comptonization of soft photons on hot electrons gives a consistent fit to the spectra of both luminosity classes. In the higher luminosity systems a second blackbody component, probably from the boundary layer is also required. This blackbody component is absent in the lower luminosity systems, suggesting they contain neutron stars that have been spun up to sub-millisecond periods.

I. INTRODUCTION

While considerable progress has been made towards understanding the spectral properties of the X-ray pulsars and the black hole candidate Cygnus X-1, the spectra of the non-pulsing LMXRB have, until recently, recieved little attention. Early observations of Sco X-1, the prototype LMXRB, had suggested that its entire spectrum from the infrared through to the X-ray could be modelled as a single thermal bremsstralung spectrum (Chodil et al 1968, Neugebauer et al 1969). However, recent higher quality X-ray spectra of Sco X-1, and other related LMXRB, do not give a good fit to this model (White et al 1986). In addition, the observation of non-simultaneous optical/X-ray variations (Canizares et al 1975, Ilovaisky et al 1980), an iron line with a width that is constant irrespective of source state (White, Peacock and Taylor 1985) and the detection of quasi-periodic oscillations (Middleditch and Priedhorsky 1986) are also inconsistent with a single component bremsstralung model.

Over the past 15 years our understanding of the nature of the LMXRB has increased considerably. A number of key observational results (e.g. the detection of orbital periods, X-ray bursts etc.), directly show that these systems are binary systems containing a compact object, usually a neutron star, that is accreting material from a late type companion that fills its Roche lobe. The high specific angular momentum of the gas stream from the inner Lagrangian point results in the flow being mediated by an accretion disk. X-ray illumination of this disk will dominate the optical light. The failure to detect X-ray pulsations from most LMXRB is taken as evidence that the neutron star magnetic dipole field has decayed to a low value and that the accretion disk penetrates close to the neutron star surface. These considerations have resulted in several groups considering the possibility that the spectra of these LMXRB are formed in the inner region of an accretion disk surrounding a neutron star (White, Peacock and Taylor 1985; White et al 1986; Mitsuda et al 1984; Czerny, Czerny and Grindlay 1986).

The structure of the inner region of an accretion disk surrounding a neutron star is not well understood and a number of different possibilities have been considered. One common feature to these models is a two component spectrum, with one component a blackbody spectrum from the boundary layer between the neutron surface and the accretion disk, and the other component from the inner accretion disk. The accretion disk emission is assumed by Mitsuda et al (1984) to be simply the sum of blackbody spectra whose temperature, T, increases as the radius, r, with $T \propto r^{-3/4}$. Czerny, Czerny and Grindlay (1986) follow a similar approach, but also include reprocessing of the boundary layer emission in the disk. It is, however, not clear that such a simple spectral model is applicable in this case, e.g. the 1-100 kev power law spectrum ($\alpha \sim 0.7$) of the black hole candidate Cyg X-1 cannot be described by such a model. The spectrum of Cyg X-1 can be understood if Comptonization dominates the radiative losses (Shapiro, Lightman and Eardley 1976). White, Peacock and Taylor (1985) and White et al (1986) show that a similar model also provides a good description of the spectrum of Sco X-1 and several other related LMXRB.

This paper reviews recent developments in the study and understanding of the spectra of the non-pulsing LMXRB. In §II the 1-20 keV spectral properties of these objects are summarized. A number of different accretion disk models are compared with the data in §III. This includes two new disk models not previously considered for these systems. These results are discussed in §IV. A full account of this work can be found in White, Stella and Parmar (1987).

II. THE OBSERVED SPECTRAL PROPERTIES

The LMXRB can be divided into two groups based on the luminosity of the system. The first group have a luminosity of $\sim 10^{37}$ erg/s and are usually X-ray burst sources with X-ray detected orbital periods of between 50 min and 7 hr. A second group of ~ 10 systems have higher luminosities of $\sim 10^{38}$ erg/s and are often referred to as the bright galactic bulge sources. The failure to detect X-ray orbital modulations in these systems may suggest that the periods are quite long, i.e. greater than a few days (cf. White 1985). This view is supported by the similarity of their properties to those of Cyg X-2 and Sco X-1, which have orbital periods established from optical spectroscopy and photometry of 9 days and 0.8 days repectively. A late type star that fills its Roche lobe and has an orbital period of a few days or longer must be a giant. The nuclear evolution of such a star naturally gives the required mass loss rate (Webbink, Rappaport and Savonije 1983). The factor of ten lower mass transfer rates in the shorter orbital period systems is consistent with the orbits decaying via angular momentum loss (Rappaport, Verbunt and Joss 1983).

The spectral properties of the non-pulsing LMXRB will be described in this section using simple analytic models. While these have little foundation in a physical model, they give an acceptable fit to the data (cf. Swank and Serlemitsos 1985; White, Peacock and Taylor 1985; White and Mason 1985) and allow an unbiased comparison of the various spectra.

The spectra of the persistent emission from the low luminosity class of LMXRB, the X-ray burst sources, are in the majority of cases well fit by power law spectra with a high energy exponential cutoff. This is illustrated in Figure 1 where the spectra from four different burst sources are shown. The details of the EXOSAT observations from which these were taken are given in White, Stella and Parmar (1987). In the case of EXO0748-676 two spectra are given for two different intensity states. The power law photon index Γ was in the range 1 - 2

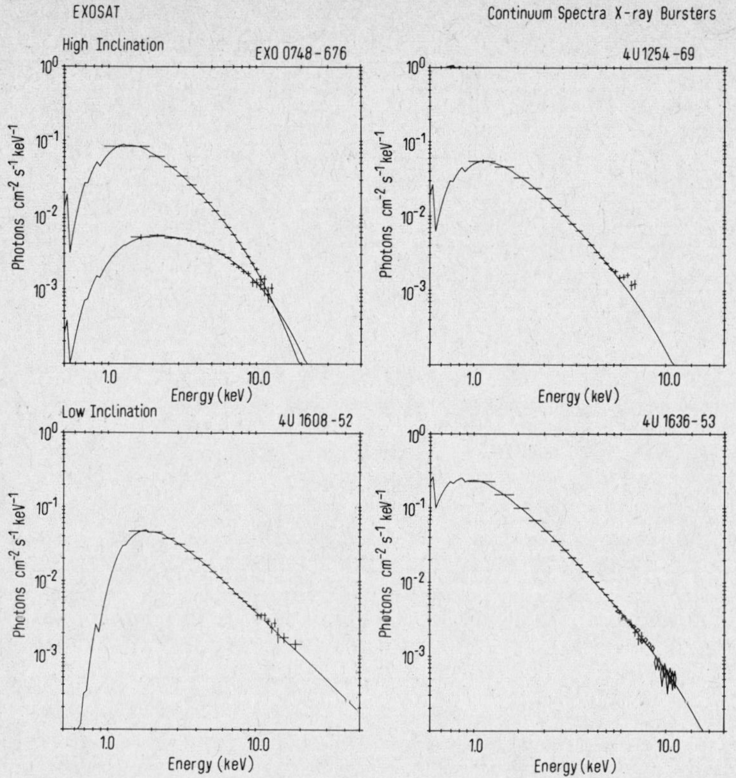

Figure 1. The incident spectra observed by EXOSAT from four low luminosity LMXRB. All four sources are burst sources. The top two show X-ray orbital modulations (periodic dips), indicating their inclinations to be high (~75°). The bottom two do not show any evidence for an X-ray modulation. For EXO0748-676 two spectra are shown taken at different intensity levels. The solid lines represent the best fitting power law times an exponential cutoff model.

for all except the low state spectrum of EXO0748-676 where it decreased to Γ ~ 0. The high energy cutoff varied between 5 and >20 Kev.

Both EXO0748-676 and XB1254-69 show periodic absorption dips in their flux that are caused by material at the outer edge of the accretion disk (cf. Mason 1987). The other two do not show these dips. Eclipses by the companion star are also seen from EXO0748-676, which for a main sequence companion give an estimate for the inclination of ~75° (Parmar et al 1986). The inclination of the two bursters that do not show dips is unknown, however it seems reasonable to assume that it is substantially less than 70°. There is no obvious difference between the spectrum of the persistent emission from a dipper, and that from a non-dipping source. This indicates that the spectral properties of these sources is not strongly dependent on inclination.

The high luminosity LMXRB can be well described by the same power law with an exponential cutoff model, but only if a second, blackbody component is included. The blackbody emission typically accounts for between 10 and 40% of the total flux, and has an equivalent radius for a spherical emitter of ~4-10 km, i.e. similar to or less than the radius of a neutron star (White et al 1986). The characteristic Sco X-1 like flaring activity that is seen from several of these

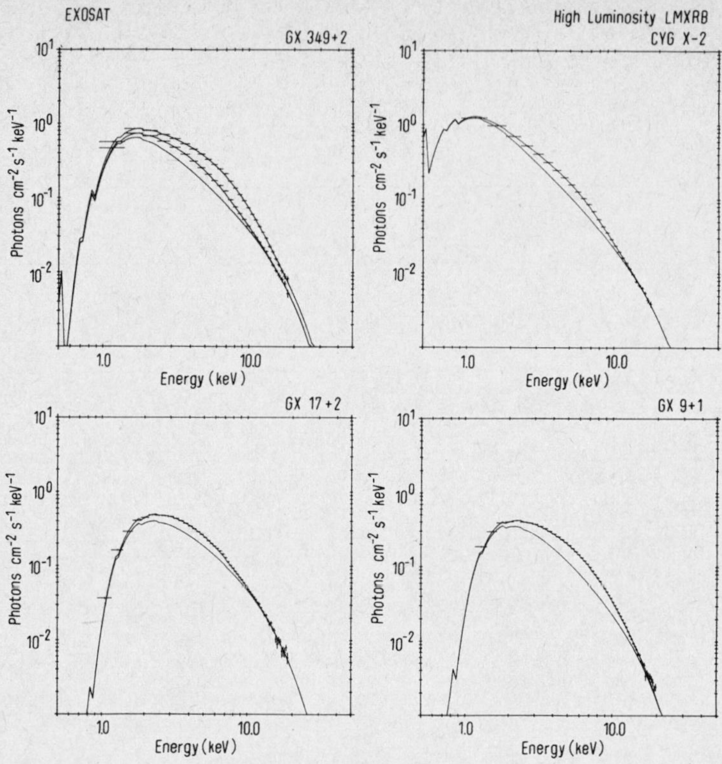

Figure 2: The incident spectra of four highly luminous ($\sim10^{38}$ erg/s) X-ray LMXRB. Two spectra are shown for GX349+2 from the mimimum and maximum of a flare. The solid lines represent the best fitting Comptonization model both with and without the contribution of the blackbody.

sources (cf. Mason et al 1976) is caused by factor of two (or more) increases in the luminosity of the blackbody component. A selection of spectra of these sources is shown in Figure 2. In the case of GX349+2 (Sco X-2, XB1702-363) two spectra are shown from the minimum and maximum of a flare. The contribution of the blackbody component is indicated by the solid lines, with the lower lines indicating the contribution of the power law component. The upper limits to the presence of a blackbody component in the lower luminosity systems range between 5 and 15%, typically a factor of three lower than the blackbody component detected from the higher luminosity LMXRB.

While both the low and high luminosity LMXRB discussed above show factor of 2-3 variations in luminosity, the luminosity distributions do not usually overlap. Because of this it is not immediately obvious whether the additional blackbody component found in the high luminosity systems is present because of an overall evolution of the spectrum with luminosity, or because of a difference in the emission region in the two luminosity classes. There are a number of transient X-ray sources, and a few persistent sources with long term quasi-cyclic intensity variations (see e.g. Terrell and Priedhorsky 1984) that extend over both luminosity classes. These sources can be used to compare the spectrum at both low and high luminosities from a single system. 4U1820-30, in the globular cluster NGC6624, is one such system that varies between a luminosity of 2×10^{37}

erg/s and 6×10^{37} erg/s, with a quasi-period of 176 day. The maximum luminosity seen from this source, corresponds to the minimum of the range found from the bright galactic center sources. X-ray bursts are seen from 4U1820-30 when it is in a low state and the luminosity is similar to that seen from the low luminosity LMXRB. In Figure 3 the spectra recorded in both the low and the high state are shown (taken from Stella, White and Priedhorsky 1987). In the high state the spectrum is similar to that of the bright bulge sources, with blackbody component contributing ~25% the total luminosity. In the low state the blackbody component is still present, with the luminosity increasing to ~50% the total. The spectrum in the low state is quite different from any other burst sources at a similar luminosity level (cf. Figure 1 and 3). This indicates that the presence of the blackbody component in the bright galactic center systems, is not caused by changes in the emission region resulting from the difference in luminosity.

III. ACCRETION DISK MODELS

There exist a number of different theories for the inner region of an accretion disk surrounding a neutron star (or black hole). The simple α disk model considered by Shakura and Sunyaev (1973), where the viscous stress scales as the total gas plus radiation pressure predicts spectra that, for a given accretion rate, can be compared with the observations. For luminosities in excess of 10^{32} erg/s, the opacity of the inner disk is dominated by electron scattering, and the emission will be a modified blackbody spectrum. Because the radiative efficiency is decreased the surface temperature at a given radius will be higher. The inner radiation dominated disk is unstable to secular and thermal instabilities which will probably increase the scale height and it will become optically thin. It has been suggested (see e.g. Stella and Rosner 1984) that in the radiation dominated region the viscous stress should scale with only the gas pressure. In this case the inner disk is stable and optically thick at all radii.

Because the spectrum of the black hole candidate Cyg X-1 is not well described by the simple optically thick disk models, it seems likely that, in that case at least, the energy loss is dominated by Compton cooling of electrons by a source of uv (or longer wavelength) photons (Shapiro, Lightman and Eardley 1976). Depending on the value of the Comptonization parameter $y=4kT_e/m_e c^2$, the spectrum can range between a power law with a high energy cutoff, to a Wein spectrum (cf. Sunyaev and Titarchuk 1980). Such a spectrum may be the signature of an inner disk that has puffed up because of an instability and/or because it is optically thin in the inner region.

The spectral models can be divided as follows into four different types that depend on the assumed structure of the inner disk:

Model A: An optically thick disk radiating with a blackbody spectrum up to the neutron star surface, or the point where the inner disk becomes optically thin. This is the model adopted by Mitsuda et al (1984) and later by Czerny, Czerny and Grindlay (1986). In general it is not relevant to the inner accretion disk surrounding a neutron star in an LMXRB, because electron scattering will dominate the opacity.

Models B and C: An optically thick disk where electron scattering dominates so that it radiates with a modified blackbody spectrum. Model B is for a Shakura and Sunyaev (1973) disk where the viscosity scales as the total pressure and Model C is for a disk where the viscosity is only a function of the gas pressure (Stella and Rosner 1984). The predicted spectra from these two models are similar. For model B an optically thin region will develop at small radii (cf Shakura and Sunyaev 1973) when $\alpha > 10^{-3}$; in this model α was fixed at 10^{-3}, for model C α was fixed at 1.

Model D: Comptonization dominates the spectral formation. This uses the solution to the Kompaneets equation given by Sunyaev and Titarchuk (1980) where soft photons are scattered on hot electrons. For the case of unsaturated Comptonization (y~1), the spectrum from this model can be approximated as a power law with an exponential cutoff, the same as the analytic function used in the previous section (cf. Eardley et al 1978).

In models A, B and C the free parameters are the mass accretion rate, the viscosity α (B and C only), and the distance to the source scaled by cosine of the inclination (to account for limb darkening effects). For model D the free parameters are the normalization, the temperature of the scattering electrons, kT_e, and their optical depth, τ. It is important to note that while the blackbody disk models A-C uniquely predict the observed spectrum for a given mass acceretion rate, distance and inclination, whereas model D only gives a measure of the optical depth and temperature of the scaterring region, without information on the mass accretion rate or the geometry. While model A is not appropriate it will nonetheless be included to provide a reference point with earlier work.

A blackbody component was included to account for any emission from the boundary layer. The energy released is equal to the difference between the kinetic energy of the orbital velocity of the Keplerian disk and the surface rotational velocity of the neutron star. Both the temperature and the normalization were kept as free parameters. Figure 4 shows how the fraction of the blackbody component relative to that from the disk depends on the rotation period of the neutron star. For slow rotation periods (>1 ms) one half of the total luminosity is emitted from the boundary layer. As the neutron star rotation period approaches its limiting break up value, the fractional luminosity decreases to zero. Viewing geometry is another factor that influences the relative strength of the blackbody versus the disk component. The height of the boundary layer will be small relative to the radius of the neutron star and the emission will be preferentially radiated back into the disk plane. In the blackbody disk models A-C limb darkening will enhance this effect. Thus the blackbody component should be dominant in systems viewed at high inclinations. The relative strength of the blackbody component was included as a free parameter.

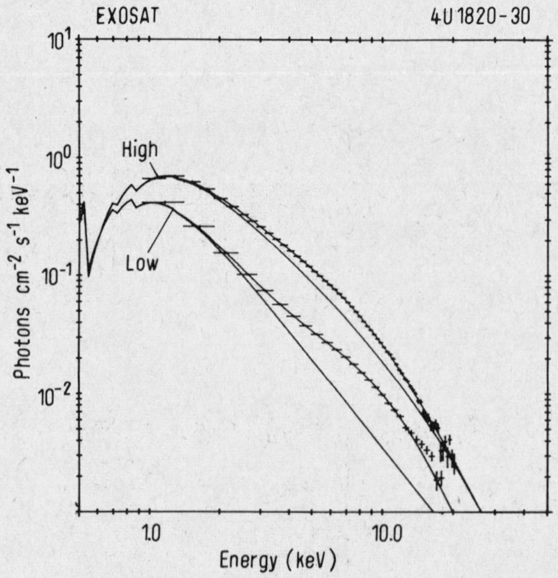

Figure 3: The high and low state spectra of 4U1820-30. The solid lines show the best fitting power law plus blackbody model, both with and without the blackbody.

Models A-C cannot reproduce the spectra of the lower luminosity LMXRB because the predicted spectra decay too rapidly with energy. In Figure 5 the spectrum of XB1608-52 is compared with model A, both with and without a blackbody component. As can be seen, the model spectra fall off much to rapidly. This is analogous to the problem found with modelling the spectrum of Cyg X-1. In contrast Model D gives a good fit to the data with typical values of $kT_e \sim 3$ keV and $\tau \sim 12$, which gives a value of $y \sim 3$.

For the bright bulge sources models A - C all give reasonable fits, although the decomposition of the contribution of the blackbody and disk components was very different in the two cases. In Figure 6 the fits to models A and C are illustrated. For model A the spectrum at high energies is dominated by the blackbody component, whereas for the other models the blackbody is only responsible for a small part of the flux in the 2-10 kev band. The problem with these models is that for all three models the derived mass accretion rates are in excess of the Eddington limit by a factor of 3 or more. Such high accretion rates would cause the accretion disk to undergo strong mass loss from the central regions, which will limit the X-ray luminosity to approximately the Eddington limit (Shakura and Sunyaev 1973). As with the lower luminosity systems model D again provides a good fit to the data. The derived Comptonization parameters are similar to those obtained for the low luminosity LMXRB.

The blackbody component was only required in the high luminosity cases. In the lower luminosity systems it was not required with a luminosity greater than ~10% the total. The luminosities found for the high luminosity systems were similar to those found for the simple analytic model given in the previous section.

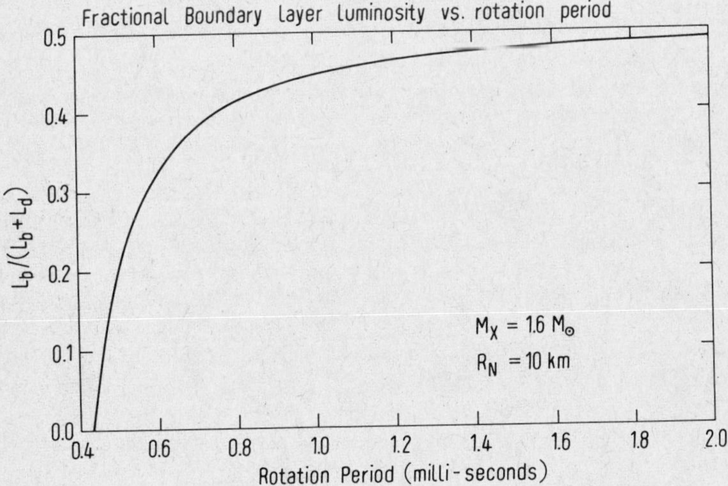

Figure 4: The fraction of the accretion luminosity released in the boundary layer as a function of the rotation period of the neutron star.

IV. DISCUSSION.

The spectra of the non-pulsing LMXRB are best understood if the energy loss in the inner accretion disk is dominated by the Comptonization of low energy photons on hot electrons, similar to the mechanism proposed to explain the spectrum of Cyg X-1. Accretion disk models where the spectrum is the sum of multi-temperature blackbody are not correct because electron scattering dominates the opacity. Blackbody disk spectra, modified to take into account electron scattering, do not work either. In the low luminosity LMXRB the blackbody disk models give rather soft spectra and cannot reproduce the observed power law spectra. In the case of the higher luminosity galactic center sources the accretion rates required to reproduce the observed spectra are well in excess of the Eddington limit and are inconsistent with the model (this would also be the case for an unmodified blackbody disk model if it was appropriate).

The value of the Comptonization parameter $y \sim 3$ found for the LMXRB is similar to that obtained for the spectrum of Cyg X-1 by Sunyaev and Trumper (1979). This is in spite of the fact that the the temperature and τ differ considerably, with Cyg X-1 having values of 27 kev and 5, compared to typical values of 3 kev and 12 for the LMXRB. These differences may reflect the smaller size of the emission region in the neutron star systems increasing the optical depth, or the change in the X-ray luminosity relative to the Eddington limit.

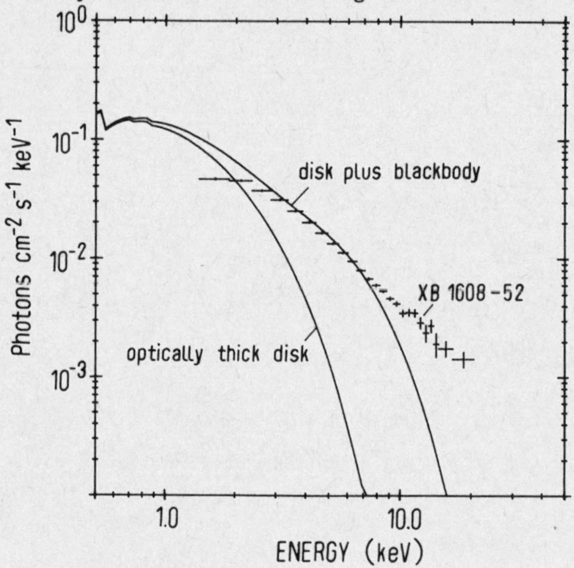

Figure 5: A comparison of the incident spectrum of XB1608-52 with model A, the multi-temperature blackbody accretion disk model (first suggested by Mitsuda et al 1984). The predicted spectrum is given both including and excluding a contribution from the boundary layer with a luminosity of 50% the total. This model does not provide a good desciption of the data.

Only the high luminosity bright galactic center sources require the presence of a second, blackbody component that would represent emission from the boundary layer. The fact that a strong blackbody component is not detected from the lower luminosity LMXRB is puzzling. It is unlikely that the boundary layer is hidden because the neutron star surface is clearly seen when bursts occur. Also in 4U1820-30 where the luminosity cycles between high and low values, the blackbody emission from the boundary layer is clearly detected in the low state (with a higher relative luminosity than in the high state). This indicates that a blackbody component should also be detected in the other low luminosity systems. This leads to the conclusion that the blackbody is absent in the majority of the low luminosity LMXRB because the neutron star has been spun up to close to its break up period, and that little energy is released in the boundary layer (cf. Figure 4). This would require that the neutron stars in these systems have extremely rapid rotation periods of ~0.5 ms. While this is very fast, recent work by Friedman et al (1986) indicates that in principle it may be possible to spin a neutron star up to breakup, rather than the gravitational radiation limited value of 1.5 ms suggested earlier by Papaloizou and Pringle (1978). How close to break up the neutron star will get is not clear, however the X-ray

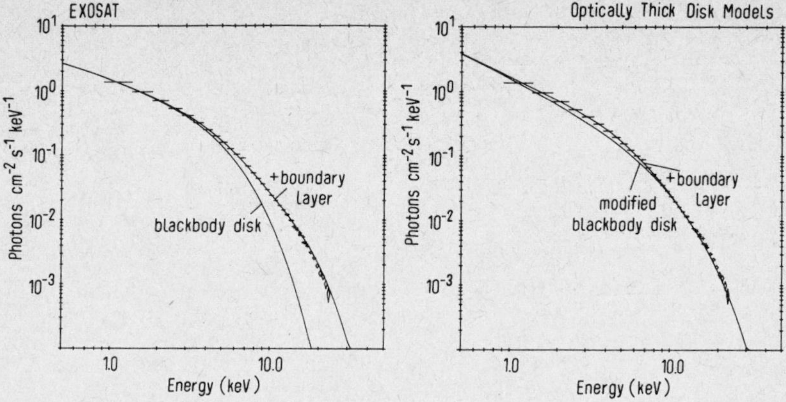

Figure 6: A comparison with the data of the best fitting models B and C. The contribution of the blackbody is shown. While these two models can give an acceptable fit to the data, they require mass accretion rates well in excess of the Eddington limit, which are inconsistent with the model.

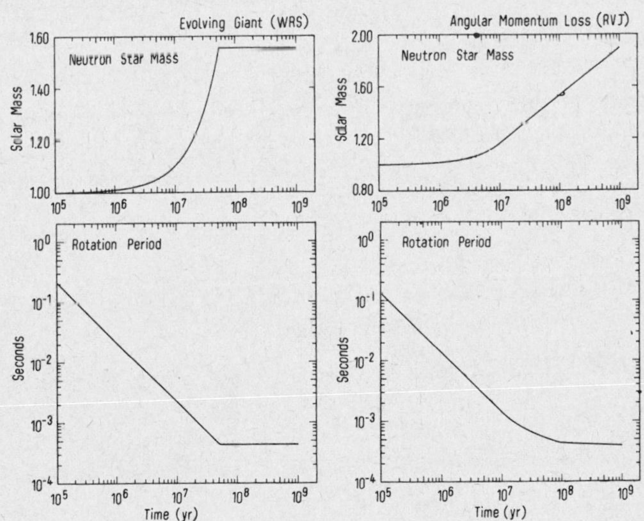

Figure 7: The spin and mass history of a neutron star in a LMXRB for the evolutionary models considered by Rappaport, Verbunt and Joss (1983) - RVJ, and Webbink, Rappaport and Savonije (1983) - WRS. The initial spin period of the neutron star was assumed to be slow (700 s) and the initial mass 1 M$_\odot$. The spin up is assumed to stop when the neutron star reaches its limiting break up value. In the WRS model the mass accretion rate stops after a few 10^7 yr.

spectra indicate that in most cases in all cases some sort of equilibrium has been reached where the energy released in the boundary layer is minimal.

The above discussion requires that in the bright galactic center sources the neutron stars are still spinning relatively slowly, compared to their counterparts in the X-ray lower luminosity systems. This may be consistant with the two different evolutionary models suggested by the difference in accretion rate of the two luminosity classes. In Figure 7 the spin and mass history of an accreting neutron star is shown for the evolutionary model of Rappaport, Verbunt and Joss (1983) and Webbink, Rappaport and Savonije (1983). Because the lifetime of the evolving giant scenario is relatively short the neutron star still falls short of reaching its limiting breakup value at the end of the mass exchange. In the angular momentum loss scenario the much longer time over which accretion occurs, combined with an initial phase of higher mass transfer, spins the neutron star up to its limiting spin period in the first 10% of the lifetime of the system.

Lately, there has been considerable interest recently in the possibility that the magnetic dipole moment of a neutron star may decay to a limiting value of order 10^8-10^9 Gauss (Kulkarni 1986; van den Heuvel, van Paradijs and Taam 1986). Such a field may cause a small magnetosphere to form with a radius of a few tens of kilometers. The interaction and penetration of the accretion flow into such a magnetosphere has been suggested as responsible for the quasi-periodic oscillations discovered by EXOSAT from many of the high luminosity LMXRB (cf. van der Klis 1987; Lewin 1987). Any disruption of the accretion flow prior to the neutron star must increase the relative contribution of the blackbody component to that from the disk. The blackbody component will rapidly increase to dominate the total, unless the interaction of the magnetosphere with the disk transports neutron star spin energy into the disk (Priedhorsky 1986). The lack of any strong blackbody component from the lower luminosity sources argues that if there is a residual magnetic field on the neutron stars in these systems, then it does not disrupt the accretion disk. The strength of the blackbody component in the high luminosity systems is similar to or less than that expected from the simple boundary layer model and again argues against any major disruption of the accretion disk, unless a bloated inner disk obscures the central region (Lamb 1985; van der Klis et al 1987).

ACKNOWLEDGEMENTS

Luigi Stella and Arvind Parmar are thanked for their contributions to this work.

REFERENCES.

Canizares, C.R., et al. 1975. Ap.J. 197, 457.
Chodil, G., et al 1968, Ap.J., 154, 645.
Czerny, B., Czerny, M. and Grindlay, J.E. 1986. Ap.J. in the press.
Eardley, D.M., Lightman, A.P., Payne, D.G., and Shapiro, S.L. 1978, Ap.J.
    224, 53.
Friedman, J.L., Ipser, J.R. and Parker, L. 1984. Nature, 312, 255.
Ilovaisky, S.A., Chevalier, C., White, N.E., Mason, K.O., Sanford, P.W.,
    Delvaille, J., and Schnopper, H.W. 1980, Ap.J., 191, 81.
Kulkarni, S.R., 1986, Ap.J., 306, L85.
Lamb, F.K. 1986. in in The Evolution of X-ray Binaries, [Reidel:ed. J. Truemper,
    W.H.G. Lewin and W. Brinkman]. page 151.
Lewin, W.H.G. 1987. these proceedings.
Mason, K.O. 1987. these proceedings.
Mason, K.O., Charles, P.A., White, N.E., Culhane, J.L., Sanford, P.W., and
    Strong, K.T.,1976. M.N.R.A.S., 177, 513.
Middleditch, J. and Priedhorsky, W.C. 1986, 306, 230.
Mitsuda, K. et al 1984, P.A.S.J., 36, 741.
Neugebauer, G., Oke, J.B., Becklin, E., and Garmire, G. 1969. Ap.J., 155, 1.
Papaloizou, J. and Pringle, J.E. 1978. M.N.R.A.S., 184, 501.
Parmar, A.N., White, N.E., Giommi, P. and Gottwald, M. 1986, Ap.J. in press.
Priedhorsky, W.C. 1986, Ap.J. 306, L97.
Rappaport, S.A., Verbunt, F. and Joss, P.C. 1983. Ap.J. 275, 713.
Shapiro, S.L., Lightman, A.P. and Eardley, D.M. 1976, Ap.J., 204,187.
Shakura, N.I. and Sunyaev, R.A. 1973, Astr.Ap., 24, 337.
Stella, L. and Rosner, R. 1984. Ap.J., 277, 312.
Stella, L., White, N.E. and Priedhorsky, W.C., 1987. Ap.J., in press.
Sunyaev, R.A., and Titarchuk, L.G., 1980, Astr. Ap. 86, 121.
Sunyaev, R.A., and Trumper, J. 1979, Nature, 279, 506
Swank, J. and Serlemitsos, P.J. 1985. in Galactic and Extragalactic Compact
    X-ray Sources [ISAS:ed. Y. Tanaka and W.H.G. Lewin]. page 175.
Terrell, J. and Priedhorsky, W.C. 1984. Ap.J., 285, L15.
van den Heuvel, E.P.J., van Paradijs, and Taam, R.E., 1986. Nature, 322, 153.
van der Klis, M.K. 1987, these proceedings.
van der Klis, M.K., Stella, L., White, N., Jansen, F. and Parmar, A.N.
    1987. Ap.J. in press.
Webbink, R.F., Rappaport, S.A. and Savonije, G.J. 1983, Ap.J. 270, 678.
White, N.E., 1985. in The Evolution of X-ray Binaries, [Reidel:ed. J. Truemper,
    W.H.G. Lewin and W. Brinkman]. page 227.
White, N.E. and Mason, K.O. 1985. Sp. Sci. Rev. 40, 167.
White, N.E., Peacock, A. and Taylor, B.G. 1985. Ap.J. 296, 475.
White, N.E., Peacock, A., Hasinger, G., Mason, K.O., Manzo, G., Taylor, B.G.,
    and Branduardi-Raymont, G. 1986. M.N.R.A.S., 218, 129.
White, N.E., Stella, L. and Parmar, A.N. 1987 in preperation.

CONTINUUM FEATURES IN QUASARS

Martin Elvis,
Bożena Czerny[1],
and
Belinda J. Wilkes
Harvard-Smithsonian Center for Astrophysics
Cambridge, MA 02138, USA

1. The Quasar Continuum

The quasar phenomenon is an essentially broad-band one. Quasars emit almost constant power per decade of frequency from 100μm to at least 100 keV. Although this is more than a factor of 10^6 in frequency it is possible to plot clearly the power in the continuum of the quasar 3C273 on a <u>linear</u> scale throughout this range (Figure 1, Carleton et al. 1987). The nature of the physical

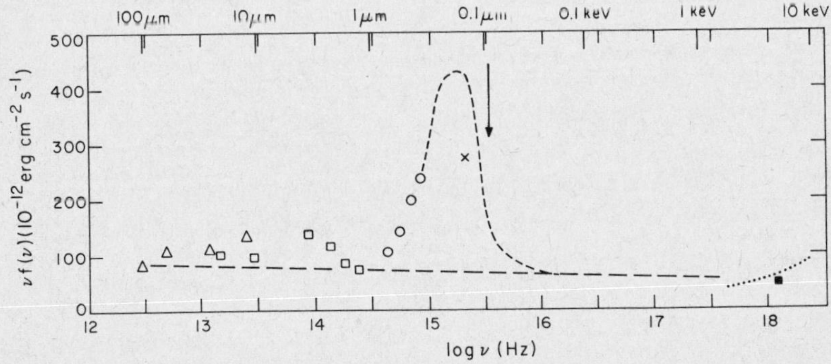

Figure 1.   Linear   energy distribution (νf_ν vs. log ν) for 3C273.

processes that produce this highly luminous continuum is one of the major problems, perhaps even the central problem, of quasar research. Energetics arguments as well as observed rapid variability tell us that the continuum originates closer to the central object than any other component of the emission. In principle then the continuum can give us information about the region of greatest interest in a quasar. Unfortunately, as its name implies, the continuum has very few features that we can measure and so provides us with few

---

1. Present address: Astronomy Dept., Univ. of Leicester, Leicester, England. On leave from N. Copernicus Astronomical Institute, Warsaw, Poland.

quantities that can constrain models. The situation could hardly be more different from emission line studies where there are so many parameters to measure that it is unclear which are important and which peripheral. In this paper we discuss the features which are seen in the broad-band continua of 'normal' (non-OVV) quasars.

The democratic distribution of quasar luminosity almost equally amongst frequency ranges (Figure 1 again) has led to their discovery in every waveband surveyed. As each such discovery was made the claim was made by waveband partisans that "at last we have found where the bulk of a quasar's luminosity is emitted".   We were all wrong. The different discovery techniques have led to a confusing variety of active galaxy names. (We recently counted 21 names.) Only in the last three years or so has it become possible to construct the overall continuum energy distribution for a reasonable sample of quasars by using a combination of satellite (<u>Einstein</u>, IUE, IRAS) and ground-based (optical, $1-20\mu$m infrared, radio) data. From these studies a far simpler view of the underlying continuum is emerging.

At low luminosities ($\lesssim 10^{44.5}$ erg s^{-1}, 2-10 keV, $M_v \gtrsim -23$) many complications exist. Contamination of the nuclear spectrum by host galaxy starlight becomes important at these luminosities (Stein, Tokunaga, and Rudy 1985, Edelson and Malkan 1986, Ward et al. 1987) as do the modifying effects of dust, both in emission and absorption (Lawrence and Elvis 1982, Ward et al. 1987, Carleton et al. 1987). At high luminosities however there seem to be only two major distinctions to be made: between the highly variable, highly polarized 'Blazars' and other quasars; and, among the other quasars, between radio-loud and radio-quiet objects.   [The 'red quasars' (Rieke, Lebofsky and Wisniewski 1982, and references therein) may require that a third distinction be made. However, their broad-band continua are not yet well studied.] The distinctions that we see easily at high luminosities probably carry over into low luminosity objects once allowance is made for the contamination mentioned above. The effects of dust may tell us a great deal about the outer structure of active nuclei (Lawrence and Elvis 1982, Carleton et al. 1987) but here we are concerned with the primary underlying 'non-thermal' continuum.

'Blazars' (BL Lacs and Optically Violent Variables, OVVs, Angel and Stockman 1980) form a distinct class. While they are important for the physical clues they provide the Blazars are rare. They make-up only 2% of all known active galaxies (Vèron-Cetty and Vèron 1985). Blazar properties are covered in this volume by M. Urry. Here we discuss the continuum features of the less dramatic, garden variety, quasars. These are much less variable and all have strong emission lines.

In Figure 2a the radio to x-ray energy distributions (log νf_ν vs. log ν) of two typical quasars, 3C273 and Mkn 509 are shown. In the 1-100μm infrared they are both almost flat (Ward et al. 1987). A single nearly horizontal power-law fits the IR points reasonably well and intersects the x-ray point at about 1 keV. In a number of radio-quiet quasars the same power-law seems to continue through into the 0.1-3 keV x-ray range (light line in fig 2b, see also fig 7,

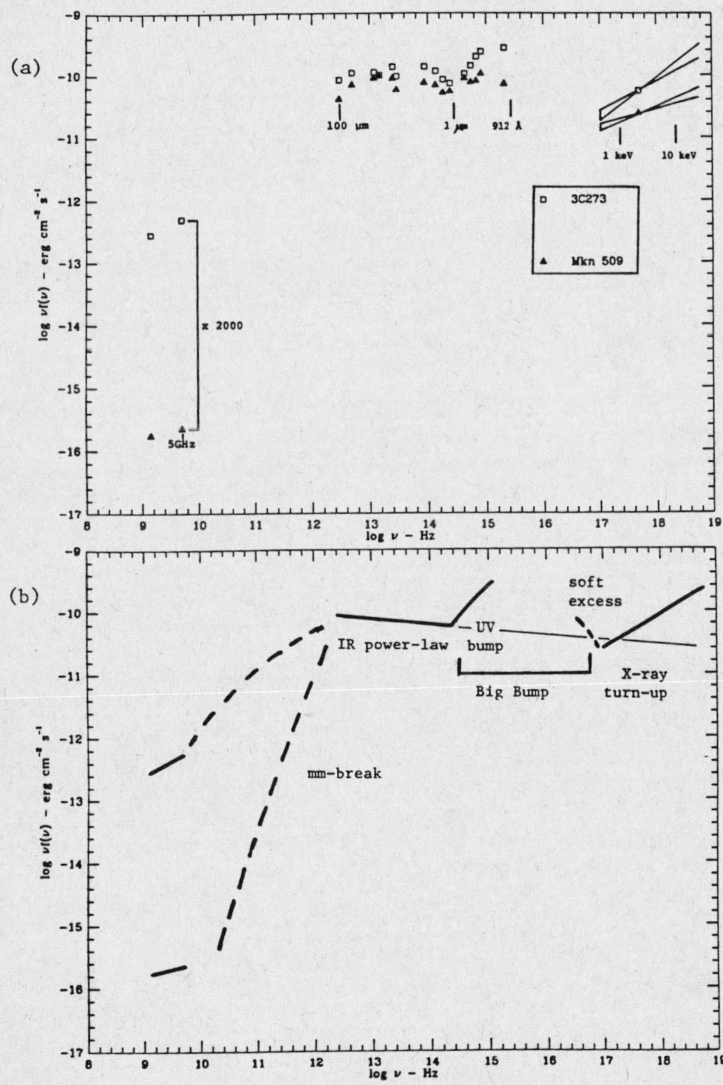

Figure 2. (a) radio-to-x-ray energy distributions for a typical radio-loud quasar (3C273) and a typical radio-quiet quasar (or Seyfert galaxy, Mkn 509).
(b) Schematic representation of 2(a) marking the continuum features referred to in this article.

Elvis et al. 1986). Deviations from this baseline power-law are what we will describe as continuum features. Altogether three continuum features can be identified with certainty. We show a schematic quasar spectrum in figure 2b. Here we have marked the features that we will discuss in the rest of this article.

The infrared baseline power-law always drops in the millimeter band ("mm-break", fig 2b) but the size of the drop varies dramatically from object to object. Quasars in which the drop is only 2 decades are called "radio-loud" (e.g., 3C273). The majority of quasars however drop by 5 or even 6 decades and are called "radio-quiet" (e.g., Mkn 509). The size of the mm-break is the major feature in normal quasar continua. Its size does not seem to affect the infrared to ultraviolet continuum shape.

The optical-ultraviolet continuum rises above the horizontal infrared power-law and forms a "uv-bump" (discussed by M. Malkan, this volume). A corresponding rise is occasionally seen in the lowest energy x-rays above the higher energy x-ray power-law (Elvis, Wilkes, and Tananbaum 1985, Bechtold et al. 1987, Arnaud et al. 1985, Singh et al. 1985). It is natural to suggest that the "soft x-ray excess" is the high energy extension of the "uv-bump" and that together they form a "big bump" which is second major feature in normal quasar continua. This big bump is most often discussed in terms of accretion disk emission (M. Malkan, this volume).

X-ray spectra of quasars (or, strictly, of high luminosity Seyfert 1 galaxies) often rise to high frequencies in log νf_ν vs. log ν energy distributions (Mushotzky 1984). They cannot then be an extension of the flat, or slightly falling, infrared power-law. A new emission component must be emerging in the x-rays in these objects. This "x-ray turn-up" near 1 keV is the third clear feature of normal quasar continua. It is not always present (Elvis et al. 1986).

Malkan (this volume) discusses a fourth, smaller, feature that is prominent around 3 to 5 microns, the "5μm hump" (Edelson and Malkan 1986). Since he covers this feature fully we will not consider it further here.

The next sections of this article will discuss the uv/x-ray bump in terms of accretion disk models and a connection between the x-ray turn-up and the mm-break.

2. UV/X-ray Bump and Accretion Disks[2]

a. Difficulties with sum-of-black-body accretion disk models

The idea that the optical/UV/soft x-ray big bump is thermal emission from an accretion disk is appealing since it immediately defines a geometry for the nucleus and allows us to make estimates of the central mass and accretion rate. However this interpretation has some difficulties with the data. We show below that two of these difficulties are the result of using disk models that are two simplistic.

The quasars to which disk models have been fitted (Malkan, this volume, Malkan and Sargent 1982, Malkan 1983) all show a turn-down in the UV at around 1500 - 2000 Å. This is readily interpreted as indicating a maximum temperature around 25-30,000 K. Such a uniform maximum temperature is surprising given the large range of accretion rates, central masses and inclination angles expected in quasars. It is also inconsistent with the discovery of AGN with strong soft x-ray excesses (e.g., PG1211+143, Elvis, Wilkes and Tananbaum 1985, Bechtold et al. 1987) if these are due to extensions of the accretion disk spectrum. More evidence against a single maximum temperature comes from Bechtold et al. (1984) who have shown that the UV spectra of high redshift quasars extend shortwards of 600 Å without dropping exponentially as would be expected for a maximum temperature of 30,000 K. Furthermore a simple sum-of-black-bodies model for the 'soft excess' quasar PG1211+143 gives an accretion rate about 20 times the Eddington limit (Bechtold et al., 1986). This is not only physically implausible, it is also inconsistent with the thin disk assumption of the models since radiation pressure puffs up the inner regions of disks when they go super-Eddington. A simple uniform maximum temperature does not seem to be a sufficient description of the big bump, but simply to allow higher, soft x-ray, temperatures lead to physically inconsistent models.

We have explored the possibility that opacity effects modify the disk spectrum in an important manner away from the simplest sum-of-black-bodies spectrum (Czerny and Elvis, 1987). These effects have the hoped for properties of predicting a turndown in the UV and of greatly reducing the super-Eddington problem. We describe these models in the next section and then look at the effect of different viewing angles on the emergent spectrum when the disks are not completely geometrically thin.

---

2. This section has also been published by M. Elvis and B. Czerny in the proceedings of the NOAO Workshop on Quasar Continua, ed. M. Sitko (Tucson: NOAO), 1986.

b. Opacity Effects

In real disks, opacity effects will be important just as they are in stars with similar temperatures (the expected surface gravities in the disks are similar to stars). To make a complete model taking this into account would be a large task. We can, however, use some approximations that keep the task simple while showing the size of the opacity effects.

Figure 3 shows a cross-section of the disk in terms of its total optical depth (measured from its surface) as a function of frequency. The free-free absorption coefficient goes as ν^{-3} (Rybicki and Lightman 1979) so that electron scattering becomes relatively more important at high frequencies. The surface of last absorption from which a black body spectrum is emitted (hatched line in Figure 3) drops toward the disk equator at high frequencies. The shape of the spectrum is modified above a frequency ν_0 at which $\tau_{absorption} = \tau_{electron\ scattering}$. ν_0 may be calculated from the disk density and temperature. Shakura and Sunyaev (1973)

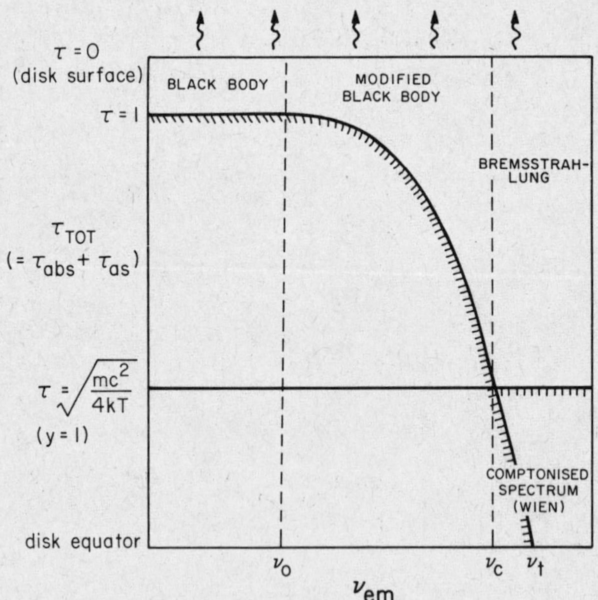

Figure 3. Accretion disk cross-section (in optical depth) versus frequency.

give formulae for disk density and temperature at a given radius and for these quantities in turn as a function of the basic disk parameters: M, central mass; ṁ, accretion rate (both in solar units) and α, the viscosity parameter. We have had to modify the Shakura and Sunyaev formulae when calculating an emergent spectrum to give continuity between the different regions (gas-dominated, radiation-dominated, ...; see Czerny and Elvis 1987).

Applying these formulae (Shakura and Sunyaev case (a), i.e., $P_R \gg P_{gas}$, $\kappa_{es} \gg \kappa_{abs}$) results in an expression for T_o, the temperature at which $\tau_{abs} = \tau_{es}$:

$$T_o = 1.6 \times 10^6 \cdot \frac{1}{\dot{m}^{4/11}} \cdot \frac{1}{M^{3/11}} \cdot \frac{1}{\alpha^{3/11}} \qquad (1)$$

The low powers in equation (1) makes the temperature T_o only weakly dependent on the disk parameters. For plausible quasar parameters ($M = 10^7$, $\dot{m} = 10$, $\alpha = 0.1$) $T_o = 1.6 \times 10^4$ K. Changing M, $\dot{m}$ or α by a factor 10 only changes T_o by a factor 2.

The effect of this opacity on the disk spectrum is illustrated in Figure 4. The dashed line is a normal sum-of-black-bodies disk spectrum. The solid curve is the modified black body produced by our method. At a frequency ν_o our disk spectrum starts to drop below the pure sum-of-black-bodies model. This happens in the ultraviolet for a wide range of model parameters. The temperature T_o corresponds to a frequency $\nu_o = 3.9kT_o/h$, so log $\nu_o = 15.47$ (1025 Å). (h$\nu = 3.9$ kT corresponds to the peak of a Planck function in νf_ν space, i.e., the peak energy input.) This value of ν_o includes non-Hydrogen bound-free opacity via a Rosseland mean opacity approximation. Opacity thus provides a natural explanation for the UV turn-down seen by Malkan.

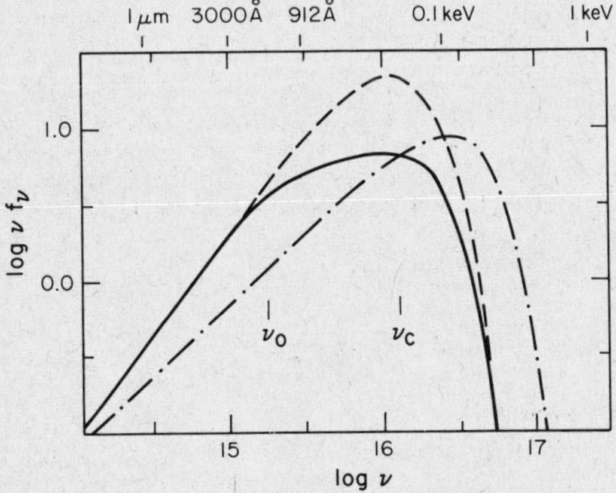

Figure 4. Accretion disk spectra with different opacities: solid line, includes $\bar{e}$-scattering and Comptonization; dashed line, pure sum of black bodies; dot-dashed line, grey-body opacity.

This interpretation effect removes the apparent conflict between Malkan (1983) and Bechtold et al. (1984,1986).

The modified disk model has another advantage. Because it departs from a black body, it is a less efficient radiator. It thus reaches higher temperatures at lower luminosities. This allows even a disk which emits soft x-rays to avoid being highly super-Eddington. For PG1211+143 (Figure 5) the modified disk model lies at only twice L_{Edd}(point 'C') instead of 20 times L_{Edd}(point 'B').

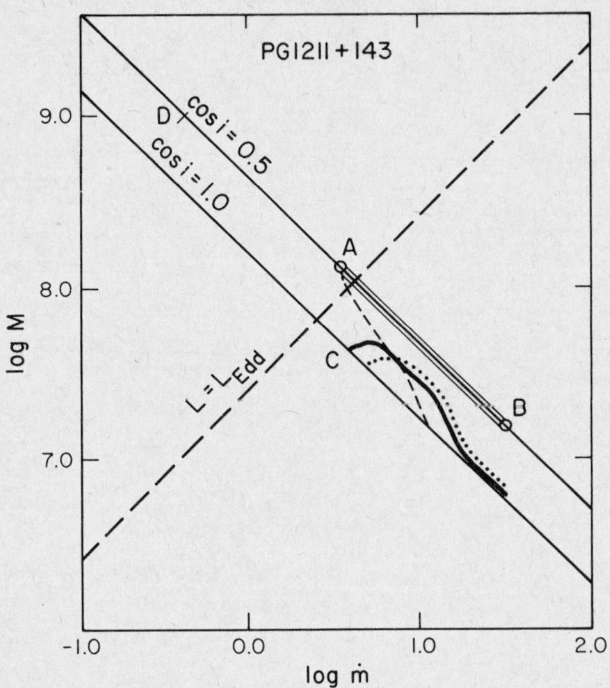

Figure 5.   Constraints on M and ṁ for disk models of PG1211+143.   A. black body disk extending through UV;   B. black body disk extending to soft x-rays for cos i = 0.5; · C.   face-on e⁻ scattering disk model;   solid curve from C-effect of changing inclination angle;   D. Wandel and Mushotzky M.

Figure 4 also shows a grey body disk spectrum (dot-dashed curve) which has been used by some authors to improve on the simplest black-body model (e.g., Arnaud et al., 1985).   It seems that for quasar disks, a grey body is a poorer approximation than a sum-of-black-bodies and should be avoided.

c.   Inclination Effects

Any disk near the Eddington limit will be puffed-up in its inner regions by radiation pressure.   The disk can then be self-occulting for observers at a wide range of inclination angles in the sense that this puffed-up region will hide radiation from parts of the disk that are even closer to the center.   Since the soft X-ray emission comes from closest in to the black hole it will be most affected by this self-occultation.

Figure 6 (solid line) shows the highest frequency ν_{ex} to which

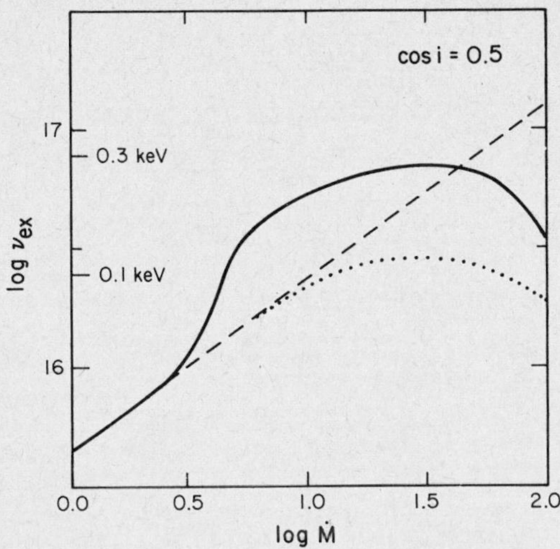

Figure 6. ν_{ex}, high frequency extent of disk spectrum for given $\dot{m}$ (cos i = 0.5).

radiation will be visible for an average observer at cos i = 0.5 as a function of accretion rate (ν_{ex} is defined by $[\nu f_\nu]_{\nu_{ex}} = 0.1 [\nu f_\nu]_{10^{15}Hz}$). The dashed line is the prediction for a black body model with no occultation; and the dotted line is the prediction for the black body model including occultation. For a wide range of $\dot{m}$, the end of the high frequency tail of the disk spectrum will lie in the softest X-ray range. For the particular case of PG1211+143, however, the soft X-ray luminosity is so large that only by making cos i > 0.7 can sufficient soft X-rays be made visible. (Only a spherical corona of substantial electron scattering optical depth ($\sim$ 0.5-1.0) could mitigate these occultation effects).

This result implies that substantial soft X-ray excesses from accretion disks should not be visible except in the minority (10-20%) of face-on disks. This percentage is consistent with the numbers of soft X-ray excesses so far reported (e.g., Pounds 1985). When these objects are found, however, they have the advantage that their inclination is determined. This should greatly simplify their understanding.

d. Comparison of M from disk models and from L_x-relationship

If one assumes that x-ray variability is due to processes occuring at a few (say 5) Schwarzchild radii then one can estimate a mass of the central black hole (Barr and Mushotzky 1986). These mass estimates agree strikingly with those based on [OIII] line widths and luminosity assuming Keplerian velocities (Wandel and Mushotzky 1986). The results imply relatively large black holes accretion at only $\sim 1\%$ L_{Edd}. This contrasts strongly with the results of accretion disk fits to the big bump which, as seen above for the case of PG1211+143, tend to give accretion rates near L_{Edd}.

We can make an explicit comparison of the results for PG1211+143.   For this quasar $L_x(2-10$ keV$)$ = 3×10^{44} erg s^{-1} which implies a mass M = $1 \times 10^9 M_\odot$ using the relationships of Barr and Mushotzky (1986) and Wandel and Mushotzky (1986).   In order to be consistent with this large a mass and with the constraint imposed by the luminosity of the $\nu^{1/3}$ part of the big bump disk spectrum (diagonal line from top left to bottom right of Figure 5) the quasar must be accreting at only 0.5 $M_\odot$ yr^{-1} (point D in Figure 5).   This implies a high efficiency of conversion of gravitational potential energy into radiation of $\sim 20\%$.   This is possible, but unlikely. Moreover, such an $(M,\dot{m})$ combination would produce only a cool disk with a peak in the visible band at ~ 5000Å (log ν_{max} ~ 14.8).   The big bump from this disk would drop strongly in the uv, which is inconsistent with the IUE observations.

Clearly one (or both) of these mass estimates must be wrong. The big bump may not be emission from an accretion disk after all; or the x-ray timescale may not be equated with a $5r_s$ radius.   One surprise of the Wandel and Mushotzky (1986) result is that the [OIII] derived mass is not several times larger than the x-ray variability derived mass even though the former is a measurement out on a 100 pc scale and so includes a significant galaxian component.   The [OIII] mass and the x-ray mass may scale together but the absolute normalization of the x-ray mass may be a factor 5 or so too large.

Resolving the differences between these two mass estimates is clearly important.   Our comparison was made _via_ a two step process which introduces some scatter in the estimate.   Also R. Warwick (this volume) has reported observations that seem to increase the scatter in the Barr and Mushotzky relationship.

3. A Connection Between the X-ray Turn-up and the mm-break[3]

We now move to the features at either end of the quasar continuum: the mm-break and the x-ray turn-up. A survey of quasar spectra with <u>Einstein</u> IPC shows that there is a link between these two features.

The x-ray turn-up is not always seen and it has not been obvious what property of a quasar controls whether or not an x-ray turn-up is present. The lack of a turn-up was first noted by Elvis et al. (1986). They presented a sample of 8 optically-selected ($\approx$radio-quiet) quasars that seem to show a continuous power-law of slope $\sim$1-1.2 (in log f_ν) in both the x-ray and infrared regions that join smoothly from the one region to the other (Figure 7). This strongly suggests (but does not prove) that a single underlying power-law extends across the whole infrared to x-ray region. Preliminary IPC results for a larger sample (Elvis, Wilkes, and Tananbaum 1985; Elvis 1985) suggested that there was a systematic difference between radio-loud and radio-quiet quasars in their x-ray spectra. This suggests that the <u>presence</u> of an x-ray turn-up implies the <u>absence</u> of a mm-break.

We have now made a survey of IPC spectra including all quasars in the <u>Einstein</u> data bank, excluding OVV's (Wilkes and Elvis 1987). The total sample includes 33 QSOs all of which were first discovered through either optical or radio search techniques.. This sample is equivalent in size to that found by combining all the previous studies of emission-line AGN that gave well-determined x-ray spectra. In addition our sample covers a wider range of radio properties, redshift and luminosity. We can now begin to investigate how the x-ray spectral index depends on other quasar properties — in particular the strength of the mm-break.

The spread of energy spectral index, α_E, present in the sample is wide. In Figure 8 a histogram of the best fit α_E is displayed. The typical 90% confidence error margin is also shown for guidance. The distribution is broad, with extreme values of -0.3 and 1.8, and is possibly bi-modal with peaks at $\sim$0.5 and $\sim$1.0 and a slight dip at $\sim$0.7.

There is a clear trend for radio-quiet quasars to have steeper x-ray spectra than radio-loud quasars. In Figure 8 the solid line denotes radio-loud objects while the dashed line denotes radio-quiet

---

3.   A full acount of this work will be submitted to Ap.J. by B.J. Wilkes and M. Elvis.

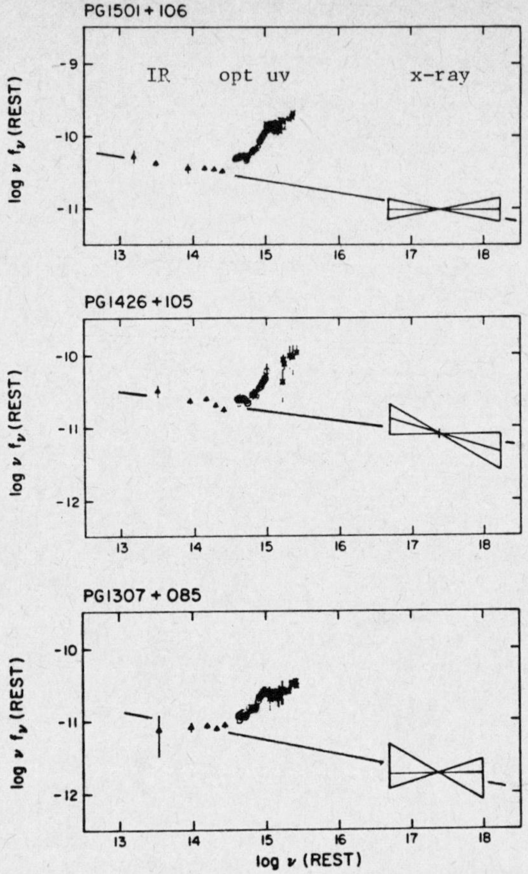

Figure 7. Three IR-x-ray energy distributions for optically-selected quasars illustrating the apparent continuity between the IR and x-ray continua.

objects (We define radio-loud as having $R_L (= \log(f_r(5\ GHz)/f_o(B)) > 1$; see below.) The radio-loud quasars cluster around $\alpha_E = 0.5$ while the radio-quiet quasars are grouped around $\alpha_E = 1.0$, with a rather distinct division of the two types. There is convincing evidence in the literature that the radio properties of quasars influence their x-ray luminosity. A very strong link between 90 GHz and x-ray emission was found for strong mm sources by Owen, Helfand, and Spangler (1981). Zamorani et al (1981) and Worrall et al (1987) have shown that radio-loud quasars, especially those with compact flat spectra, are relatively more luminous in x-rays. These authors suggested that radio-loud quasars possess an additional x-ray production mechanism linked to the radio emission such as synchrotron or synchrotron-self-Compton. The dominant x-ray component may be different depending upon a quasar's

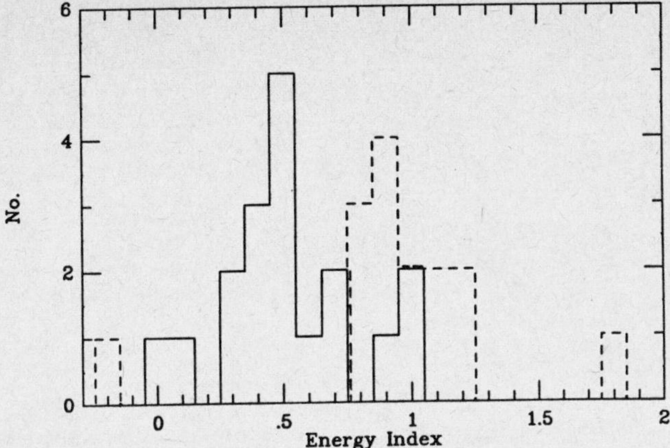

Figure 8. Double histogram of IPC (0.2-3.5 keV) energy spectral
indices for radio-quiet (dashed histogram) and radio-loud
(solid histogram) quasars.

radio luminosity since different components may have different x-ray
production mechanisms, we might also expect the x-ray slope to change
with radio-loudness, R_L.

However, in our IPC sample those quasars with high radio
luminosity are also those with high x-ray luminosity and high
redshift. Any of these three variables — radio-loudness, luminosity
or redshift — could be the controlling variable that determines the
x-ray spectral slope. In the case of redshift this change of slope
could simply be the redshifting of a soft excess out of the low
energy carbon window in the IPC response.

In order to distinguish between these different possibilities we
have collected from the literature 5 GHz radio core fluxes for all
the quasars in our sample. The fluxes for the PG quasars were kindly
supplied by K. Kellerman in advance of publication (Kellerman et
al. 1987). We have selected as far as possible observations taken
with the same physical beam size, typically a few tens of
kiloparsecs. When the beam sizes do not match it is usually because
the radio-loud quasars were observed at the highest VLA resolution.
Any correction to a larger beam could only make these objects more
radio-loud. The ratio of 5 GHZ radio core flux density to B-band
flux density, log (f_R/f_o), forms an index of radio-loudness, R_L. With
this index we can now treat radio-loudness as a continuous variable.

Figure 9 shows x-ray energy slope versus radio-loudness, R_L.
Radio-loudness is clearly correlated with x-ray slope. Because of
the interconnectedness of ℓ_o, z, and R_L we have checked that this
correlation is fundamental. We have performed a non-parametric

partial Spearman Rank correlation analysis on the four variables. It is clear from this that the only significant correlation with x-ray slope is that of radio-loudness, R_L. Radio-loud quasars do have flatter x-ray spectra, so two of the features in the quasar continuum are linked. When a mm-break is weak the x-ray turn up is strong and vice versa.

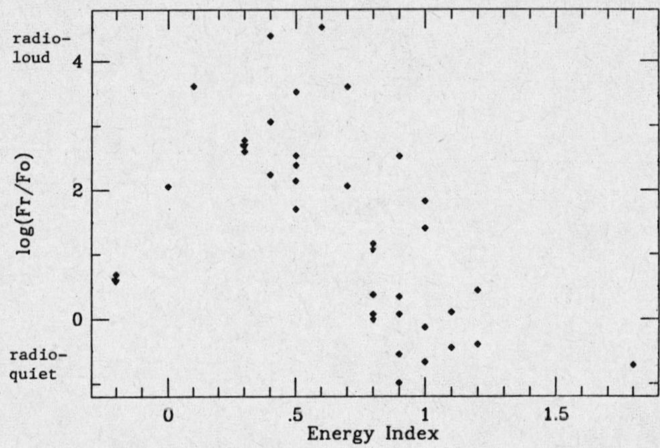

Figure 9. Correlation of x-ray spectral index with radio-loudness for IPC measured quasars.

The apparently smooth change of x-ray spectral index α_E with radio-loudness, R_L (Figure 9) can be understood if we are seeing two emission mechanisms, one of which is linked to radio emission, which are mixed in different proportions along the correlation. With the data in hand there is no way though to rule out the possibility that we are instead witnessing an intrinsic change of slope in a single mechanism. The work of Zamorani et al. and Worrall et al. (discussed above) suggested two components. The apparent continuity of the x-ray and infrared continua for radio-quiet quasars also suggests that any flattening is due to another component (Elvis et al. 1986). The reality of this important continuity needs to be checked with better data and with a larger sample of quasars.

If we are seeing a mix of two processes the slope of the radio-linked one is usually ~0.5 and of the other one ~1.0. Given the errors on the IPC slope estimation these could be 'universal' values but this is certainly not required.

One problem with invoking two components lies in explaining why the break point between one or the other mechanism dominating the observed emission should lie within the kilovolt IPC band. This seems to imply a surprising balance between the two processes. In radio-loud quasars VLBI data suggest that the flat radio spectra arise from the superposition of the output from numerous physically separate emission regions (e.g., Wittels, Shapiro and Cotton, 1982). There would be no obvious physical reason to expect a balance between synchrotron emission in one component and Compton emission in another.

A single physical component might explain this balance. Zdziarski (1986b) shows that in a single component synchrotron self-Compton model for radio-quiet AGN a kilovolt break point corresponds to equipartition betwen magnetic field and synchrotron photon energy densities. This model could be extended to radio loud objects by reducing the magnetic field and thus the self-absorption frequency. In this case though the radio spectrum should be a simple straight line continuation of the infrared instead of dropping away as is observed (Figure 2).

A single inhomogeneous synchrotron source (Marscher 1977) would explain this dropping away in the mm region with a suitable choice of radial density profile. However each part of this region is assumed to produce the same electron distribution and this produces the flat infrared power-law. There is no initial synchrotron spectrum that matches the rising x-ray slope in this model. The x-rays could not then come from Compton scattering of any of the radio components.

Otherwise two component models seem to be necessary in a synchrotron or synchrotron-self-Compton picture.

The infrared to x-ray power-law may not be synchrotron emission at all. If not then two main alternatives have been discussed: Non-thermal pair production related processes (Zdziarski and Lightman 1985, Svensson 1986, Fabian et al. 1986a, Fabian, this volume); and cyclotron self-Compton emission from thermal electrons (Maraschi, Roasio and Treves 1982, Zdziarski 1986a). Pair production models usually require a soft photon source. The soft source has to be at longer wavelengths than the continuum. When an infrared-to-x-ray continuum has to be explained and if the source is thermal this implies a large (pc-scale) soft source, because of the black body limit. This large size is probably inconsistent with variability constraints. These pair production models also have difficulty producing slopes of 1.0 or slightly greater as is required. A slope of ~ 0.9 seems to be a maximum. The other class of models employing Compton-scattered cyclotron radiation does not suffer from the black body size limit problem since the electron temperatures involved can be much larger than for thermal emission at the same wavelength. This type of model is a viable competitor for synchrotron self-Compton models.

The last possibility, of course, is that the single infrared-to-x-ray power law is an illusion and that given better data on more objects it will go away. We are gathering data to test this possibility.

4. Conclusions

We have discussed the major continuum features seen in normal (non-OVV) quasars from the point-of-view provided by the new Einstein IPC spectral data. We have focussed especially on accretion disk interpretations of the optical/UV/soft x-ray 'big bump' and on the connection between radio-loudness and the presence or absence of a turn-up in the x-ray spectrum toward higher energies.

In accretion disks opacity effects offer an alternative explanation for the observed UV turn-down. They simultaneously explain the short-UV and soft x-ray extensions of the big bump and avoid the highly super-Eddington luminosities implied by simple black-body models of soft X-ray excesses. Our models predict far-UV slopes of ~ 1, should these become observable. Inclination effects due to self-occultation of near-Eddington accretion disks lead to a limit on the observed high frequency extension of disk spectra in the lowest energy X-ray range. This leads to the prediction that only in the minority of disks viewed almost face-on will soft X-ray excesses be seen. Calculations of inner disk structure are in such a formative stage however that the magnitude of the self-occultation effects can only be roughly estimated. Self-occultation effects can be important for the state of, and even the existence of, the broad emission line region in quasars (Fabian et al. 1986b, Bechtold et al. 1987). The accretion disk models explored here are certainly still much simpler than the real disks. A fuller model should consider: absorption edges; photoionization of the disk by the non-thermal power-law; radial energy transfer and self-irradiation of the thick disk. The disk models we have made are also thermally unstable. It is not known whether this instability would completely disrupt such disks or merely lead to variability. All these issues need to be explored. The hypothesis that the big bump in quasars is due to an accretion disk has, however, survived a first consistency challenge.

Radio-loud quasars are found to have systematically flatter x-ray spectra than radio-quiet quasars in our Einstein IPC survey. This could be due to a single component of changing slope or, perhaps more probably, to a mix of two components. The radio-associated component would have a flat (α ~ 0.5) slope and the optical-associated component would have a steeper slope (α ~1.0). Thermal cyclo-Compton and non-thermal synchrotron self-Compton models remain possible.   The reality of the single infrared-to-x-ray power-law continuum needs greater testing.

The broad-band continua of normal quasars do possess features that can be used to explore the innermost regions of a quasar nucleus.   The first multi-wavelength studies are beginning to explore these possibilities.   The inclusion of x-ray data in these studies provides important constraints on models, from testing the viability of accretion disk models to finding regularities in the 'non-thermal' continuum features.

We thank Nat Carleton, Meg Urry and especially Andrzej Zdziarski for helpful and stimulating discussions.   This work was supported in part by NASA Contract NAS8-30751 and by the European Space Agency.

References

Angel, J.R.P. and Stockman, H.S. 1980, Ann.Rev.Astr.Ap., **18**, 321.

Arnaud, K.A., Branduardi-Raymont, G., Culhane, J.L., Fabian, A.C., Hazard, C., McGlynn, T.A., Shafer, R.A., Tennant, A.F., and Ward, M.J., 1985, MNRAS, **217**, 105.

Barr, P. and Mushotzky, R.F. 1986, Nature, **320**, 421.

Bechtold, J. et al., 1984, Ap.J., **281**, 76.

Bechtold, J., Czerny, B., Elvis, M., Fabbiano, G., and Green, R.F., 1986, Ap.J., submitted.

Carleton, N.P., Elvis, M., Fabbiano, G., Lawrence, A., Ward, M.J., and Willner, S.P., 1987, Ap.J., submitted.

Czerny, B. and Elvis, M., 1987, in preparation.

Edelson, R.A., and Malkan, M.A., 1986, Ap.J., in press.

Elvis, M., 1985, in "Compact Galactic and Extragalactic X-ray Sources", Japan-U.S. Seminar (eds. Y. Tanaka and W.H.G. Lewin), [Tokyo, ISAS] p.291.

Elvis, M., Wilkes, B.J. and Tananbaum, H., 1985 Ap.J., **292**, 357.

Elvis, M., et al. 1986, Ap.J., **210** (Nov. 1) in press.

Fabian , A.C., Blandford, R.D., Guilbert, P.W., Phinney, E.S., and Cuellar, L. 1986a, MNRAS, in press.

Fabian, A.C., Guilbert, P.W., Arnaud, K.A., Shafer, R.A., Tennant, A.F., and Ward, M.J. 1986b, MNRAS, **218**, 457.

Kellerman, K.I., Sramek, R., Shaffer, D., Green, R., and Schmidt, M., 1986, IAU Symposium No. 119, "Quasars" eds. G. Swarup and V.K. Kapahi (Reidel, Dordrecht).

Lawrence, A., and Elvis, M., 1982, Ap.J., **256**, 410.

Malkan, M.A., and Sargent, W.L.W., 1982, Ap.J., **254**, 22.

Malkan, M.A. 1983, Ap.J., **268**, 582.

Malkan, M.A., this volume.

Maraschi, L., Roasio, R., and Treves, A. 1982, Ap.J., **253**, 312.

Mushotzky, R.F. 1984, Advances in Space Research, **3**, no. 10-13, 157.

Owen, F.N., Helfand, D.J., and Spangler, S.R., 1981, Ap.J.(Letters), **250**, L55.

Pounds, K.A., 1985, in "Galactic and Extragalactic Compact X-ray Sources", eds. Y. Tanaka, W.H.G. Lewin [ISAS, Tokyo].

Rieke, G.H., Lebofsky, M.J., and Wisniewski, W.Z. 1982, Ap.J., **263**, 73.

Rybicki, G.B. and Lightman, A.P., 1979, "Radiative Processes in Astrophysics" [Wiley, New York].

Shakura, N.I. and Sunyaev, R.A., 1973, Astron.Astrophys., **24**, 337.

Stein, W.A., Tokunaga, A.T., and Rudy, R.J., 1986, PASP, submitted.

Svensson, R. 1986, in "Radiative Hydrodynamics in Stars and Compact Objects" (Heidelberg: Springer), in press.

Urry, C.M., 1984, Ph.D. Thesis, John Hopkins University.

Véron-Cetty, M.-P., and Véron, P., 1986, Astr.Ap.Suppl., in press.

Wandel, A. and Mushotzky, R.F. 1986, Ap.J.(Letters), in press.

Ward, M.J., Elvis, M., Fabbiano, G., Carleton, N.P., Willner, S.P., and Lawrence, A., 1987, Ap.J., submitted.

Wilkes, B.J. and Elvis, M., 1987, Ap.J., submitted.

Wittels J.J., Shapiro I.I., and Cotton W.D., 1982, Ap.J.,**262**, L27.

Worrall, D.M., Giommi, P., Tananbaum, H., and Zamorani, G., 1987, Ap.J., submitted.

Zamorani, G., et al., 1981, Ap.J., **245**, 357.

Zdziarski, A.A. 1986a, Ap.J. **303**, 94.

Zdziarski, A.A. 1986b, Ap.J., **305**, 45.

Zdziarski, A.A. and Lightman, A.P. 1985, Ap.J.(Letters), **294**, L79.

The Soft X-ray Spectra of Active Galactic Nuclei

G. Branduardi-Raymont

Mullard Space Science Laboratory
University College London
Holmbury St.Mary, Dorking
Surrey, U.K.

ABSTRACT

EXOSAT LE and ME data on Seyfert 1 galaxies and quasars are reviewed: the observations indicate that a strong flux of soft X-rays, in excess of the power-law continuum that extends over most of the spectrum, is a common feature of the emission of Active Galactic Nuclei (AGN). This excess is seen in both Seyfert galaxies and quasars, irrespective of whether they are X-ray or optically selected. EXOSAT observations reveal that this low-energy X-ray spectral component, which is often identified with the emission of an optically thick accretion disc surrounding a central massive black hole, is highly variable, with changes in flux by more than a factor of ten over a few months. This new result and the observation of spectral variability correlated with the changes in flux give clues to the mechanisms of X-ray production in AGN and to the structure of the innermost parts of the accretion disc in these objects.

1. INTRODUCTION

Attention has recently been drawn to an excess of soft X-rays above the power-law continuum fit to the medium-energy X-ray spectrum of a number of Seyfert 1 galaxies: these are Mkn 509 (Singh et al. 1985), Mkn 841 (Arnaud et al. 1985), NGC 4051 (Lawrence et al. 1985), NGC 4151 and Mkn 335 (Pounds 1985), Fairall 9 (Morini et al. 1986) and NGC 5548 (Branduardi-Raymont 1986). The soft X-ray flux is generally interpreted as the high-energy tail of the emission of an optically thick accretion disc surrounding a central massive black hole, with the bulk of the radiative output being in the ultraviolet (the 'big bump'). A spectral fit of combined EXOSAT and IUE data has been attempted for only two objects: Mkn 841 (Arnaud et al. 1985) and NGC 5548 (Hanson et al. 1986). In section 2 we review the EXOSAT observations of NGC 5548, which have revealed a further new aspect of the soft X-ray emission of AGN: its extreme variability.

Early work suggested that the X-ray spectrum of all Seyfert 1 galaxies is a power-law with energy index 0.7 (Mushotzky 1984). Quasars in general, though, do not conform to a single slope power-law spectrum (but see Worrall and Marshall 1984). For a number of years the only available X-ray spectrum of a quasar, that of 3C 273, was known to have an energy slope roughly consistent with 0.4 (Worrall et al. 1979). More recently, Pravdo and Marshall (1984) discovered a quasar with a steep X-ray spectrum (energy index of 1.3). Elvis et al. (1985, 1986) found a diversity of spectra in quasars observed with the Einstein IPC (0.2 - 4.0 keV): their energy slopes range from 0.7 to 2.2. In the case of the quasar PG1211+143, which has the steepest spectral slope observed with the IPC, Bechtold et al. (1986) have found evidence for a large excess of soft X-rays superposed on the overall power-law continuum of energy slope 1.25 which extends from 10 μm to 2 keV. A discussion of the continuum properties of quasars appears in these proceedings (Elvis et al.).

Branduardi-Raymont et al. (1985, Paper I) have reported the serendipitous discovery of an unexpectedly large number (4) of AGN, 3 of which are quasars, in an EXOSAT CMA exposure (0.02 - 2.5 keV) of a 1 square degree field near the Coma cluster. They have attributed this result to the combination of two facts: these AGN have a bright, soft component in their X-ray spectrum and their position on the sky coincides with a 'hole' in our Galaxy's obscuration, which allows us to see this component. The EXOSAT observations thus provide evidence that quasars in general possess a strong flux of soft X-rays.

Quasars are thought to be important contributors of the X-ray background: the size of their contribution is obviously dependent on the spectral characteristics of their emission. Because of the importance of determining the spectral properties of quasars and of the unique character of the EXOSAT results in Coma, section 3 presents a review of EXOSAT follow-up observations of the Coma field, which reinforce the original conclusion of Paper I. Moreover, the new data also indicate that the soft X-ray emission of the AGN in Coma is subject to extreme variability, as is the emission from NGC 5548: this is reported in section 4. The implications of the EXOSAT observations are discussed in section 5.

2. X-RAY VARIABILITY OF THE SEYFERT 1 GALAXY NGC 5548

NGC 5548 is an X-ray bright ($L_x \sim 10^{43.5}$ erg s^{-1}) type 1 Seyfert galaxy which was suspected, on the basis of Einstein Observatory combined MPC and IPC data, to possess a complex spectrum (Halpern 1982). The EXOSAT observations of NGC 5548 cover a period of one and a half years (a complete account of the observations will be found in Branduardi-Raymont 1986). Figure 1 shows the 2 - 6 keV lightcurve observed with the ME array of proportional counters (top panel) and those obtained simultaneously with the LE CMA (0.02 - 2.5 keV) and filters (Boron, Aluminium/Parylene and Lexan 3000Å, from top to bottom). NGC 5548 underwent a dramatic flare at both soft and medium X-ray energies between February and March 1984: the CMA flux more than doubled in all filters, while the ME flux only increased by about 30%. During the following year the X-ray flux of NGC 5548 declined by more than a factor of 10 and 3 in the CMA and the ME respectively.

The variability of NGC 5548 at soft X-ray energies is at times strongly correlated with that at medium energies: Figure 2 shows the CMA and filters countrates plotted against the simultaneous ME measurements. Interestingly, the correlation is very clear when the ME flux is above 2 counts s^{-1}; below this threshold the CMA and ME fluxes are uncorrelated (although the countrate range sampled is only about one third of that in the high state where correlation is observed). Figure 2 suggests that the low-energy

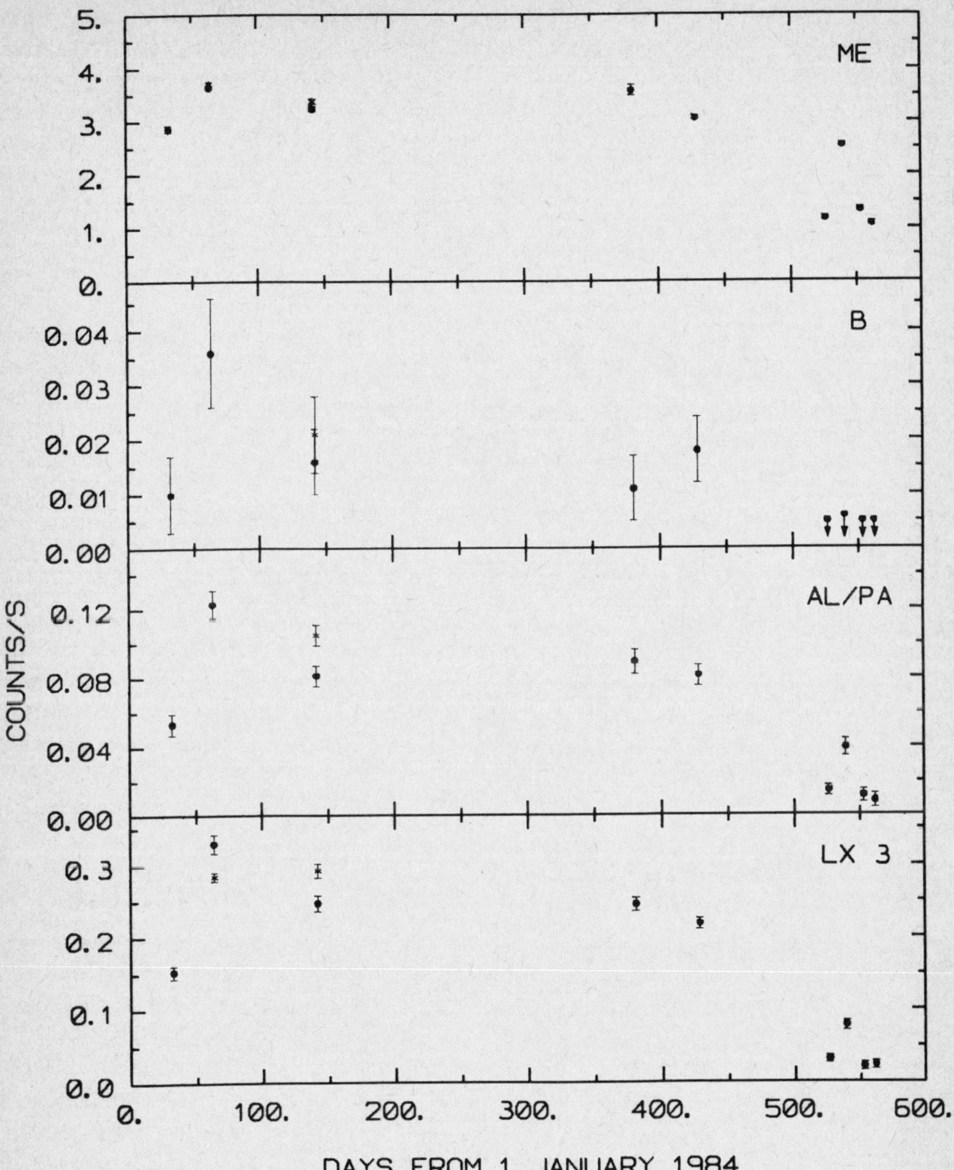

Figure 1 EXOSAT lightcurves of NGC 5548; *top panel:* ME (2 - 6 keV); *bottom three panels:* LE CMA (0.02 - 2.5 keV) with Boron, Aluminium/Parylene, Lexan 3000Å filters.

X-ray flux of NGC 5548 in the high state pertains to a spectral component separate from that producing the medium-energy emission, but in some way physically related to it.

Analysis of the individual ME observations shows that changes in the ME flux are associated with statistically significant spectral variability: the 2 - 6 keV spectrum of

NGC 5548 is softer when the source is brighter, with the best fit power-law energy index varying from 0.3 to 0.8 as the flux varies from 1 to 3.7 counts s^{-1}. This is to be compared with observations of other objects: spectral softening with increasing X-ray brightness has recently been observed in 3C 120 (Halpern 1985), NGC 4051 (Lawrence et al. 1985), NGC 4151 (Perola et al. 1986) and Fairall 9 (Morini et al. 1986).

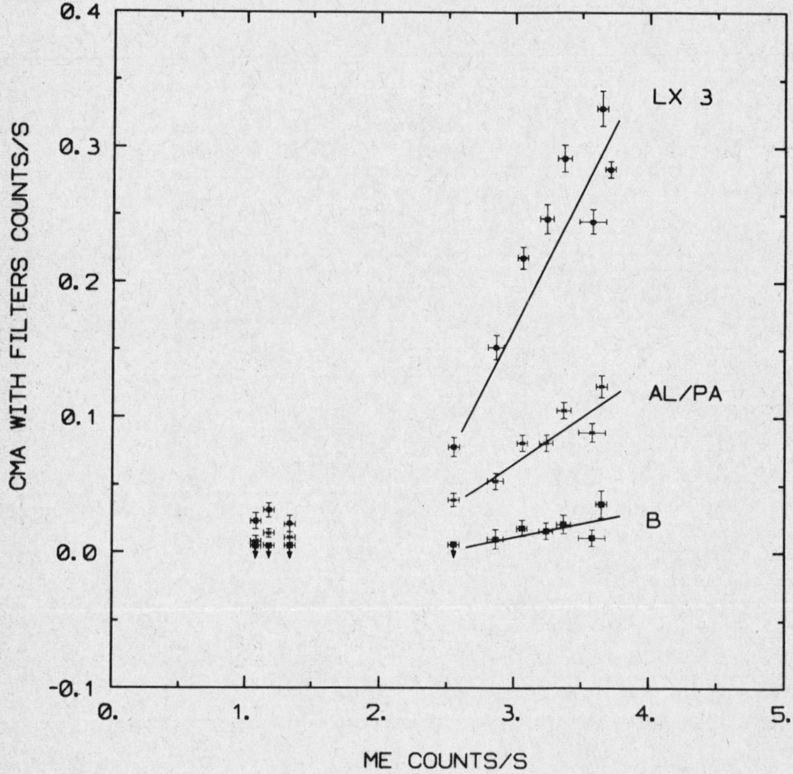

Figure 2 NGC 5548 countrates in the LE CMA with various filters plotted against the simultaneous ME measurements (the solid lines are there to guide the eye and are not formal fits to the data).

The combined ME and LE spectra of NGC 5548 show that the best fit ME power-law extrapolates below 1 keV only when the ME flux is below 2 counts s^{-1}. Above this threshold, extrapolation of the power-law into the LE range requires an amount of low-energy absorption about 4 times smaller than indicated by recent 21 cm Hydrogen absorption measurements (1.65 x 10^{20} cm^{-2}; F.J.Lockman, L.Danly and B.Savage, personal communication). Fixing N_H at this value in fitting the combined ME and LE data, the presence of a soft X-ray excess in the spectrum of NGC 5548 is unequivocally identified, although only when the source is brightest. The excess can be fitted by a blackbody component with a temperature of $\sim 3 \times 10^5$ K. The blackbody temperature does not vary significantly with the CMA flux, but its normalization is highly uncertain: the broad-band measurements made with the CMA and its various filters do not provide sufficient spectral resolution to pose stringent constraints on the shape of the spectrum at low energies. For all observations, the index of the medium-energy power-law from the combined ME and LE spectral fits is identical (within the errors) to that obtained by fitting the ME data on their own. The steepening of the ME spectrum with increasing flux is not a consequence of the variable strength of the low-energy excess in NGC 5548: the blackbody component is far too cool to make any contribution to the flux in the ME band. One is led to conclude that, as Figure 2 qualitatively suggests, the soft X-ray excess in NGC 5548 is a separate component from that producing the medium-energy X-ray flux, although their correlated variability indicates a physical link between the two.

A model for the low-energy X-ray flux of NGC 5548 that is more realistic than a single temperature blackbody is emission from an optically thick accretion disc: the characteristic disc spectrum (Pringle 1981) is flatter than that of a blackbody but decays exponentially at high frequencies like a blackbody. It would then be indistinguishable from the latter with the 3-colour CMA and filters measurements. Hanson et al. (1986) have tested this possibility by combining the EXOSAT data from the May 1984 observation with simultaneous IUE data: indeed the single temperature blackbody spectrum fitted to the EXOSAT data falls short of explaining the UV flux by more than an order of magnitude, but a disc spectrum with a characteristic temperature of $\sim 6 \times 10^5$ K (according to the nomenclature of Pringle 1981) fits both the IUE and EXOSAT data.

3. THE 1985 OBSERVATIONS OF THE COMA CLUSTER REGION

In June 1985 Branduardi-Raymont et al. (1986) obtained a second EXOSAT CMA exposure of the Coma cluster region 3 times longer than that described in Paper I and centred 40 arcmin further East. All the original serendipitous sources contained in the new CMA field are detected, but the 1985 image also reveals 4 new sources. As one would expect if we are looking through a small window of low Galactic obscuration, 3 of the new sources are concentrated on the sky in the vicinity of the original AGN detections. A search for blue objects on the Palomar Sky Survey prints at the positions of the new CMA sources and subsequent optical spectroscopy with the INT at the Observatorio del Roque de los Muchachos on La Palma reveal that the counterparts of the 4 new X-ray sources are all background quasars: their redshifts are 0.28, 0.51, 0.68 and 1.88 respectively.

This result brings the total number of soft X-ray selected, background AGN in the field of the Coma cluster up to 8; 7 are contained within a projected area of 1 square degree on the sky; of the 7, one has redshift larger than 1. The total number of AGN known to be present in this 1 square degree area (11) is still consistent with the number (10.3) of objects of this type, brighter than 19.5 mag, expected from optical counts (Braccesi et al. 1980): thus the AGN in the Coma field are probably not unusual and their X-ray properties may be representative of those of the entire quasar population. The CMA detection of 8 AGN in an area of small angular size on the sky then reinforces the earlier suggestion that these objects are intrinsically bright in soft X-rays and that they are observed through a window of low Galactic obscuration near the North Galactic Pole.

4. VARIABILITY OF THE SOFT X-RAY FLUX OF THE COMA AGN

In 1985 Branduardi-Raymont et al. also obtained a CMA exposure at the Coma cluster centre, with the same pointing position of that taken in 1983 and the same observing time ($\sim$ 7 hours), in order to investigate how the X-ray flux of the original serendipitous sources evolves with time.

Table 1 lists the 1983 and 1985 countrates of the original 5 sources reported in Paper I; the nature of the optical counterpart is also indicated. There are striking differences in the fluxes two years apart, even when the $\sim$ 10% lower throughput of the EXOSAT LE1 CMA (used in 1985) with respect to LE2 (1983) is taken into account. Source 4 is not detected in the 1985 exposure centred on the Coma cluster, although it is detected in the longer, offset exposure at a level which is more than a factor of 5 down with respect to 1983; Table 1 shows a 3σ upper limit from the short exposure. The 1985 flux of source 2, the type 1 Seyfert X Comae, is less than 20% of that in 1983. Source 1 is also weaker in 1985, at about half the strength of 1983 (its flux is very uncertain because the source is embedded in the diffuse emission of the Coma Cluster). Sources 3 and 5 have remained approximately constant between 1983 and 1985 and constitute a check on the stability of the CMA response.

TABLE 1: EXOSAT CMA Countrates of Sources in the Coma Field

Source number	Type of identification	1983, LE2 counts min^{-1}	1985, LE1 counts min^{-1}
1	QSO	0.27 ± 0.10	0.12 ± 0.10
2	Seyfert 1 galaxy	0.91 ± 0.09	0.23 ± 0.09
3	QSO	0.26 ± 0.08	0.20 ± 0.04
4	QSO	0.34 ± 0.07	≤ 0.08
5	QSO	0.31 ± 0.07	0.24 ± 0.11

5. DISCUSSION

With the exception of the highly unusual object PG1211+143 (Bechtold et al. 1986), it is only very recently that examination of Einstein IPC data has led to the conclusion that a soft excess is indeed present in the X-ray spectrum of quasars (Elvis, personal communication). The softer response of the EXOSAT CMA and the large area of the EXOSAT ME experiment have offered the opportunity of investigating this issue in more detail.

The EXOSAT results in the region of Coma are indeed evidence that a large fraction of the luminosity of quasars is emitted in the UV/soft X-ray part of the spectrum. Moreover, soft X-ray excesses have been discovered in several Seyfert galaxies, and in particular in NGC 5548, although only when the source is brightest. If the soft X-ray flux is the high-energy tail of the emission of an optically thick accretion disc surrounding a central massive black hole, indeed one expects the disc spectrum to extend to high energies for smaller black hole masses (thus in the case of Seyferts, like NGC 5548, rather than quasars) and higher accretion rates (i.e. at higher luminosities, such as in the brightest states of NGC 5548). In addition, the 'big bump' shifts more and more to longer wavelengths with increasing redshift. All of this suggests that EUV/soft X-ray observations, such as those which will be possible with the ROSAT Wide Field Camera, could produce a wealth of AGN detections, if the Galactic obscuration is sufficiently small over a sufficiently large fraction of the sky. One could expect a selection effect against detecting high redshift AGN in the EUV/soft X-ray band, and thus a break in the AGN redshift distribution, since the 'big bump' in their spectra would occur at longer wavelengths than those covered by the ROSAT Wide Field Camera.

As well as demonstrating that the presence of a strong flux of soft X-rays is a common feature of the spectra of AGN in general, the EXOSAT data of NGC 5548 and of the Coma AGN reveal a new property of this emission: its strong variability. The size and timescale of this variability, coupled with the correlated spectral changes at medium X-ray energies, are clues to the mechanisms of X-ray production and to the inner structure of accretion discs in AGN. Czerny and Elvis (1987) have investigated AGN accretion disc models incorporating opacity and inclination effects: although they are able to avoid violating the Eddington limit in this way, only 10 - 20% of AGN are predicted to possess soft X-ray excesses from the inner edges of accretion discs. Models of this type may require modification (possibly the presence of some form of hot corona above the disc) if the total fraction of Seyferts and quasars observed to display a soft X-ray excess and/or the rate of AGN detections in the EUV band are found to be high.

REFERENCES

Arnaud, K.A., Branduardi-Raymont, G., Culhane, J.L., Fabian, A.C., Hazard, C., McGlynn, T.A., Shafer, R.A., Tennant, A.F., Ward, M.J. 1985. M.N.R.A.S., **217**, 105.

Bechtold, J., Czerny, B., Elvis, M., Fabbiano, G. and Green, R.F. 1986. Ap. J., submitted.

Braccesi, A., Zitelli, V., Bonoli, F. and Formiggini, L. 1980. Astr. Astrophys., **85**, 80.

Branduardi-Raymont, G., Mason, K.O., Murdin, P.G. and Martin, C. 1985. M.N.R.A.S., **216**, 1043 (Paper I).

Branduardi-Raymont, G. 1986, in preparation.

Branduardi-Raymont, G., Mason, K.O., Smale, A.P., Murdin, P.G. and Martin, C. 1986, in preparation.

Czerny, B. and Elvis, M. 1987, in preparation.

Elvis, M., Wilkes, B.J. and Tananbaum, H. 1985. Ap. J., **292**, 357.

Elvis, M., Green, R.F., Bechtold, J., Schmidt, M., Neugebauer, G., Soifer, B.T., Matthews, K. and Fabbiano, G. 1986. Ap. J., in press.

Halpern, J.P. 1982. Ph.D. Thesis, Harvard University.

Halpern, J.P. 1985. Ap.J., **290**, 130.

Hanson, C.G., Branduardi-Raymont, G., Coe, M.J., Elsmore, B., Wamsteker, W., Wills, B. and Wills, D. 1986, in preparation.

Lawrence, A., Watson, M.G., Pounds, K.A. and Elvis, M. 1985. M.N.R.A.S., **217**, 685.

Morini, M., Scarsi, L., Molteni, D., Salvati, M., Perola, G.C., Piro, L., Simari, G., Boksenberg, A., Penston, M.V., Snijders, M.A.J., Bromage, G.E., Clavel, J., Elvius, A. and Ulrich, M.H. 1986. Ap. J., **307**, 486.

Mushotzky, R.F., 1984. Advances in Space Research, **3**, no. 10-13, 157.

Perola, G.C., Piro, L., Altamore, A., Fiore, F., Boksenberg, A., Penston, M.V., Snijders, M.A.J., Bromage, G.E., Clavel, J., Elvius, A. and Ulrich, M.H. 1986. Ap. J., **306**, 508.

Pounds, K.A. 1985. In "Galactic and Extragalactic Compact X-ray Sources", Ed.s Y. Tanaka and W.H.G. Lewin. ISAS, Tokyo.

Pravdo, S.H. and Marshall, F.E. 1984. Ap. J., **281**, 570.

Pringle, J.E. 1981. Ann. Rev. Astr. Astrophys., **19**, 137.

Singh, K.P., Garmire, G.P. and Nousek, J. 1985. Ap. J., **297**, 633.

Worrall, D.M., Mushotzky, R.F., Boldt, E.A., Holt, S.S. and Serlemitsos, P.J. 1979. Ap. J., **232**, 683.

Worrall, D.M. and Marshall, F.E. 1984. Ap. J., **276**, 434.

SUBJECT INDEX

Lecture Notes in Physics